Handbook of Engineering Mathematics Formulae

Handbook of Engineering Mathematics Formulae

Harish C Rai
PhD (Electrical Engg, IIT Delhi), FIE (India), FIETE, MISTE, MAeSI

Pro Vice Chancellor
Galgotias University, Greater Noida, UP

Former
Professor, Department of Electrical and Electronics Engineering
Chhotu Ram State College of Engineering
(Presently Deenbandhu Chhotu Ram University of Science and Technology)
Murthal, Haryana 131039

Controller of Examinations
Director, Academic Affairs
Director, Research Project Monitoring Cell
Director, Organization and Development
GGS Indraprastha University, Delhi

Advisor I, All India Council of Technical Education
(Ministry of Human Resource and Development, Delhi)

Himanshu Rai
Programme Manager
Digital Media, Cognizant Global

CBS Publishers & Distributors Pvt Ltd

New Delhi • Bengaluru • Chennai • Kochi • Kolkata • Mumbai
Hyderabad • Jharkhand • Nagpur • Patna • Pune • Uttarakhand

Disclaimer

Science and technology are constantly changing fields. New research and experience broaden the scope of information and knowledge. The authors have tried their best in giving information available to them while preparing the material for this book. Although, all efforts have been made to ensure optimum accuracy of the material, yet it is quite possible some errors might have been left uncorrected. The publisher, the printer and the authors will not be held responsible for any inadvertent errors or inaccuracies.

Handbook of
Engineering Mathematics Formulae

ISBN: 978-93-86478-21-4

First Edition: 2018

Published by Satish Kumar Jain and produced by Varun Jain for
CBS Publishers & Distributors Pvt Ltd
4819/XI Prahlad Street, 24 Ansari Road, Daryaganj, New Delhi 110 002, India.
Ph: 23289259, 23266861, 23266867 Website: www.cbspd.com
Fax: 011-23243014 e-mail: delhi@cbspd.com; cbspubs@airtelmail.in.
Corporate Office: 204 FIE, Industrial Area, Patparganj, Delhi 110 092
Ph: 4934 4934 Fax: 4934 4935 e-mail: publishing@cbspd.com; publicity@cbspd.com

Branches

- **Bengaluru:** Seema House 2975, 17th Cross, K.R. Road, Banasankari 2nd Stage, Bengaluru 560 070, Karnataka
  Ph: +91-80-26771678/79 Fax: +91-80-26771680 e-mail: bangalore@cbspd.com
- **Chennai:** 7, Subbaraya Street, Shenoy Nagar, Chennai 600 030, Tamil Nadu
  Ph: +91-44-26680620, 26681266 Fax: +91-44-42032115 e-mail: chennai@cbspd.com
- **Kochi:** Ashana House, 39/1904, AM Thomas Road, Valanjambalam, Ernakulam 682 016, Kochi, Kerala
  Ph: +91-484-4059061-62-64-65 Fax: +91-484-4059065 e-mail: kochi@cbspd.com
- **Kolkata:** 6/B, Ground Floor, Rameswar Shaw Road, Kolkata-700 014, West Bengal
  Ph: +91-33-22891126, 22891127, 22891128 e-mail: kolkata@cbspd.com
- **Mumbai:** 83-C, Dr E Moses Road, Worli, Mumbai-400018, Maharashtra
  Ph: +91-22-24902340/41 Fax: +91-22-24902342 e-mail: mumbai@cbspd.com

Representatives

• **Hyderabad**	0-9885175004	• **Jharkhand**	0-9811541605
• **Nagpur**	0-9021734563	• **Patna**	0-9334159340
• **Pune**	0-9623451994	• **Uttarakhand**	0-9716462459

Printed at:

to

My respected parents
Late Sh Balraj
Late Smt Ram Devi

My lively children, Shivanshu and Himanshu
My loving wife, Sangeeta; and
Shipra and grandson Vivaan, whose
affection is always appreciated

Preface

This book introduces students of engineering, physics, mathematics, and computer science to those areas of mathematics which from a modern point of view are the most important domains in connection with practical problems.

The contents and characters of mathematics needed in application are changing rapidly. Linear algebra, especially matrices and numerical methods for computers are gaining importance, statistics and graph theory play more prominent roles. Real analysis and complex analysis remain indispensable.

The purpose of the book is to provide a concise summary of the major tools of engineering mathematics in a single volume, a desktop-reference for practising engineers in industry, scientists, architects and academics. The material has been compiled so as to serve the needs of students and professionals, who wish to have a ready-reference to formulas, equations, methods, concepts and their mathematical formulations. The material in this book is arranged accordingly in twenty three chapters and appendices.

Chapters 1–7 cover set theory, algebra, geometry, trigonometry, trigonometric functions, analytic geometry, and space coordinate geometry with major emphasis on topics frequently occurring in the solution of physical problems.

Chapters 8–13 present the differential calculus, sequence and series, integral calculus, vector analysis, functions of complex variables, Fourier series, special functions and their applications encountered in applied sciences and engineering analysis.

Chapters 14–16 are the extensive summary of higher transcendent functions, special cases of ordinary and partial differential equations and related topics.

Chapter 17–20 give a summary of the terminologies and the major formulas of Laplace transforms, numerical methods, probability and statistics and related tables of numerical coefficients.

Chapters 21–23 contain tables of indefinite integrals, digital logic, and linear programming.

Finally, tables of fundamental, mechanical, electrical and magnetic units, trigonometric and other relations and tables of numerical values of the most important functions are assembled in the appendices.

The graphical arrangement offers the possibility of using this handbook as a pictorial dictionary of engineering mathematics, a review outline for examinations, and a manual of comparative study.

Students and professionals alike will find this book a valuable supplement to standard textbooks, a source of review and a handy-reference. The book summarizes the applicable mathematical symbols and physical constants.

We hope that this handbook will effectively satisfy the professional requirements, at the end, we would like to request the esteemed readers to kindly send us their valuable suggestions for improvement of the book and notify for any error they may come across while going through the book.

Harish C Rai
Himanshu Rai

Acknowledgements

I thank all my undergraduate students who suggested that I should write this book and indeed, all those who have encouraged me in this venture. I derive immense pleasure in expressing my sincere thanks to Prof Yogesh Singh, Vice Chancellor, and Prof Annu Singh Lather, Pro Vice Chancellor, Delhi Technological University (DTU), for the invaluable encouragement throughout this work. I am indebted to their guidance and invaluable suggestions.

I express my gratitude to Prof SS Murthy, former Vice Chancellor, Central University of Karnataka; Prof ZH Zaidi, former Vice Chancellor, MJP Rohilkhand University, Bareilly; Prof BP Singh, Prof Bhim Singh, Department of Electrical Engineering, IIT Delhi, for sparing their valuable time and providing useful guidance on various chapters.

I thank my colleagues, Prof Alok Mittal, Member Secretary, AICTE, New Delhi; Prof JRP Gupta (NSIT, Delhi); Prof DR Bhaskar (DTU); Prof SS Inamdar (Vishwaniketan, Mumbai); Prof VK Sharma (NIT, Uttarakhand); Prof Rominder Randhwa, Director, Guru Tegh Bahadur Institute of Technology; Prof SS Tyagi, Director, BSA Institute of Technology, Haryana, and Prof Lajpat Rai, IIT Delhi, with whom I have discussed power electronics while teaching courses on this subject.

I express my gratitude to my brother Dr Mahesh Popli (Income Tax Department), Rajasthan; Dr Vikas Gupta (DU); Mr Pankaj Munjal, Director, Training and Development, RVIT, Bijnore; Brig Pradeep Upmanu; Dr Nitin Malik, GGS Indraprastha University, and Mr Ankit Popli for their immense help and constructive criticism on the manuscript.

My special thanks are due to Sh RC Taneja and late Sh KR Munjal for their moral support which has enabled me to complete this work. I am grateful to Sh Satish Kumar Jain (Mataji), CMD, and Sh Varun Jain, Director, CBS Publishers & Distributors, New Delhi, for their patience, goodwill and cooperation. I express my gratitude to Mr YN Arjuna (Senior Vice President Publishing, Editorial and Publicity); Mrs Ritu Chawla (AGM Production); Mr Sumit Behl; Ms Sanjubala Tripathy (Copy Editor) and Mr Parmod Kumar, for their skillful service and immense help.

Finally, I appreciate the patience and solid support of my family—my wife Sangeeta Rai; children Shivanshu, Shipra and Himanshu.

Harish C Rai

Contents

CHAPTER 1

Set Theory

1.1 INTRODUCTION

A set is a well defined collection of objects. If a is an element of set A, then $a \in A$ or a is a member of A. The collection of vowels in English alphabet comprises five elements, namely a, e, i, o, u.

Types of Sets

i. *Empty set*: A set is said to be an empty or null or void set if it has no element. It is denoted by ϕ.

ii. *Equal sets*: Two sets A and B are said to be equal if every element of A is a member of B, and every element of B is a member of A.

iii. *Subset*: Let A and B be two sets. If every element of A is an element of B then A is called a subset of B, can be written as

   $A \subseteq B$, which is read as 'A is a subset of B' or 'A is contained in B'.

iv. *Universal set*: A set that contains all sets in a given context is called the universal set, denoted by U.

1.2 VENN DIAGRAMS

The diagrams drawn to represent sets are called Venn diagrams. In Venn diagrams, the universal set U is represented by points within in a rectangle and its subsets are represented by points in a closed circle within the rectangle. If a set A is a subset of B than the circle representing A is drawn inside the circle representing B. If A and B are not equal but have some common elements can be represented by two intersecting circles. Two disjoint sets are represented by two nonintersecting circles.

1.3 OPERATIONS OF SETS

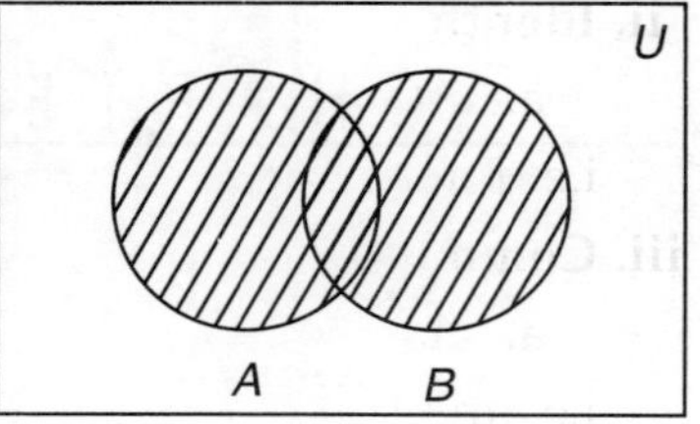

Fig. 1.1

i. Union of Sets: Let A and B be two sets. The union of A and B is the set of all those elements which belong either to A or B or both A and B.

The notation $A \cup B$ (read as 'A union B') to denote the union of A and B. Thus,

$$A \cup B = \{x : x \in A \quad \text{or} \quad x \in B\}$$

In Fig. 1.1, the shaded portion represents $A \cup B$

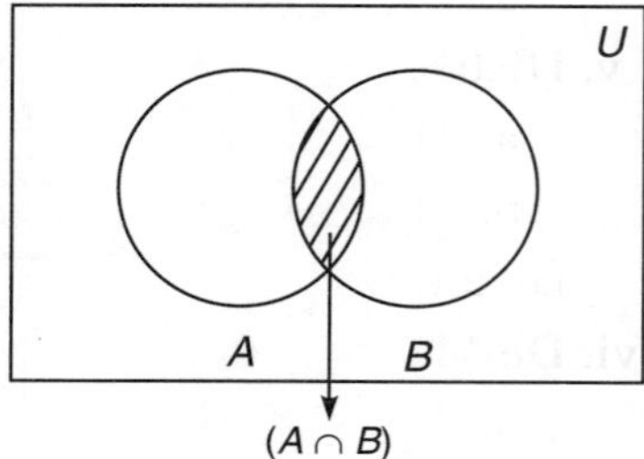

Fig. 1.2

ii. Intersection of Sets: Let A and B be two sets. The intersection of A and B is the set of all those elements belong to both A and B. The intersection of A and B is denoted by $A \cap B$ (read as 'A intersection B'). Thus,

$$A \cap B = \{x : x \in A \text{ and } x \in B\}$$

In Fig. 1.2, the shaded portion represents $A \cap B$.

iii. Disjoint Sets: Two sets A and B are said to be disjoint, if $A \cap B = \phi$. If $A \cap B \neq \phi$, then A and B are said to be intersecting or overlapping sets.

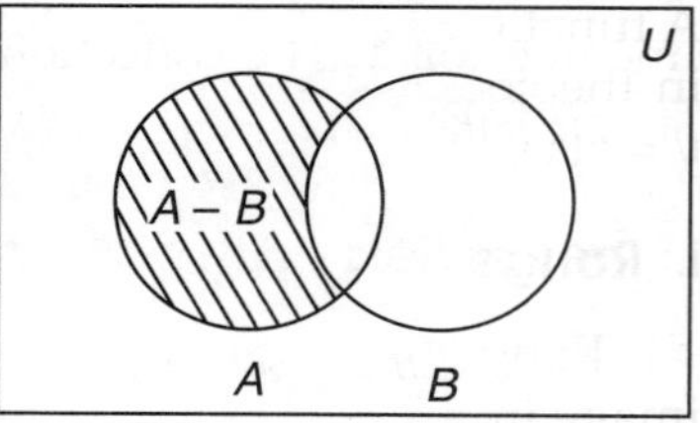

Fig. 1.3

iv. Difference of Sets: Let A and B be two sets. The difference of sets A and B written as $A - B$, is the set of all those elements in A that are not in B. Thus,

$$A - B = \{x : x \in A \text{ and } x \notin B\}$$

In Fig. 1.3, the shaded portion represents $A - B$

v. Symmetric Difference of Two Sets: Set A and B be two sets. The symmetric difference of two sets A and B is the set $(A - B) \cup (B - A)$ and is denoted by $A \Delta B$. Thus,

$$A \Delta B = (A - B) \cup (B - A) = \{x : x \notin A \cap B\}$$

The shaded portion in Fig. 1.4 represents $A \Delta B$.

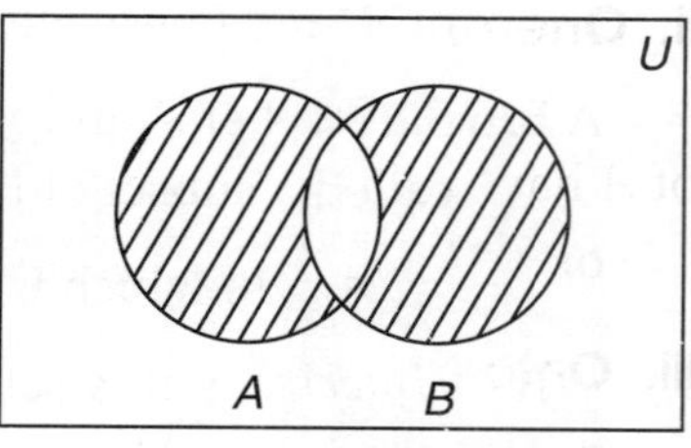

Fig. 1.4

vi. Complement of Set: Let U be the universal set and let A be a set such that $A \subset U$. Then the complement of A, with respect to U is denoted by A' or A^C or $U - A$ and is defined as the set of all elements of U which are not in A [refer to Fig. 1.5].

Thus, $$A' = \{x \in U : x \notin A\}$$

Clearly, $$x \in A' \Leftrightarrow x \notin A.$$

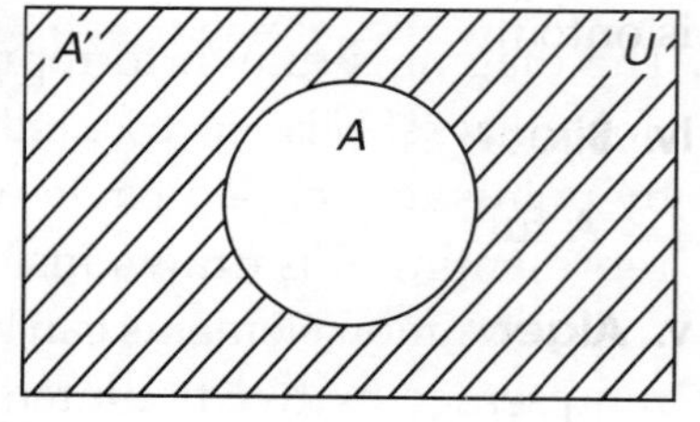

Fig. 1.5

vii. Cartesian Product of Sets: Let A and B be two nonempty sets, then the cartesian product of A and B is the set of ordered pairs (a, b), where $a \in A$ and $b \in B$. It is denoted as $A \times B$.

$$A \times B = \{(a, b): a \in A, b \in B\}$$

1.4 LAWS OF ALGEBRA OF SETS

i. Idempotent Laws: For any set A;

a. $A \cup A = A$ b. $A \cap A = A$

ii. Identity Laws: For any set A;

a. $A \cup \phi = A$ b. $A \cap U = A$

i.e. ϕ and U are identity elements for union and intersection respectively.

iii. Commutative Laws: For any two sets A and B

a. $A \cup B = B \cup A$ b. $A \cap B = B \cap A$

i.e. union and intersection are commutative.

iv. Associative Laws: If A, B and C are any three sets, then

a. $(A \cup B) \cup C = A \cup (B \cup C)$ b. $A \cap (B \cap C) = (A \cap B) \cap C$

v. Distributive Laws:

a. $A \cup (B \cap C) = (A \cup B) \cap (A \cup C)$

b. $A \cap (B \cup C) = (A \cap B) \cup (A \cap C)$

i.e. union and intersection are distributive over intersection and union respectively.

vi. De-Morgan's Laws: If A and B are any two sets, then

a. $(A \cup B)' = A' \cap B'$ b. $(A \cap B)' = A' \cup B'$

1.5 FUNCTION

A function, $f: A \rightarrow B$ is a rule that assigns to every element of domain of A, a unique element in the codomain B. In other words, for every $x \in A$, there exist a unique $y \in B$ such that $y = f(x)$. Here, y is said to be the image of x and x is the preimage y.

i. Range of a Function

Range of $f: A \rightarrow B$ is a subset of B that contains all the elements of B that have a pre-image in A.

$$R_{ng} f = \{y \in B \mid y = f(x) \text{ for some } x \in A\}$$

ii. One-one Function/Injection

A function $f: A \rightarrow B$ is said to be an injection or a one-one function if distinct elements of A have distinct images. If $x_1, x_2 \in A$ such that $x_1 \neq x_2$ then $f(x_1) \neq f(x_2)$

or $f(x_1) = f(x_2)$ implies $x_1 = x_2$

iii. Onto Function/Surjection

A function $f: A \rightarrow B$ is said to be a surjection or an onto function if every element of B has a preimage in A.

In other words, for every $y \in B$, there exist some $x \in A$ such that $y = f(x)$. If a function f is onto then $R_{ng} f =$ codomain of f.

iv. Bijection

A function $f: A \rightarrow B$ is said to be a bijection if it is both an injection and a surjection.

v. Algebra of Continuous Functions

1. If f and g are continuous functions then $f + g, f - g, f \cdot g, f / g$ are all continuous.
2. If f is a continuous function and c is any scalar then cf is a continuous function.

vi. Algebra of Differentiable Function

1. If f and g are differentiable functions on (a, b) then $f \pm g, f \cdot g, f / g$ are all differentiable on $[a, b]$.
2. If f is a differentiable functions on (a, b) and c is any scalar then cf is also differentiable on (a, b).

vii. Intervals

Closed interval: Let a and b be two given real numbers such that $a < b$. Then the set of real numbers x such that $a \leq x \leq b$ is called a closed interval and is denoted by $[a, b]$.

$$[a, b] = [x \in R \mid a \leq x \leq b]$$

Open interval: Let a and b be two given real numbers such that $a < b$. Then the set of all real numbers x such that $a \leq x \leq b$ is called a closed loop interval and is denoted by (a, b) (set of all real point numbers excluding the end points).

viii Real Function

Some standard real functions which are frequently met with calculus:

a. Constant function
b. Modulus function
c. Logarithmic function
d. Polynomial function
e. Inverse trignometrical functions
f. Reciprocal function
g. Identify function
h. Signum function
i. Exponential functions
j. Trignometrical function
k. Rational function

1.6 RELATION

Let A and B be two nonempty sets. A relation R on $A \times B$ is a subset of $A \times B$

$$A \times B = \{(a, b) \mid a \in A, b \in B\}$$

$$R \subseteq A \times B$$

If $(a, b) \in R$ then it is said that a is R, related to b and is written as aRb

If $A = B$, then R is said to be a relation on A.

i. Reflexive Relation

Let R be a relation defined on A. R is said to be a reflexive relation if $(a, a) \in R$ for every $a \in A$. In other words, aRa for every $a \in A$.

ii. Symmetric Relation

Let R be a relation defined on A. R is said to be symmetric relation if $(a, b) \in R$ implies $(b, a) \in R$, i.e. aRb implies bRa.

iii. Antisymmetric Relation

Let R be a relation defined on A. R is said to be of antisymmetric relation if $(a, b) \in R$ and $(b, a) \in R$ implies $A = b$, i.e. aRb and bRa implies $a = b$.

iv. Transitive Relation

A relation R defined on A, is said to be transitive if $(a, b) \in R$ and $(b, c) \in R$ implies $(a, c) \in R$, i.e. aRb and bRc implies aRc.

v. Equivalence Relation

A relation R defined on A, is an equivalence relation if it is reflexive, symmetric and transitive.

vi. Partial Order Relation

A relation R defined on A, is a partial order relation, if it is reflexive, anti-symmetric and transitive.

1.7 LIMIT OF A FUNCTION

Let f be a function defined on the interval $[a, b]$. Let $x_0 \in [a, b]$ then the limit of f at $x = x_0$ is denoted by $\lim\limits_{x \to x_0} f(x)$ and the limit is said to exist, if the left hand limit at x_0 is equal to the right hand limit at x_0. If the left hand limit and the right hand limit both exist but are not equal at $x = x_0$, then it is said that $\lim\limits_{x \to x_0} f(x)$ does not exist.

ε–δ Definition of Limit

If for given $\varepsilon > 0$, there exist a $\delta > 0$ such that

$|f(x) - L| < \varepsilon$ whenever $0 < |x - x_0| < \delta$ then

$$\lim_{x \to x_0} f(x) = L$$

Continuity of a function at a point. A function $f(x)$ is said to be continuous at a point $x = x_0$ if $\lim\limits_{x \to x_0} f(x) = f(x_0)$.

In other words, a function f defined on $[a, b]$ is said to be continuous at $x = x_0$ if the following three conditions hold:

1. Both left hand limit and right hand limit exist at $x = x_0$.
2. Left hand limit equals to the right hand limit at $x = x_0$ (say they are both L).
3. $L = f(x_0)$, i.e. the value of the function at $x = x_0$.

i. Continuous Function

Let f be a function defined on $[a, b]$. Then, f is said to be continuous on $[a, b]$ if it is continuous everywhere in the interval $[a, b]$.

Differentiability of a function at a point: A continuous function f defined on the interval $[a, b]$ is said to be differentiable at a point $x_0 \in (a, b)$ if

$$\lim_{h \to 0} \frac{f(x_0 + h) - f(x_0)}{h} \quad \text{exists}$$

and, the value of this limit is called the derivative of f at $x = x_0$.

$$f'(x_0) = \lim_{h \to 0} \frac{f(x_0 + h) - f(x_0)}{h}$$

ii. Algebra of Limits

1. $\lim\limits_{x \to c}[f(x) \pm g(x)] = \lim\limits_{x \to c} f(x) \pm \lim\limits_{x \to c} g(x)$
2. $\lim\limits_{x \to c}[f(x) \cdot g(x)] = \left[\lim\limits_{x \to c} f(x)\right] \cdot \left[\lim\limits_{x \to c} g(x)\right]$
3. $\lim\limits_{x \to c}[f(x) / g(x)] = \left[\lim\limits_{x \to c} f(x)\right] \Big/ \left[\lim\limits_{x \to c} g(x)\right];\ \lim\limits_{x \to x_0} g(x) \neq 0$
4. $\lim\limits_{x \to c}[c\, f(x)] = c \lim\limits_{x \to c} f(x)$
5. $\lim\limits_{x \to c}[f(x)]^n = \left[\lim\limits_{x \to c} f(x)\right]^n$

iii. A Few Theorems on Limits

1. $\lim\limits_{x \to a} \dfrac{x^n - a^n}{x - a} = n\, a^{n-1}$, $a > 0$ and n is an integer
2. $\lim\limits_{x \to 0} e^x = 1$
3. $\lim\limits_{x \to 0} \dfrac{e^x - 1}{x} = 1$
4. $\lim\limits_{x \to 0} \dfrac{a^x - 1}{x} = \log a$
5. $\lim\limits_{x \to 0} (1 + x)^{1/x} = e$
6. $\lim\limits_{x \to 0} \dfrac{1}{x}[\log(1 + x)] = 1$

CHAPTER 2

Algebra

2.1 ALGEBRA

Algebra is the branch of mathematics that involves in the study of mathematical symbols and the laws to manipulate these symbols. The basic parts of algebra, such as solution of equations, addition, subtraction, multiplication, division, etc. form the elementary algebra, whereas the more abstract form, such as groups, rings, fields, vector spaces, etc. form abstract algebra or modern algebra.

i. Linear Algebra

It is a part of elementary algebra. It includes the theory and application of linear systems of equation (briefly called linear system), linear transformation and eigen value problems, frameworks in mechanics, curve fitting and other optimization problems, systems of differential equation, and processes in statistics.

Linear algebra makes systematic use of vectors, matrices and determinants. This requires the study of properties of matrices as a central task by itself.

ii. Laws of Algebra

Fundamental properties (real numbers)

Commutative law

$a + b = b + a$

$ab = ba$

Distributive law

$a(b + c) = ab + ac$

$(a + b)c = ac + bc$

Associative law

$a + (b + c) = (a + b) + c$

$a(bc) = (ab)c$

Division law

If $ab = 0$, then $a = 0$ or $b = 0$

Law of additive identity

$a + 0 = 0 + a$

Law of additive inverse

$a + (-a) = (-a) + a = 0$

Law of multiplication inverse

$a\,(1/a) = (1/\,a)\,a = 1, a \neq 0$

Law of multiplication identity

$(a)(1) = (1)\,(a) = a$

iii. Algebraic Rules of Signs

Summation	Multiplication	Division
$a + (+b) = a + b$	$(+a)(+b) = +ab$	$\frac{+a}{+b} = +\frac{a}{b}$
$a + (-b) = a - b$	$(+a)(-b) = -ab$	$\frac{+a}{-b} = -\frac{a}{b}$
$a - (+b) = a - b$	$(-a)(+b) = -ab$	$\frac{-a}{+b} = -\frac{a}{b}$
$a - (-b) = a + b$	$(-a)(-b) = +ab$	$\frac{-a}{-b} = +\frac{a}{b}$

iv. Algebraic Powers and Roots

$a^k = aa\ldots a$ (k times)	$a^0 = 1$*	$a^1 = a$
$a^{-k} = \frac{1}{a^k}$	$\frac{1}{a^{-k}} = a^k$	
$a^m a^n = a^{m+n}$	$(ab)^k = a^k b^k$	
$\frac{a^m}{a^n} = a^{m-n}$	$\left(\frac{a}{b}\right)^k = \frac{a^k}{b^k}$	
$(a^m)^n = +a^{mn}$	$(a^m)^{-n} = \frac{1}{a^{mn}}$	
$(\pm a)^{2k} = +a^{2k}$	$(\pm a)^{2k+1} = \pm a^{2k+1}$	

$a^0 = 1$, for all a

v. Roots

$\sqrt[k]{a} = a^{1/k}$	$\sqrt[0]{a} = \infty$*	$\sqrt[1]{a} = a$
$\sqrt[-k]{a} = \frac{1}{\sqrt[k]{a}}$	$\frac{1}{\sqrt[-k]{a}} = \sqrt[k]{a}$	
$\sqrt[m]{a}\sqrt[n]{a} = a^{1/mn}$	$\sqrt[k]{ab} = \sqrt[k]{a}\sqrt[k]{b}$	
$\frac{\sqrt[m]{a}}{\sqrt[n]{a}} = a^{(n-m)/mn}$	$\sqrt[k]{a/b} = \frac{\sqrt[k]{a}}{\sqrt[k]{b}}$	
$\sqrt[n]{a^m} = a^{m/n}$	$\sqrt[-n]{a^m} = \frac{1}{a^{m/n}}$	
$\sqrt[2k]{a^{2k}} = a; a > 0$	$\sqrt[2k+1]{a^{2k+1}} = a; a > 0$	

* If $a > 1$

2.2 LOGARITHMS

i. Definition

If $\log_a x$ is a number N such that

$$a^N = x, \text{ i.e. } \log_a x = N$$

$$\Leftrightarrow \qquad x = a^N$$

Here a is known as the base of the logarithm. Generally the letter e is taken as the base of the logarithm unless otherwise stated. Logarithms of numbers calculated to the base e are called *Naperian logarithms* or *natural logarithms* and logarithms to the base 10 are known as common logarithms.

ii. Formulas of Logarithm

Logarithm to the base 10 ($\log_{10} x$)		Logarithm to the base e ($\ln_e x$)	
$10^{\infty} = \infty$	$\log \infty = \infty$	$\ln \infty = \infty$	$e^{\infty} = \infty$
$10^2 = 100$	$\log 100 = 2$	$\ln e^2 = 2$	$e^2 = e^2$
$10^1 = 10$	$\log 10 = 1$	$\ln e = 1$	$e^1 = e$
$10^0 = 1$	$\log 1 = 0$	$\ln 1 = 0$	$e^0 = 1$
$10^{-1} = \frac{1}{10}$	$\log \frac{1}{10} = -1$	$\ln \frac{1}{e} = -1$	$e^{-1} = \frac{1}{e}$
$10^{-2} = \frac{1}{100}$	$\log \frac{1}{100} = -2$	$\ln \frac{1}{e^2} = -2$	$e^{-2} = \frac{1}{e^2}$
$10^{-\infty} = 0$	$\log 0 = -\infty$	$\ln 0 = -\infty$	$e^{-\infty} = 0$

iii. Transformations

Change of base ($a \neq 1$): $\log_b x = \log_a x \cdot \log_b a$

$$\log x = \frac{\ln x}{\ln 10} = (0.43429....) \ln x \qquad \ln x = \frac{\log x}{\log e} = (2.30259...) \log x$$

iv. Operations of Logarithms

$\log_b xy = \log_b x + \log_b y$

$\log_b \frac{x}{y} = \log_b x - \log_b y$

$\log_b x^k = k \log_b x$

$\log_b \sqrt[k]{x} = \frac{1}{k} \log_b x$

$\log_b 10^k = k$

$\log_b (1/x) = -\log_b (x)$

$\log_b b = 1$

$\log_b 1 = 0$

$\log_b e = 0.43429...$

$\ln xy = \ln x + \ln y$

$\ln \frac{x}{y} = \ln x - \ln y$

$\ln x^k = k \ln x$

$\ln \sqrt[k]{x} = \frac{1}{k} \ln x$

$\ln e^k = k$

$\ln (1/x) = -\ln (x)$

$\ln e = 1$

$\ln 10 = 2.30259...$

v. Characteristic of Common Logarithm

A common logarithm consists of an integer, called the *characteristic*, and a decimal called the *mantissa*. The mantissa is a tabulated value. The characteristic is determined from the following.

$\log 100 = 2.00$	$\log 0.1 = -1$
$\log 10 = 1.00$	$\log 0.01 = -2$
$\log 1 = 0.00$	$\log 0.001 = -3$

2.3 BINOMIAL THEORY

i. Factorial Function

$$n! = n(n-1)\,(n-2).....(2)(1)$$
$$n! = n(n-1)!$$
$$n! = n(n-1)\,(n-2)!$$

....

$$0! = 1$$

ii. Table of Factorial n!

Index of binomial n	$n!$						
1							1
2							2
3							6
4							24
5							120
6							720
7						5	040
8						40	320
9						362	880
10					3	628	800
11					39	916	800
12					479	001	600
13				6	227	020	800
14				87	178	291	200
15			1	307	674	368	000
16			20	922	789	888	000
17			355	687	428	096	000
18		6	402	373	705	728	000
19		121	645	100	408	832	000
20	2	432	902	008	176	640	000

iii. For Large Value of *n*

For large n: the function becomes very large.

A convenient approximation for large n is the Stirling formula expressed as

$$n! \sim \sqrt{2\pi n}\left(\frac{n}{e}\right)^n \qquad \ldots (e = 2.718)$$

$$n! = \left(\frac{n}{e}\right)^2 \sqrt{2n\pi}\left(1+\frac{1}{12n}+\frac{1}{288n^2}+\frac{139}{51{,}840n^3}+\ldots.\right)$$

$$\ln n! = \left(n+\frac{1}{2}\right)\ln n - n + \ln\sqrt{2\pi}$$

n may be a fraction. The error for $n = 1$ is less than 10% and decreases as n increases.

iv. Binomial Coefficients

The binomial coefficients are defined by the following formula:

$${}^nC_k \text{ or } \binom{n}{k} = \frac{n(n-1)(n-2)\ldots(n-k+1)}{(1)(2)(3)\ldots.k} = \frac{n!}{(n-k)!\,k!} \qquad \ldots (k \geq 0, \text{ integer})$$

$$\binom{n+1}{k+1} = \binom{n}{k+1} + \binom{n}{k}, \quad \binom{n+1}{k} = \binom{n}{k}\frac{n+1}{n-k+1} \qquad \ldots (k \geq 0, \text{ integer})$$

$$\binom{n}{k} = \binom{n}{n-k}, \quad \binom{n}{0} = 1, \quad \binom{n}{n-1} = n, \quad \binom{n}{l} = n.$$

v. Factors and Expansion

$$(a + b)^2 = a^2 + b^2 + 2ab$$

$$(a - b)^2 = a^2 + b^2 - 2ab$$

$$(a + b)^3 = a^3 + b^3 + 3a^2b + 3ab^2$$

$$(a - b)^3 = a^3 - b^3 - 3a^2b + 3ab^2$$

$$(a^2 - b^2) = (a - b)(a + b)$$

$$(a^3 - b^3) = (a - b)(a^2 + ab + b^2)$$

$$(a^3 + b^3) = (a + b)(a^2 - ab + b^2)$$

vi. Binomial Theorem

For positive integer n, if x and y are real numbers, than for all $n \in N$

$$(x + y)^n = x^n + n\,x^{-1}.\,y + \frac{n(n-1)}{2!}x^{n-2}.\,y^2 + \frac{n(n-1)(n-2)}{3!}x^{n-3}.\,y^3 + n\,x\,y^{n-1} + y^n$$

It can be expressed as:

$$(a+b)^n = a^n + \binom{n}{1} a^{n-1} b + \binom{n}{2} a^{n-2} b^2 + \ldots + b^n$$

$$= \sum_{k=0}^{n} \binom{n}{k} a^{n-k} b^k$$

vii. Pascal's Triangle of $\binom{n}{k}$

Index of binomial	Coefficient of various terms
$(a+b)^0 \rightarrow$	1
$(a+b)^1 \rightarrow$	1 1
$(a+b)^2 \rightarrow$	1 2 1
$(a+b)^3 \rightarrow$	1 3 3 1
$(a+b)^4 \rightarrow$	1 4 6 4 1

. . . .

viii. Binomial Sums

$$\binom{n}{k} + \binom{n-1}{k} + \binom{n-2}{k} + \ldots + \binom{k}{k} = \binom{n+1}{k+1}$$

$$\binom{n}{0} + \binom{n}{1} + \binom{n}{2} + \ldots + \binom{n}{n} = 2^n$$

$$\binom{n}{0} - \binom{n}{1} + \binom{n}{2} - \ldots + (-1)^n \binom{n}{n} = 0$$

ix. Special Cases

$$(1 \pm x)^n = 1 \pm \binom{n}{1} x + \binom{n}{2} x^2 \pm \ldots$$

If n is a positive integer, the series is finite.

$$\left(1 \pm \frac{1}{n}\right)^n = 1 \pm \binom{n}{l} n^{-1} + \binom{n}{2} n^{-2} \pm \ldots$$

If n is a negative integer or fraction, the series is infinite.

If $n \rightarrow \infty$, then $\left(1 + \frac{1}{n}\right)^n = 1 + \frac{1}{1!} + \frac{1}{2!} + \frac{1}{3!} + \ldots = e = 2.718281\ldots$ (Euler's number).

x. Progression

An arithmetic progression is a sequence in which the difference between only term and the proceeding term is a constant (d).

$$a, a+d, a+2d, \ldots, a+(n-1)d$$

If the last term is denoted by $l = [a + (n-1)d]$, then the sum $S = \frac{n}{2}(a + l)$

A geometric progression is a sequence in which the ratio of any term to the proceeding term is a constant r, for n terms

$$a, ar, ar^2, \ldots, ar^{n-1}$$

$$\text{then the sum } S = \frac{a - ar^n}{1 - r}$$

2.4 RATIONAL ALGEBRAIC FUNCTIONS

i. Integral Function

a. Every integral rational algebraic function of independent variable x is given by

$$f(x) = a_0 + a_1 x + a_2 x^2 + \ldots + a_n x^n = \sum_{k=0}^{n} a_k x^k$$

can be expressed as product of real linear factors of the form $c + dx$ and of real irreducible polynomials.

b. If two of these functions of the same degree are equal for all values of x, then the coefficients of like powers are equal.

$$\sum_{k=0}^{n} a_k x^k = \sum_{k=0}^{n} b_k x^k \quad \Rightarrow \quad a_0 = b_0, a_1 = b_1, \ldots, a_n = b_n$$

ii. Fractional Function

a. A rational algebraic function of independent variable x, where $g(x)$ and $h(x)$ are integral rational algebraic function; is called a *rational algebraic fraction*.

$$f(x) = \frac{g(x)}{h(x)}$$

b. If the degree of $g(x)$ is less than the degree of $h(x)$, $f(x)$ is called *proper*. If the opposite is true, $f(x)$ is called *improper*.

c. Every proper, rational algebraic fraction can be resolved into a sum of simpler fractions, whose denominators are of the form $(c + dx)^k$ and $(e + fx + gx^2)^l$, k and l being positive integers.

$$\frac{g(x)}{h(x)} = \sum_k \left[\frac{b_{k1}}{(x - x_k)} + \frac{b_{k2}}{(x - x_k)^2} + \frac{b_{km}}{(x - x_k)^m} \right]$$

iii. Coefficients

The coefficients b_{kj} are obtained by one of the following methods:

a. If $m = 1$ (x_k = distinct root), then

$$b_{kj} = \frac{g(x_k)}{f'(x_k)}$$

in which $f'(x_k)$ is the first derivative of $f'(x)$ with respect to x evaluated for $x = x_k$.

b. Multiply both sides of $h(x)$, and use the above theorem.

c. Multiply both sides by $h(x)$, and differentiate successively. Solve this set of equations for $b_{km}, b_{km-1}, \ldots, b_{k1}$.

The partial fractions corresponding to any pair of complex-conjugate roots $a_k + i\alpha_k$, $a_k - i\alpha_k$ of order m may be combined into

$$c_{kj} \frac{x + d_{kj}}{[(x - a_k)^2 + \alpha_k^2]^j}$$

2.5 EQUATIONS OF HIGHER DEGREE

Consider equation of the type

$$ax^2 + bx + c = 0, \qquad a \neq 0$$

The roots of this equation are given as follows:

$$x_{1,2} = \frac{-b \pm \sqrt{b^2 - 4ac}}{2a}$$

If a, b, c are real and if

$b^2 - 4ac > 0$, the roots are real and unequal;

$b^2 - 4ac = 0$, the roots are real and equal;

$b^2 - 4ac < 0$, the roots are imaginary and unequal.

To get the roots of a polynomial of type: $ax^4 + bx^2 + c = 0$,

substitute $y = x^2$ to derive $y^2 + py + q = 0; \qquad a \neq 0$

This reduces by the substitution

$y = x^2$ to $y^2 + py + q = 0$

in which $p = b/a$; and $q = c/a$

$$x_{1,2,3,4} = \pm\sqrt{-\frac{p}{2} \pm \sqrt{\frac{p^2}{4} - q}}$$

Polynomial $x^n - a = 0$, $a \neq 0$ has the following roots

$$x_k = \sqrt[n]{a}\left(\frac{\cos 2k\pi}{n} + \frac{i \sin 2k\pi}{n}\right) \quad \text{where, } k = 0, 1, 2, \ldots n - 1$$

Polynomial $x^n + a = 0$, $a \neq 0$ has the following roots;

$$x_k = \sqrt[n]{a}\left(\frac{\cos (2k+1)\pi}{n} + i \sin \frac{(2k+1)\pi}{n}\right)$$

where, $k = 0, 1, 2, \ldots, n - 1$

In particular,

roots of $x^3 \pm a = 0; \quad a \neq 0$

$$x_1 = \sqrt[3]{a}\,; \qquad x_{2,3} = -\frac{(1 \pm i\sqrt{3})\,\sqrt[3]{a}}{2}$$

roots of $x^4 + a = 0; \quad a \neq 0$: are $x_{1,2,3,4}$

$$= \pm \frac{(1 \pm i)\sqrt[4]{a}}{\sqrt{2}}$$

Roots of $x^4 - a = 0$; $a \neq 0$: are $x_{1,2}$

$$= \pm \sqrt[4]{a}; \quad x_{3,4} = \pm i \sqrt[4]{a}$$

Consider a cubic equation of the form

$$ax^3 + bx^2 + cx + d = 0 \quad (a \neq 0)$$

This reduces by the substitution

$$x = y - \frac{b}{3a} \quad \text{to} \quad y^3 + py + q = 0$$

where,

$$p = \frac{1}{3}\left[3\left(\frac{c}{a}\right) - \left(\frac{b}{a}\right)^2\right]$$

$$q = \frac{1}{27}\left[2\left(\frac{b}{a}\right)^3 - 9\left(\frac{b}{a}\right)\left(\frac{c}{a}\right) + 27\left(\frac{d}{a}\right)\right]$$

$$y_1 = u + v; \quad y_2 = -\frac{u+v}{2} + \frac{u-v}{2} i\sqrt{3}$$

$$y_3 = -\frac{u+v}{2} - \frac{u-v}{2} i\sqrt{3}$$

If $D < 0$, a trigonometric formulation is useful.

$$y_1 = 2\sqrt{\frac{|p|}{3}} \cos\frac{\phi}{3}$$

$$y_2 = -2\sqrt{\frac{|p|}{3}} \cos\frac{\phi + \pi}{3}$$

$$y_3 = -2\sqrt{\frac{|p|}{3}} \cos\frac{\phi - \pi}{3}$$

$$D = \left(\frac{p}{3}\right)^3 + \left(\frac{q}{2}\right)^2$$

$$u = \sqrt[3]{-\frac{q}{2} + \sqrt{D}}; \quad v = \sqrt[3]{-\frac{q}{2} - \sqrt{D}}$$

If a, b, c, d are real and if

$D > 0$, there are one real and two conjugate imaginary roots;

$D = 0$, there are three real roots of which at least two are equal;

$D < 0$, there are three real unequal roots.

The value of ϕ is calculated from the expression

$$\cos\phi = \frac{-q/2}{\sqrt{|p|^3/27}}$$

2.6 DETERMINANTS

i. Definition

a. A *determinant* of the nth order contains $n \times n$ elements, arranged in n rows and n columns. If $A = [a_{ij}]$ is a square matrix of order n, then the determinant A is denoted by determine A or $|A|$.

$$\text{Det } A = \begin{vmatrix} a_{11} & a_{12} & a_{1n} \\ a_{21} & a_{22} & a_{24} \\ \cdots & \cdots & \cdots \\ a_{n_1} & a_{n_2} & a_{n_n} \end{vmatrix}$$

A number equal to $\sum(\pm 1)\, a_{1i}\, a_{2j}\, a_{3k} \ldots a_{nl}$, where $i, j, k, \ldots l$ is a permutation of the n integers $1, 2, 3 \ldots n$ in some order. The sign is plus if the permutation is even and in minus if the permutation is odd. The 2×2 determinant $\begin{vmatrix} a_{11} & a_{12} \\ a_{21} & a_{22} \end{vmatrix}$ has the value $a_{11}a_{22} - a_{12}a_{21}$, since the permutation (1, 2) is even and (2, 1) is odd. For 3×3 determinants; permutation are as follows:

$$\text{Det } A = \begin{vmatrix} a_{11} & a_{12} & a_{13} \\ a_{21} & a_{22} & a_{23} \\ a_{31} & a_{32} & a_{33} \end{vmatrix}$$

$$= a_{11}(a_{22}a_{33} - a_{32}a_{23} - a_{12}\,(a_{21}a_{33} - a_{31}a_{23}) + a_{13}\,(a_{21}a_{32} - a_{22}a_{31})$$

A determinant of order n is seen to be the sum of ϕ $n!$ signed product.

b. A *minor* M_{jk} of the element a_{jk} in the nth order determinant is the $(n - 1)$ th order determinant obtained from A by deleting the jth row and the kth column.

ii. Evaluation by Cofactors

Each element a_{ij} has determinant of order $(n-1)$ called a minor (M_{ij}) obtained by suppresing all elements in row i and column j. For example, the minor of element a_{22} in the 3×3 determinant above is

$$\begin{vmatrix} a_{11} & a_{13} \\ a_{31} & a_{33} \end{vmatrix}$$

The cofactor of element a_{ij}, denoted A_{ij} is defined as $\pm M_{ij}$, where the sign is determined from i and j.

$$A_{ij} = (-1)^{i+j} M_{ij}$$

The value of the $n \times n$ determinant equals the sum of products of elements of any row or (column) and their respective cofactors. Thus for the 3×3 determinant, i.e.

$$\text{det. } A = a_{11}A_{11} + a_{12}A_{12} + a_{13}A_{13} \text{ (first row)}$$

or $$= a_{11}A_{11} + a_{21}A_{21} + a_{31}A_{31} \text{ (first column) etc.}$$

2.7 MATRICES

i. Definition

A *matrix* is a rectangular array of *elements* arranged in rows and columns.

m = number of rows; a_{jk} = any element of the matrix in row j and column k

n = number of columns; $m \times n$ = dimension of matrix, the order of matrix is $m \times n$ ('m' by 'n')

$$[A] = \begin{bmatrix} a_{11}\ a_{12} \dots a_{1n} \\ a_{21}\ a_{22} \dots a_{2n} \\ \dots\dots\dots\dots \\ a_{m1}\ a_{m2} \dots a_{mn} \end{bmatrix} = [a_{jk}]$$

ii. Types of Matrices

a. *Square matrix.* A matrix A in which number of rows is equal to the number of columns is called a square matrix.

b. *Upper triangular matrix.* A matrix A in which entries below the main diagonal are zero.

c. *Lower triangular matrix.* A matrix A in which the entries above the main diagonal are zero.

d. *Unit matrix*

$$\begin{bmatrix} 1 & 0 & 0 \\ 0 & 1 & 0 \\ 0 & 0 & 1 \end{bmatrix}$$

e. *Diagonal matrix*

$$\begin{bmatrix} a & 0 & 0 \\ 0 & b & 0 \\ 0 & 0 & c \end{bmatrix}$$

f. *Zero matrix*

$$\begin{bmatrix} 0 & 0 & 0 \\ 0 & 0 & 0 \\ 0 & 0 & 0 \end{bmatrix}$$

g. *Symmetrical matrix*

$$A = A^T$$

$$\begin{bmatrix} a & d & l \\ d & b & f \\ l & f & c \end{bmatrix}$$

h. *Skew symmetric matrix*

$$A = -A^T$$

$$\begin{bmatrix} 0 & +d & -l \\ -d & 0 & +f \\ +l & -f & 0 \end{bmatrix}$$

i. *Point symmetrical matrix*

$$\begin{bmatrix} a & d & c \\ d & b & d \\ c & d & a \end{bmatrix}$$

iii. Operations of Matrices

a. *Equal matrices*: Two matrices $[A]$ and $[B]$ are equal, if they have the same dimensions and their corresponding elements are equal.

$$a_{jk} = b_{ik}$$

b. *Sum of matrices*: The sum of two or several $m \times n$ matrices is an $m \times n$ matrix, each of whose elements is equal to the sum of the corresponding elements of the initial matrices, i.e. matrices A, B or C of the same order be added as

$$A + B + C = [a_{jk} + b_{jk} + c_{jk} + \dots]$$

c. *Scalar–matrix multiplication*: The product of a scalar k and an $m \times n$ matrix $[A]$ is an $m \times n$ matrix, each of whose elements is equal to the product of the scalar and the corresponding element of $[A]$.

$$k[A] = [kA]$$

d. *Matrix–matrix multiplication*: A product of two rectangular conformable matrices of dimensions $m_1 \times n_1$ and $m_2 \times n_2$ is a rectangular (or square) matrix of dimensions $m_1 \times n_2$ whose elements are equal to the sum of products of the inner elements.

$$\begin{bmatrix} a_{11} & a_{12} & a_{13} \\ a_{21} & a_{22} & a_{23} \end{bmatrix} \begin{bmatrix} b_{11} & b_{12} \\ b_{21} & b_{22} \\ b_{31} & b_{32} \end{bmatrix} = \begin{bmatrix} a_{11}b_{11} + a_{12}b_{21} + a_{13}b_{31} & a_{11}b_{12} + a_{12}b_{22} + a_{13}b_{32} \\ a_{21}b_{11} + a_{22}b_{21} + a_{23}b_{31} & a_{21}b_{12} + a_{22}b_{22} + a_{23}b_{32} \end{bmatrix}$$

Two rectangular matrices are conformable, if $n_1 = m_2$.

2.8 MATRIX TRANSPOSE

a. *Definition*: The transpose $[A]^T$ of a matrix $[A]$ has each row identical with the corresponding column of $[A]$.

$$[A] = \begin{bmatrix} a \\ b \end{bmatrix} \qquad [B] = \begin{bmatrix} a & b \\ c & d \end{bmatrix} \qquad [C] = [a\ b]$$

$$[A]^T = [a\ b] \qquad [B]^T = \begin{bmatrix} a & c \\ b & d \end{bmatrix} \qquad [C]^T = \begin{bmatrix} a \\ b \end{bmatrix}$$

b. *Properties of transpose*:

Transpose of transpose

$[[A]^T]^T = [A]$

Transpose of unit matrix

$[I]^T = [I]$

Transpose of zero matrix

$[0]^T = [0]$

Transpose of diagonal matrix

$$\begin{bmatrix} a & 0 \\ 0 & d \end{bmatrix}^T = \begin{bmatrix} a & 0 \\ 0 & d \end{bmatrix}$$

$[D]^T = [D]$

Transpose of symmetrical matrix

$$\begin{bmatrix} a & b \\ b & c \end{bmatrix}^T = \begin{bmatrix} a & b \\ b & c \end{bmatrix}$$

$[E]^T = [E]$

Transpose of skew symmetric matrix

$$\begin{bmatrix} 0 & b \\ -b & 0 \end{bmatrix}^T = \begin{bmatrix} 0 & -b \\ b & 0 \end{bmatrix}^T$$

$[F]^T = -[F]$

c. *Transpose of product*: The transpose of product of two or more matrices is equal to the product of their transposes in reverse order.

$$[[A]\,[B]\,[C]\,[D]]^T = [D]^T\,[C]^T\,[B]^T\,[A]^T$$

d. *Matrix transpose—sum and difference*: The sum of a square matrix and its transpose is a symmetrical matrix. The difference of a square matrix and its transpose is a skew symmetric matrix.

$$\underset{[B]}{\begin{bmatrix} a & b \\ c & d \end{bmatrix}} + \underset{[B]^T}{\begin{bmatrix} a & c \\ b & d \end{bmatrix}} = \underset{\text{Symmetrical}}{\begin{bmatrix} 2a & b+c \\ c+b & 2d \end{bmatrix}} \qquad \underset{[B]}{\begin{bmatrix} a & b \\ c & d \end{bmatrix}} - \underset{[B]^T}{\begin{bmatrix} a & c \\ c & d \end{bmatrix}} = \underset{\text{Skew symmetric}}{\begin{bmatrix} 0 & b-c \\ c-b & 0 \end{bmatrix}}$$

e. *Matrix transpose—product*: The product of a square matrix and its transpose, or vice versa, is a symmetrical matrix.

$$\underset{[B]}{\begin{bmatrix} a & b \\ c & d \end{bmatrix}} \underset{[B]^T}{\begin{bmatrix} a & c \\ b & d \end{bmatrix}} = \underset{\text{Symmetrical}}{\begin{bmatrix} a^2+b^2 & ac+bd \\ ac+bd & c^2+d^2 \end{bmatrix}} \qquad \underset{[B]^T}{\begin{bmatrix} a & c \\ b & d \end{bmatrix}} \underset{[B]}{\begin{bmatrix} a & b \\ c & d \end{bmatrix}} = \underset{\text{Symmetrical}}{\begin{bmatrix} a^2+c^2 & ab+cd \\ ab+cd & b^2+d^2 \end{bmatrix}}$$

f. *Matrix resolution*: Every unsymmetrical square matrix can be expressed as the sum of a symmetrical matrix and a skew symmetric matrix.

g. *Scalar matrix*: A matrix in which, each element of the main diagonal is the same constant a and all other elements zero is called a scalar matrix.

$$\begin{bmatrix} a & 0 & 0 & \cdots & 0 \\ 0 & a & 0 & \cdots & 0 \\ 0 & 0 & a & \cdots & 0 \\ \cdots & \cdots & \cdots & \cdots & \cdots \\ 0 & 0 & 0 & \cdots & a \end{bmatrix}$$

A scalar matrix with diagonal elements I is called the identity or unit matrix and is denoted by I. Thus for any n the order matrix A, the identity matrix of order n has the property $AI = IA = A$.

2.9 MATRIX INVERSE

a. *Definition*

The inverse $[A]^{-1}$ of a square matrix $[A]$ is uniquely defined by the conditions

$[A]^{-1}[A] = [I] = [A][A]^{-1}$ and $|A| \neq 0$

If $|A| = 0$, the inverse of $[A]$ does not exist, and $[A]$ is said to be a *singular matrix*. Only nonsingular square matrices have inverses.

$$[A]^{-1} = \begin{bmatrix} a_{11} & a_{12} & \ldots & a_{1n} \\ a_{21} & a_{22} & \ldots & a_{2n} \\ \ldots & \ldots\ldots & & \\ a_{n1} & a_{n2} & \ldots & a_{nn} \end{bmatrix}^{-1} = \frac{1}{|A|}\begin{bmatrix} A_{11} & A_{21} & \ldots & A_{n1} \\ A_{12} & A_{22} & \ldots & A_{n2} \\ \ldots & \ldots\ldots & & \\ A_{1n} & A_{2n} & \ldots & A_{nn} \end{bmatrix} = \frac{[A_{jk}]}{|A|}$$

in which $|A|$ = determinant of $[A]$, A_{jk} = minor jk of $|A|$, and $[A_{jk}]$ = adjoint matrix of minors of $|A|$.

b. *Special cases of matrix inverse*

Inverse of inverse

$[[A]^{-1}]^{-1} = [A]$

Initial matrix

Inverse of unit matrix

$[I]^{-1} = [I]$

Unit matrix

Inverse of zero matrix

$[0]^{-1} = [0]$

Zero matrix

Inverse of diagonal matrix

$$\begin{bmatrix} a & 0 \\ 0 & d \end{bmatrix}^{-1} = \begin{bmatrix} \frac{1}{a} & 0 \\ 0 & \frac{1}{d} \end{bmatrix}$$

Reciprocal diagonal matrix

Inverse of symmetrical matrix

$$\begin{bmatrix} a & b \\ b & c \end{bmatrix}^{-1} = \frac{1}{ac-b^2}\begin{bmatrix} c & -b \\ -b & a \end{bmatrix}$$

Symmetrical matrix

Inverse of antisymmetrical matrix

$$\begin{bmatrix} 0 & b \\ -b & 0 \end{bmatrix}^{-1} = \frac{1}{b^2}\begin{bmatrix} 0 & -b \\ b & 0 \end{bmatrix}$$

Antisymmetrical matrix

c. *Inverse of product*

The inverse of the product of two or more matrices is equal to the product of their inverses in reverse order.

$$[[A]\,[B]\,[C]\,[D]]^{-1} = [D]^{-1}\,[C]^{-1}\,[B]^{-1}\,[A]^{-1}$$

d. *Normal matrix*

A square matrix is said to be normal if it is equal to its transpose. All symmetrical matrices are normal.

If $[A] = [A]^T$ then $[A]\,[A]^T = [A]^T\,[A] = [A]^2$

e. *Orthogonal matrix*

A square matrix is said to be orthogonal if its transpose is equal to its inverse.

If $[A]^T = [A]^{-1}$ then $[A]\,[A]^T = [A]^T\,[A] = [I]$

f. *Adjoint*

If A is an nth-order square matrix and A_{ij} the cofactor of element a_{ij}, the transpose of $[A_{ij}]$ is called the adjoint of A.

$$\text{adj } A = [A_{ij}]^T$$

g. *Properties of matrices*:

If A, B and C are three matrices of the same order. Then

$A + B = B + A$	Commutative property
$A + (B + C) = (A + B) + C$	Associative property
$(C_1 + C_2)\,A = C_1A + C_2A$	Scalar multiplication property
$C(A + B) = CA + CB$	Scalar multiplication property
$C_1\,(C_2A) = (C_1C_2)\,A$	Matrix multiplication is associative
$(AB)\,C = A\,(BC)$	Distribution matrix multiplication
$(A + B)\,(C) = AC + BC$	Matrix multiplication is distributive
$AB \neq BA$ (in general)	Matrix multiplication is not commutative (in general)

2.10 COMPLEX MATRICES

Definition

a. *Complex matrix*

A matrix $A_{m\times n}$ is said to be a complex matrix if the entries of the matrix are complex numbers.

$$C = \begin{bmatrix} a_1 + ib_1 & a_2 + ib_2 \\ a_3 + ib_3 & a_4 + ib_4 \end{bmatrix}$$

Conjugate of a matrix: Let A be an $m \times n$ matrix with complex entries. Conjugate of A, denoted as A^θ is an $m \times n$ matrix with entries that are conjugates of the corresponding complex entries of A.

b. *Hermitian matrix*

A complex matrix A is said to be Hermitian if its tranjugate (transpose of conjugate matrix) is equal to A.

$$A = (A^\theta)^T$$

c. *Skew–hermitian matrix*

A complex matrix A is said to be skew–hermitian if its tranjugate (transpose of conjugate matrix) is equal to A.

$$A = (A^{\theta})^T$$

d. *Unitary matrix*

A complex matrix A is said to be unitary if tranjugate of A is equal to inverse of A.

$$(A^{\theta})^T = (A)^{-1} \quad \text{or} \quad (A^{\theta})^T \cdot A = A \cdot (A^{\theta})^T = I.$$

e. *Involutory matrix*

A square conjugate-complex matrix which is hermitian and unitary is called *involutory*. The zero matrix and the unit matrix are special cases of the involutory matrix.

If $[A^{\theta}] = [A^{\theta}]^T = [A^{\theta}]^{-1}$ then $[A^{\theta}]\,[A^{\theta}] = [I]$

f. *Table for real and conjugate—complex matrices*

Real matrix		**Conjugate–complex matrix**	
$A = B$		$A^{\theta} = B - Ci$	
Normal	$A = A^T$	Hermitian	$A^{\theta} = A^{\theta T}$
Antinormal	$A = -A^T$	Alternating	$A^{\theta} = -A^{\theta T}$
Orthogonal	$A^T = A^{-1}$	Unitary	$A^{\theta T} = A^{\theta - 1}$
Antiorthogonal	$A^T = -A^{-1}$	Antiunitary	$A^{\theta T} = -A^{\theta - 1}$
Involutory	$A = A^T = A^{-1}$	Involutory	$A^{\theta} = A^{\theta T} = A^{\theta - 1}$
Anti-involutory	$A = -A^T = -A^{-1}$	Anti-involutory	$A^{\theta} = -A^{\theta T} = -A^{\theta - 1}$

2.11 PROPERTIES OF MATRICES

Commutative Property

$A + B = B + A$

$[A + B]^T = [B + A]^T$

$[A + B]^{-1} = [B + A]^{-1}$

$AB \neq BA$

$[AB]^T \neq [BA]^T$

$[AB]^{-1} \neq [BA]^{-1}$

Associative Property

$A + (B + C) = (A + B) + C$

$[A + (B + C)]^T = [(A + B) + C]^T$

$[A + (B + C)]^{-1} = [(A + B) + C]^{-1}$

$A(BC) = (AB)C$

$[A(BC)]^T = [(AB)C]^T$

$[A(BC)]^{-1} = [(AB)C]^{-1}$

Distributive Property

$A[B + C] = AB + AC$

$[A + B]C = AC + BC$

Division Property

If $AB = 0$, then A and/or B may or may not be zero.

i. Property of Determinant Matrices $n \times n$

$\text{Det}\,(A^T) = \text{Det}\,(A)$	$\text{Det}\,(AB) = \text{Det}\,(A)\,\text{Det}\,(B)$
$\text{Det}\,(A^{-1}) = \dfrac{1}{\text{Det}\,(A)}$	$\text{Det}\,(AB^{-1}) = \dfrac{\text{Det}\,(A)}{\text{Det}\,(B)}$
$\text{Det}\,(I) = 1$	$\text{Det}\,(A^{-1}B^{-1}) = \dfrac{1}{\text{Det}\,(A)\,\text{Det}\,(B)}$
$\text{Det}\,[Adj\,(A)] = [\text{Det}\,(A)]^{n-1}$	$Adj\,(AB) = Adj\,(B)\,Adj\,(A)$

ii. Characteristics of Matrices

a. The **rank of a matrix** is the order of the largest nonzero determinant that can be obtained from the elements of the matrix. The matrix whose order exceeds its rank is singular.

b. The **trace of a matrix** is the sum of its diagonal elements.

iii. Relationships of Two Matrices

a. *Equivalence*

The square matrix A is equivalent to another square matrix B if there exist nonsingular matrices P and Q such that $A = PBQ$

b. *Congruence*

The square matrix A is congruent to another square matrix B if there exists a nonsingular matrix Q such that $A = Q^T BQ$

c. *Similarity*

The square matrix A is similar to another square matrix B if there exists a nonsingular matrix Q such that $A = Q^{-1} BQ$

2.12 PROPERTIES OF DETERMINANTS

a. The *value* of a determinant is *unchanged* if the corresponding rows and columns are interchanged.

$$\begin{vmatrix} a_{11} & a_{12} & a_{13} \\ a_{21} & a_{22} & a_{23} \\ a_{31} & a_{32} & a_{33} \end{vmatrix} = \begin{vmatrix} a_{11} & a_{21} & a_{31} \\ a_{12} & a_{22} & a_{32} \\ a_{13} & a_{23} & a_{33} \end{vmatrix} = D$$

b. The *sign* of a determinant is *changed* (unchanged) if an odd (even) number of interchanges of any two rows or of any two columns is introduced.

c. If a determinant has *two identical rows* (or columns) or if all the elements of one row (or column) are zero, then the value of the determinant is zero.

d. If each element of a row (or column) is *multiplied by m*, the *new determinant is equal to mD*.

$$\begin{vmatrix} a_{11} & a_{12} & a_{13} \\ ma_{21} & ma_{22} & ma_{23} \\ a_{31} & a_{32} & a_{33} \end{vmatrix} = m \begin{vmatrix} a_{11} & a_{12} & a_{13} \\ a_{12} & a_{22} & a_{23} \\ a_{31} & a_{32} & a_{33} \end{vmatrix} = mD$$

e. If each element of a row (or column) is *added m times* to the corresponding element in another row (or column), the value of the determinant is unchanged.

f. If the elements of any row (or column) are *linear combinations* of the corresponding elements of the other rows (or columns), the value of the determinant is zero.

$$\begin{vmatrix} ba_{21} + ca_{31} & ba_{22} + ca_{32} & ba_{23} + ca_{33} \\ a_{21} & a_{22} & a_{23} \\ a_{31} & a_{32} & a_{33} \end{vmatrix} = 0$$

g. If two *determinants differ* from each other only *in the elements of any one row* (or column), they may be added as follows:

$$\begin{vmatrix} a_{11} & a_{12} & a_{13} \\ a_{21} & a_{22} & a_{23} \\ a_{31} & a_{32} & a_{33} \end{vmatrix} + \begin{vmatrix} b_{11} & b_{12} & b_{13} \\ a_{21} & a_{22} & a_{23} \\ a_{31} & a_{32} & a_{33} \end{vmatrix} = \begin{vmatrix} a_{11}+b_{11} & b_{12}+b_{12} & a_{13}+b_{13} \\ a_{21} & a_{22} & a_{23} \\ a_{31} & a_{32} & a_{33} \end{vmatrix}$$

2.13 SYSTEMS OF SIMULTANEOUS LINEAR EQUATIONS

a. A system of n simultaneous nonhomogeneous linear equations

$$\begin{aligned} a_{11}x_1 + a_{12}x_2 + \ldots + a_{1n}x_n &= b_1 \\ a_{12}x_1 + a_{22}x_2 + \ldots + a_{2n}x_n &= b_2 \\ \ldots\ldots\ldots\ldots\ldots\ldots\ldots\ldots\ldots \\ a_{n1}x_1 + a_{n2}x_2 + \ldots + a_{nn}x_n &= b_n \end{aligned}$$

has a *unique solution* for the unknowns $x_1, x_2, \ldots, x_n$ if

$$D = \begin{bmatrix} a_{11} & a_{12} & \cdots & a_{1n} \\ a_{21} & a_{22} & \cdots & a_{2n} \\ \ldots & \ldots & \ldots & \ldots \\ a_{n1} & a_{n2} & \cdots & a_{nn} \end{bmatrix} \neq 0$$

and at least one of the absolute terms $b_1, b_2, \ldots, b_n$ is different from zero.

b. *The unique solution* by determinants (Cramer's rule) for the unknowns is

$$x_1 = \frac{D_1}{D},\ x_2 = \frac{D_2}{D}, \ldots,\ x_n = \frac{D_n}{D}$$

in which the augmented determinants are

$$D_1 = \begin{bmatrix} b_1 & a_{12} & \cdots & a_{1n} \\ b_2 & a_{22} & \cdots & a_{2n} \\ \ldots & \ldots & \ldots & \ldots \\ b_n & a_{n2} & \cdots & a_{nn} \end{bmatrix} \quad D_2 = \begin{bmatrix} a_{11} & b_1 & \cdots & a_{1n} \\ a_{12} & b_2 & \cdots & a_{2n} \\ \ldots & \ldots & \ldots & \ldots \\ a_{1n} & b_n & \cdots & a_{nn} \end{bmatrix} \quad D_n = \begin{bmatrix} a_{11} & b_{12} & \cdots & b_1 \\ a_{12} & a_{22} & \cdots & b_2 \\ \ldots & \ldots & \ldots & \ldots \\ a_{n1} & a_{n2} & \cdots & a_n \end{bmatrix}$$

c. *The matrix solution*

The linear system may be written in matrix form.

$AX = B$, where A is the matrix of coefficients $[a_{ij}]$ and X and B are

$$X = \begin{bmatrix} x_1 \\ x_2 \\ \vdots \\ x_n \end{bmatrix} \qquad B = \begin{bmatrix} b_1 \\ b_2 \\ \vdots \\ b_n \end{bmatrix}$$

If det. $A \neq 0$, A^{-1} exists and a unique solution exists given by $X = A^{-1}B$.

d. The system of n simultaneous homogeneous linear equations $AX = 0$ in n unknowns has a solution different from

$$x_1 = x_2 = \ldots = x_n = 0$$

if and only if

$$D = 0$$

e. A system of n linear equations in m variables, $AX = B$, given as

$$\begin{aligned} a_{11}x_1 + a_{12}x_2 + \ldots + a_{1m}x_m &= b_1 \\ a_{21}x_1 + a_{22}x_2 + \ldots + a_{2m}x_m &= b_2 \\ &\vdots \\ a_{n_1}x_1 + an_2x_2 + \ldots + a_{nm}x_m &= b_m \end{aligned}$$

- has unique solution if rank (A) = rank $([A \mid B])$ = order (A)
- has infinite solutions if rank (A) = rank $[A \mid B]$ < order (A)
- has no solution if rank $(A) \neq$ rank $[A \mid B]$.

2.14 PERMUTATION, VARIATION, COMBINATION

i. Permutation

a. A permutation is an ordered arrangement (sequence) of n elements. The number of all possible permutations of n different elements is

$${}^nP_n = n(n-1)(n-2) \ldots (3)(2)(1) = n!$$

b. The *number* of all different permutations of n elements, among which there are a elements of equal value is

$${}^aP_n = \frac{n(n-1)(n-2)\ldots(3)(2)(1)}{a(a-1)(a-2)\ldots(3)(2)(1)} = \frac{n!}{a!b!}$$

c. The number of all different permutations of n elements, among which there are a elements of one equal value and b elements of another equal value is

$${}^{a,b}P_n = \frac{n(n-1)(n-2)\ldots(3)(2)(1)}{a(a-1)(a-2)\ldots(3)(2)(1)\,b(b-1)(b-2)\ldots(3)(2)(1)} = \frac{n!}{a!b!}$$

ii. Variation

A variation is an arrangement of n elements into a sequence of k terms. The number of all possible variations is

$${}^kV_n = \frac{n(n-1)(n-2)\ldots(3)(2)(1)}{(n-k)(n-k-l)(n-k-2)\ldots(3)(2)(1)} = \frac{n!}{(n-k)!k} = \binom{n}{k}$$

iii. Combination

A combination is an arrangement (without repetition) of n elements into a sequence of k terms. The number of all possible combinations is

$$^{k}C_{n} = \frac{n(n-1)(n-2)\ldots(n-k+2)(n-k+1)}{(n-k)(n-k-l)(n-k-2)\ldots(3)(2)(1)} = \frac{n!}{(n-k)!k!} = \binom{n}{k}$$

Permutation, Variation and Combination

Permutations	**Elements** A, B, C	$n = 3$
	ABC BCA CAB ACB BAC CBA	$P_3 = 234(3)(2)(1) = 6$
	Elements A, A, C	$n = 3$ $a = 2$
	AAC ACA CAA	$^{2}P_{3} = \frac{(3)(2)(1)}{(2)(1)} = 3$
Variations	**Elements** A, B, C	$n = 3$ $k = 2$
	AB BC CA BA CB AC	$^{2}V_{3} = \frac{(3)(2)(1)}{1} = 6$
Combinations	**Elements** A, B, C	$n = 3$ $k = 2$
	AB BC CA	$^{2}C_{3} = \frac{(3)(2)(1)}{(2)(1)} = 3$

CHAPTER

3

Geometry

3.1 SIMPLE TWO-DIMENSIONAL FIGURES

The following is a collection of *common simple* geometric figures. Area (A), volume (V) and other measurable features, i.e. a, b, h, R, ϕ, θ are indicated in the following figures.

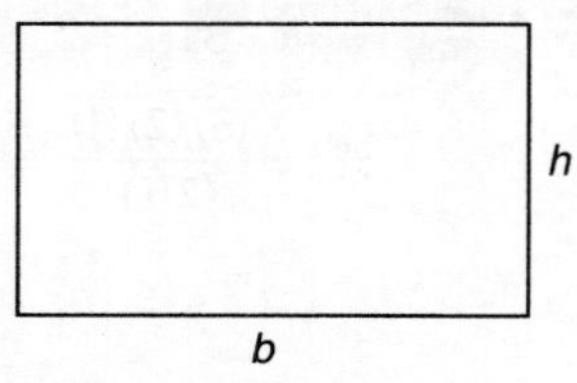

$A = bh$

Rectangle

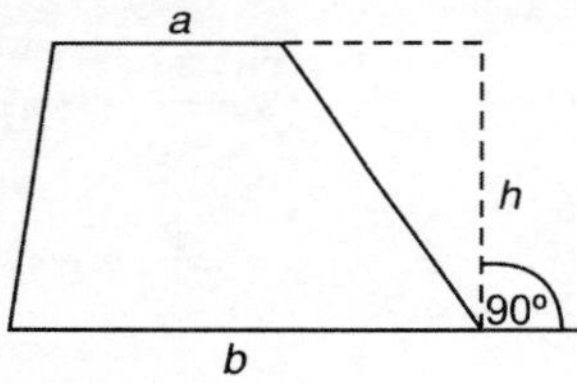

$A = \frac{1}{2}(a + b)h$

Trapezium

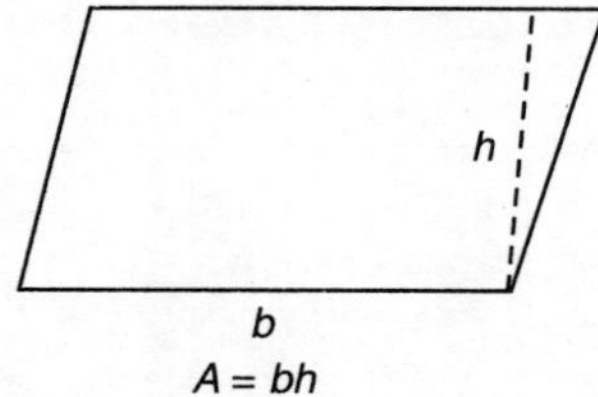

$A = bh$

Parallelogram

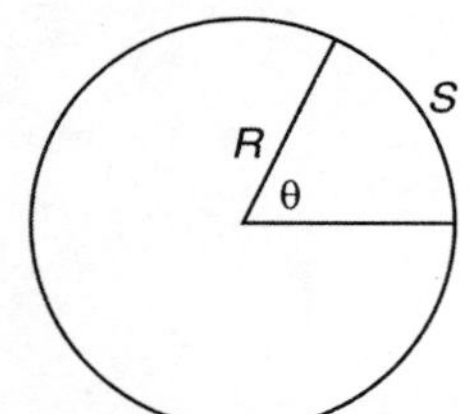

$A = \pi R^2$; circumference $= 2\pi R$;
arc length $S = R\theta$ (θ in radians)

Circle

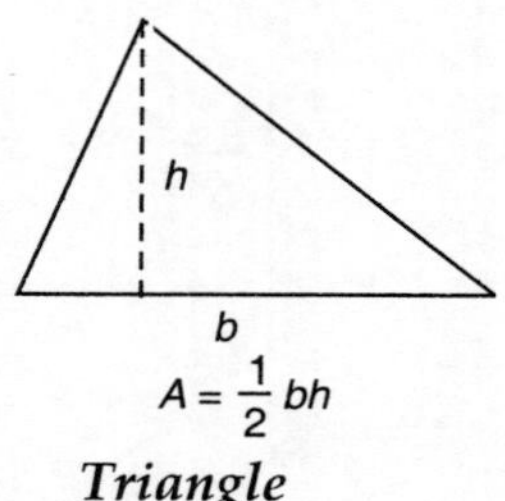

$A = \frac{1}{2}bh$

Triangle

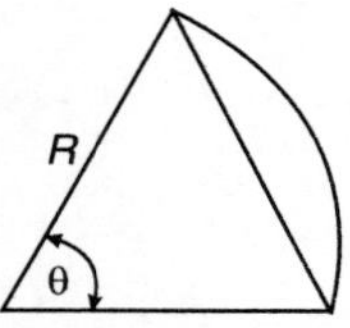

$A_{sector} = \frac{1}{2}R^2\theta$; $A_{segment} = \frac{1}{2}R^2(\theta - \sin\theta)$

Sector

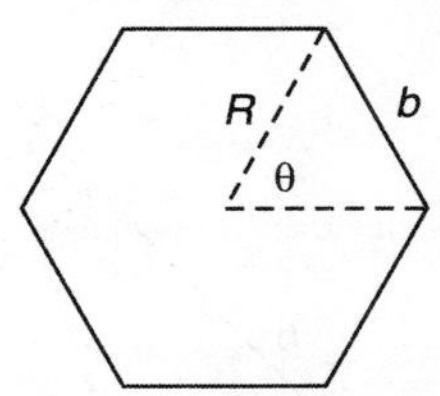

Regular polygon of n sides

$A = \frac{\pi}{4} b^2 \text{ ctn } \frac{\pi}{n}; \; R = \frac{\pi}{7} \csc \frac{\pi}{n}$

Polygon

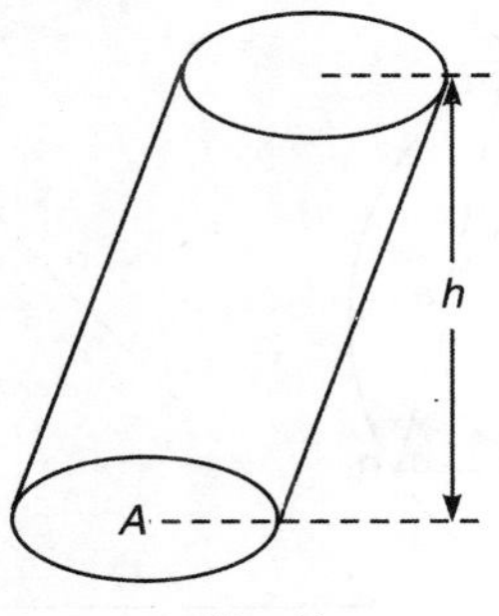

$V = Ah$

Prism

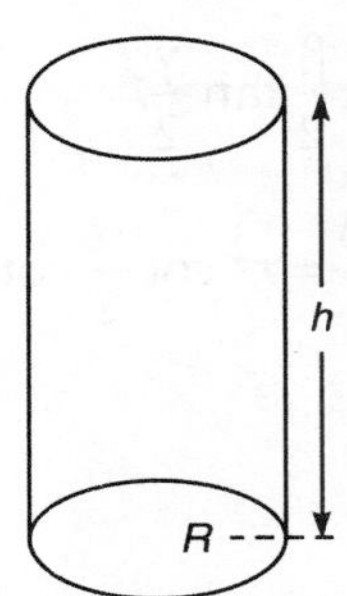

$V = \pi R^2 h$; lateral surface area = $2\pi Rh$

Right circular cylinder

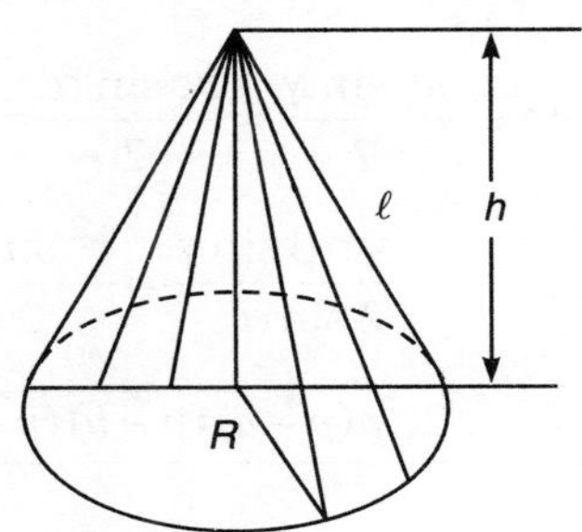

$V = \frac{1}{3}\pi R^2 h$; lateral surface area

$= \pi Rl = \pi R\sqrt{R^2 + h^2}$

Right circular cone

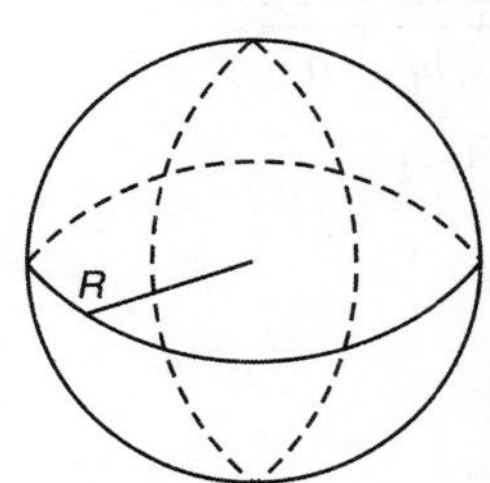

$V = \frac{4}{3}\pi R^2 h$; surface area = $4\pi R^2$

Sphere

3.2 TRIANGLES

In triangle, different parameters are:

a, b, c = sides	A = area	h = altitude
α, β, γ = angles	R = circumradius	m = median
$2p = a + b + c$	r = radius	t = bisector

i. *Oblique triangle*: $(\alpha + \beta + \gamma = 180°)$

$$A = \sqrt{p(p-a)(p-b)(p-c)} = \frac{abc}{4R} = pr$$

$$A = \frac{ah_a}{2} = \frac{bh_b}{2} = \frac{ch_c}{2} = 2R^2 \sin\alpha \sin\beta \sin\gamma$$

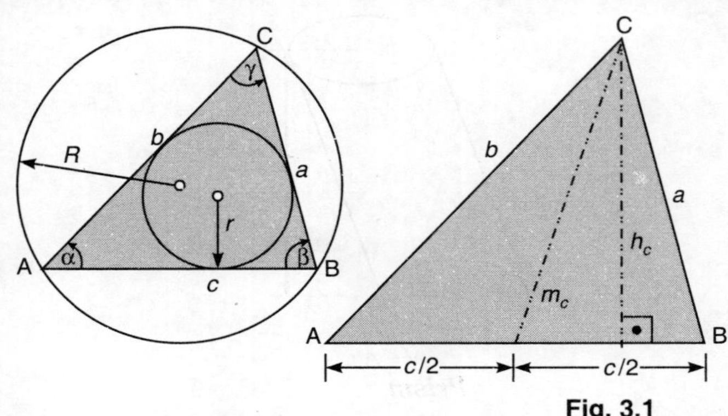

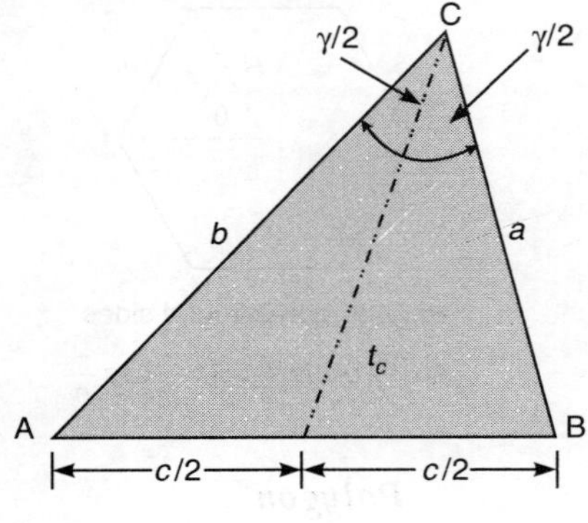

Fig. 3.1

Area $\quad A = \dfrac{ab\sin\gamma}{2} = \dfrac{bc\sin\alpha}{2} = \dfrac{ac\sin\beta}{2} = p^2\tan\dfrac{\alpha}{2}\tan\dfrac{\beta}{2}\tan\dfrac{\gamma}{2}$

or $\quad A = \dfrac{a^2\sin\beta\sin\gamma}{2\sin\alpha} = \dfrac{b^2\sin\alpha\sin\gamma}{2\sin\beta} = \dfrac{c^2\sin\alpha\sin\beta}{2\sin\gamma} = r^2\cot\dfrac{\alpha}{2}\cot\dfrac{\beta}{2}\cot\dfrac{\gamma}{2}$

Altitude $h_c = \dfrac{2\sqrt{p(p-a)(p-b)(p-c)}}{c}$ $\qquad R = \dfrac{abc}{4A}$

Median $m_c = \dfrac{\sqrt{2(a^2+b^2)-c^2}}{2}$ $\qquad r = \dfrac{A}{p}$

Bisector $t_c = \dfrac{\sqrt{ab(a+b)-c^2}}{a+b}$ $\qquad \dfrac{1}{r} = \dfrac{1}{h_a} + \dfrac{1}{h_b} + \dfrac{1}{h_c}$

$a:b:c = \dfrac{1}{h_a}:\dfrac{1}{h_b}:\dfrac{1}{h_c}$ $\qquad h_a:h_b:h_c = \dfrac{1}{a}:\dfrac{1}{b}:\dfrac{1}{c}$

ii. *Right angled triangle*: $\quad (\alpha + \beta = 90°, \gamma = 90°)$

$$A = \frac{ab}{2} = \frac{h_c}{2}$$

$$h = \frac{ab}{c} \qquad R = \frac{c}{2}$$

$$r = \frac{a+b-c}{2}$$

$$a^2 + b^2 = c^2$$

$$p = \frac{b^2}{c} \qquad q = \frac{a^2}{c}$$

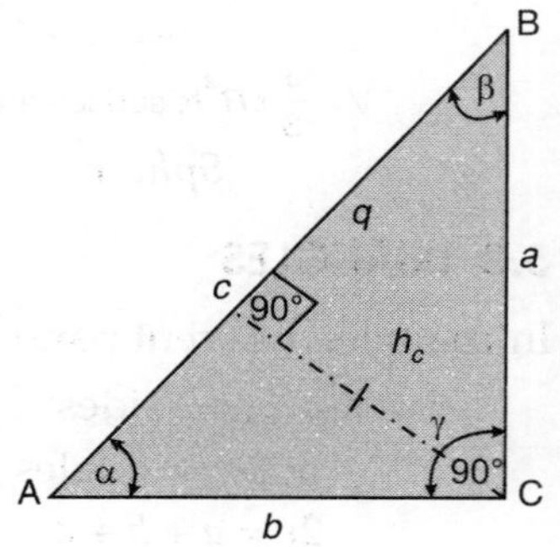

Fig. 3.2

iii. *Equilateral triangle*: $\quad (\alpha = \beta = \gamma = 60°)$

$$A = \frac{a^2}{4}\sqrt{3} = \frac{h^2}{3}\sqrt{3}$$

$$h = m = t = \frac{a}{2}\sqrt{3}$$

$$R = \frac{a}{3}\sqrt{3}$$

$$r = \frac{a}{6}\sqrt{3}$$

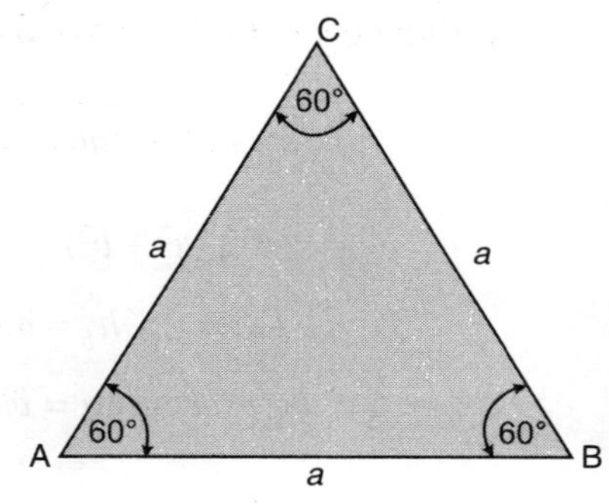

Fig. 3.3

3.3 PARALLELOGRAMS

In parallelograms, different parameters are:

a, b, c, d = sides $\quad 2s = a + b + c + d \quad A$ = area

e, f = diagonals $\quad h$ = altitude $\quad R$ = circumradius

$\alpha, \beta, \gamma, \delta$ = angles $\quad r$ = radius

i. *Square*: $(\alpha = \beta = \gamma = \delta = 90°)$

$$e = \sqrt{2}a = 1.4142a \qquad R = \frac{a}{2}\sqrt{2} = 0.7071a$$

$$A = a^2 \qquad r = \frac{a}{2}$$

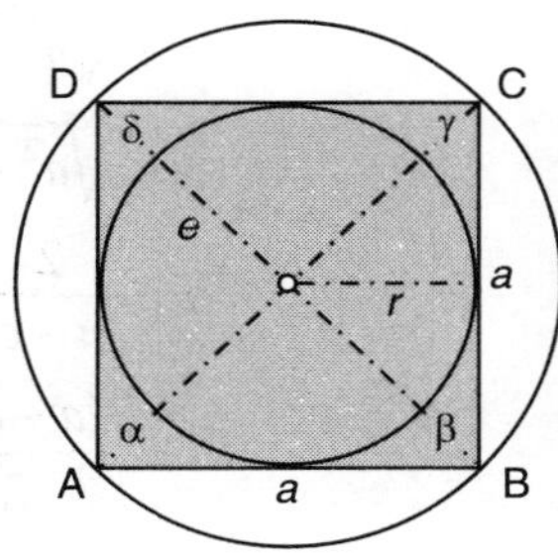

Fig. 3.4

ii. *Rectangle*: $(\alpha = \beta = \gamma = \delta = 90°)$

$$e = f = \sqrt{a^2 + b^2} \qquad R = \frac{e}{2} = \frac{\sqrt{a^2 + b^2}}{2} \qquad A = ab$$

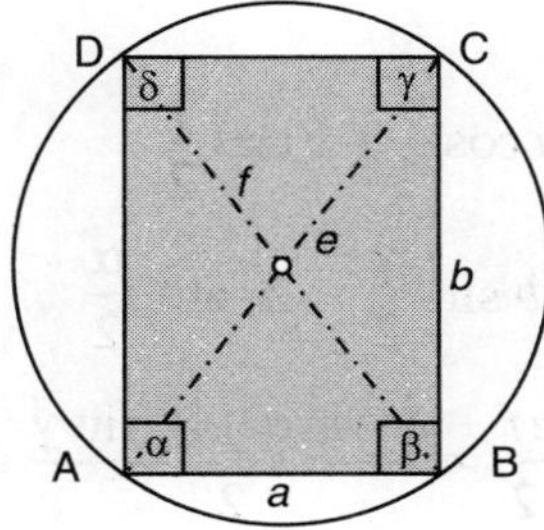

Fig. 3.5

iii. *Rhombus*: $(\alpha + \beta = \gamma + \delta = 180°) \quad (\alpha = \gamma, \beta = \delta)$

$$e = 2a \cos \frac{\alpha}{2} \qquad f = 2a \sin \frac{\alpha}{2}$$

$$e^2 + f^2 = 4a^2$$

$$h = a \sin \alpha \qquad r = \frac{a}{2} \sin \alpha$$

$$A = ah = a^2 \sin \alpha = \frac{ef}{2}$$

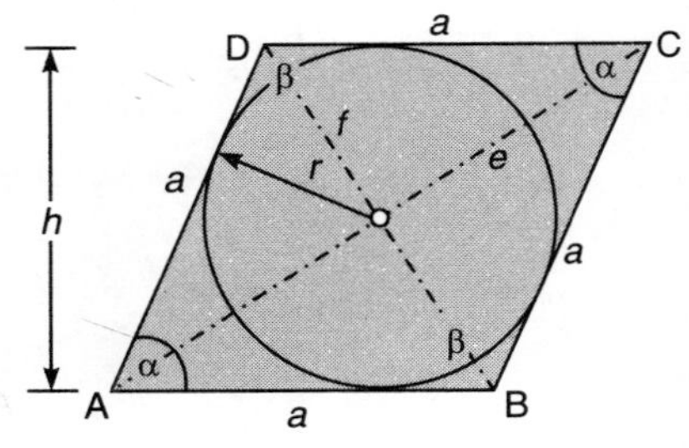

Fig. 3.6

iv. *Rhomboid*: $(\alpha + \beta = \gamma + \delta = 180°)$ $(\alpha = \gamma, \beta = \delta)$

$$e = \sqrt{a^2 + b^2 - 2ab\cos\beta} \qquad f = \sqrt{a^2 + b^2 - 2ab\cos\alpha}$$

$$e^2 + f^2 = 2(a^2 + b^2)$$

$$h_a = b\sin\alpha, \quad h_b = a\sin\alpha$$

$$A = ah_a = ab\sin\alpha = bh_b$$

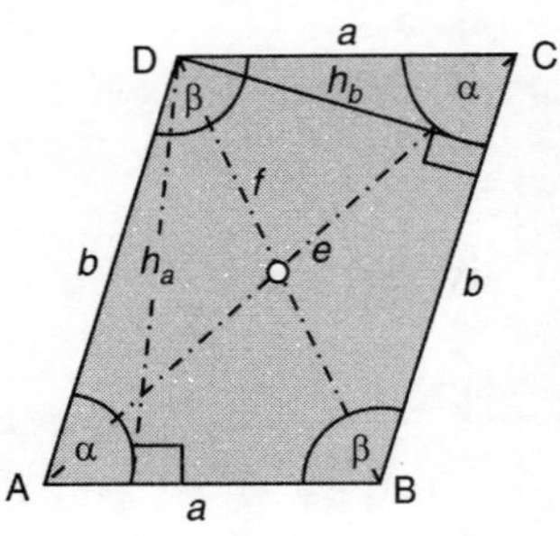

Fig. 3.7

3.4 QUADRILATERALS

In quadrilaterals, different parameters are:

a, b, c, d = sides	$2s = a + b + c + d$	A = area
e, f = diagonals	h = altitude	R = circumradius
$\alpha, \beta, \gamma, \delta$ = angles	r = radius	

i. *Trapezoid*: $(\alpha + \beta = \beta + \gamma = 180°)$

$$e = \sqrt{a^2 + b^2 - 2ab\cos\beta}$$

$$f = \sqrt{a^2 + d^2 - 2ad\cos\alpha}$$

$$h = \frac{2}{a-c}\sqrt{s(s-a+c)(s-b)(s-d)}$$

$$A = \frac{(a+c)h}{2}$$

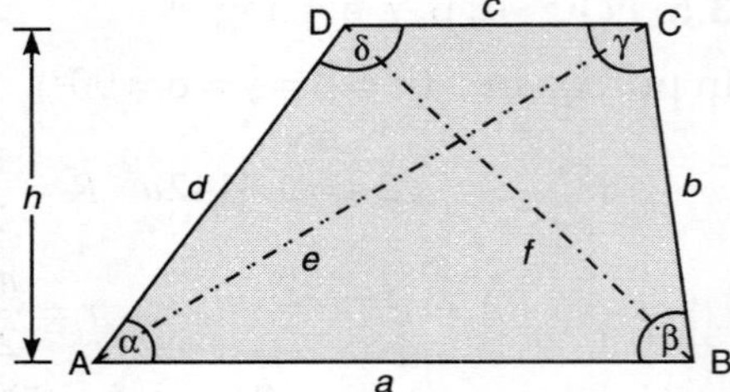

Fig. 3.8

ii. *Deltoid*: $\left(\frac{\alpha}{2} + \beta + \frac{\gamma}{2} = 180°\right)$

$$e = a\cos\frac{\alpha}{2} + b\cos\frac{\gamma}{2}$$

$$f = 2b\sin\frac{\gamma}{2} = 2a\sin\frac{\alpha}{2}$$

$$A = \frac{ef}{2} = \frac{a^2\sin\alpha + b^2\sin\gamma}{2}$$

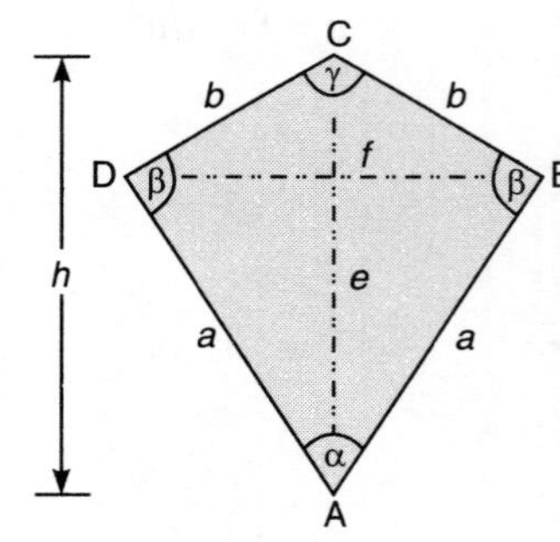

Fig. 3.9

iii. *Tangent-quadrilateral*: $(a + c = b + d)$

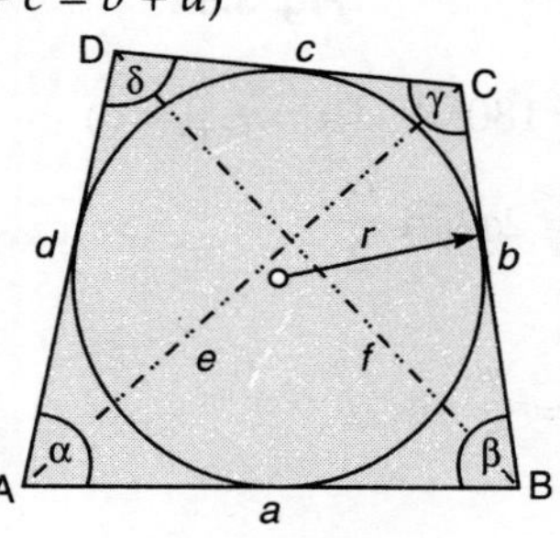

Fig. 3.10

$$A = sr$$

If $\alpha + \gamma = \beta + \delta = 180°$,

$$A = \sqrt{abcd}; \quad r = \frac{\sqrt{abcd}}{s}$$

iv. *Secant–quadrilateral*: $(\alpha + \gamma = \beta + \delta = 180°)$

$$ef = ac + bd = g$$

$$e = \sqrt{\frac{(ad + bc)g}{ab + cd}}$$

$$f = \sqrt{\frac{(ab + cd)g}{ad + bc}}; \quad \sin \omega = \frac{2A}{a}$$

$$A = \sqrt{(s-a)(s-b)(s-c)(s-d)}$$

$$R = \frac{\sqrt{(ab + cd)(ad - bc)g}}{4A}$$

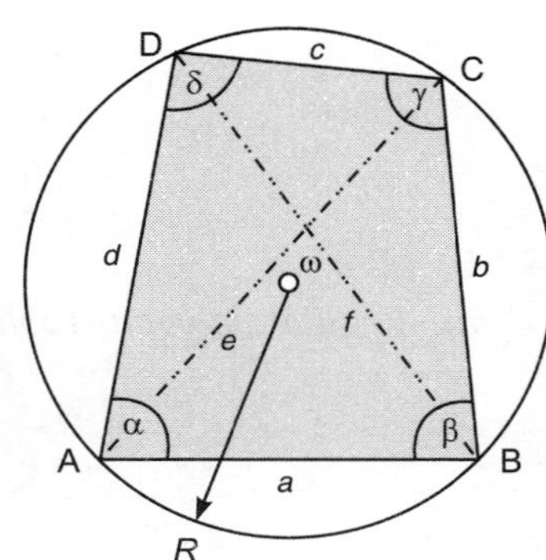

Fig. 3.11

3.5 POLYGONS

In polygons, different parameters are:

α = central angle	a = side	A = area
β = interior angle	n = number of sides	R = circumradius
γ = exterior angle	r = radius	

i. *General polygon*:

$$\sum_{1}^{n} \beta_j = (n-2)\, 180°$$

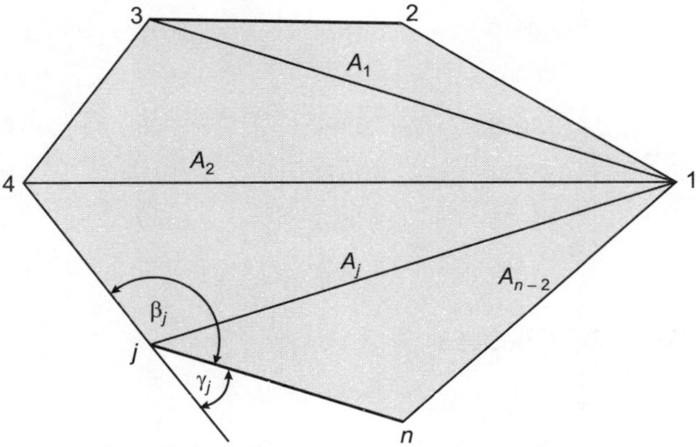

Fig. 3.12

$$\sum_{1}^{n} \gamma_j = 360°$$

$$\sum_{1}^{n-2} A_j = A$$

ii. *Regular polygon*: $[(n\alpha = 360°, n\beta = (n-2)\,180°, n\gamma = 360°)]$

$$A = \frac{na^2}{4}\cot\frac{\pi}{n} = \frac{nar}{2} = \frac{nR^2}{2}\sin\frac{2\pi}{n}$$

$$R = \frac{a}{2}\operatorname{cosec}\frac{\pi}{n} \qquad r = \frac{a}{2}\cot\frac{\pi}{n}$$

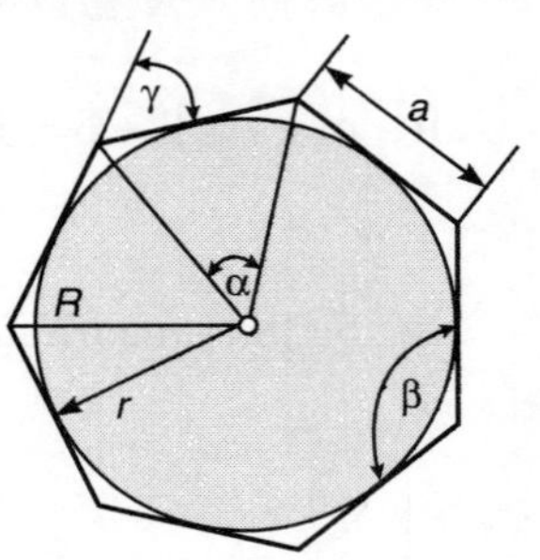

Fig. 3.13

iii. *Regular polygon–Table of coefficients*

n	$\frac{A}{a^2}$	$\frac{A}{R^2}$	$\frac{A}{r^2}$	$\frac{R}{a}$	$\frac{r}{a}$	$\frac{R}{r}$
3	0.4330	1.2990	5.1962	0.5774	0.2887	2.0000
4	1.0000	2.0000	4.0000	0.7071	0.5000	1.4142
5	1.7205	2.3776	3.6327	0.8507	0.6882	1.2361
6	2.5981	2.5981	3.4641	1.0000	0.8660	1.1547
7	3.6339	2.7364	3.3710	1.1524	1.0383	1.1099
8	4.8284	2.8284	3.3137	1.3066	1.2071	1.0824
9	6.1818	2.8925	3.2757	1.4619	1.3737	1.0642
10	7.6942	2.9389	3.2492	1.6180	1.5388	1.0515
12	11.196	3.0000	3.2154	1.9319	1.8660	1.0353
15	17.642	3.0505	3.1883	2.4049	2.3523	1.0223
16	20.109	3.0615	3.1826	2.5629	1.5137	1.0196
20	31.569	3.0902	3.1677	3.1962	3.1569	1.0125
24	45.575	3.1058	3.1597	3.8306	3.7979	1.0086
32	81.225	3.1214	3.1517	5.1011	5.0766	1.0048
48	183.08	3.1326	3.1461	7.6449	7.6285	1.0021
64	325.67	3.1365	3.1441	10.190	10.178	1.0012

3.6 CIRCLES AND PARTS OF CIRCLES

In circle, different parameters are:

C = circumference	S = length of arc	$2l$ = chord
A = area	α = angle	h = altitude
R = radius	D = diameter	

Arc 1° $= \dfrac{\pi}{180}$ $= 0.017453293$ radians

Arc 1 minute $= \dfrac{\pi}{10{,}800}$ $= 0.000290888$ radians

Arc 1 second $= \dfrac{\pi}{648{,}000}$ $= 0.000004848$ radians

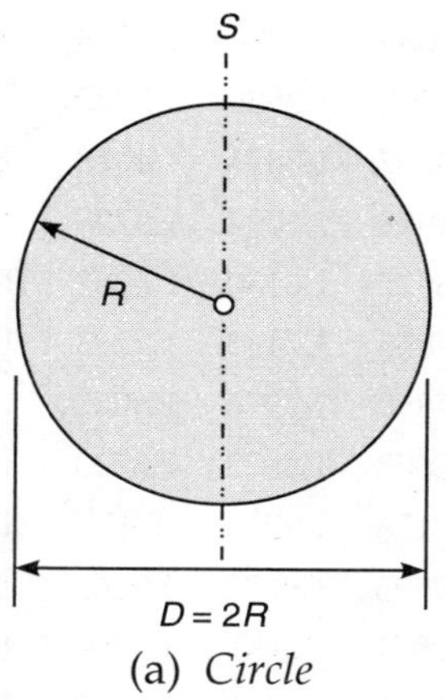

(a) *Circle*

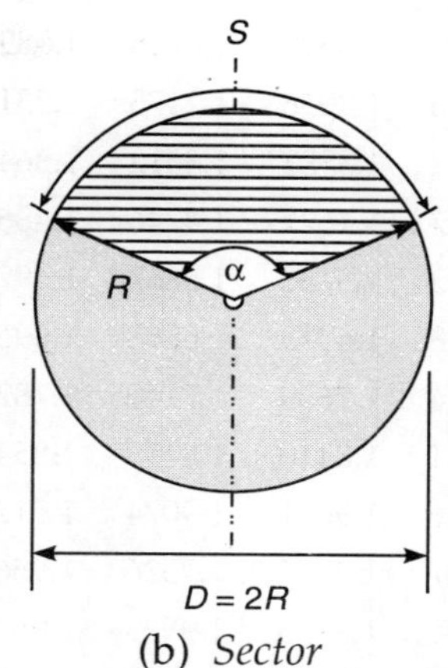

(b) *Sector*

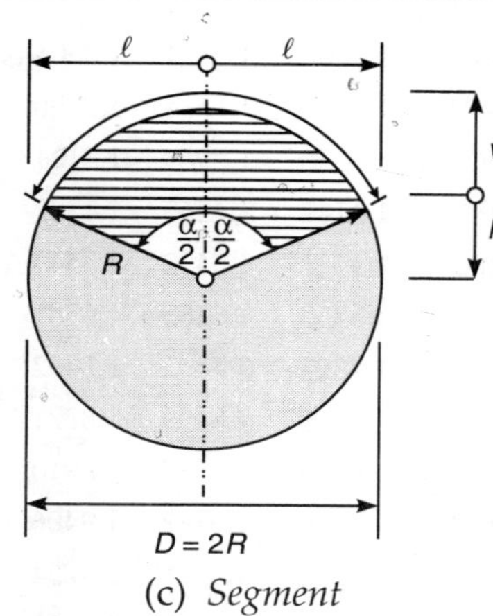

(c) *Segment*

Fig. 3.14

$C = 2\pi R = \pi D$

$S = \dfrac{\pi R\alpha^\circ}{180^\circ} = R\alpha = \dfrac{D\alpha}{2}$

$l = R\sin\dfrac{\alpha}{2}$

$h = R\cos\dfrac{\alpha}{2}$

$A_0 = \pi R^2 = \dfrac{\pi D^2}{4}$

$A = \dfrac{\pi R^2\alpha^\circ}{360^\circ} = \dfrac{R^2\alpha}{2}$

$A = \dfrac{R^2}{2}\left(\dfrac{\pi\alpha^\circ}{180^\circ} - \sin\alpha\right)$

Values of π constants

n	$n\pi$	$\dfrac{1}{n\pi}$	$\dfrac{\pi}{n}$	$\dfrac{n}{\pi}$
1	3.14159 26536	0.31830 98862	3.14159 26536	0.31830 98862
2	6.28318 53072	0.15915 49431	1.57079 63268	0.63661 97724
3	9.42477 79608	0.10610 32954	1.04719 75512	0.95492 95686
4	12.56637 06144	0.07957 74715	0.78539 81634	1.27323 95447
5	15.70796 32679	0.06366 19772	0.62831 85307	1.59154 94309
6	18.84955 59215	0.05305 16477	0.52359 87756	1.90985 93171
7	21.99114 85751	0.04547 28409	0.44879 89505	2.22816 92033
8	25.13274 12287	0.03978 87358	0.39269 90817	2.54647 90895
9	28.27433 88823	0.03536 77651	0.34906 58504	2.86478 89757

Table of Coefficients of Circles $\left[l = R\sin\dfrac{\alpha}{2}\right]$

i. *Chord length in terms of R*

α	0°	1°	2°	3°	4°	5°	6°	7°	8°	9°	α
0°	0.0000	0.0175	0.0349	0.0524	0.0698	0.0872	0.1047	0.1221	0.1395	0.1569	0°
10°	0.1743	0.1917	0.2091	0.2264	0.2437	0.2611	0.2783	0.2956	0.3129	0.3301	10°
20°	0.3473	0.3645	0.3816	0.3987	0.4158	0.4329	0.4499	0.4669	0.4838	0.5008	20°
30°	0.5176	0.5345	0.5513	0.5680	0.5947	0.6014	0.6180	0.6346	0.6511	0.6676	30°
40°	0.6840	0.7004	0.7167	0.7330	0.7492	0.7654	0.7815	0.7975	0.8135	0.8294	40°
50°	0.8452	0.8610	0.8767	0.8924	0.90080	0.9235	0.9389	0.9534	0.9696	0.9848	50°

60°	1.0000	1.0151	1.0301	1.0450	1.0598	1.0746	1.0893	1.1039	1.1184	1.1328	60°
70°	1.1472	1.1614	1.1756	1.1896	1.2036	1.2175	1.2313	1.2450	1.2586	1.2722	70°
80°	1.2856	1.2989	1.3121	1.3252	1.3383	1.3512	1.3640	1.3767	1.3893	1.4018	80°
90°	1.4142	1.4265	1.4387	1.4507	1.4627	1.4746	1.4863	1.4979	1.5094	1.5208	90°
100°	1.5321	1.5432	1.5543	1.5652	1.5760	1.5867	1.5973	1.6077	1.6180	1.6282	100°
110°	1.6383	1.6483	1.6581	1.6678	1.6773	1.6868	1.6961	1.7053	1.7143	1.7233	110°
120°	1.7321	1.7407	1.7492	1.7576	1.7659	1.7740	1.7820	1.7899	1.7976	1.8052	120°
130°	1.8126	1.8199	1.8271	1.8341	1.8410	1.8478	1.8544	1.8608	1.8672	1.8733	130°
140°	1.8794	1.8853	1.8910	1.8966	1.9021	1.9074	1.9126	1.9176	1.9225	1.927	140°
150°	1.9319	1.9363	1.9406	1.9447	1.9487	1.9526	1.9563	1.9598	1.9633	1.9665	150°
160°	1.9696	1.9726	1.9754	1.9780	1.9805	1.9829	1.9831	1.9871	1.890	1.9908	160°
170°	1.9924	1.9938	1.9951	1.9963	1.9973	1.9981	1.9988	1.9993	1.9997	1.9999	170°
180°	2.0000										180°
α	0°	1°	2°	3°	4°	5°	6°	7°	8°	9°	α

ii. *Segment height in terms of* $R\left[h = R\cos\left(\frac{\alpha}{2}\right)\right]$

α	0°	1°	2°	3°	4°	5°	6°	7°	8°	9°	α
0°	0.0000	0.0000	0.0002	0.0003	0.0006	0.0010	0.0014	0.0019	0.0024	0.0031	0°
10°	0.0038	0.0046	0.0055	0.0064	0.0075	0.0086	0.0097	0.0110	0.0123	0.0137	10°
20°	0.0152	0.0167	0.0184	0.0201	0.0219	0.0237	0.0256	0.0276	0.0297	0.0319	20°
30°	0.0341	0.0364	0.0387	0.0696	0.0728	0.0761	0.0795	0.0829	0.0865	0.0900	30°
40°	0.0603	0.0633	0.0664	0.0694	0.0728	0.0761	0.0795	0.0829	0.0865	0.0900	40°
50°	0.0937	0.0974	0.1012	0.1051	0.1090	0.1130	0.1171	0.1212	0.1254	0.1296	50°
60°	0.1340	0.1384	0.1428	0.1474	0.1520	0.1566	0.1613	0.1661	0.1710	0.1759	60°
70°	0.1808	0.1859	0.1910	0.1961	0.2014	0.2066	0.2120	0.2174	0.2229	0.2284	70°
80°	0.2340	0.2396	0.2453	0.2510	0.2569	0.2627	0.2686	0.2746	0.2807	0.2867	80°
90°	0.2929	0.2991	0.3053	0.3116	0.3180	0.3244	0.3309	0.3374	0.3439	0.3506	90°
100°	0.3572	0.3639	0.3707	0.3775	0.3843	0.391	0.3982	0.4052	0.4122	0.4193	100°
110°	0.4264	0.4336	0.4408	0.4481	0.4554	0.4627	0.4701	0.4775	0.4850	0.4925	110°
120°	0.5000	0.5076	0.5152	0.5228	0.5305	0.5383	0.5460	0.5538	0.5616	0.5695	120°
130°	0.5774	0.5853	0.5933	0.6013	0.6093	0.6173	0.6254	0.6335	0.6416	0.498	130°
140°	0.6580	0.6662	0.6744	0.6827	0.6910	0.693	0.7076	0.7160	0.7244	0.7328	140°
150°	0.7412	0.7496	0.7581	0.7666	0.7750	0.7836	0.7921	0.8006	0.8092	0.8178	150°
160°	0.8264	0.8350	0.8436	0.8522	0.8608	0.8695	0.8781	0.8868	0.8955	0.9042	160°
170°	0.9128	0.9215	0.9302	0.9390	0.9477	0.964	0.9651	0.9738	0.9825	0.9913	170°
180°	1.0000										180°
α	0°	1°	2°	3°	4°	5°	6°	7°	8°	9°	α

3.7 CONICAL SECTIONS

In conical sections, different parameters are:

a, b = semiaxis $\qquad S$ = length of curve $\qquad A$ = area

$\dfrac{1}{r_a}$ = curvature at $A \qquad \dfrac{1}{r_b}$ = curvature at $B \qquad C$ = circumference

i. *Ellipse*:

$$r_b = \frac{a^2}{b} \qquad r_a = \frac{b^2}{a} \qquad m = \frac{a-b}{a+b}$$

$$C = \pi\,(a+b)\left(1 + \frac{m^2}{4} + \frac{m^4}{64} + \frac{m^6}{256} + \dots\right)$$

$$C_{\text{approx}} = \pi\left[1.5\,(a+b) - \sqrt{ab}\right]$$

$$= \pi\,(a+b)\frac{64-3m^4}{64-16m^2}$$

$$A_{\ominus} = \pi ab \qquad A_{\text{D}} = \frac{\pi ab}{4}$$

$$A_{\triangle} = \frac{ab}{2}\cos^{-1}\frac{x}{a} \qquad A_{\triangleright} = \frac{ab}{2}\sin^{-1}\frac{x}{a}$$

$$A_{\text{D}} = -xy + ab\cos^{-1}\frac{x}{a}$$

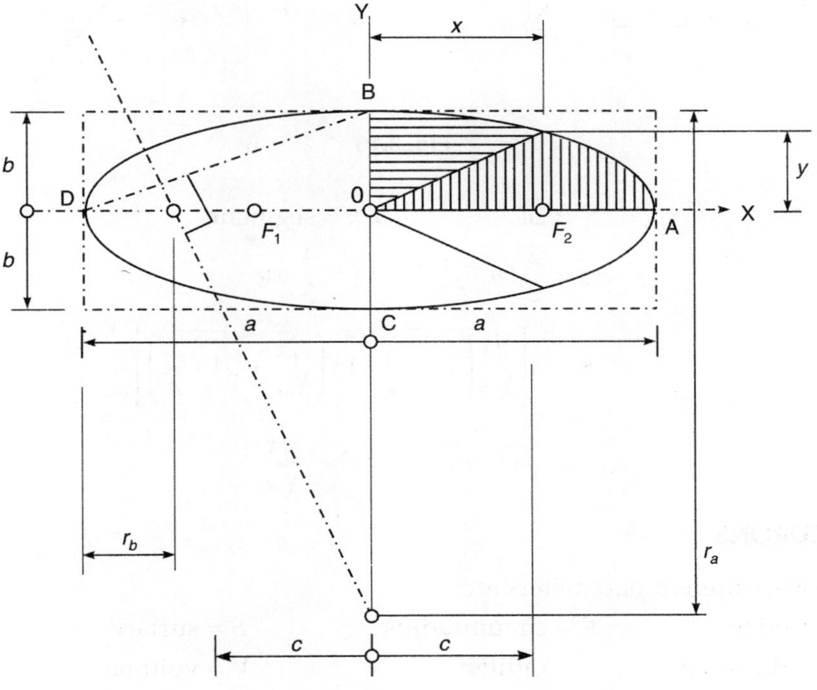

Fig. 3.15

ii. *Hyperbola*:

$$r_a = r_b = \frac{b^2}{a}$$

$$\alpha = \tan^{-1} \frac{b}{a}$$

$$A_{\measuredangle} = ab \ln\left(\frac{x}{a} + \frac{y}{b}\right) = ab \cosh^{-1} \frac{x}{a}$$

$$A_{\triangleleft} = xy - ab \ln\left(\frac{x}{a} + \frac{y}{b}\right)$$

$$= xy - ab \cosh^{-1} \frac{x}{a}$$

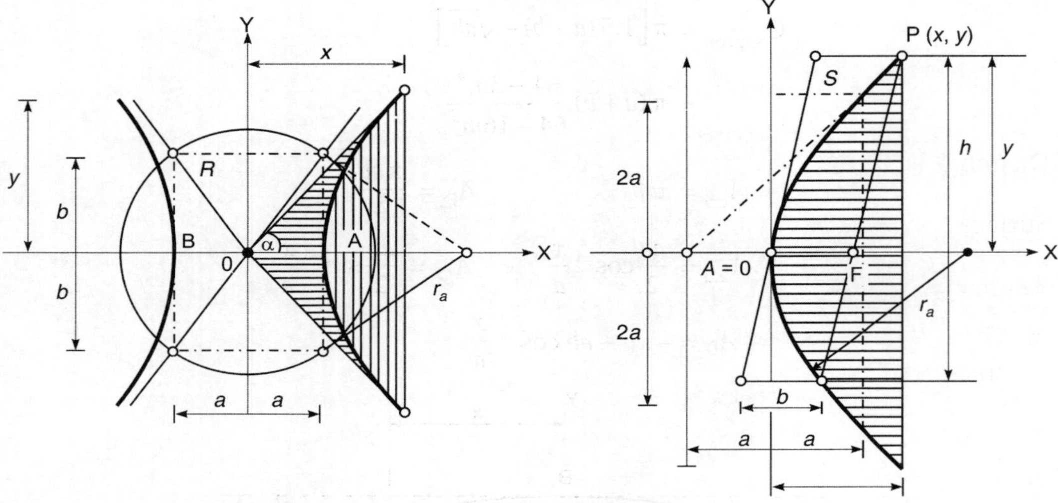

Fig. 3.16

iii. *Parabola*:

$$r_a = 2a \qquad b, h = \text{segments}$$

$$A_{\triangleleft} = \frac{2}{3}\, bh \qquad A_{\triangleleft} = \frac{4}{3}\, xy$$

$$S = a\left[\sqrt{\frac{x}{a}\left(1+\frac{x}{a}\right)} + \ln\left(\sqrt{\frac{x}{a}} + \sqrt{1+\frac{x}{a}}\right)\right]$$

$$= \sqrt{x(x+a)} + a \sinh^{-1} \sqrt{\frac{x}{a}}$$

3.8 POLYHEDRONS

In polyhedrons, different parameters are:

a = edge	R = circumradius	S = surface
e = diagonal	r = radius	V = volume
v = altitude	ω = dihedral angle	

i. *Tetrahedron* contains 4 triangles, 6 edges, 4 vertices ($\omega = 70° 31' 44''$):

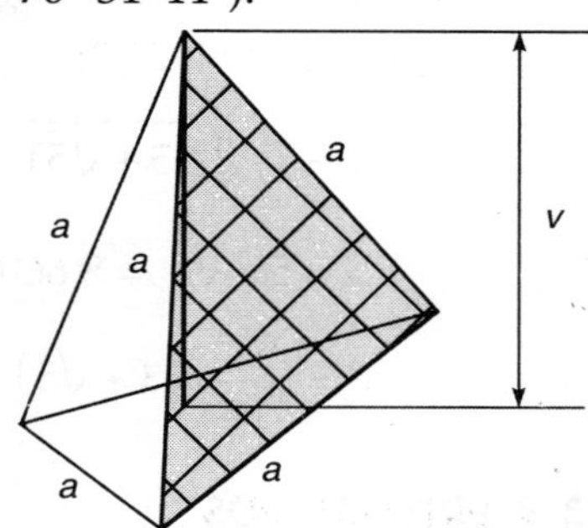

Fig. 3.17

$$R = \frac{a}{4}\sqrt{6} \qquad r = \frac{a}{12}\sqrt{6}$$

Altitude $v = \frac{a}{3}\sqrt{6}$

Surface $S = a^2\sqrt{3} = 1.7321\, a^2$

Volume $V = \frac{a^3\sqrt{2}}{12} = 0.1179\, a^3$

ii. *Cube* contains 6 squares, 12 edges, 8 vertices ($\omega = 90°$):

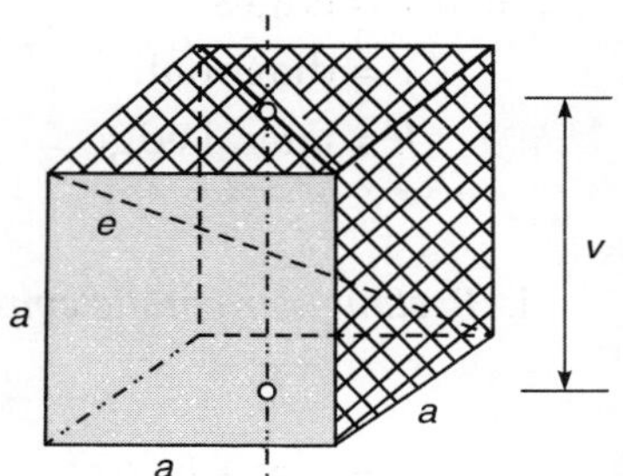

Fig. 3.18

$$R = \frac{a}{2}\sqrt{3} \qquad r = \frac{a}{2}$$

Diagonal $e = a\sqrt{3}$

Surface $S = 6a^2$

Volume $V = a_3$

iii. *Octahedron* contains 8 triangles, 12 edges, 6 vertices ($\omega = 109° 28' 16''$):

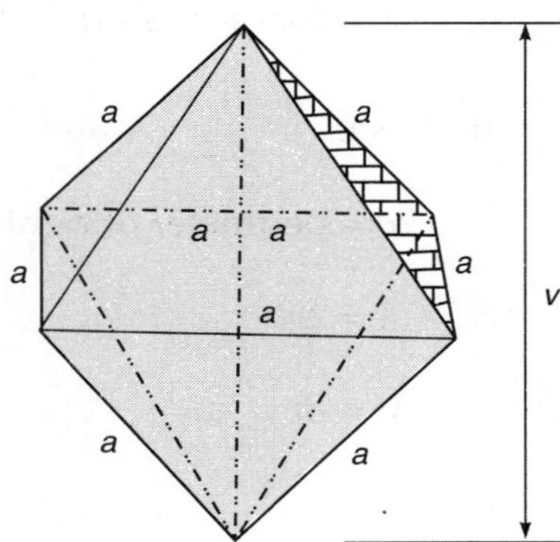

Fig. 3.19

$$R = \frac{a}{2}\sqrt{2}$$

$$r = \frac{a}{6}\sqrt{6}$$

$$S = 2a^2\sqrt{3} = 3.4641\, a^2$$

$$V = \frac{a^3}{3}\sqrt{2}\ 10.4714\, a^3$$

iv. *Dodecahedron* contains 12 pentagons, 30 edges, 20 vertices ($\omega = 116° 33' 54''$):

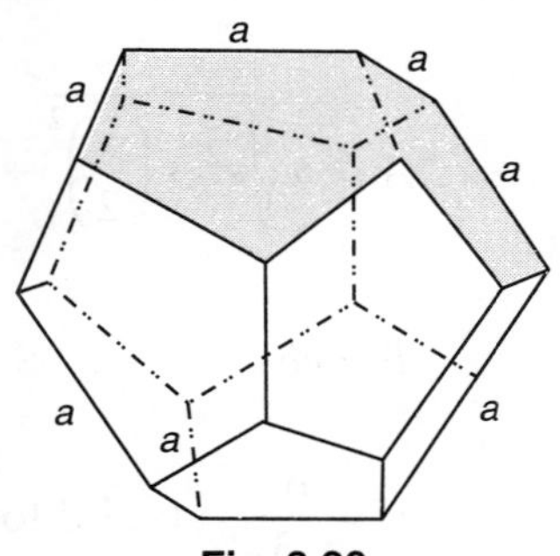

Fig. 3.20

$$R = \frac{a(1+\sqrt{5})\sqrt{3}}{4} \qquad r = \frac{a}{4}\sqrt{\frac{50+22\sqrt{5}}{5}}$$

$$S = 3a^2\sqrt{5(5+2\sqrt{5})} = 20.6457\, a^2$$

$$V = \frac{a^3}{4}(15+7\sqrt{5}) \approx 7.6631\, a^3$$

v. *Icosahedron* contains 20 triangles, 30 edges, 12 vertices ($\omega = 138° 11' 23''$):

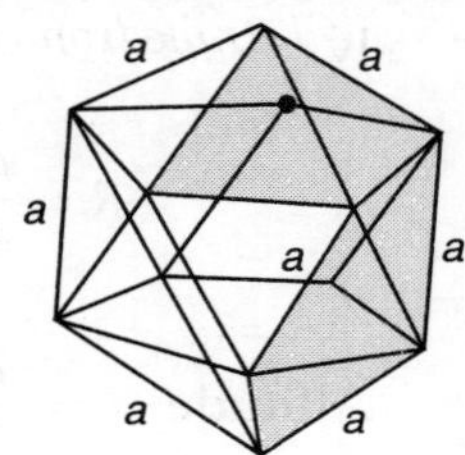

Fig. 3.21

$$R = \frac{a}{4}\sqrt{2(5+\sqrt{5})} \qquad r = \frac{a}{2}\sqrt{\frac{7+3\sqrt{5}}{6}}$$

$$S = 5a^2\sqrt{3} = 8.6603\, a^2$$

$$V = \frac{5a^3}{12}\left(3+\sqrt{5}\right) \approx 2.1817\, a^2$$

3.9 PRISMATOIDS

In prismatoids, different parameters are:

a, b, c = edges	R = circumradius	A = lateral area
e = diagonal	B = area of base	S = surface
h = lateral edge	v = altitude	V = volume

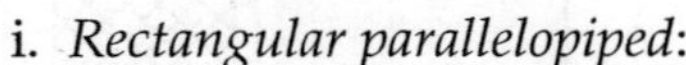

i. *Rectangular parallelopiped*:

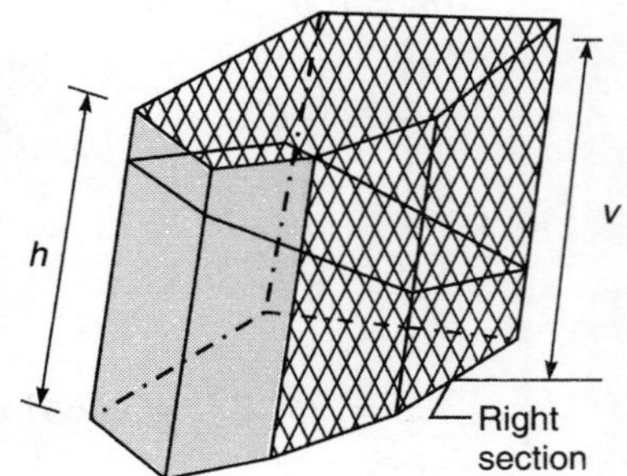

Fig. 3.22

$$e = \sqrt{a^2+b^2+c^2} \qquad R = \frac{\sqrt{a^2+b^2+c^2}}{2}$$

$$S = 2(ab+bc+ca) \qquad V = abc$$

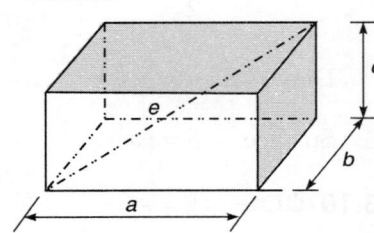

ii. *Prism*:

$2p$ = perimeter of right section

$$A = 2ph$$

$$V = Bv$$

Fig. 3.23

iii. *Right pyramid*:

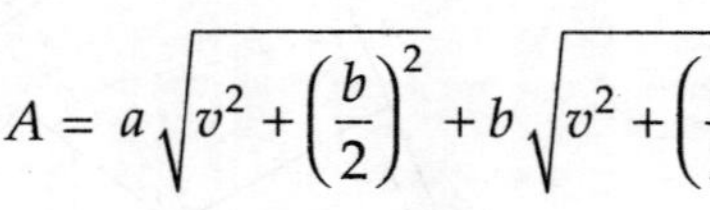

$$A = a\sqrt{v^2+\left(\frac{b}{2}\right)^2} + b\sqrt{v^2+\left(\frac{a}{2}\right)^2}$$

$$B = ab \qquad S = A + B$$

$$V = \frac{Bv}{3} \quad \text{(valid for any pyramid)}$$

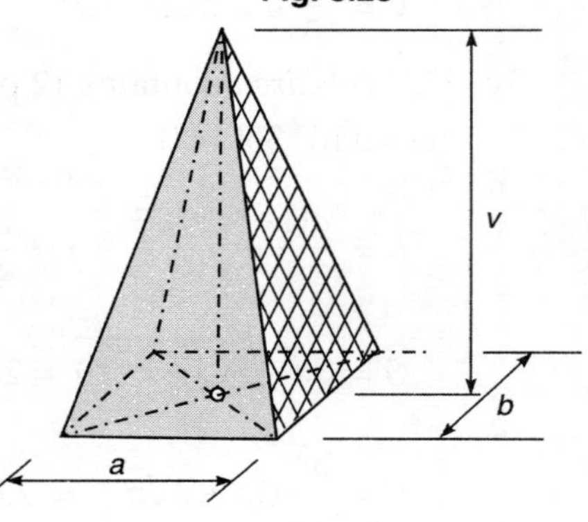

Fig. 3.24

iv. *Frustum of right pyramid*:

(subscript b = bottom; t = top)

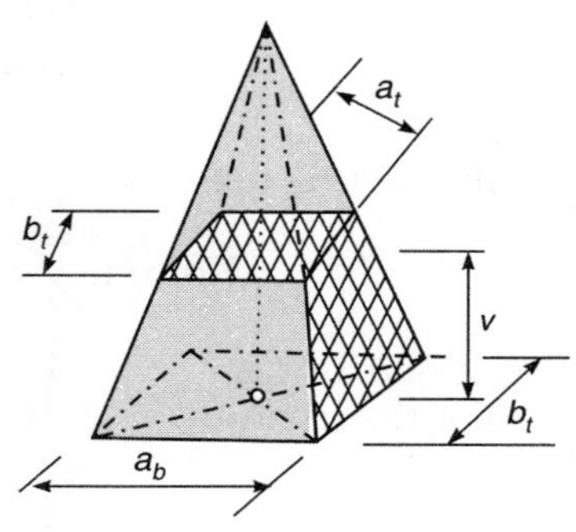

Fig. 3.25

$$A = (a_b + a_t)\sqrt{v^2 + \left(\frac{b_b - b_t}{2}\right)^2} + (b_b + b_t)\sqrt{v^2 + \left(\frac{a_b - a_t}{2}\right)^2}$$

$$S = A + a_b b_b + a_t b_t$$

$$V = \frac{v}{3}\left(B_b + B_t + \sqrt{(B_b B_t)}\right) \quad \text{(valid for any frustum)}$$

v. *Right wedge:*

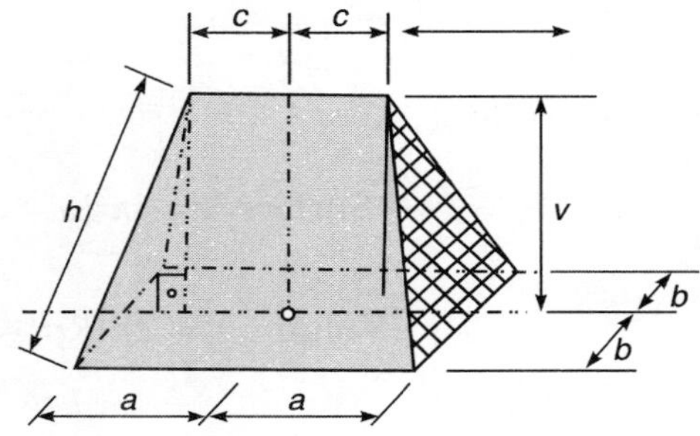

Fig. 3.26

$$A = 2(a + c)\sqrt{v^2 + b^2} + 2b\sqrt{v^2 + (a - c)^2}$$

$$S = A + 4ab$$

$$V = \frac{2bv}{3}\,(2a + c)$$

3.10 CONES, TORUS, BARREL

Different parameters are:

π = 3.14159...	R = radius	A = lateral area
r = radius	B = area of base	S = surface
h = slant height	v = altitude	V = volume

i. *Circular right cone*:

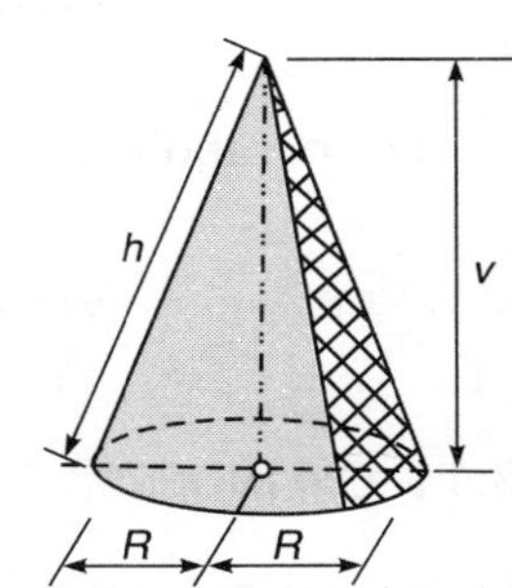

Fig. 3.27

$$A = \pi R\sqrt{v^2 + R^2} = \pi Rh$$

$$B = \pi R^2$$

$$S = \pi R\,(R + h)$$

$$V = \frac{\pi R^2 v}{3}$$

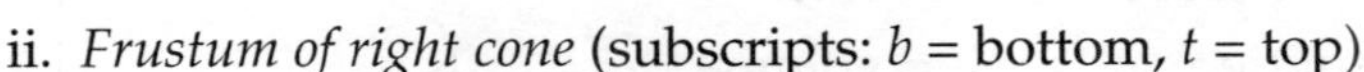

ii. *Frustum of right cone* (subscripts: b = bottom, t = top):

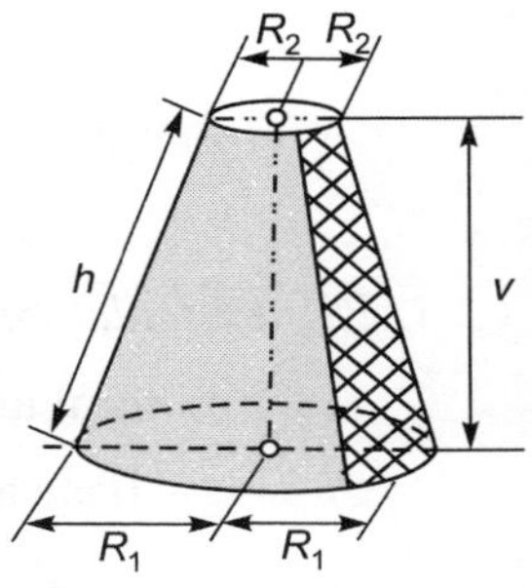

Fig. 3.28

$$A = \pi(R_1 + R_2)\sqrt{v^2 + (R_1 - R_2)^2} = \pi(R_1 + R_2)h$$

$$B_b = \pi R_1^2; \qquad B_t = \pi R_2^2$$

$$S = \pi[R_1^2 + (R_1 + R_2)\,h + R_2^2]$$

$$V = \frac{\pi v}{3}\,(R_1^2 + R_1R_2 + R_2^2)$$

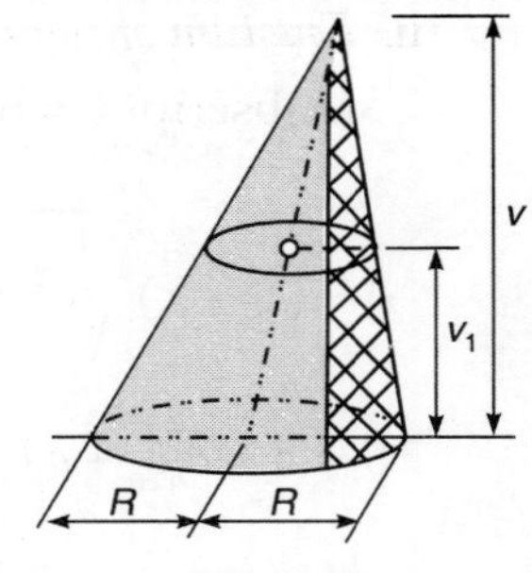

Fig. 3.29

iii. *General cone*:

$$\text{Volume } V = \frac{Bv}{3}$$

For the frustum,

$$\text{Volume } V = \frac{v_1}{3}\,(B_b + B_t + \sqrt{B_b B_t}\,)$$

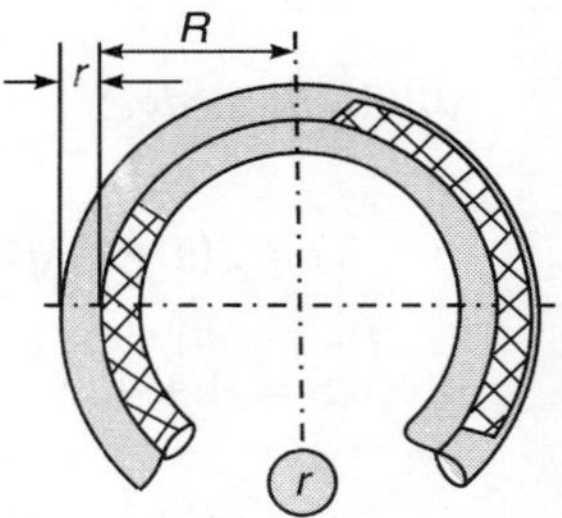

Fig. 3.30

iv. *Torus*:

$$\text{Surface } S = 4\pi^2 Rr = 39.4784Rr$$

$$\text{Volume } V = 2\pi^2 Rr^2 \approx 19.7392Rr^2$$

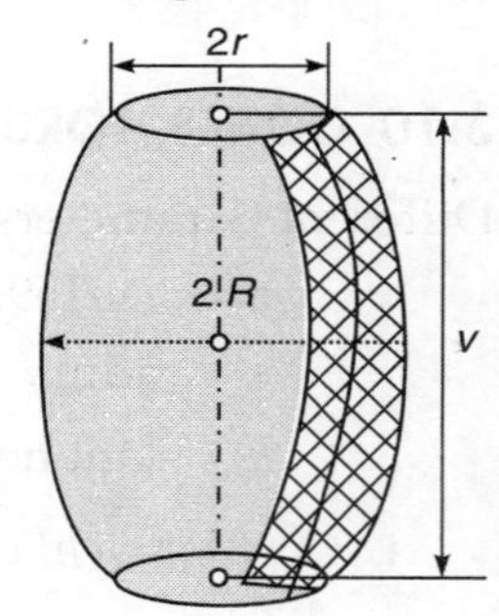

Fig. 3.31

v. *Circular barrel*:

For circular curvature,

$$\text{Volume } V = \frac{1}{3}\,\pi v\,(2R^2 + r^2)$$

For parabolic curvature,

$$\text{Volume } V = \frac{1}{15}\,\pi v\,(8R^2 + 4Rr + 3r^2)$$

3.11 CYLINDERS

Different parameters are:

$\pi = 3.14159...$	R = radius	A = lateral area
r = radius	B = area of base	S = surface
h = height	v = altitude	V = volume

i. *Right circular cylinder*:

Lateral area $A = 2\pi Rv$

Base area $B = \pi R^2$ $\qquad S = \pi R\,(R + 2v)$

Volume $V = \pi R^2 v$

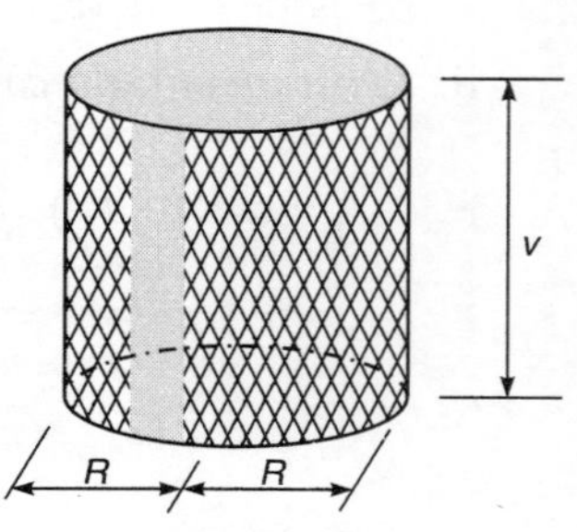

Fig. 3.32

ii. *Truncated frustum of right circular cylinder*:

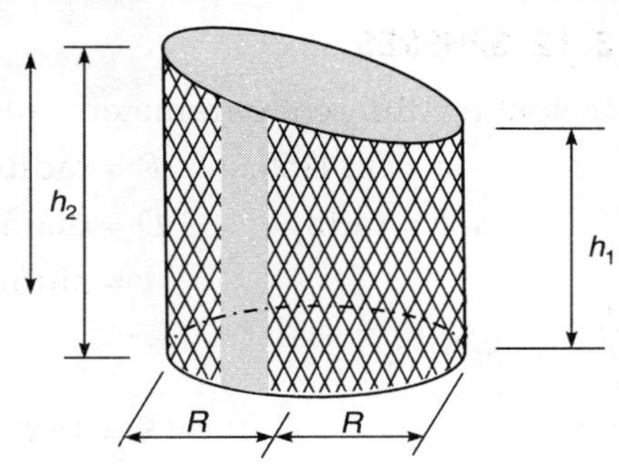

Fig. 3.33

$$A = \pi R \ (h_1 + h_2)$$

$$S = \pi R \left[h_1 + h_2 + R + \sqrt{R^2 + \left(\frac{h_2 - h_1}{2} \right)^2} \right]$$

$$V = \pi R^2 \frac{h_1 + h_2}{2}$$

iii. *Ungula of right circular cylinder* (2ω = central angle):

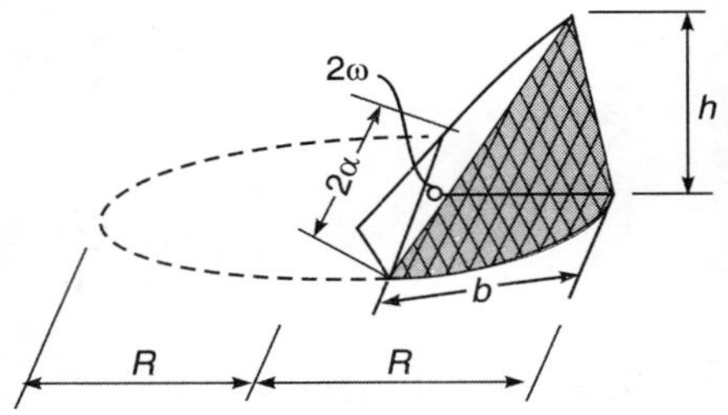

Fig. 3.34

$$A = \frac{2Rh}{b} \ [(b - R)\ \omega + a]$$

$$V = \frac{h}{3b} \ [a(3R^2 - a^2) + 3R^2 \ (b - R)\ \omega]$$

If $a = b = R$, then

$$A = 2Rh \qquad V = \frac{2}{3} R^2 h$$

iv. *Hollow right circular cylinder*:

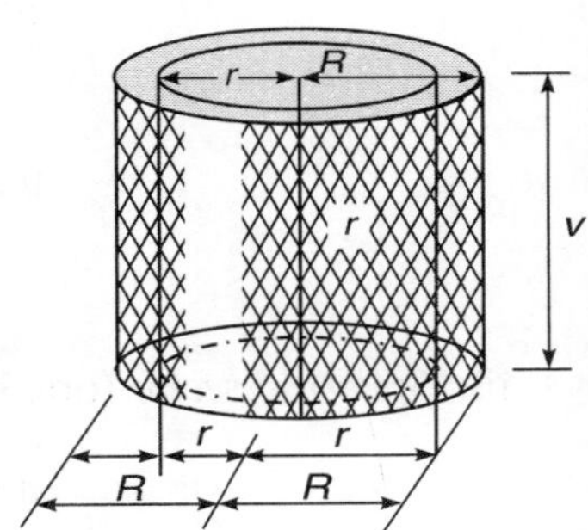

Fig. 3.35

$$t = R - r \qquad \rho = \frac{R + r}{2}$$

$$A = 2\pi R v$$

$$B = \pi \ (R^2 - r^2)$$

$$V = \pi v \ (R^2 - r^2) = 2\pi v t \rho$$

v. *General circular cylinder*:

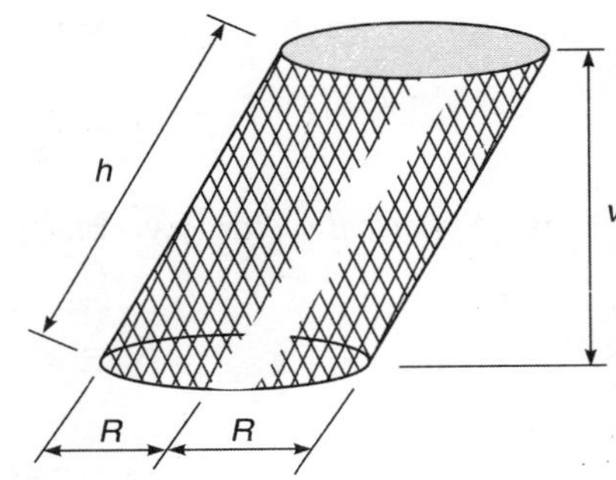

Fig. 3.36

C = circumference of right section

Lateral area $A = Ch$

Base area $B = \pi R^2$

$$V = Bv$$

3.12 SPHERES

In sphere, different parameters are:

$\pi = 3.14159...$	R = radius	A = lateral area
a, b = radii	D = diameter	S = surface
	v = altitude	V = volume

i. *Sphere*:

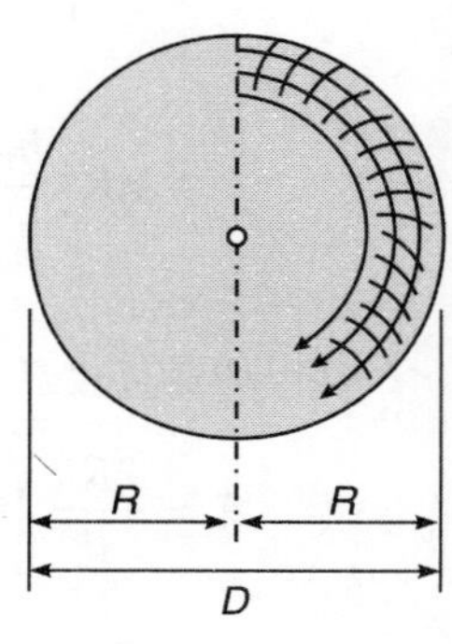

Fig. 3.37

$$S = 4\pi R^2 = 12.5663\ R^2$$

$$= \pi D^2 \approx \sqrt[3]{36\pi V^2} = 4.8362\ \sqrt[3]{V^2}$$

$$V = \frac{4}{3}\ \pi R^2 \approx 4.1888 R^3$$

$$= \frac{\pi D^3}{6} = \frac{1}{6}\sqrt{\frac{S^3}{\pi}} \approx 0.0940\sqrt{S^3}$$

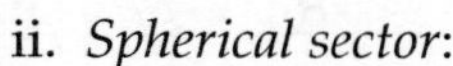

ii. *Spherical sector*:

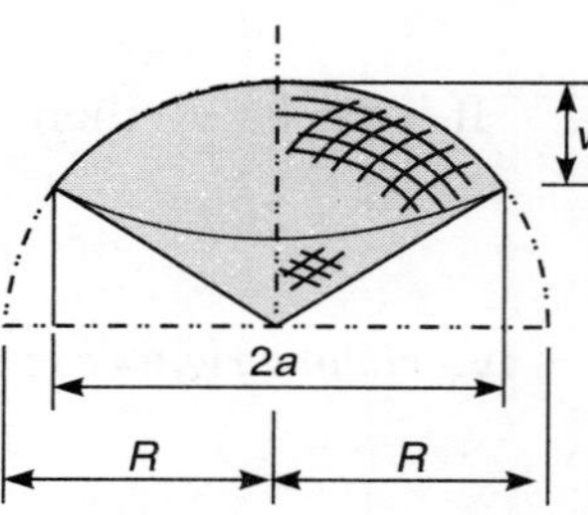

Fig. 3.38

$$S = \pi R\ (2v + a)$$

$$V = \frac{2\pi}{3}\ R^2 v = 2.0943\ R^2 v$$

iii. *Spherical sector* (one base):

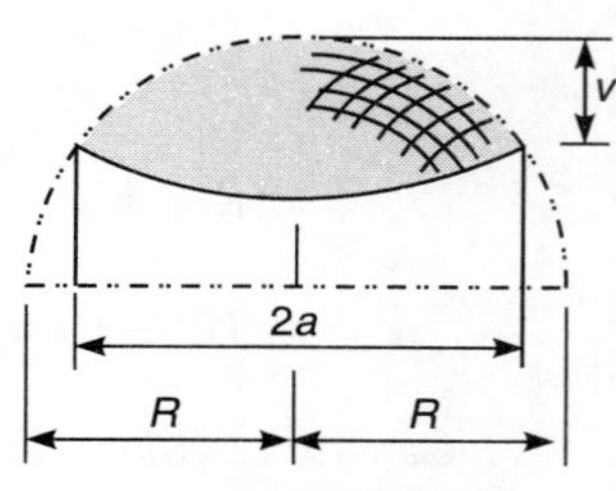

Fig. 3.39

$$a = \sqrt{v(2R - v)}$$

$$A = 2\pi R v = 6.2832 R v$$

$$S = \pi v\ (4R - v)$$

$$V = \frac{\pi}{3}\ v^2 (3R - v) = 1.0972\ v^2\ (3R - v)$$

iv. *Spherical sector* (two bases):

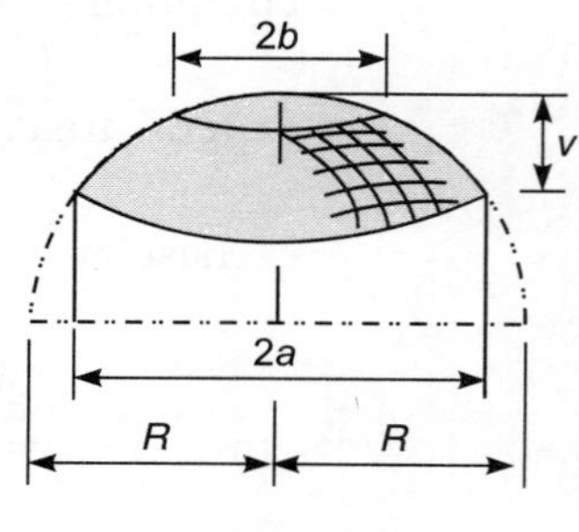

Fig. 3.40

$$R^2 = a^2 + \left(\frac{a^2 - b^2 - v^2}{2v}\right)^2$$

$$A = 2\pi R v$$

$$S = \pi(2Rv + a^2 + b^2)$$

$$V = \frac{\pi v}{6} (3a^2 + 3b^2 + v^2)$$

v. *Conical ring*:

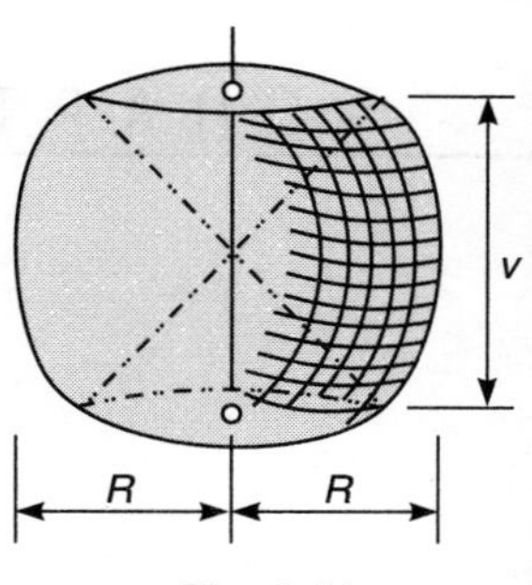

Fig. 3.41

$$S = 2\pi R \left(v + \sqrt{R^2 - \frac{v^2}{4}} \right)$$

$$V = \frac{2\pi}{3} R^2 v = 2.0943 R^2 v$$

CHAPTER

4

Trigonometry

4.1 RELATIONSHIPS AND SOLUTIONS OF RIGHT ANGLED TRIANGLE

In triangle:

a, b = legs	c = hypotenuse	A = area
A, B, C = vertices	p, q = segment of c	R = circumradius
α, β, γ = angles	h = height	r = radius

i. Trigonometric Functions of an Angle

With reference to Fig. 4.1, $P(x, y)$ is a point in either one of the four quadrants and A is an angle whose initial side is coincident with the positive x-axis and whose terminal side contains the point $P(x, y)$ is denoted by r and is positive. The trigonometric functions of the angle A are defined as:

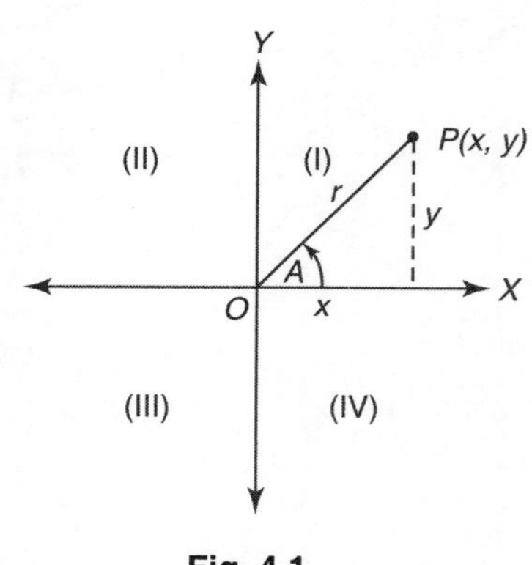

Fig. 4.1

$\sin A$ = sine A $= y / r$

$\cos A$ = cosine A $= x / r$

$\tan A$ = tangent A $= y / x$

ctn A = cotangent A $= x / y$

$\sec A$ = secant A $= r / x$

$\csc A$ = cosecant A $= r / y$

Angles are measured in degree or radian;

$$180° = \pi \text{ radians};$$

$$1 \text{ radian} = 180° / \pi \text{ degrees}.$$

Fig. 4.1: From the trigonometric point, angle A is taken to be positive when the rotation is counterclockwise and negative when the rotation is clockwise. The plane is divided into 4 quadrants.

The trigonometric function of 0°, 30°, 45°, and integer multiples of these are directly computed.

	0°	30°	45°	60°	90°	120°	135°	150°	180°
sin	0	$\frac{1}{2}$	$\frac{\sqrt{2}}{2}$	$\frac{\sqrt{3}}{2}$	1	$\frac{\sqrt{3}}{2}$	$\frac{\sqrt{2}}{2}$	$\frac{1}{2}$	0
cos	1	$\frac{\sqrt{3}}{2}$	$\frac{\sqrt{2}}{2}$	$\frac{1}{2}$	0	$-\frac{1}{2}$	$-\frac{\sqrt{2}}{2}$	$-\frac{\sqrt{3}}{2}$	-1
tan	0	$\frac{\sqrt{3}}{3}$	1	$\sqrt{3}$	∞	$-\sqrt{3}$	-1	$-\frac{\sqrt{3}}{3}$	0
ctn	∞	$\sqrt{3}$	1	$\frac{\sqrt{3}}{3}$	0	$-\frac{\sqrt{3}}{3}$	-1	$-\sqrt{3}$	∞
sec	1	$\frac{2\sqrt{3}}{3}$	$\sqrt{2}$	2	∞	-2	$-\sqrt{2}$	$-\frac{2\sqrt{3}}{3}$	-1
csc	∞	2	$\sqrt{2}$	$\frac{2\sqrt{3}}{3}$	1	$\frac{2\sqrt{3}}{3}$	$\sqrt{2}$	2	∞

ii. Trigonometric Identities and Basic Formulae

$$\sin A = \frac{1}{\csc A}$$

$$\cos A = \frac{1}{\sec A}$$

$$\tan A = \frac{1}{\operatorname{ctn} A} = \frac{\sin A}{\cos A}$$

$$\csc A = \frac{1}{\sin A}$$

$$\sec A = \frac{1}{\cos A}$$

$$\operatorname{ctn} A = \frac{1}{\tan A} = \frac{\cos A}{\sin A}$$

$$\sin^2 A + \cos^2 A = 1$$

$$1 + \tan^2 A = \sec^2 A$$

$$1 + \operatorname{ctn}^2 A = \csc^2 A$$

iii. Sum and Difference Formulae

$\sin(A \pm B) = \sin A \cos B \pm \cos A \sin B$

$\cos(A \pm B) = \cos A \cos B \pm \sin A \sin B$

$$\tan(A \pm B) = \frac{\tan A \pm \tan B}{1 \mp \tan A \tan B}$$

$$\sin 2A = 2 \sin A \cos As$$

$$\sin 3A = 3 \sin A - 4 \sin^3 A$$

$$\sin nA = 2 \sin(n-1) A \cos A - \sin(n-2) A$$

$$\cos 2A = 2\cos^2 A - 1 = 1 - 2\sin^2 A$$

$$\cos 3A = 4\cos^3 A - 3\cos A$$

$$\cos nA = 2\cos(n-1) A \cos A - \cos(n-2) A$$

iv. Sum and Difference into Products

$$\sin A + \sin B = 2 \sin \frac{1}{2}(A+B) \cos \frac{1}{2}(A-B)$$

$$\sin A - \sin B = 2 \cos \frac{1}{2}(A+B) \sin \frac{1}{2}(A-B)$$

$$\cos A + \cos B = -2 \cos \frac{1}{2}(A+B) \cos \frac{1}{2}(A-B)$$

$$\cos A - \cos B = -2 \sin \frac{1}{2}(A+B) \sin \frac{1}{2}(A-B)$$

$$\tan A \pm \tan B = \frac{\sin(A \pm B)}{\cos A \cos B}$$

$$\text{ctn } A \pm \text{ctn } B = M \frac{\sin(A \pm B)}{\sin A \sin B}$$

v. Product into Sum or Difference

$$\sin A \sin B = \frac{1}{2}\cos(A-B) - \frac{1}{2}\cos(A+B)$$

$$\cos A \cos B = \frac{1}{2}\cos(A-B) + \frac{1}{2}\cos(A+B)$$

$$\sin A \cos B = \frac{1}{2}\sin(A+B) + \frac{1}{2}\sin(A-B)$$

$$\sin \frac{A}{2} = \pm\sqrt{\frac{1-\cos A}{2}}$$

$$\cos \frac{A}{2} = \pm\sqrt{\frac{1+\cos A}{2}}$$

$$\tan\frac{A}{2}=\frac{1-\cos A}{\sin A}=\frac{\sin A}{1+\cos A}=\pm\sqrt{\frac{1-\cos A}{1+\cos A}}$$

$$\sin^2 A=\frac{1}{2}(1-2A)$$

$$\cos^2 A=\frac{1}{2}(1+\cos 2A)$$

$$\sin^3 A=\frac{1}{4}(3\sin A-\sin 3A)$$

$$\cos^3 A=\frac{1}{4}(\cos 3A+3\cos A)$$

$$\sin ix=\frac{1}{2}i(e^x-e^{-x})=i\sinh x$$

$$\cos ix=\frac{1}{2}(e^x+e^{-x})=\cosh x$$

$$\tan ix=\frac{i(e^x-e^{-x})}{e^x+e^{-x}}=i\tanh x$$

$$e^{x+iy}=e^x(\cos y+i\sin y)$$

$$(\cos x\pm i\sin x)^n=\cos nx\pm i\sin nx$$

vi. Inverse Trigonometric Functions

The inverse trigonometric functions are multiple valued and this should be taken into account in the use of the following formulas.

$$\sin^{-1}x=\cos^{-1}\sqrt{1-x^2}$$

$$=\tan^{-1}\frac{x}{\sqrt{1+x^2}}=\operatorname{ctn}^{-1}\frac{\sqrt{1-x^2}}{x}$$

$$=\sec^{-1}\frac{1}{\sqrt{1-x^2}}=\csc^{-1}\frac{1}{x}$$

$$=-\sin^{-1}(-x)$$

$$\cos^{-1}x=\sin^{-1}\sqrt{1-x^2}$$

$$=\tan^{-1}\frac{\sqrt{1-x^2}}{x}=\operatorname{ctn}^{-1}\frac{x}{\sqrt{1-x^2}}$$

$$=\sec^{-1}\frac{1}{x}=\csc^{-1}\frac{1}{\sqrt{1-x^2}}$$

$$=\pi-\cos^{-1}(-x)$$

$$\tan^{-1} x = \text{ctn}^{-1} \frac{1}{x}$$

$$= \sin^{-1} \frac{x}{\sqrt{1+x^2}} = \cos^{-1} \frac{1}{\sqrt{1+x^2}}$$

$$= \sec^{-1} \sqrt{1+x^2} = \csc^{-1} \frac{\sqrt{1+x^2}}{x}$$

$$= -\tan^{-1}(-x)$$

vii. Rules of Triangles

In any triangle ΔABC (in a plane), with sides a, b, and c and corresponding opposite angles A, B, C.

$$\frac{a}{\sin A} = \frac{b}{\sin B} = \frac{c}{\sin C} \qquad \text{(sine rule)}$$

$$a^2 = b^2 + c^2 - 2cb \cos A \qquad \text{(cosine rule)}$$

$$\frac{a+b}{a-b} = \frac{\tan \frac{1}{2}(A+B)}{\tan \frac{1}{2}(A-B)} \qquad \text{(tangent rule)}$$

viii. Half of the Angle of a Triangle

$$\sin \frac{1}{2} A = \sqrt{\frac{(s-b)(s-c)}{bc}}, \quad \text{where } s = \frac{1}{2}(a+b+c)$$

$$\cos \frac{1}{2} A = \sqrt{\frac{s(s-a)}{bc}}$$

$$\tan \frac{1}{2} A = \sqrt{\frac{(s-b)(s-c)}{s(s-a)}}$$

ix. Area of a Triangle

$$\text{Area} = \frac{1}{2} bc \sin A$$

$$= \sqrt{s(s-a)(s-b)(s-c)}$$

If the vertices have coordinates (x_1, y_1), (x_2, y_2), (x_3, y_3), the area is the *absolute* value of the expression;

$$\frac{1}{2}\begin{vmatrix} x_1 & y_1 & 1 \\ x_2 & y_2 & 1 \\ x_3 & y_3 & 1 \end{vmatrix}$$

x. Relationships of Right Angled Triangle

$a^2 + b^2 = c^2$ $\qquad \alpha + \beta = 90^\circ$

$\sin \alpha = \dfrac{a}{c}$ $\qquad \sin \beta = \dfrac{b}{c}$

$\cos \alpha = \dfrac{b}{c}$ $\qquad \cos \beta = \dfrac{a}{c}$

$\tan \alpha = \dfrac{a}{b}$ $\qquad \tan \beta = \dfrac{b}{a}$

$\cot \alpha = \dfrac{b}{a}$ $\qquad \cot \beta = \dfrac{a}{b}$

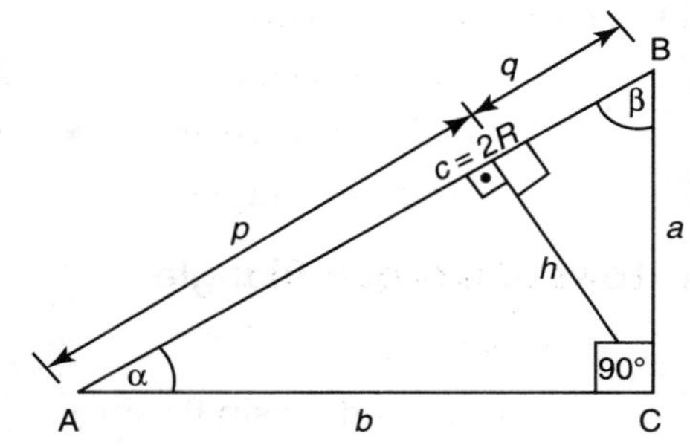

Fig. 4.2

xi. Formulas of Right Angled Triangle

$$h = a \sin \beta = b \sin \alpha \qquad p = b \cos \alpha = h \cot \alpha \qquad q = a \cos \beta = h \cot \beta$$

$$h = \sqrt{ab \cos \alpha \cos \beta} \qquad c = b \cos \alpha + a \cos \beta$$

$$R = \frac{c}{2} \qquad r = \frac{a+b-c}{2} = \frac{c(\sin\alpha + \sin\beta - 1)}{2}$$

$$A = \frac{ab}{2} = \frac{c^2}{4} \sin 2\alpha = \frac{a^2}{2} \cot \alpha = \frac{b^2}{2} \cot \beta$$

xii. Solutions of Right Angled Triangle

Variable	*Results*					
	a	b	c	α	β	A
a, b			$\sqrt{a^2 + b^2}$	$\tan \alpha = \dfrac{a}{b}$	$\tan \beta = \dfrac{b}{a}$	$\dfrac{ab}{2}$
a, c		$\sqrt{c^2 - a^2}$		$\sin \alpha = \dfrac{a}{c}$	$\cos \beta = \dfrac{a}{c}$	$\dfrac{a\sqrt{c^2 - a^2}}{2}$
a, α		$a \cot \alpha$	$\dfrac{a}{\sin \alpha}$		$90^\circ - \alpha$	$\dfrac{a^2 \cot\alpha}{2}$
b, α	$b \tan \alpha$		$\dfrac{b}{\cos \alpha}$		$90^\circ - \alpha$	$\dfrac{b^2 \tan \alpha}{2}$
c, α	$c \sin \alpha$	$c \cos \alpha$			$90^\circ - \alpha$	$\dfrac{c^2 \sin 2\alpha}{4}$

4.2 RELATIONSHIPS OF OBLIQUE TRIANGLE

Different parameters are:

a, b, c = sides	h = altitude	A = area
A, B, C = vertices	m = median	R = circumradius
α, β, γ = angles	t = bisector	r = radius

i. Laws of Oblique Triangle

a. *Law of sine*

$a : b : c = \sin\alpha : \sin\beta : \sin\gamma$

b. *Law of cosine*

$a^2 = b^2 + c^2 - 2bc\cos\alpha$

c. *Law of tangent*

$(a + b) : (a - b) = \tan\frac{\alpha+\beta}{2} : \tan\frac{\alpha-\beta}{2}$

d. *Projection formula*

$a = b\cos\gamma + c\cos\beta$

e. *Law of angles*

$\sin(\alpha + \beta) = \sin\gamma, \quad \cos(\alpha + \beta) = -\cos\gamma$

$\tan(\alpha + \beta) = -\tan\gamma, \quad \cot(\alpha + \beta) = -\cot\gamma$

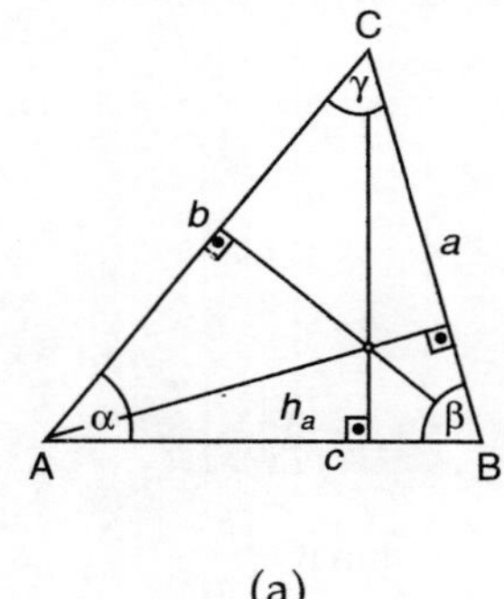

(a)

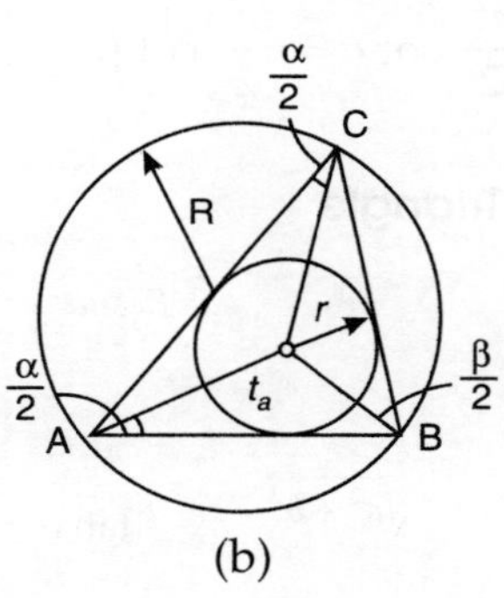

(b)

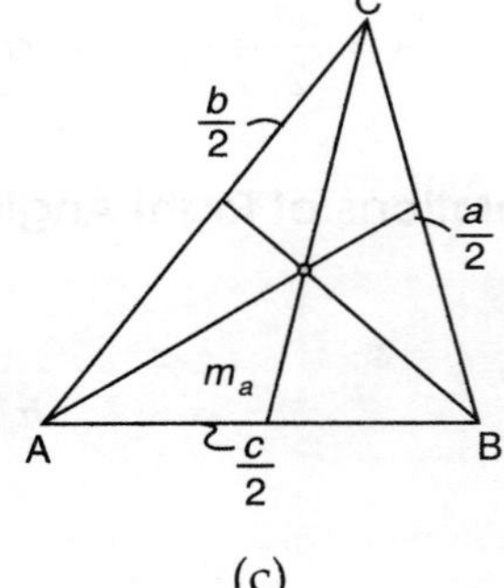

(c)

Fig. 4.3

ii. Formulas of a Triangle (2*p* = *a* + *b* + *c*)

In ΔABC:

a. *For half angles*:

$$\sin\frac{\alpha}{2} = \sqrt{\frac{(p-b)(p-c)}{bc}}$$

$$\cos\frac{\alpha}{2} = \sqrt{\frac{p(p-a)}{bc}}$$

$$\tan\frac{\alpha}{2} = \sqrt{\frac{(p-b)(p-c)}{p(p-a)}}$$

b. *For R, r*:

$$R = \frac{a}{2\sin\alpha}$$

$$r = (p-a)\tan\frac{\alpha}{2} = 4R\sin\frac{\alpha}{2}\sin\frac{\beta}{2}\sin\frac{\gamma}{2}$$

c. *For h, m, t*:

$$h_a = b\sin\gamma = c\sin\beta$$

$$m_a = \frac{1}{2}\sqrt{b^2 + c^2 + 2bc\cos\alpha}$$

$$t_a = \frac{2bc\cos(\alpha/2)}{b+c}$$

d. *For area*:

$$A = \frac{ah_a}{2} = \frac{ab\sin\gamma}{2}$$

$$= \frac{abc}{4R} = \frac{a^2\sin\beta\sin\gamma}{2\sin\alpha}$$

$$= \sqrt{p(p-a)(p-b)(p-c)}$$

4.3 PLANE OBLIQUE TRIANGLE—SOLUTIONS

Variable(s)	*Result* $(2p = a + b + c)$
a, b, c	$\cos\alpha = \frac{b^2 + c^2 - a^2}{2bc}$ or $\cos\frac{\alpha}{2} = \sqrt{\frac{p(p-a)}{bc}}$ $\cos\beta = \frac{a^2 + c^2 - b^2}{2ac}$ or $\cos\frac{\beta}{2} = \sqrt{\frac{p(p-b)}{ac}}$ $\gamma = 180° - (\alpha + \beta)$ *Area,* $A = \sqrt{p(p-a)(p-b)(p-c)}$
a, b, α	$\sin\beta = \frac{b\sin\alpha}{a}$ $\gamma = 180° - (\alpha + \beta)$ $c = \frac{a\sin\gamma}{\sin\alpha}$ Area, $A = \frac{ab}{2}\sin\gamma$

Variable(s)	*Result* $(2p = a + b + c)$
a, b, γ	$\frac{\alpha+\beta}{2} = 90° - \frac{\gamma}{2}$ $\tan\frac{\alpha-\beta}{2} = \frac{a-b}{a+b}\cot\frac{\gamma}{2}$ $\alpha = \frac{\alpha+\beta}{2} + \frac{\alpha-\beta}{2}$ $\beta = \frac{\alpha+\beta}{2} - \frac{\alpha-\beta}{2}$ $c = \sqrt{a^2 + b^2 - 2ab\cos\gamma}$ Area, $A = \frac{ab}{2}\sin\gamma$
a, α, β	$\gamma = 180° - (\alpha + \beta)$ $b = \frac{a\sin\beta}{\sin\alpha}$ $c = \frac{a\sin\gamma}{\sin\alpha} = \frac{a\sin(\alpha+\beta)}{\sin\alpha}$ Area, $A = \frac{ab}{2}\sin\gamma$
a, β, γ	$\alpha = 180° - (\beta + \gamma)$ $b = \frac{a\sin\beta}{\sin\alpha} = \frac{a\sin\beta}{\sin(\beta+\gamma)}$ $c = \frac{a\sin\gamma}{\sin\alpha} = \frac{a\sin\gamma}{\sin(\beta+\gamma)}$ Area, $A = \frac{a^2\sin\beta\sin\gamma}{2\sin(\beta+\gamma)}$

4.4 SPHERICAL TRIANGLES

In spherical triangles different parameters are:

a, b, c = sides $\quad$ ε = spherical excess $\quad$ A = area

α, β, γ = angles $\quad$ d = spherical defect $\quad$ R = radius of sphere

i. In Right Angled Triangle

Napier's formulas

$\sin a = \tan b \cot\beta \quad \sin a = \sin\alpha \sin c$

$\sin b = \tan a \cot\alpha \quad \sin b = \sin\beta \sin c$

$\cos c = \cot\alpha \cot\beta \quad \cos c = \cos a \sin b \qquad +90° < \alpha + \beta < +270°$

$\cos\alpha = \tan b \cot c \quad \cos\alpha = \cos a \sin\beta \qquad -90° < \alpha - \beta < +90°$

$\cos\beta = \tan\alpha \cot c \quad \cos\beta = \sin\alpha \cos b$

ii. In Oblique Triangle $(2p = a + b + c)$

a. *Law of sine*

$$\sin a : \sin b : \sin c = \sin \alpha : \sin \beta : \sin \gamma$$

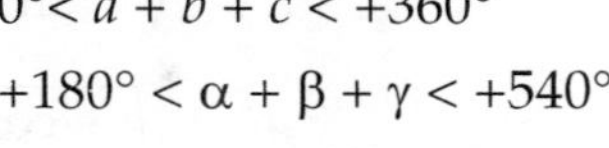

$0° < a + b + c < +360°$

$+180° < \alpha + \beta + \gamma < +540°$

b. *Law of cosine I*

$$\cos a = \cos b \cos c + \sin b \sin c \cos \alpha$$

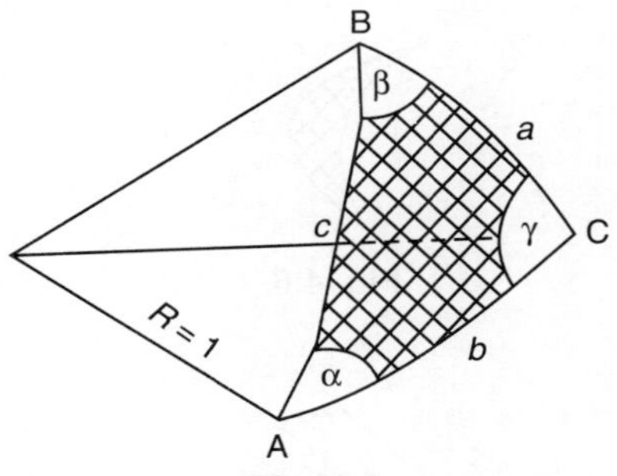
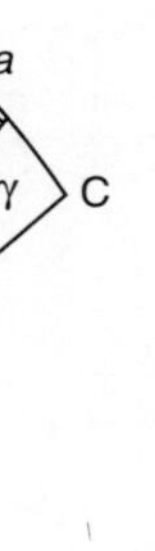

Fig. 4.4

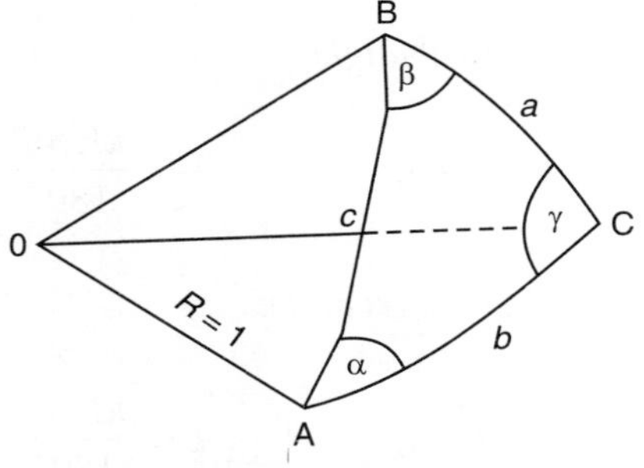

Fig. 4.5

c. *Law of cosine II*

$$\cos \alpha = -\cos \beta \cos \gamma + \sin \beta \sin \gamma \cos a$$

d. *Delambre's equations*

$$\sin\frac{\alpha+\beta}{2}\cos\frac{c}{2} = \cos\frac{a-b}{2}\cos\frac{\gamma}{2}$$

$$\sin\frac{\alpha-\beta}{2}\sin\frac{c}{2} = \sin\frac{a-b}{2}\cos\frac{\gamma}{2}$$

$$\cos\frac{\alpha+\beta}{2}\cos\frac{c}{2} = \cos\frac{a+b}{2}\sin\frac{\gamma}{2}$$

$$\cos\frac{\alpha-\beta}{2}\sin\frac{c}{2} = \sin\frac{a+b}{2}\sin\frac{\gamma}{2}$$

e. *Napier's equations*

$$\tan\frac{\alpha+\beta}{2}\cos\frac{a+b}{2} = \cos\frac{a-b}{2}\cot\frac{\gamma}{2}$$

$$\tan\frac{\alpha-\beta}{2}\sin\frac{a+b}{2} = \sin\frac{a-b}{2}\cot\frac{\gamma}{2}$$

$$\tan\frac{a+b}{2}\cos\frac{\alpha+\beta}{2} = \cos\frac{\alpha-\beta}{2}\tan\frac{c}{2}$$

$$\tan\frac{a-b}{2}\sin\frac{\alpha+\beta}{2} = \sin\frac{\alpha-\beta}{2}\tan\frac{c}{2}$$

f. *In angles of spherical triangles:*

$$\tan\frac{\alpha}{2} = \sqrt{\frac{\sin(p-b)\sin(p-c)}{\sin p \sin(p-a)}}$$

$$\tan\frac{a}{2} = \sqrt{\frac{-\cos\sigma\cos(\sigma-\alpha)}{\cos(\sigma-\beta)\cos(\alpha-\gamma)}}$$

g. *For spherical angle:*

Area, $$A = \frac{\pi R^2 \alpha^\circ}{180^\circ}$$

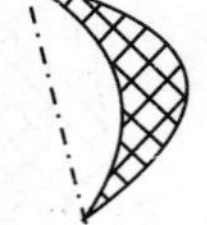

Fig. 4.6

h. *For spherical triangle:*

Area, $$A = \frac{\pi R^2 \varepsilon^\circ}{180^\circ}$$

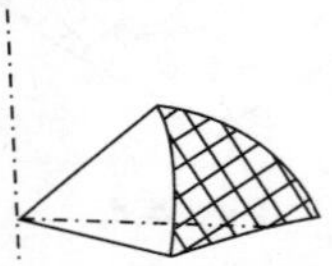

Fig. 4.7

i. *For spherical excess:*

$$\varepsilon = (\alpha + \beta + \gamma) - 180^\circ$$

$$\tan\frac{\varepsilon}{4} = \sqrt{\tan\frac{p}{2}\tan\frac{p-a}{2}\tan\frac{p-b}{2}\tan\frac{p-c}{2}}$$

j. *For spherical defect:*

$$d = 360^\circ - (a + b + c)$$

$$= \tan\left(\frac{a}{2} - \frac{d}{4}\right)$$

$$= \sqrt{\cot\frac{\sigma}{2}\cot\frac{\sigma-\alpha}{2}\tan\frac{\sigma-\beta}{2}\tan\frac{\sigma-\gamma}{2}}$$

CHAPTER

5

Trigonometric Functions

5.1 BASIC CONCEPTS OF TRIGONOMETRIC FUNCTIONS

i. Angle

The independent variable A is measured in radians.

$$180° = \pi \text{ radian} = 3.1415926535 \text{ radians} \qquad 1° = \frac{\pi}{180°} = 0.0174532925 \text{ radians} \qquad 1 \text{ radian} = \frac{180°}{\pi} = 57.2957795130°$$

ii. Trigonometric Functions

Trigonometric functions in terms of angle ω.

$$\sin^2 \omega + \cos^2 \omega = 1$$

$$\tan^2 \omega + 1 = \sec^2 \omega$$

$$\cot^2 \omega + 1 = \csc^2 \omega$$

$$\sin \omega \cdot \csc \omega = 1$$

$$\cos \omega \cdot \sec \omega = 1$$

$$\tan \omega \cdot \cot \omega = 1$$

$$\sin \omega + \text{covers } \omega = +1$$

$$\cos \omega + \text{vers } \omega = +1$$

$$\tan \omega = \frac{\sin \omega}{\cos \omega}$$

$$\cot \omega = \frac{\cos \omega}{\sin \omega}$$

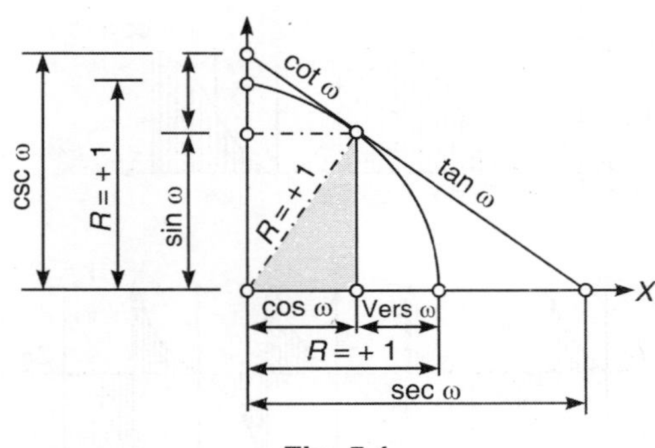

Fig. 5.1

iii. Reductions (k = 0, 1, 2, ...)

Function	$\pm \alpha$	$\frac{\pi}{2} \pm \alpha$	$2k\pi \pm \alpha$	$(4k+1)\frac{\pi}{2} \pm \alpha$	$(4k+2)\frac{\pi}{2} \pm \alpha$	$(4k+3)\frac{\pi}{2} \pm \alpha$
sin	$\pm \sin \alpha$	$+ \cos \alpha$	$\pm \sin \alpha$	$+ \cos \alpha$	$\mp \sin \alpha$	$- \cos \alpha$
cos	$+ \cos \alpha$	$\mp \sin \alpha$	$+ \cos \alpha$	$\mp \sin \alpha$	$- \cos \alpha$	$\mp \sin \alpha$
tan	$\pm \tan \alpha$	$\mp \cot \alpha$	$\pm \tan \alpha$	$\mp \cot \alpha$	$\pm \tan \alpha$	$\mp \cot \alpha$
cot	$\pm \cot \alpha$	$\mp \tan \alpha$	$\pm \cot \alpha$	$\mp \tan \alpha$	$\pm \cot \alpha$	$\mp \tan \alpha$
sec	$+ \sec \alpha$	$\mp \csc \alpha$	$+ \sec \alpha$	$\mp \csc \alpha$	$- \sec \alpha$	$+ \csc \alpha$
csc	$\pm \csc \alpha$	$+ \sec \alpha$	$\pm \csc \alpha$	$+ \sec \alpha$	$\mp \csc \alpha$	$- \sec \alpha$

5.2 PROPERTIES OF TRIGONOMETRIC FUNCTIONS

i. Signs in Quadrants of Trigonometric Functions

Quadrant	*sin*	*cos*	*tan*	*cot*	*sec*	*csc*	*vers*	*covers*
I	+	+	+	+	+	+	+	+
II	+	−	−	−	−	+	+	+
III	−	−	+	+	−	−	+	+
IV	−	+	−	−	+	−	+	+

ii. Graphs

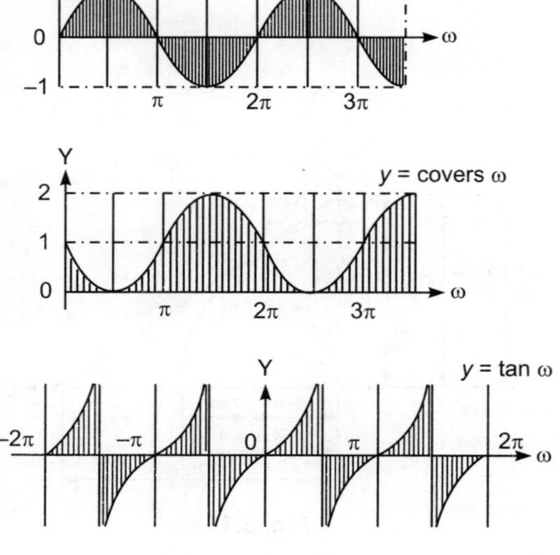

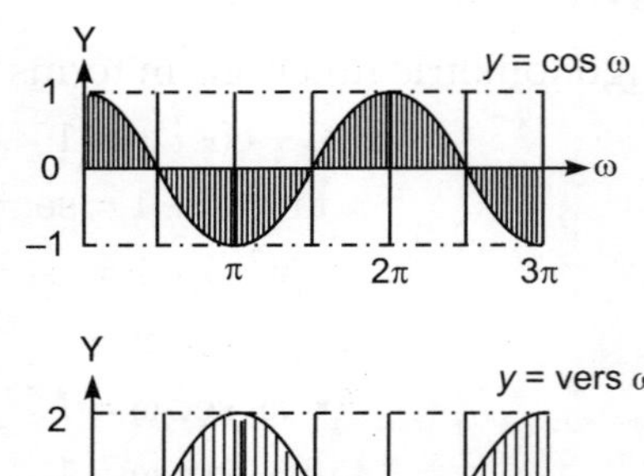

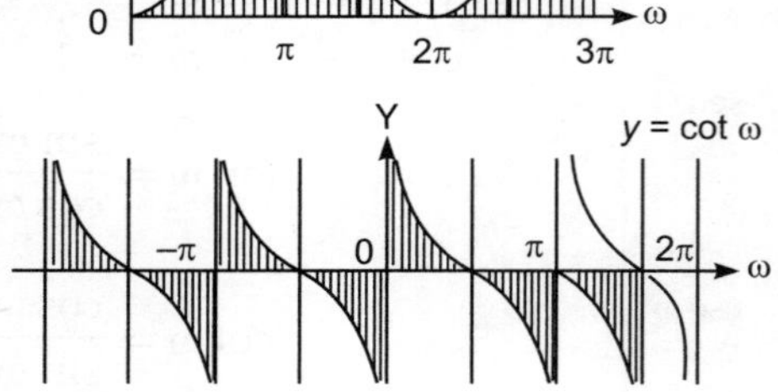

Fig. 5.2

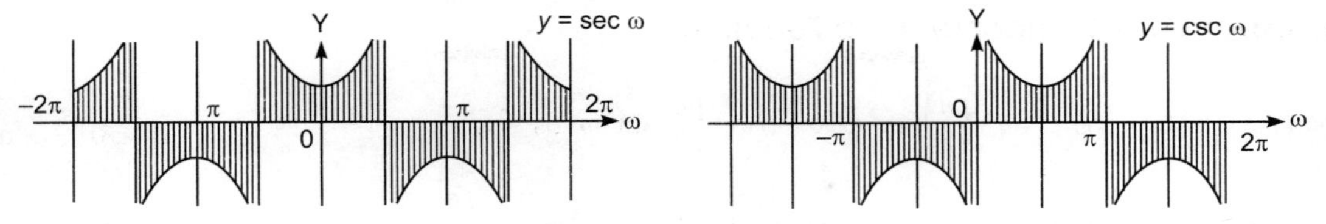

Fig. 5.2a

iii. Limit Values

Degree	*Radians*	*sin ω*	*cos ω*	*tan ω*	*cot ω*	*sec ω*	*csc ω*	*vers ω*	*covers ω*
0°	0	0	+1	0	$\mp\infty$	+1	$\mp\infty$	0	+1
90°	$\pi/2$	+1	0	$\pm\infty$	0	$\pm\infty$	+1	+1	0
180°	π	0	−1	0	$\mp\infty$	−1	$\pm\infty$	+2	+1
270°	$3\pi/2$	−1	0	$\pm\infty$	0	$\mp\infty$	−1	+1	+2
360°	2π	0	+1	0	$\mp\infty$	+1	$\mp\infty$	0	+1

5.3 GENERAL FORMULAS OF TRIGONOMETRIC FUNCTIONS

i. Transformations

Functions	sin ω	cos ω	tan ω	cot ω	sec ω	csc ω
sin ω	$\sin\omega$	$\pm\sqrt{1-\cos^2\omega}$	$\dfrac{\tan\omega}{\pm\sqrt{1+\tan^2\omega}}$	$\dfrac{1}{\pm\sqrt{1+\cot^2\omega}}$	$\dfrac{\pm\sqrt{\sec^2\omega-1}}{\sec\omega}$	$\dfrac{1}{\csc\omega}$
cos ω	$\pm\sqrt{1-\sin^2\omega}$	$\cos\omega$	$\dfrac{1}{\pm\sqrt{1+\tan^2\omega}}$	$\dfrac{\cot\omega}{\pm\sqrt{1+\cot^2\omega}}$	$\dfrac{1}{\sec\omega}$	$\dfrac{\pm\sqrt{\csc^2\omega-1}}{-\csc\omega}$
tan ω	$\dfrac{\sin\omega}{\pm\sqrt{1-\sin^2\omega}}$	$\dfrac{\pm\sqrt{1-\cos^2\omega}}{\cos\omega}$	$\tan\omega$	$\dfrac{1}{\cot\omega}$	$\pm\sqrt{\sec^2\omega-1}$	$\dfrac{1}{\pm\sqrt{\csc^2\omega-1}}$
cot ω	$\dfrac{\pm\sqrt{1-\sin^2\omega}}{\sin\omega}$	$\dfrac{\cos\omega}{\pm\sqrt{1-\cos^2\omega}}$	$\dfrac{1}{\tan\omega}$	$\cot\omega$	$\dfrac{1}{\pm\sqrt{\sec^2\omega-1}}$	$\pm\sqrt{\csc^2\omega-1}$
sec ω	$\dfrac{1}{\pm\sqrt{1-\sin^2\omega}}$	$\dfrac{1}{\cos\omega}$	$\pm\sqrt{1+\tan^2\omega}$	$\pm\dfrac{\sqrt{1+\cot^2\omega}}{\cot\omega}$	$\sec\omega$	$\dfrac{\csc\omega}{\pm\sqrt{\csc^2\omega-1}}$
csc ω	$\dfrac{1}{\sin\omega}$	$\dfrac{1}{\pm\sqrt{1-\cos^2\omega}}$	$\dfrac{\pm\sqrt{1+\tan^2\omega}}{\tan\omega}$	$\pm\sqrt{1+\cot^2\omega}$	$\dfrac{\sec\omega}{\pm\sqrt{\sec^2\omega-1}}$	$\csc\omega$

ii. Expansions of Trigonometric Functions

$$\sin n\omega = n \sin \omega \cos^{n-1} \omega - \binom{n}{3} \sin^3 \omega \cos^{n-3} \omega + \binom{n}{5} \sin^5 \omega \cos^{n-5} \omega - \ldots$$

$$\cos n\omega = \cos^n \omega - \omega - \binom{n}{2} \sin^2 \omega \cos^{n-2} \omega + \binom{n}{4} \sin^4 \omega \cos^{n-4} \omega - \ldots$$

$$\sin^{2n} \omega = \frac{(-1)^n}{2^{2n-1}} \left[\cos 2n\omega - \binom{2n}{1} \cos (2n-2)\,\omega + \ldots (-1)^{n-1} \binom{2n}{n-1} \cos 2\omega \right] + \binom{2n}{n} \frac{1}{2^{2n}}$$

$$\sin^{2n-1} \omega = \frac{(-1)^{n-1}}{2^{2n-2}} \left[\sin (2n-1)\,\omega - \binom{2n-1}{1} \sin (2n-3)\,\omega + \ldots (-1)^{n-1} \binom{2n-1}{n-1} \sin \omega \right]$$

$$\cos^{2n} \omega = \frac{1}{2^{2n-1}} \left[\cos^2 n\omega + \binom{2n}{1} \cos (2n-2)\,\omega + \ldots + \binom{2n}{n-1} \cos 2\omega \right] + \binom{2n}{n} \frac{1}{2^{2n}}$$

$$\cos^{2n-1} \omega = \frac{1}{2^{2n-2}} \left[\cos (2n-1)\,\omega + \binom{2n-1}{1} \cos (2n-3)\,\omega + \ldots + \binom{2n-1}{n-1} \cos \omega \right]$$

5.4 SUMS

i. Sums of Angles of Trigonometric Functions

$\sin (\alpha + \beta) = \sin \alpha \cos \beta \pm \cos \alpha \sin \beta$

$\cos (\alpha \pm \beta) = \cos \alpha \cos \beta \mp \sin \alpha \sin \beta$

$\tan (\alpha \pm \beta) \ \dfrac{\tan \alpha \pm \tan \beta}{1 \mp \tan \alpha \tan \beta}$

$\cot (\alpha \pm \beta) = \dfrac{\cot \alpha \cot \beta \mp 1}{\cot \beta + \cot \alpha}$

ii. Sums of Functions of Trigonometric Functions

$\sin \alpha + \sin \beta = 2 \sin \dfrac{\alpha + \beta}{2} \cos \dfrac{\alpha - \beta}{2}$

$\cos \alpha + \cos \beta = 2 \cos \dfrac{\alpha + \beta}{2} \cos \dfrac{\alpha - \beta}{2}$

$\sin \alpha - \sin \beta = 2 \sin \dfrac{\alpha - \beta}{2} \cos \dfrac{\alpha + \beta}{2}$

$\cos \alpha - \cos \beta = -2 \sin \dfrac{\alpha - \beta}{2} \sin \dfrac{\alpha + \beta}{2}$

$\tan \alpha \pm \tan \beta = \dfrac{\sin (\alpha \pm \beta)}{\cos \alpha \cos \beta}$

$\cot \alpha \pm \cot \beta = \dfrac{\sin (\beta \pm \alpha)}{\sin \alpha \sin \beta}$

$\sin \alpha + \cos \alpha = \sqrt{2} \sin \left(\dfrac{\pi}{4} + \alpha \right)$

$\sin \alpha - \cos \alpha = -\sqrt{2} \cos \left(\dfrac{\pi}{4} + \alpha \right)$

$\tan \alpha + \cot \alpha = 2 \csc 2\alpha$

$\tan \alpha - \cot \alpha = -2 \cot 2\,\alpha$

$\dfrac{1 + \tan \alpha}{1 - \tan \alpha} = \tan \left(\dfrac{\pi}{4} + \alpha \right)$

$\dfrac{1 + \cot \alpha}{1 - \cot \alpha} = -\cot \left(\dfrac{\pi}{4} - \alpha \right)$

iii. Sums of Sums

$\sin(\alpha+\beta)+\sin(\alpha-\beta) = 2\sin\alpha\cos\beta$

$\cos(\alpha+\beta)+\cos(\alpha-\beta) = 2\cos\alpha\cos\beta$

$\sin(\alpha+\beta)-\sin(\alpha-\beta) = 2\sin\beta\cos\alpha$

$\cos(\alpha+\beta)-\cos(\alpha-\beta) = -2\sin\alpha\sin\beta$

$$\tan(\alpha+\beta)+\tan(\alpha-\beta) = \frac{\sin 2\alpha}{\cos(\alpha+\beta)\cos(\alpha-\beta)}$$

$$\cot(\alpha+\beta)+\cot(\alpha-\beta) = \frac{\sin 2\alpha}{\sin(\alpha+\beta)\sin(\alpha-\beta)}$$

$$\tan(\alpha+\beta)-\tan(\alpha-\beta) = \frac{\sin 2\beta}{\cos(\alpha+\beta)\cos(\alpha-\beta)}$$

$$\cot(\alpha+\beta)-\cot(\alpha-\beta) = -\frac{\sin 2\beta}{\sin(\alpha+\beta)\sin(\alpha-\beta)}$$

5.5 MULTIPLE AND HALF ANGLES OF TRIGONOMETRIC FUNCTIONS

i. Multiple Angles of Trigonometric Functions

$\sin 2\alpha = 2\sin\alpha\cos\alpha$

$\cos 2\alpha = 1-2\sin^2\alpha$

$\sin 3\alpha = (\sin\alpha)(3-4\sin^2\alpha)$

$\cos 3\alpha = 4\cos^3\alpha - 3\cos\alpha$

$\sin 4\alpha = (4\sin\alpha\cos\alpha)(1-2\sin^2\alpha)$

$\cos 4\alpha = 8\cos^4\alpha - 8\cos^2\alpha + 1$

$\sin 5\alpha = (\sin\alpha)(5-20\sin^2\alpha+16\sin^4\alpha)$

$\cos 5\alpha = 16\cos^5\alpha - 20\cos^3\alpha + 5\cos\alpha$

$$\tan 2\alpha = \frac{2\tan\alpha}{1-\tan^2\alpha}$$

$$\cot 2\alpha = \frac{\cot^2\alpha-1}{2\cot\alpha}$$

$$\tan 3\alpha = \frac{3\tan\alpha-\tan^3\alpha}{1-3\tan^2\alpha}$$

$$\cot 3\alpha = \frac{3\cot\alpha-\cot^3\alpha}{1-3\cot^2\alpha}$$

$$\tan 4\alpha = \frac{4\tan\alpha-4\tan^3\alpha}{1-6\tan^2\alpha+\tan^4\alpha}$$

$$\cot 4\alpha = \frac{3\cot^4\alpha-6\cot^2\alpha+1}{4\cot^3\alpha-4\cot\alpha}$$

$$\tan 5\alpha = \frac{5\tan\alpha-10\tan^3\alpha+\tan^5\alpha}{1-10\tan^2\alpha+5\tan^4\alpha}$$

$$\cot 5\alpha = \frac{5\cot\alpha-10\cot^3\alpha+\cot^5\alpha}{1-10\cot^2\alpha+5\cot^4\alpha}$$

ii. Half Angles of Trigonometric Functions

$$\sin\frac{\alpha}{2} = \sqrt{\frac{1-\cos\alpha}{2}}$$

$$\cos\frac{\alpha}{2} = \sqrt{\frac{1+\cos\alpha}{2}}$$

$$\tan\frac{\alpha}{2} = \sqrt{\frac{1-\cos\alpha}{1+\cos\alpha}} = \frac{\sin\alpha}{1+\cos\alpha} = \frac{1-\cos\alpha}{\sin\alpha}$$

$$\cot\frac{\alpha}{2} = \sqrt{\frac{1+\cos\alpha}{1-\cos\alpha}} = \frac{1+\cos\alpha}{\sin\alpha} = \frac{\sin\alpha}{1-\cos\alpha}$$

iii. Relations of Trigonometric Functions

$$\sin\alpha = 2\sin\frac{\alpha}{2}\cos\frac{\alpha}{2}$$

$$\cos\alpha = \cos^2\frac{\alpha}{2} - \sin^2\frac{\alpha}{2}$$

$$\tan\alpha = \frac{2\tan(\alpha/2)}{1-\tan^2(\alpha/2)} = \frac{2\sin(\alpha/2)\cos(\alpha/2)}{\cos^2(\alpha/2)-\sin^2(\alpha/2)}$$

$$\cot\alpha = \frac{\cot^2(\alpha/2)-1}{2\cot(\alpha/2)} = \frac{\cos^2(\alpha/2)-\sin^2(\alpha/2)}{2\sin(\alpha/2)\cos(\alpha/2)}$$

$$\sin\alpha = \sqrt{\frac{1-\cos 2\alpha}{2}}$$

$$\cos\alpha = \sqrt{\frac{1+\cos 2\alpha}{2}}$$

$\tan\alpha = \sqrt{\dfrac{1-\cos 2\alpha}{1+\cos 2\alpha}}$ $\cot\alpha = \sqrt{\dfrac{1+\cos 2\alpha}{1-\cos 2\alpha}}$

$= \dfrac{\sin 2\alpha}{1+\cos 2\alpha} = \dfrac{1-\cos 2\alpha}{\sin 2\alpha}$ $= \dfrac{1+\cos 2\alpha}{\sin 2\alpha} = \dfrac{\sin 2\alpha}{1-\cos 2\alpha}$

5.6 POWERS AND PRODUCTS OF TRIGONOMETRIC FUNCTIONS

i. Powers of Trigonometric Functions

$\sin^2\alpha = \dfrac{1}{2}(-\cos 2\alpha + 1)$ $\cos^2\alpha = \dfrac{1}{2}(\cos 2\alpha + 1)$

$\sin^3\alpha = \dfrac{1}{4}(-\sin 3\alpha + 3\sin\alpha)$ $\cos^3\alpha = \dfrac{1}{4}(\cos 3\alpha + 3\cos\alpha)$

$\sin^4\alpha = \dfrac{1}{8}(\cos 4\alpha - 4\cos 2\alpha + 3)$ $\cos^4\alpha = \dfrac{1}{8}(\cos 4\alpha + 4\cos 2\alpha + 3)$

$\sin^5\alpha = \dfrac{1}{16}(\sin 5\alpha - 5\sin 3\alpha + 10\sin\alpha)$ $\cos^5\alpha = \dfrac{1}{16}(\cos 5\alpha + 5\cos 3\alpha + 10\cos\alpha)$

$\tan^2\alpha = \dfrac{1-\cos 2\alpha}{1+\cos 2\alpha}$ $\cot^2\alpha = \dfrac{1+\cos 2\alpha}{1-\cos 2\alpha}$

$\tan^3\alpha = \dfrac{-\sin 3\alpha + 3\sin\alpha}{\cos 3\alpha + 3\cos\alpha}$ $\cot^3\alpha = \dfrac{\cos 3\alpha + 3\cos\alpha}{-\sin 3\alpha + 3\sin\alpha}$

$\tan^4\alpha = \dfrac{\cos 4\alpha - 4\cos 2\alpha + 3}{\cos 4\alpha + 4\cos 2\alpha + 3}$ $\cot^4\alpha = \dfrac{\cos 4\alpha + 4\cos 2\alpha + 3}{\cos 4\alpha - 4\cos 2\alpha + 3}$

$\tan^5\alpha = \dfrac{\sin 5\alpha - 5\sin 3\alpha + 10\sin\alpha}{\cos 5\alpha + 5\cos 3\alpha + 10\cos\alpha}$ $\cot^5\alpha = \dfrac{\cos 5\alpha + 5\cos 3\alpha + 10\cos\alpha}{\sin 5\alpha - 5\sin 3\alpha + 10\sin\alpha}$

ii. Products of Trigonometric Functions

$\sin\alpha\sin\beta = \dfrac{1}{2}\cos(\alpha-\beta) - \dfrac{1}{2}\cos(\alpha+\beta)$ $\tan\alpha\tan\beta = \dfrac{\cos(\alpha-\beta) - \cos(\alpha+\beta)}{\cos(\alpha-\beta) + \cos(\alpha+\beta)}$

$\sin\alpha\cos\beta = \dfrac{1}{2}\sin(\alpha-\beta) + \dfrac{1}{2}\sin(\alpha+\beta)$ $\tan\alpha\cot\beta = \dfrac{\sin(\alpha-\beta) + \sin(\alpha+\beta)}{-\sin(\alpha-\beta) + \sin(\alpha+\beta)}$

$\cos\alpha\cos\beta = \dfrac{1}{2}\cos(\alpha-\beta) + \dfrac{1}{2}\cos(\alpha+\beta)$ $\cot\alpha\cot\beta = \dfrac{\cos(\alpha-\beta) + \cos(\alpha+\beta)}{\cos(\alpha-\beta) - \cos(\alpha+\beta)}$

If $\alpha + \beta + \gamma = \pi$, then

$4\sin\alpha\sin\beta\sin\gamma = \sin(\alpha+\beta+\gamma) + \sin(\beta+\gamma-\alpha) + \sin(\gamma+\alpha-\beta) - \sin(\alpha+\beta+\gamma)$

$4\sin\alpha\sin\beta\cos\gamma = -\cos(\alpha+\beta-\gamma) + \cos(\beta+\gamma-\alpha) + \cos(\gamma+\alpha-\beta) - \cos(\alpha+\beta+\gamma)$

$4\sin\alpha\cos\beta\cos\gamma = \sin(\alpha+\beta-\gamma) - \sin(\beta+\gamma-\alpha) + \sin(\gamma+\alpha-\beta) - \sin(\alpha+\beta+\gamma)$

$4\cos\alpha\cos\beta\cos\gamma = \cos(\alpha+\beta-\gamma) + \cos(\beta+\gamma-\alpha) + \cos(\gamma+\alpha-\beta) + \cos(\alpha+\beta+\gamma)$

5.7 INVERSE TRIGONOMETRIC FUNCTIONS

i. Definition

Inverse trigonometric functions are defined as follows:

Trigonometric functions	Inverse trigonometric functions	Principal values
$y = \sin \omega$	$\omega = \sin^{-1} y = \arcsin y$	$-\dfrac{\pi}{2} \le \omega \le +\dfrac{\pi}{2}$
$y = \cos \omega$	$\omega = \cos^{-1} y = \arccos y$	$0 \le \omega \le \pi$
$y = \tan \omega$	$\omega = \tan^{-1} y = \arctan y$	$-\dfrac{\pi}{2} < \omega < +\dfrac{\pi}{2}$
$y = \cot \omega$	$\omega = \cot^{-1} y = \text{arccot}\, y$	$0 < \omega < \pi$

which means that ω is the arc of an angle of which the trigonometric function is y.

ii. Relationship of Inverse Trigonometric Functions

$$\sin^{-1} y + \cos^{-1} y = \frac{\pi}{2} \qquad \tan^{-1} y + \cot^{-1} y = \frac{\pi}{2}$$

$$\sin^{-1}(-y) = -\sin^{-1} y \qquad \tan^{-1}(-y) = -\tan^{-1} y$$

iii. Graphs of Inverse Trigonometric Functions

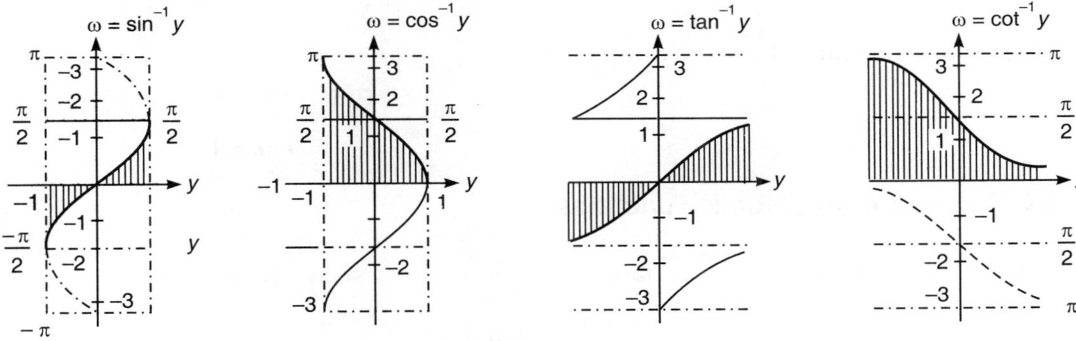

Fig. 5.3

iv. Transformation Tables

$y \le 0$	$\sin^{-1} y$	$\cos^{-1} y$	$\tan^{-1} y$	$\cot^{-1} y$
$\sin^{-1} y$	$\sin^{-1} y$	$\cos^{-1} \sqrt{1-y^2}$	$\tan^{-1} \dfrac{y}{\sqrt{1-y^2}}$	$\cot^{-1} \dfrac{\sqrt{1-y^2}}{y}$
$\cos^{-1} y$	$\sin^{-1} \sqrt{1-y^2}$	$\cos^{-1} y$	$\tan^{-1} \dfrac{\sqrt{1-y^2}}{y}$	$\cot^{-1} \dfrac{y}{\sqrt{1-y^2}}$
$\tan^{-1} y$	$\sin^{-1} \dfrac{y}{\sqrt{1+y^2}}$	$\cos^{-1} \dfrac{1}{\sqrt{1+y^2}}$	$\tan^{-1} y$	$\cot^{-1} \dfrac{1}{y}$
$\cot^{-1} y$	$\sin^{-1} \dfrac{1}{\sqrt{1+y^2}}$	$\cos^{-1} \dfrac{y}{\sqrt{1+y^2}}$	$\tan^{-1} \dfrac{1}{y}$	$\cot^{-1} y$

5.8 HYPERBOLIC FUNCTIONS

i. Definition

A hyperbolic function is a combination of e^x and e^{-x} and is introduced as follows:

Hyperbolic sine of x is $\sinh x = \dfrac{e^x - e^{-x}}{2}$

Hyperbolic cosine of x is $\cosh x = \dfrac{e^x + e^{-x}}{2}$

Hyperbolic tangent of x is $\tanh x = \dfrac{e^x - e^{-x}}{e^x + e^{-x}}$

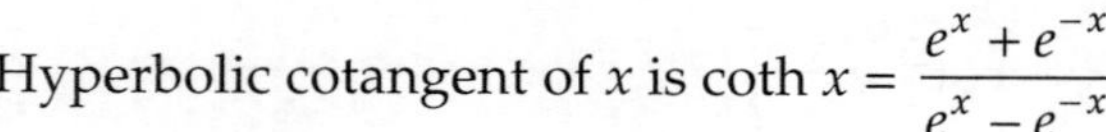

Hyperbolic cotangent of x is $\coth x = \dfrac{e^x + e^{-x}}{e^x - e^{-x}}$

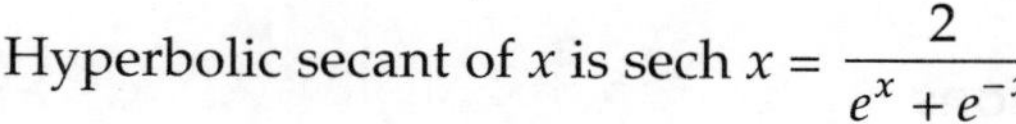

Hyperbolic secant of x is $\text{sech}\, x = \dfrac{2}{e^x + e^{-x}}$

Hyperbolic cosecant of x is $\text{csch}\, x = \dfrac{2}{e^x - e^{-x}}$

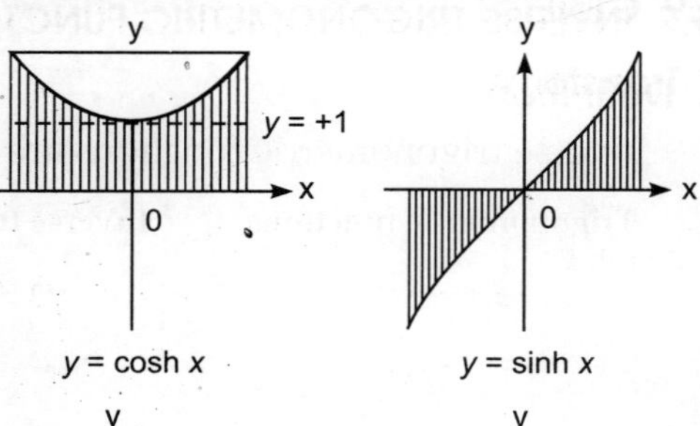

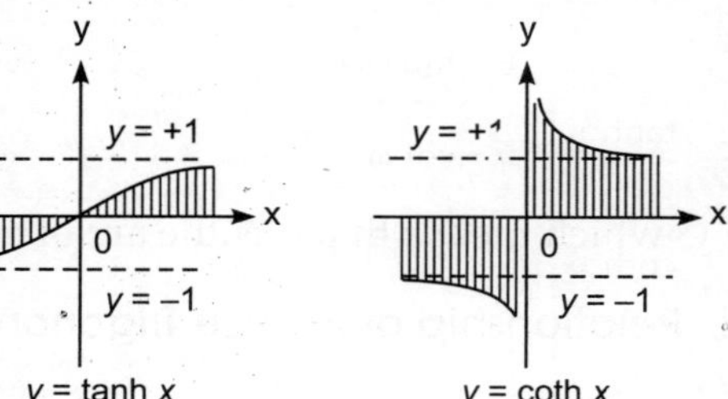

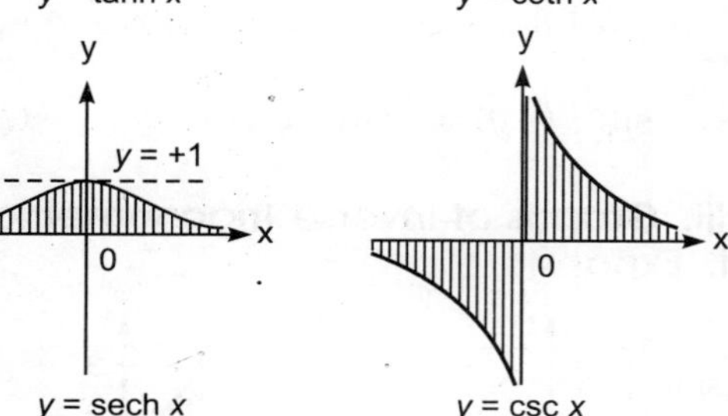

Fig. 5.4

ii. Relationships of Hyperbolic Functions

$\cosh^2 x - \sinh^2 x = 1$ $\qquad \tanh x = \dfrac{\sinh x}{\cosh x}$ $\qquad \text{sech}\, x \cosh x = 1$

$\tanh^2 x + \text{sech}^2 x = 1$ $\qquad \coth x = \dfrac{\cosh x}{\sinh x}$ $\qquad \text{csch}\, x \sinh x = 1$

$\coth^2 x - \text{csch}^2 x = 1$ $\qquad \tanh x \coth x = 1$

$\sinh(-x) = -\sinh x$ $\qquad \tanh(-x) = -\tanh x$ $\qquad \cosh(-x) = \cosh x$

$\text{sech}(-x) = \text{sech}\, x$ $\qquad \coth(-x) = -\coth x$ $\qquad \text{csch}(-x) = -\text{csch}\, x$

iii. Limit Values

x	*sinh x*	*cosh x*	*tanh x*	*coth x*	*sech x*	*csch x*
$-\infty$	$-\infty$	$+\infty$	-1	-1	0	0
-1	-1.1752	$+1.5431$	-0.7616	-1.3130	$+0.6480$	-0.8509
0	0	$+1$	0	$\mp\infty$	$+1$	$\mp\infty$
$+1$	$+1.1752$	$+1.5431$	$+0.7616$	$+1.3130$	$+0.6480$	$+0.8509$
$+\infty$	$+\infty$	$+\infty$	$+1$	$+1$	0	0

5.9 GENERAL FORMULAS OF HYPERBOLIC FUNCTIONS

i. Transformations of Hyperbolic Functions

$x > 0$	*sinh x*	*cosh x*	*tanh x*	*coth x*	*sech x*	*csch x*
$\sinh x$	$\sinh x$	$\sqrt{\cosh^2 x - 1}$	$\frac{\tanh x}{\sqrt{1-\tanh^2 x}}$	$\frac{1}{\sqrt{\coth^2 x - 1}}$	$\frac{\sqrt{1-\text{sech}^2 x}}{\text{sech}\, x}$	$\frac{1}{\text{csch}\, x}$
$\cosh x$	$\sqrt{1+\sinh^2 x}$	$\cosh x$	$\frac{1}{\sqrt{1-\tanh^2 x}}$	$\frac{\coth x}{\sqrt{\coth^2 x - 1}}$	$\frac{1}{\text{sech}\, x}$	$\frac{\sqrt{1+\text{csch}^2 x}}{\text{csch}\, x}$
$\tanh x$	$\frac{\sinh x}{\sqrt{1+\sin^2 x}}$	$\frac{\sqrt{\cosh^2 x - 1}}{\cosh x}$	$\tanh x$	$\frac{1}{\coth x}$	$\sqrt{1-\text{sech}^2 x}$	$\frac{1}{\sqrt{1+\text{csch}^2 x}}$
$\coth x$	$\frac{\sqrt{1+\sinh^2 x}}{\sinh x}$	$\frac{\cosh x}{\sqrt{\cosh^2 x - 1}}$	$\frac{1}{\tanh x}$	$\coth x$	$\frac{1}{\sqrt{1-\text{sech}^2 x}}$	$\sqrt{1+\text{csch}^2 x}$
$\text{sech}\, x$	$\frac{1}{\sqrt{1+\sinh^2 x}}$	$\frac{1}{\cosh x}$	$\sqrt{1-\tanh^2 x}$	$\frac{\sqrt{\coth^2 x - 1}}{\coth x}$	$\text{sech}\, x$	$\frac{\text{csch}\, x}{\sqrt{1+\text{csch}^2 x}}$
$\text{csch}\, x$	$\frac{1}{\sinh x}$	$\frac{1}{\sqrt{\cosh^2 x - 1}}$	$\frac{\sqrt{1-\tanh^2 x}}{\tanh x}$	$\sqrt{\coth^2 x - 1}$	$\frac{\text{sech}\, x}{\sqrt{1-\text{sech}^2 x}}$	$\text{csch}\, x$

ii. Expansions of Hyperbolic Functions

$$\sinh nx = n \sinh x \cosh^{n-1} x + \binom{n}{3} \sinh^3 x \cosh^{n-3} x + \binom{n}{5} \sinh^5 x \cosh^{n-5} x + \ldots$$

$$\cosh nx = \cosh^n x + \binom{n}{2} \sinh^2 x \cosh^{n-2} x + \binom{n}{4} \sinh^4 x \cosh^{n-4} x + \ldots$$

$$\sinh^{2n} x = \frac{1}{2^{2n-1}}\left[\cosh 2nx - \binom{2n}{1}\cosh (2n-2)x + \ldots(-1)^{n-1}\binom{2n}{n-1}\cosh 2x\right] + \binom{2n}{n}\frac{1}{2^{2n}}$$

$$\sinh^{2n-1} x = \frac{1}{2^{2n-1}}\left[\sinh (2n-1)\, x - \binom{2n-1}{1}\sinh (2n-3)\, x + \ldots(-1)^{n-1}\binom{2n-1}{n-1}\sinh x\right]$$

$$\cosh^{2n} x = \frac{1}{2^{2n-1}}\left[\cosh 2nx + \binom{2n}{1}\cosh (2n-2)\, x + \ldots + \binom{2n}{n-1}\cosh 2x\right] - \binom{2n}{n}\frac{1}{2^{2n}}$$

$$\cosh^{2n-1} = \frac{1}{2^{2n-2}}\left[\cosh (2n-1)\, x + \binom{2n-1}{1}\cosh (2n-3)\, x + \ldots + \binom{2n-1}{n-1}\cosh x\right]$$

5.10 SUMS AND PRODUCTS OF HYPERBOLIC FUNCTIONS

i. Sums of Angles of Hyperbolic Functions

$$\sinh (a \pm b) = \sinh a \cosh b \pm \cosh a \sinh b$$

$$\tanh (a \pm b) = \frac{\tanh a \pm \tanh b}{1 \pm \tanh a \tanh b}$$

$$\cosh (a \pm b) = \cosh a \cosh b \pm \sinh a \sinh b$$

$$\coth (a \pm b) = \frac{\coth a \coth b \pm 1}{\coth b \pm \coth a}$$

ii. Sums of Functions of Hyperbolic Functions

$$\sinh a + \sinh b = 2 \sinh \frac{a+b}{2} \cosh \frac{a-b}{2}$$

$$\tanh a + \tanh b = \frac{\sinh (a+b)}{\cosh a \cosh b}$$

$$\sinh a - \sinh b = 2 \cosh \frac{a+b}{2} \sinh \frac{a-b}{2}$$

$$\tanh a - \tanh b = \frac{\sinh (a-b)}{\cosh a \cosh b}$$

$$\cosh a + \cosh b = 2 \cosh \frac{a+b}{2} \cosh \frac{a-b}{2}$$

$$\coth a + \coth b = \frac{\sinh (a+b)}{\sinh a \sinh b}$$

$$\cosh a - \cosh b = 2 \sinh \frac{a+b}{2} \sinh \frac{a-b}{2}$$

$$\coth a - \coth b = \frac{\sinh (b-a)}{\sinh a \sinh b}$$

$$\sinh a + \cosh a = e^{a}$$

$$\sinh a - \cosh a = -e^{-a}$$

$$\tanh a + \coth a = 2 \coth 2a$$

$$\tanh a - \coth a = -2 \operatorname{csch} 2a$$

iii. Sums of Sums of Hyperbolic Functions

$A = a + b \qquad B = a - b$

$$\sinh A + \sinh B = 2 \sinh a \cosh b$$

$$\cosh A + \cosh B = 2 \cosh a \cosh b$$

$$\sinh A - \sinh B = 2 \cosh a \sinh b$$

$$\cosh A - \cosh B = 2 \sinh a \sin b$$

$$\tan A + \tanh B = \frac{\sinh 2a}{\cosh A \cosh B}$$

$$\coth A + \coth B = \frac{\sinh 2a}{\sinh A \sinh B}$$

$$\tanh A - \tanh B = \frac{\sinh 2b}{\cosh A \cosh B}$$

$$\coth A - \coth B = \frac{-\sinh 2b}{\sinh A \sinh B}$$

iv. Products of Hyperbolic Functions

$$\sinh a \sinh b = \frac{1}{2} \cosh A - \frac{1}{2} \cosh B$$

$$\cosh a \cosh b = \frac{1}{2} \cosh A + \frac{1}{2} \cosh B$$

$$\sinh a \cosh b = \frac{1}{2} \sinh A + \frac{1}{2} \sinh B$$

$$\tanh a \tanh b = \frac{\tanh a + \tanh b}{\coth a + \coth b}$$

$$\coth a \coth b = \frac{\coth a + \coth b}{\tanh a + \tanh b}$$

5.11 HALF ANGLES, MULTIPLE ANGLES, POWERS OF HYPERBOLIC FUNCTIONS

i. Half Angles of Hyperbolic Functions

$$\sinh \frac{a}{2} = \sqrt{\frac{1}{2}(\cosh a - 1)} = \frac{1}{2}\sqrt{\cosh a + \sinh a} - \frac{1}{2}\sqrt{\cosh a - \sinh a}$$

$$\cosh \frac{a}{2} = \sqrt{\frac{1}{2}(\cosh a + 1)} = \frac{1}{2}\sqrt{\cosh a + \sinh a} + \frac{1}{2}\sqrt{\cosh a - \sinh a}$$

$$\tanh \frac{a}{2} = \frac{\sinh a}{\cosh a + 1} = \frac{\cosh a - 1}{\sinh a} = \sqrt{\frac{\cosh a - 1}{\cosh a + 1}}$$

$$\coth \frac{a}{2} = \frac{\sinh a}{\cosh a - 1} = \frac{\cosh a + 1}{\sinh a} = \sqrt{\frac{\cosh a + 1}{\cosh a - 1}}$$

ii. Multiple Angles of Hyperbolic Functions

$\sinh 2a = 2 \sinh a \cosh a$	$\cosh 2a = \sinh^2 a + \cosh^2 a$
$\tanh 2a = \dfrac{2 \tanh a}{1 + \tanh^2 a}$	$\coth 2a = \dfrac{1 + \coth^2 a}{2 \coth a}$
$\sinh 3a = (\sinh a)\,(4 \cosh^2 a - 1)$	$\cosh 3a = (\cosh a)\,(4 \sinh^2 a - 1)$
$\tanh 3a = \dfrac{\tanh^2 a + 3 \tanh a}{3 \tanh^2 a + 1}$	$\coth 3a = \dfrac{\coth^2 a + 3 \coth a}{3 \coth^2 a + 1}$

iii. Powers of Hyperbolic Functions

$\sinh^2 a = \dfrac{1}{2}\,(\cosh 2a - 1)$	$\cosh^2 a = \dfrac{1}{2}\,(\cosh 2a + 1)$
$\tanh^2 a = \dfrac{\cosh 2a - 1}{\cosh 2a + 1}$	$\coth^2 a = \dfrac{\cosh 2a + 1}{\cosh 2a - 1}$
$\sinh^3 a = \dfrac{1}{4}\,(\sinh 3a - 3 \sinh a)$	$\cosh^3 a = \dfrac{1}{4}\,(\cosh 3a + 3 \cosh a)$
$\sinh^4 a = \dfrac{1}{8}\,(\cosh 4a - 4 \cosh 2a + 3)$	$\cosh^4 a = \dfrac{1}{8}\,(\cosh 4a + 4 \cosh 2a + 3)$
$\sinh^2 a + \cosh^2 a = \cosh 2a$ $\sinh^2 a - \cosh^2 a = -1$	$(\sinh a \pm \cosh a)^2 = \cosh 2a \pm \sinh 2a$
$\tanh^2 a + \coth^2 a = -8\,\dfrac{\cosh 2a}{\cosh 4a - 1}$	$(\tanh a + \coth a)^2 = 4\,\dfrac{\cosh 4a + 1}{\cosh 4a - 1}$
$\tanh^2 a - \coth^2 a = 2\,\dfrac{\cosh 4a + 3}{\cosh 4a - 1}$	$(\tanh a - \coth a)^2 = \dfrac{8}{\cosh 4a - 1}$
$(\sinh a \pm \cosh a)^n = \cosh na \pm \sinh na$	

5.12 INVERSE HYPERBOLIC FUNCTIONS

i. Definition

Inverse hyperbolic functions (area functions) are defined as follows :

Hyperbolic functions	Inverse hyperbolic functions	Principal values
$y = \sinh \omega$	$\omega = \sinh^{-1} y = ar \sinh y$	$-\infty \leq y \leq +\infty$
$y = \cosh \omega$	$\omega = \cosh^{-1} y = ar \cosh y$	$+1 \leq y \leq +\infty$
$y = \tanh \omega$	$\omega = \tan^{-1} y = ar \tanh y$	$-1 \leq y \leq +1$
$y = \coth \omega$	$\omega = \cot^{-1} y = ar \coth y$	$-1 \leq y \leq -\infty$ $+1 \leq y \leq +\infty$

This means that ω is the area of the hyperbolic segment of which the hyperbolic function is y.

ii. Relationships of Inverse Hyperbolic Function

$\sinh^{-1} y = ln\left(y + \sqrt{y^2+1}\right)$ $\qquad \tanh^{-1} y = \frac{1}{2} ln \frac{1+y}{1-y} \qquad | y | < 1$

$\cosh^{-1} y = \pm\, ln\left(y + \sqrt{y^2-1}\right) \qquad y \le 1$ $\qquad \coth^{-1} y = \frac{1}{2} ln \frac{y+1}{y-1} \qquad | y | > 1$

For other relationships use the logarithmic transformation, for example,

$\sinh^{-1} y + \cosh^{-1} y = ln\left(y + \sqrt{(y^2+1)}\right) \pm ln\left(y + \sqrt{y^2-1}\right)$

iii. Graphs

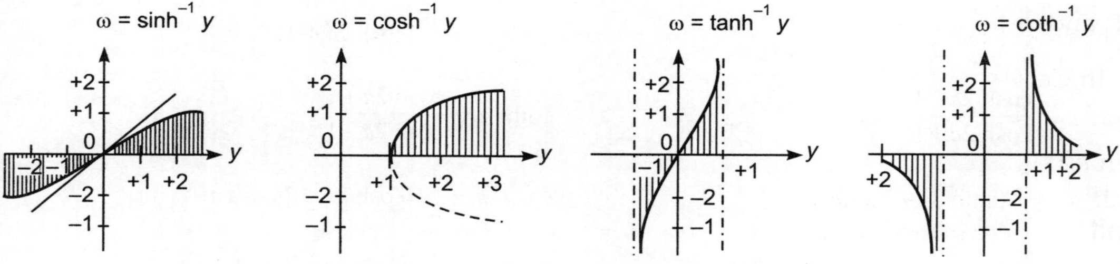

Fig. 5.5

iv. Transformation Table ($k = +1$, if $y > \theta$, or $k = -1$ if $y < \theta$)

	sinh^{-1} *y*	*cosh*$^{-1}$ *y*	*tanh*$^{-1}$ *y*	*coth*$^{-1}$ *y*
$\sinh^{-1} y$	$\sinh^{-1} y$	$k \cosh^{-1} \sqrt{y^2+1}$	$\tanh^{-1} \frac{y}{\sqrt{y^2+1}}$	$\coth^{-1} \frac{\sqrt{y^2+1}}{y}$
$\cosh^{-1} y$	$k \sinh^{-1} \sqrt{y^2-1}$	$\cosh^{-1} y$	$k \tanh^{-1} \frac{\sqrt{y^2-1}}{y}$	$k \coth^{-1} \frac{y}{\sqrt{y^2-1}}$
$\tanh^{-1} y$	$\sinh^{-1} \frac{y}{\sqrt{1-y^2}}$	$k \cosh^{-1} \frac{1}{\sqrt{1-y^2}}$	$\tanh^{-1} y$	$\coth^{-1} \frac{1}{y}$
$\coth^{-1} y$	$\sinh^{-1} \frac{1}{\sqrt{y^2-1}}$	$k \cosh^{-1} \frac{y}{\sqrt{y^2-1}}$	$\tanh^{-1} \frac{1}{y}$	$\coth^{-1} y$

Continuity at a Point

A function $f(x)$ is said to be continuous at a point $x = a$ of its domain if $\lim_{x\to a} f(x) = f(a)$.

Continuous Functions

A function $f(x)$ is said to be continuous if it continuous at each point of its domain.

Every Where Continuous Function

A function $f(x)$ is said to be every where continuous if it is continuous on the entire real line $(-\alpha\ \alpha)$.

Continuity

A vector function $v(t)$ is said to be continuous at $t = t_0$ if it is defined in some neighborhood of t_0 and $\lim_{t\to t_o} v(t) = v(t_0)$.

In Cartesian coordinate system

$$v(t) = [v_1(t), v_2(t), v_3(t)] = v_1(t)i, + v_2(t)j + v_3(t)k$$

then $v(t)$ is continuous at t_0 if and only if its three components are continuous at t_0.

Limit

A vector function $v(t)$ of a real variable t is said to have the limit l as t approaches t_0. If $v(t)$ is defined in some neighbourhood of t_0 (preferably except at t_0) and

$$\lim_{t\to t_0} \left|v(t) - t\right| = 0$$

then

$$\lim_{t\to t_0} v(t) = v(t_0)$$

Definition: Derivative of a vector function. A vector function $v(t)$ is said to be differentiable at a point t if the following limit exists:

$$v'(t) = \lim_{\Delta t\to 0} \frac{v(t+\Delta t) - v(t)}{\Delta t}$$

Results on Limits

(i) $\lim_{x\to c} [f(x) \pm g(x)] = \lim_{x\to c} f(x) \pm \lim_{x\to c} g(x)$

(ii) $\lim_{x\to c} [f(x)\cdot g(x)] = \left[\lim_{x\to c} f(x)\right]\left[\lim_{x\to c} g(x)\right]$

(iii) $\lim_{x\to c} [f(x)/g(x)] = \lim_{x\to c} f(x)/\lim_{x\to c} g(x);\ \lim_{x\to x_0} g(x) \neq 0$

(iv) If $f(x) = \dfrac{g(x)}{h(x)}$ is a rational function and c is a real number such that $h(c) \neq 0$ then

$$\lim_{x\to 0} f(x) = f(c) = \frac{g(c)}{h(c)}$$

(v) $\lim_{x \to c} [cf(x)] = c \lim_{x \to c} f(x)$

(vi) $\lim_{x \to c} [f(x)]^n = \left[\lim_{x \to c} f(x)\right]^n$

(vii) If $f(x)$ is polynomial function and c is real number than $\lim_{x \to c} f(x) = f(c)$

(viii) If $c > 0$ and n is any positive integer or if $c < 0$ and n is an odd positive integer then $\lim_{x \to c} (x)^{\frac{1}{n}} = c^{\frac{1}{n}}$

(ix) If f and g are functions such that $\lim_{x \to c} g(x) = l$ and $\lim_{x \to l} f(x) = f(l)$ then $\lim_{x \to c} f[g(x)] = f(l)$

Theorems on Limit

(i) $\lim_{x \to a} \dfrac{x^n - a^n}{x - a} = na^{n-1}$, $a > 0$ and n is an integer

(ii) $\lim_{x \to 0} e^x = 1$

(iii) $\lim_{x \to 0} \dfrac{e^x - 1}{x} = 1$

(iv) $\lim_{x \to 0} \dfrac{a^x - 1}{x} = \log a$

(v) $\lim_{x \to 0} (1 + x)^{\frac{1}{x}} = e$

(vi) $\lim_{x \to 0} \dfrac{1}{x} [\log (1 + x)] = 1$

CHAPTER

6

Analytic Geometry

6.1 COORDINATES OF A POINT IN SPACE

In two dimentional geometry, two mutually perpendicular lines divide the plane containing them into four parts which are known as quadrants and the lines are known as coordinate axes.

i. Systems of Coordinates

a. *Cartesian coordinates*

A point P is given by two mutually perpendicular distances x, y (coordinates) measured from two mutually perpendicular axes x, y (coordinate axes) intersecting at the origin O.

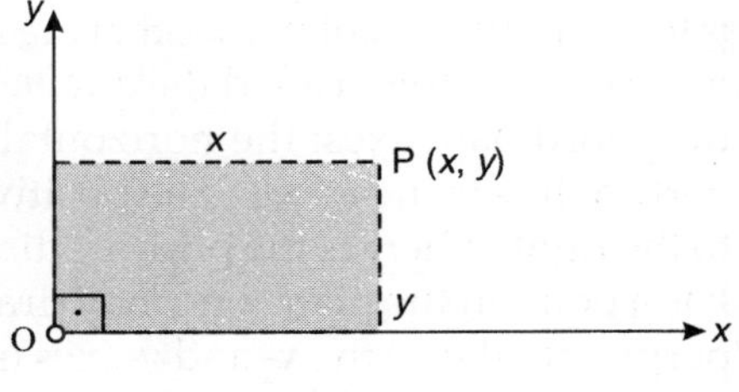

Fig. 6.1

b. *Skew coordinates*

A point P is given by the coordinates u, v parallel to two skew axes u, v intersecting at the origin O.

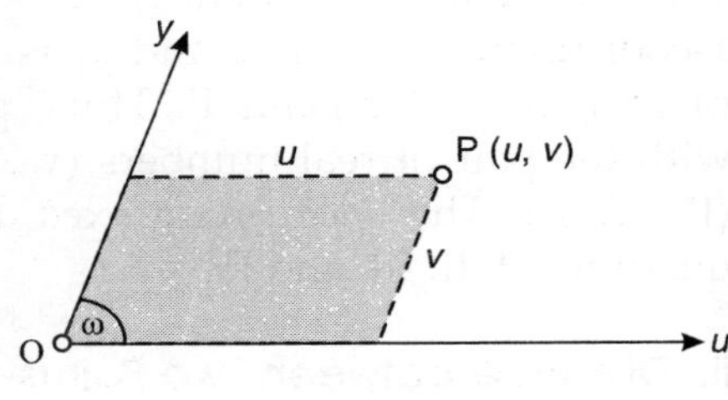

Fig. 6.2

c. *Polar coordinates*

A point P is given by two polar coordinates associated with a fixed axis-x (polar axis) and a fixed point O on this axis (pole). The first coordinate is the radius r, the distance from O to P, and the second coordinate is the position angle θ, measured from x to r.

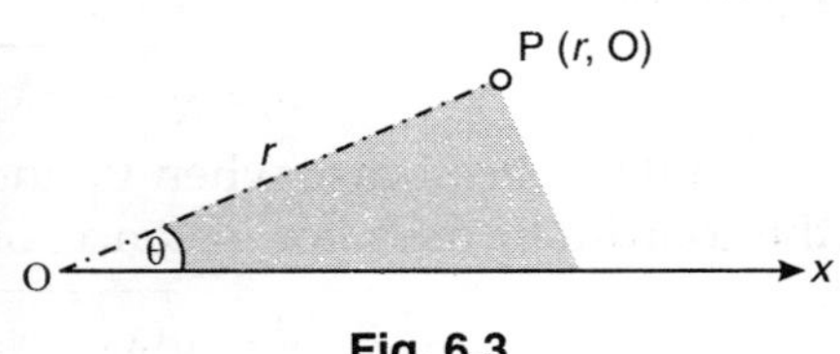

Fig. 6.3

Relationship between Coordinate Systems

	Rectangular coordinate	*Skew coordinate*	*Polar coordinate*
Rectangular coordinate	$x = x$ $y = y$	$x = u + v \cos \omega$ $y = v \sin \omega$	$x = r \cos \theta$ $y = r \sin \theta$
Skew coordinate	$u = x - y \cot \omega$ $v = y \csc \omega$	$u = v$ $v = v$	$u = r \dfrac{\sin(\omega - \theta)}{\sin \omega}$ $v = r \dfrac{\sin \theta}{\sin \omega}$
Polar coordinate	$r = \sqrt{x^2 + y^2}$ $\theta = \tan^{-1} \dfrac{y}{x}$	$r = \sqrt{u^2 + v^2 + 2uv \cos \omega}$ $\theta = \tan^{-1} \dfrac{v \sin \omega}{u + v \cos \omega}$	$r = r$ $\theta = \theta$

6.2 RECTANGULAR COORDINATES

i. Definition

The points in a plane may be placed in one-to-one correspondence with pairs of real numbers. A common method is to use perpendicular lines that are horizontal and vertical and intersect at a point called the *origin*. These two lines constitute the coordinate axes; the horizontal line is the x-axis and the vertical line is the y-axis. The positive direction of the x-axis is to the right, whereas the positive direction of the y-axis is up. If P is a point in the plane one may draw lines through it that are perpendicular to the x- and y-axes (such as the broken lines of Fig. 6.4. The lines intersect the x-axis at a point with coordinate x_1 and the y-axis at a point with coordinate y_1. We call x_1 the x-coordinate or *abscissa* and y_1 is termed the y-coordinate or *ordinate of the point* P. Thus, point P is associated with the pair of real numbers (x_1, y_1) and is denoted as (P, x_1, y_1). The coordinate axes divide the plane into quadrants I, II, III, and IV.

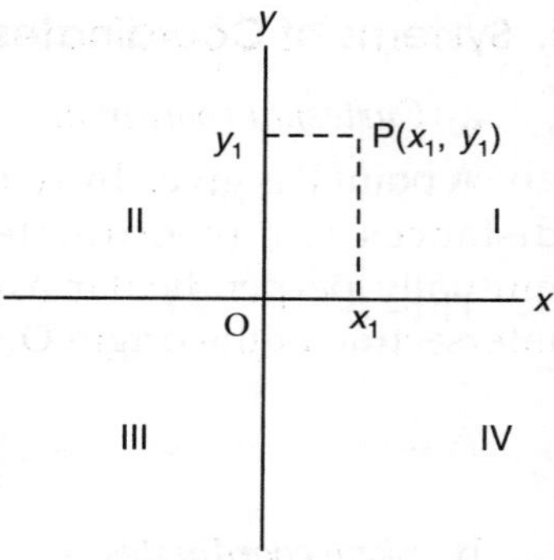

Fig. 6.4 Rectangular coordinate

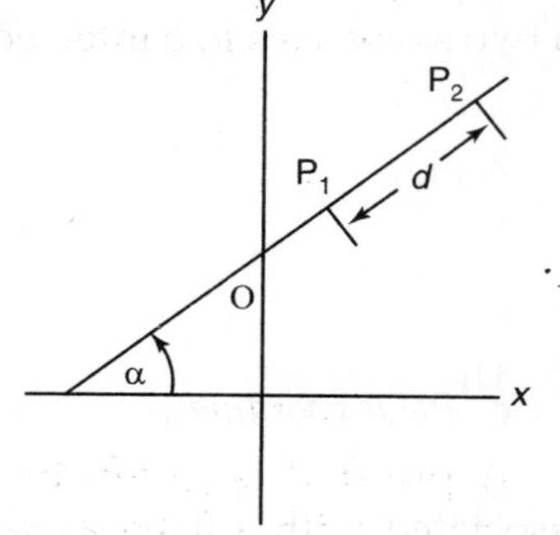

Fig. 6.5 The angle of inclination is the smallest angle measured counterclockwise from the positive x-axis to the line that contains P_1P_2

ii. Distance between Two Points–Slope

The distance d between the two points $P_1(x_1, y_1)$ and $P_2(x_2, y_2)$ is

$$d = \sqrt{(x_2 - x_1)^2 + (y_2 - y_1)^2}$$

In the special case when P_1 and P_2 are both on one of the coordinate axes, for instance, on the x-axis

$$d = \sqrt{(x_2 - x_1)^2} = |x_2 - x_1|$$

or on the y-axes,

$$d = \sqrt{(y_2 - y_1)^2} = |y_2 - y_1|$$

The midpoint of the line segment P_1P_2 is

$$\left(\frac{x_1 + x_2}{2}, \frac{y_1 + y_2}{2}\right)$$

The slope of the line segment P_1P_2, provided it is not vertical is denoted by m, and is given by

$$m = \frac{y_2 - y_1}{x_2 - x_1}$$

The slope is related to the angle of inclination α (refer to Fig. 6.5) and, is given by

$$m = \tan \alpha$$

Two lines (or line segment) with slopes m_1 and m_2 are perpendicular if

$$m = -1/m_2$$

and are parallel if $m_1 = m_2$.

iii. Equations of Straight Lines

A *vertical* line has an equation of the form

$$x = c,$$

where $(c, 0)$ is its intersection with the x-axis. A line of slope m through point (x_1, y_1) is given by

$$y - y_1 = m(x - x_1)$$

Thus a horizontal line (slope = 0) through point (x_1, y_1) is given by

$$y = y_1$$

A nonvertical line through the two points $P_1(x_1, y_1)$ and $P_2(x_2, y_2)$ is given by either

$$y - y_1 = \left(\frac{y_2 - y_1}{x_2 - x_1}\right)(x - x_1)$$

or

$$y - y_2 = \left(\frac{y_2 - y_1}{x_2 - x_1}\right)(x - x_2)$$

A line with x-intercept a and y-intercept b is given by

$$\frac{x}{a} + \frac{y}{b} = 1 \quad (a \neq 0, b \neq 0)$$

The *general equation* of a line is

$$Ax + By + C = 0$$

The normal form of the straight line equation is

$$x \cos \theta + y \sin \theta = p,$$

where p is the distance along the normal from the origin and θ is the angle that the normal makes with the x-axis (as shown in Fig. 6.6).

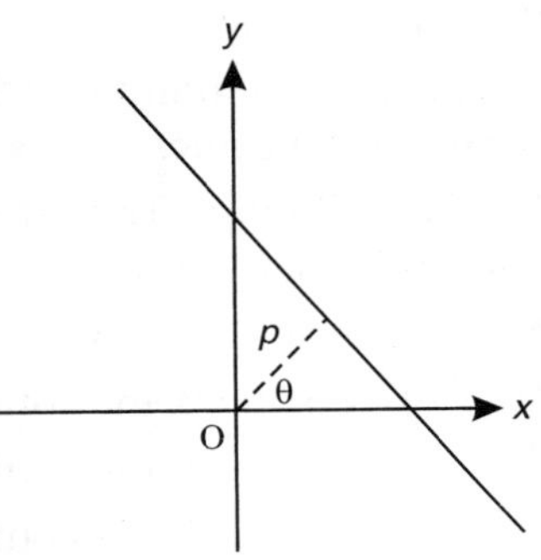

Fig. 6.6 Construction for normal form of straight line equation

The general equation of the line $Ax + By + C = 0$, may be written in normal form by dividing by $\pm\sqrt{A^2 + B^2}$, where the plus sign is used when C is negative and the minus sign is used when C is positive:

so that
$$\frac{Ax + By + C}{\pm\sqrt{A^2 + B^2}} = 0,$$

$$\cos\theta = \frac{A}{\pm\sqrt{A^2 + B^2}}, \quad \sin\theta = \frac{B}{\pm\sqrt{A^2 + B^2}}$$

and
$$p = \frac{|C|}{\sqrt{A^2 + B^2}}$$

iv. Distance from a Point to a Line

The perpendicular distance from a point $P(x_1, y_1)$ to the line $Ax + By + C = 0$ is given by

$$d = \frac{Ax_1 + By_1 + C}{\pm\sqrt{A^2 + B^2}}$$

v. Conic Section

A conic section or conic is the locus of a point P which moves in such a way that its distance form a fixed point S always bears a constant ratio to its distance, from a fixed line, all being in the same place.

Important Terms:

Focus: The fixed point is called the focus of the conic section.

Directrix: The fixed straight line is called the directrix of the conic section.

Eccentricity: The constant ratio is called the eccentricity of the conic section and is denoted by e.

Axis: The straight line passing through the focus and perpendicular to the directrix is called the axis of the conic section.

Vertex: The points of intersection of the conic section and the axis are called vertices of the conic section.

Centre: The point which bisects every chord of the conic passing through it is called the centre of the conic.

Latus-Rectum: The latus-rectum of a conic is the chord passing through the four and perpendicular to the axis.

Note: The eccentricity of a conic is generally denoted by e

i. For $e = 0$, the conic is a circle.

ii. For $e < 1$, the conic obtained is an ellipse.

iii. For $e = 1$, the conic obtained is a parabola.

iv. For $e > 1$, the conic is a hyperbola.

vi. Parabola

A parabola is the set of all points (x, y) in the plane that are equidistant from a given line called the *directrix* and a given point called the *focus*. The parabola is symmetric about a line that contains the focus and is perpendicular to the directrix. The line of symmetry intersects the parabola at its vertex (Fig. 6.7). The eccentricity $e = 1$.

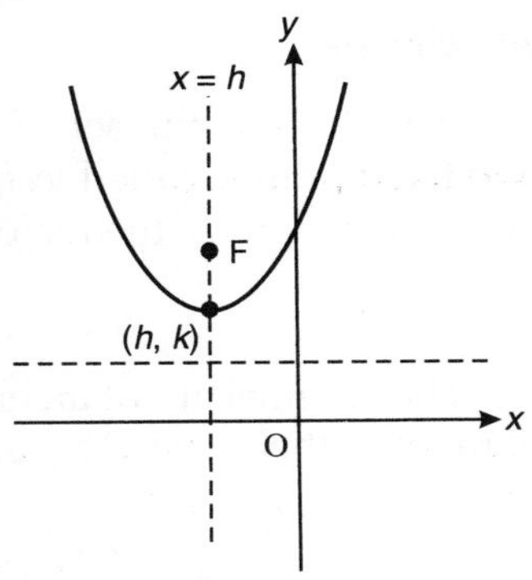

Fig. 6.7 Parabola with vertex at (h, k). F identifies the focus

The distance between the focus and the vertex, or vertex and directrix, is denoted by p (> 0) and leads to one of the following equations of a parabola with vertex at the origin (Figs 6.8 and 6.9).

$$y = \frac{x^2}{4p} \quad \text{(opens upward)}$$

$$y = -\frac{x^2}{4p} \quad \text{(opens downward)}$$

$$x = \frac{y^2}{4p} \quad \text{(opens to right)}$$

$$x = -\frac{y^2}{4p} \quad \text{(opens to left)}$$

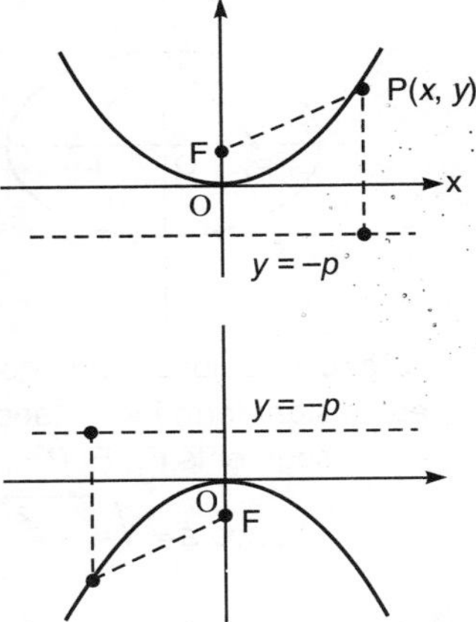

Fig. 6.8 Parabolas with y-axis as the axis of symmetry and vertex at the origin (upper) $y = \frac{x^2}{4p}$; (lower) $\frac{x^2}{4p}$

For each of the four orientations shown in Figs 6.8 and 6.9, the corresponding parabola with vertex (h, k) is obtained by replacing x by $x - h$ and y by $y - k$.

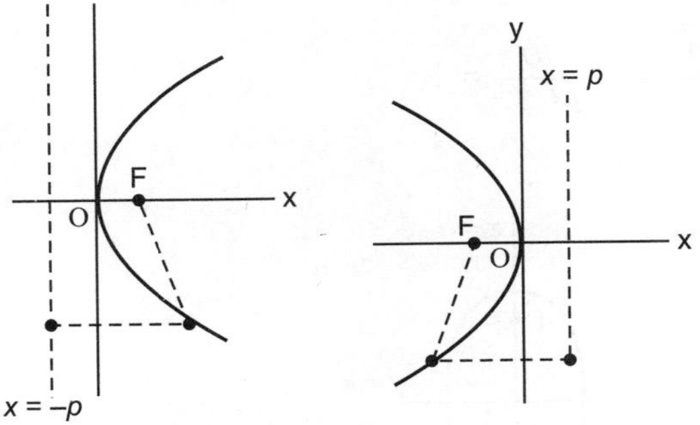

Fig. 6.9 Parabola as with x-axis as the axis of symmetry and vertex at the origin,
(upper) $x = \frac{y^2}{4p}$; (lower) $x = \frac{y^2}{4p}$

Thus, the parabola in Fig. 6.10 has the equation

$$x - h = -\frac{(y-k)^2}{4p}$$

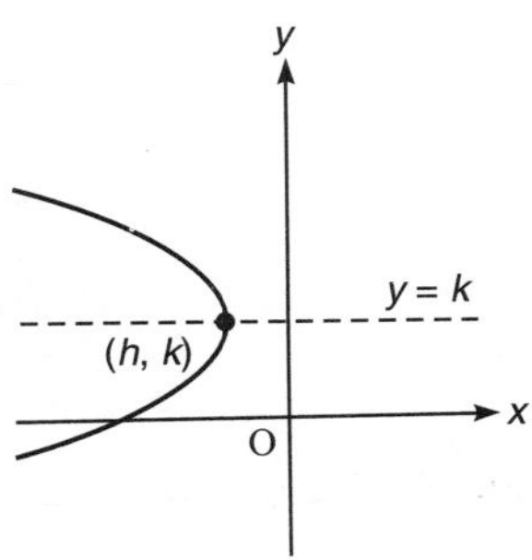

Fig. 6.10 Parabolas with *vertex at (h, k) and axis parabola to the x-axis.*

vii. Ellipse

An ellipse is the set of all point in the plane such that the sum of their distances from two fixed points, called *foci*, is a given constant $2a$. The distance between the foci is denoted as $2c$; the length of the major axis is $2a$, whereas the length of the minor axis is $2b$ (Fig. 6.11),

$$a = \sqrt{b^2 + c^2}$$

The eccentricity of an ellipse e is <1. An ellipse with center at point (h, k) and major axis parallel to the x-axis (Fig. 6.12) is given by the equation

$$\frac{(x-h)^2}{a^2} + \frac{(y-k)^2}{b^2} = 1$$

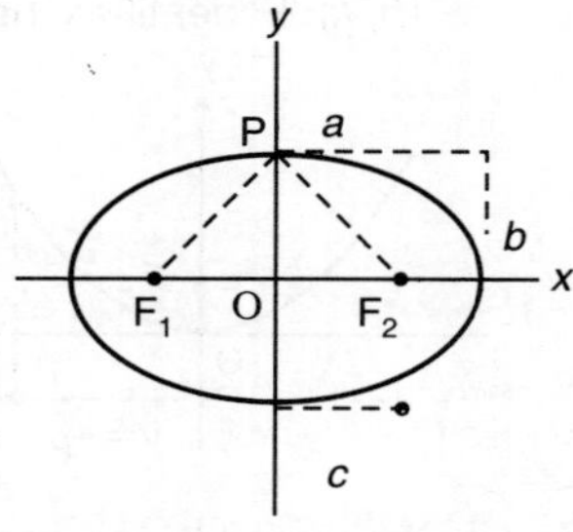

Fig. 6.11 Ellipse; since point P is equidistant form foci F_1 and F_2, the segments F_1, $F_2P = a$; hence $a = \sqrt{b^2 + c^2}$

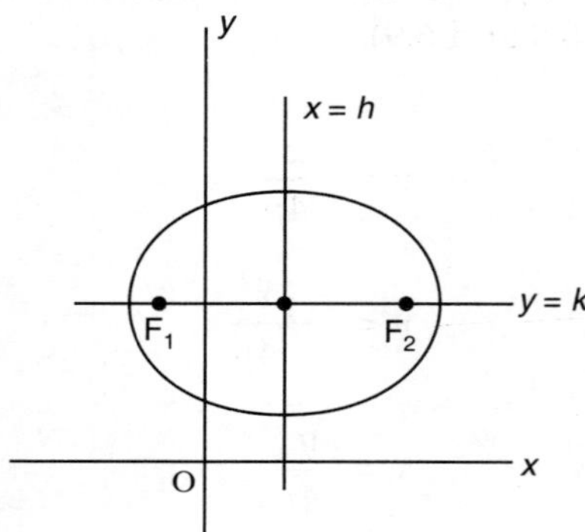

Fig. 6.12 Ellipse with major axis parallel to the x-axis, F_1 and F_2 are the foci, each a distance c from center (h, k)

An ellipse with center at (h, k) and major axis parallel to the y-axis is given by the equation (Fig. 6.13).

$$\frac{(y-k)^2}{a^2} + \frac{(x-h)^2}{b^2} = 1$$

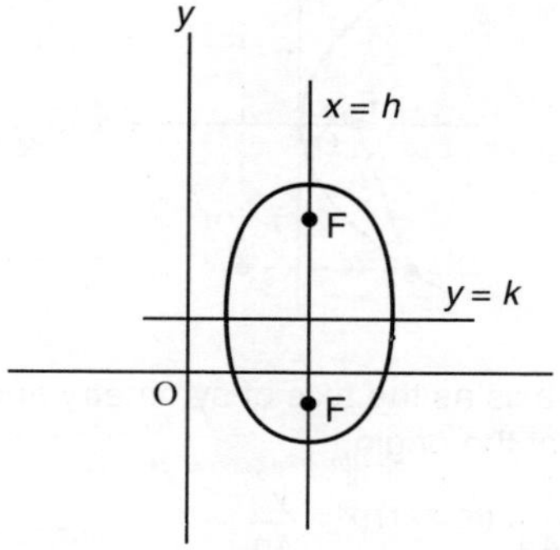

Fig. 6.13 Ellipse with major axis parallel to the y-axis. Each focus is a distance c from center (h, k).

viii. Hyperbola (e > 1)

A hyperbola is the set of all points in the plane such that the difference of its distances from two fixed points (foci) is a given positive constant denoted as $2a$. The distance between the two foci is $2c$ and that between the two vertices is $2a$. The quantity

$$b = \sqrt{c^2 - a^2}$$

and is illustrated in Fig. 6.14, which shows the construction of a hyperbola given by the equation

$$\frac{x^2}{a^2} - \frac{y^2}{b^2} = 1$$

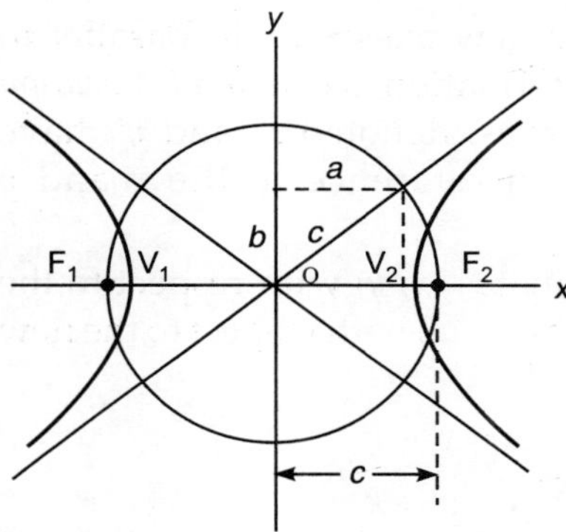

Fig. 6.14 Hyperbola, V_1, V_2 = vertices, F_1, F_2 = foot. A circle at center O with radius c contains the vertices and illustrates the relation among *a*, *b*, and *c*. Asymptotes have slopes *b*/*a* and – *b*/*a* for the orientations shown

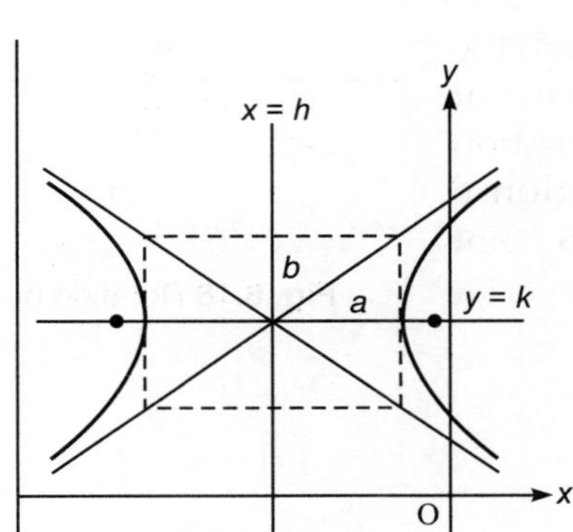

Fig. 6.15 Hyperbola with center at (*h*, *k*):
$\frac{(x-h)^2}{a^2} - \frac{(y-k)^2}{b^2} = 1$; slopes of asympototes $\pm b/a$

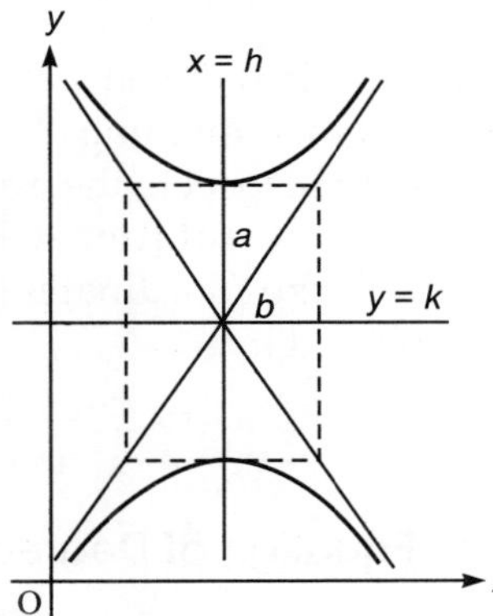

Fig. 6.16 Hyperbola with center at (*h*, *k*):
$\frac{(y-k)^2}{a^2} - \frac{(x-h)^2}{b^2} = 1$; slopes of asympototes $\pm a/b$

When the focal axis is parallel to the y-axis, the equation of the hyperbola with center (h, k) (Figs 6.15 and 6.16) is

$$\frac{(y-k)^2}{a^2} - \frac{(x-h)^2}{b^2} = 1$$

If the focal axis is parallel to the x-axis and center (h, k), then

$$\frac{(x-h)^2}{a^2} - \frac{(y-k)^2}{b^2} = 1$$

ix. Change of Axes

A change in the position of the coordinate axes will generally change the coordinates of the points in the plane. The equation of a particular curve will also generally change.

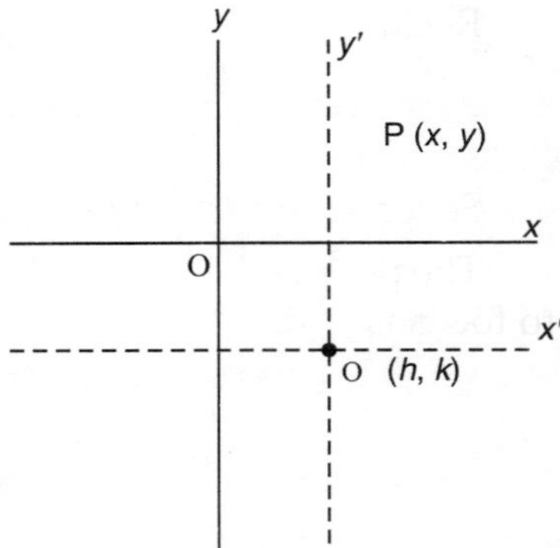

Fig. 6.17 Translation of axes

- *Translation*: When the new axes remain parallel to the original, the transformation is called a *translation* (Fig. 6.17). The new axes, denoted x' and y', have origin O at (h, k) with reference to the x and y axes.

  A point P with coordinates (x, y) with respect to the original has coordinates (x', y') with respect to the new axes. These are related by

$$x = x' + h$$
$$y = y' + k$$

  For example, the ellipse of Fig. 6.13 has the following simple equation with respect to axes x' and y' with the center at (h, k):

$$\frac{y'^2}{a^2} + \frac{x'^2}{b^2} = 1$$

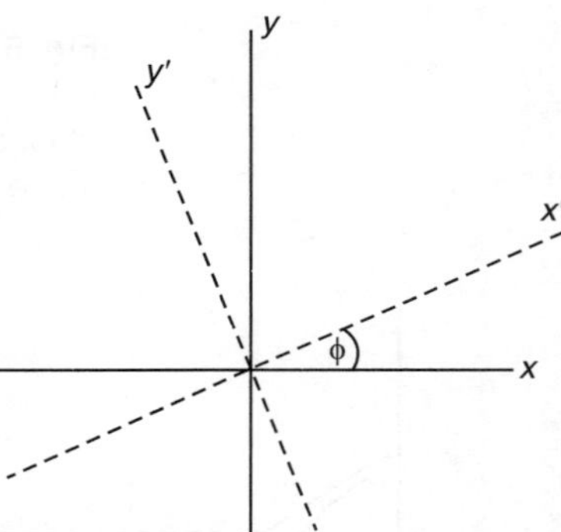

Fig. 6.18 Rotation of axes

- *Rotation*: When the new axes are drawn through the same origin, remaining mutually perpendicular, but tilted with respect to the original, the transformation is one of the rotations. For angle of rotation ϕ (Fig. 6.18), the coordinates (x, y) and (x', y') of a point P are related by

$$x = x' \cos \phi - y' \sin \phi$$
$$y = x \sin \phi + y \cos \phi$$

x. General Equation of Degree Two

$$Ax^2 + Bxy + Cy^2 + Dx + Ey + F = 0$$

Every equation of the above form defines a conic section or one of the limiting forms of a conic. By rotating the axes through a particular angle ϕ, the xy-term vanishes, yielding

$$A'x'^2 + Cy^2 + D'x' + E'y' + F' = 0$$

with respect to the axes x' and y'. The required angle ϕ (in Fig. 6.18) is calculated from

$$\tan 2\phi = \frac{B}{A-C}, \quad (\phi < 90°)$$

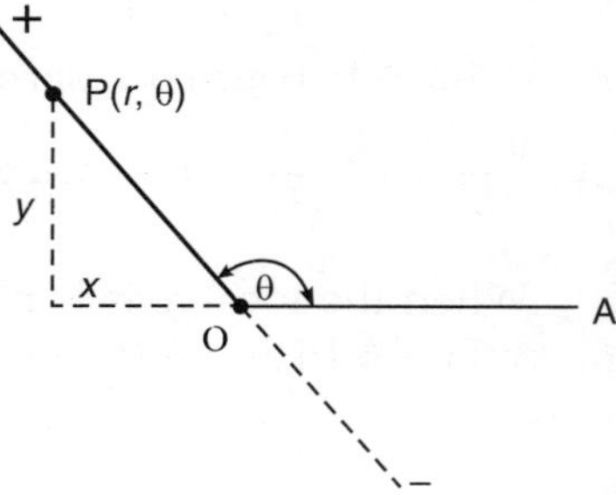

Fig. 6.19 Polar coordinates

xi. Polar Coordinates (Fig. 6.19)

The fixed point O is the origin or *pole* and a line OA drawn through it is the polar axis. A point P in the plane is determined from the distance r, measured from O, and the angle θ between OP and OA. Distances measured on the terminal line of θ from the pole are positive, whereas those measured in the opposite direction are negative.

Rectangular coordinates (x, y) and polar coordinates (r, θ) are related according to

$$x = r\cos\theta, \qquad y = r\sin\theta$$

$$r^2 = x^2 + y^2, \qquad \tan\theta = y/x$$

Several well known polar curves are shown in Fig. 6.20 to 6.24.

The polar equation of a conic section with focus at the pole and distance $2p$ from directrix to focus is either

$$r = \frac{2ep}{1 - e\cos\phi} \qquad \text{(directrix to left of pole)}$$

or

$$r = \frac{2ep}{1 + e\cos\theta} \qquad \text{(directrix to right of pole)}$$

The corresponding equations for the directrix below or above the pole are as above, except $\sin\theta$ appears instead of $\cos\theta$.

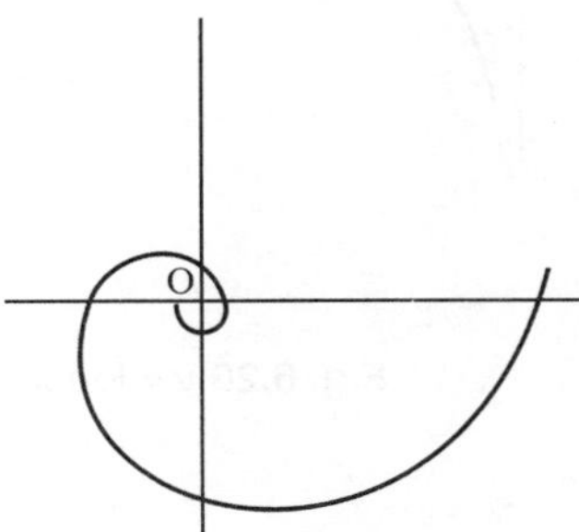

Fig. 6.20 Polar curve $r = e^{a\theta}$

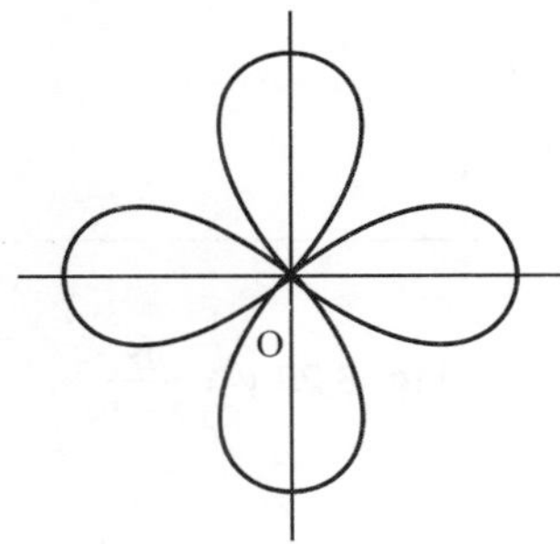

Fig. 6.21 Polar curve $r = a\cos 2\theta$

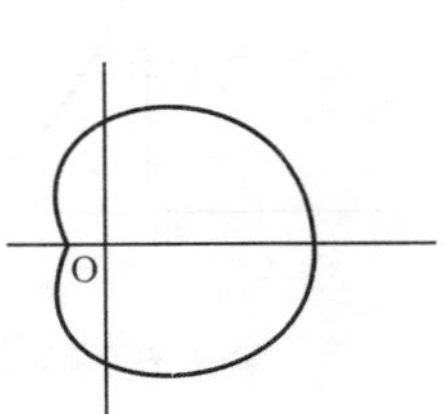

Fig. 6.22 Polar curve $r = 2a\cos\theta + b$

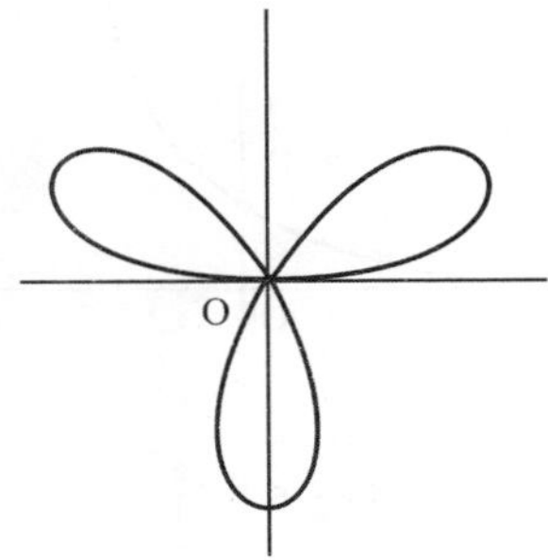

Fig. 6.23 Polar curve $r = a\sin 3\theta$

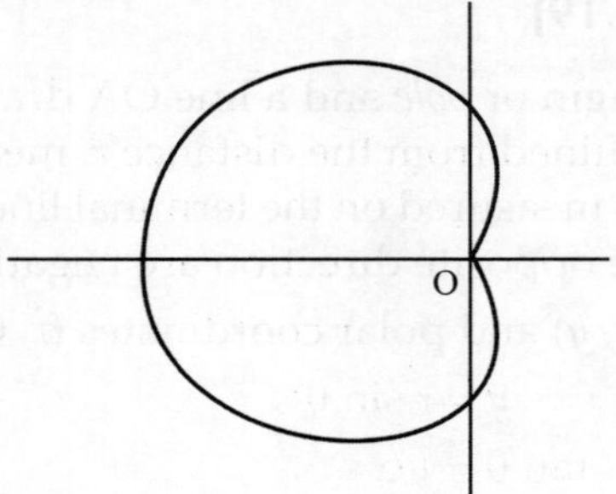

Fig. 6.24 Polar curve $r = a\,(1 - \cos\theta)$

xii. Curves and Equations

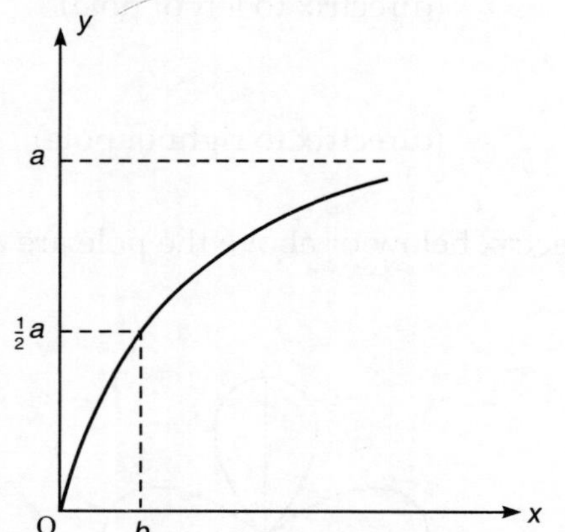

Fig. 6.25 $y = \dfrac{ax}{x + b}$

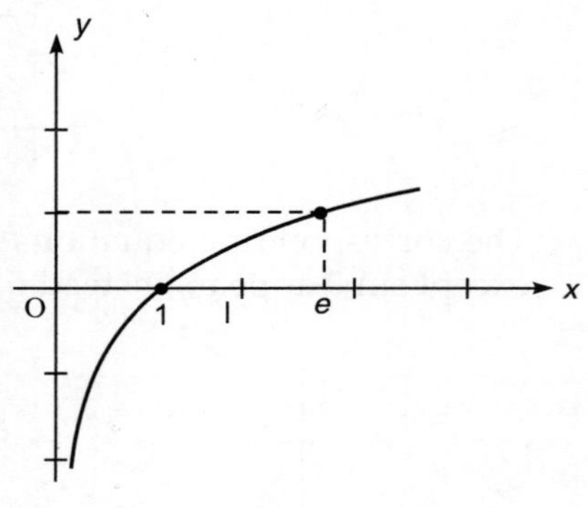

Fig. 6.26 $y = \log x$

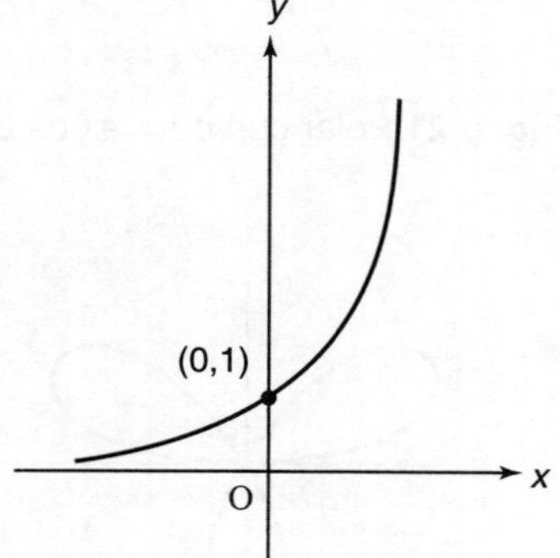

Fig. 6.27 $y = e^x$

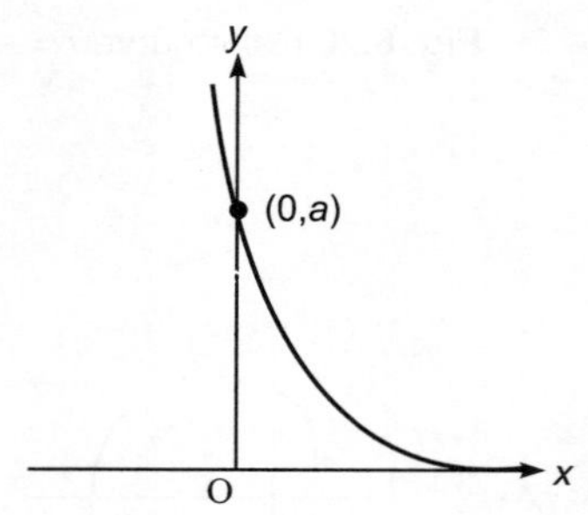

Fig. 6.28 $y = ae^{-x}$

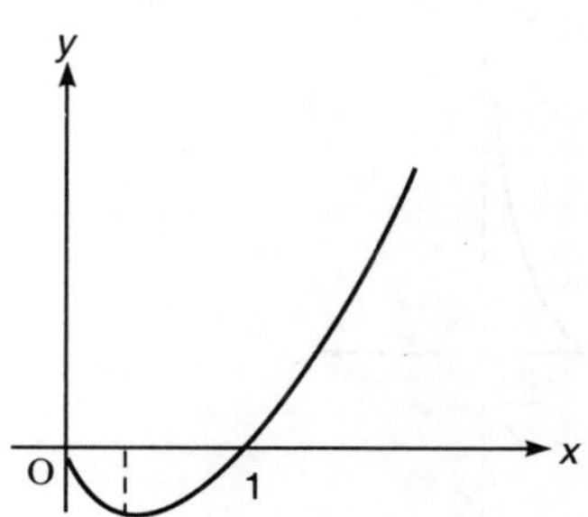

Fig. 6.29 $y = x \log x$

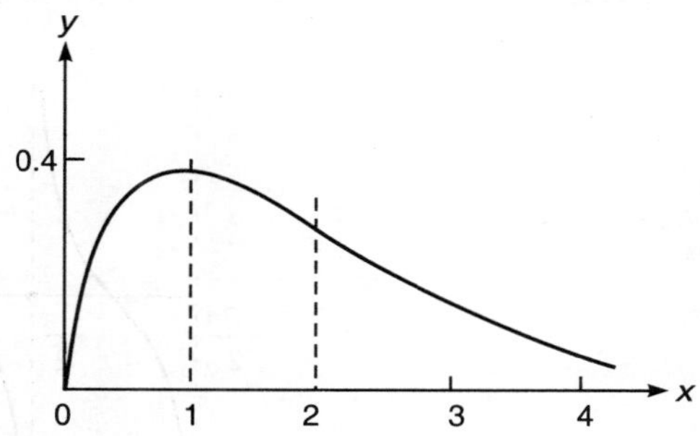

Fig. 6.30 $y = xe^{-x}$

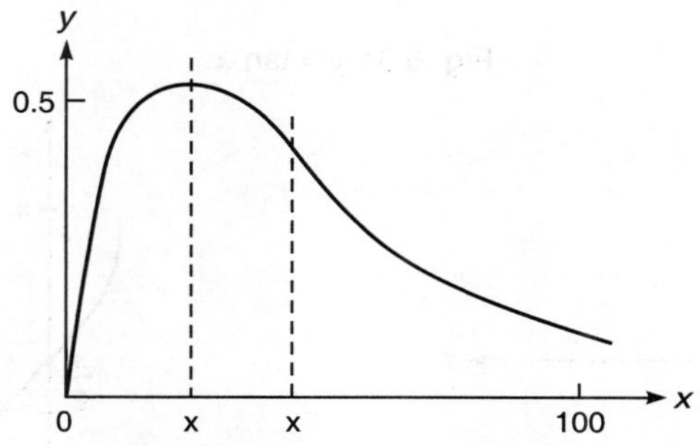

Fig. 6.31 $y = e^{-ax} - e^{-bx}$, $0 < a < b$
(drawn for $a = 0.02$, $b = 0.1$,
and showing maximum and inflection)

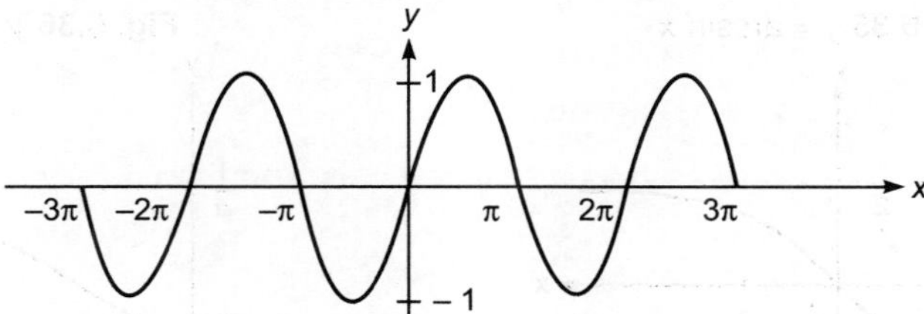

Fig. 6.32 $y = \sin x$

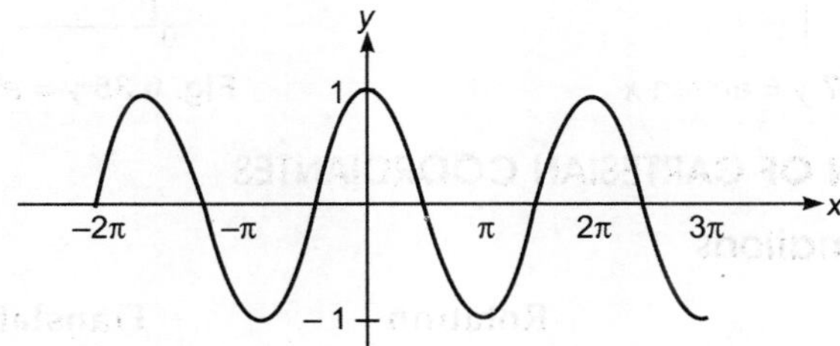

Fig. 6.33 $y = \cos x$

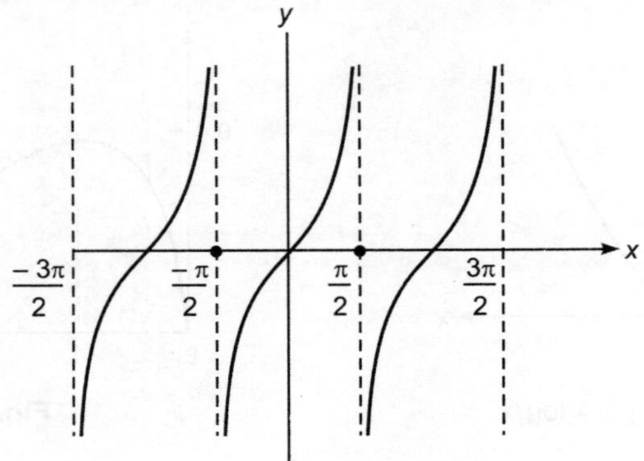

Fig. 6.34 $y = \tan x$

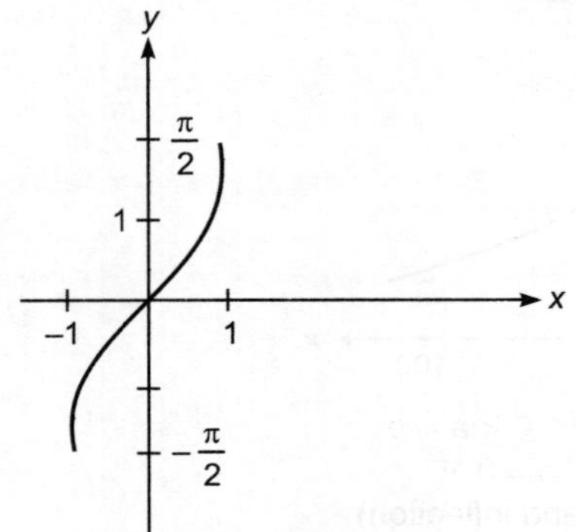

Fig. 6.35 $y = \arcsin x$

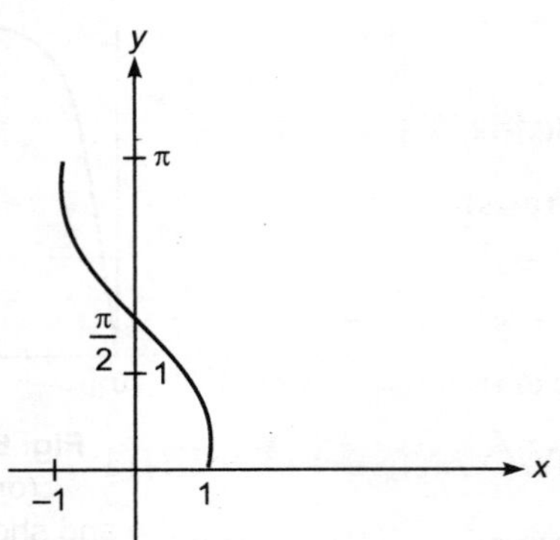

Fig. 6.36 $y = \arccos x$

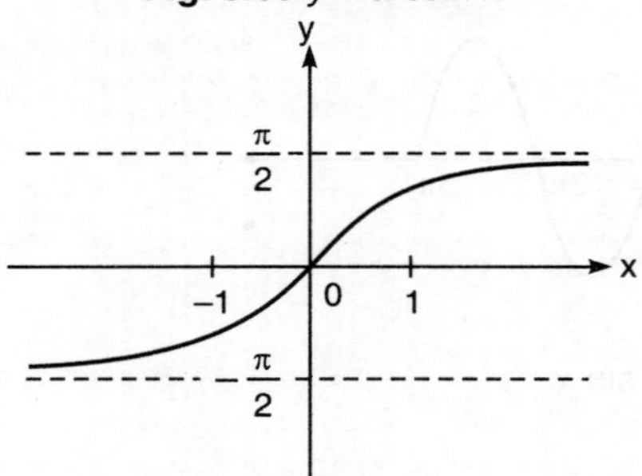

Fig. 6.37 $y = \arctan x$

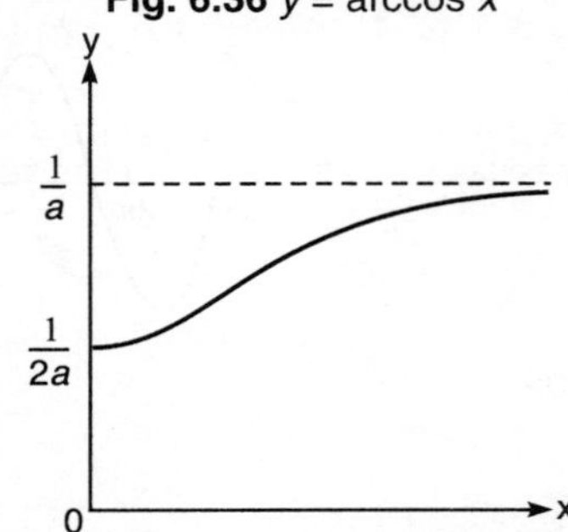

Fig. 6.38 $y = e^{bx} / a(1 + e^{bx}),\ x \geq 0$

6.3 TRANSFORMATION OF CARTESIAN COORDIANTES

i. Algebraic Transformations

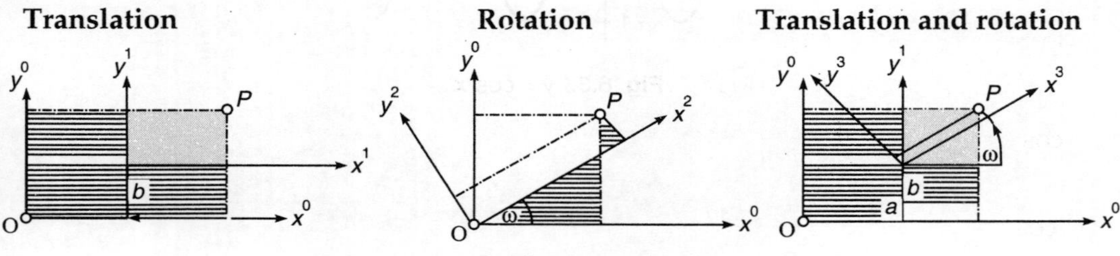

Fig. 6.39

$$x^0 = x^1 + a^0 \qquad x^0 = x^2 \cos\omega - y^2 \sin\omega \qquad x^0 = x^3 \cos\omega - y^3 \sin\omega + a^0$$
$$y^0 = y^1 + b^0 \qquad y^0 = x^2 \sin\omega + y^2 \cos\omega \qquad y^0 = x^3 \sin\omega + y^3 \cos\omega + b^0$$
$$x^1 = x^0 - a^0 \qquad x^2 = x^0 \cos\omega + y^0 \sin\omega \qquad x^3 = (x^0 - a^0)\cos\omega + (y^0 - b^0)\sin\omega$$
$$y^1 = y^0 - b^0 \qquad y^2 = -x^0 \sin\omega + y^0 \cos\omega \qquad y^3 = -(x^0 - a^0)\sin\omega + (y^0 - b^0)\cos\omega$$

Note: 0, 1, 2, 3 are superscripts and not exponents.

ii. Catalog of Matrices

$$s^0 = \begin{bmatrix} x^0 \\ y^0 \end{bmatrix} \qquad s^1 = \begin{bmatrix} x^1 \\ y^1 \end{bmatrix} \qquad s^2 = \begin{bmatrix} x^2 \\ y^2 \end{bmatrix} \qquad s^3 = \begin{bmatrix} x^3 \\ y^3 \end{bmatrix} \qquad d^0 = \begin{bmatrix} a^0 \\ b^0 \end{bmatrix}$$

$$\omega^{02} = \begin{bmatrix} \cos & -\sin \\ \sin & \cos \end{bmatrix} = \omega^{03} \qquad \omega^{20} = \begin{bmatrix} \cos & \sin \\ -\sin & \cos \end{bmatrix} = \omega^{30}$$

iii. Matrix Transformations

Translation	**Rotation**	**Translation and rotation**
$s^0 = s^1 + d^0$	$s^0 = \omega^{02} s^2$	$s^0 = \omega^{03} s^3 + d^0$
$s^1 = s^0 - d^0$	$s^2 = \omega^{20} s^0$	$s^3 = \omega^{30} (s^0 - d^0)$

Note: ω matrices are orthogonal.

6.4 TWO AND THREE POINTS

i. Distance of Two Points, Segment in Plane $\overline{P_1, P_2}$

a. *Cartesian coordiantes* $P_1(x_1, y_1)$; $P_2(x_2, y_2)$

$$d = \sqrt{(x_2 - x_1)^2 + (y_2 - y_1)^2}$$

$$\tan\alpha = \frac{y_2 - y_1}{x_2 - x_1}$$

$$\cos\alpha = \frac{x_2 - x_1}{d}$$

$$\cos\beta = \frac{y_2 - y_1}{d}$$

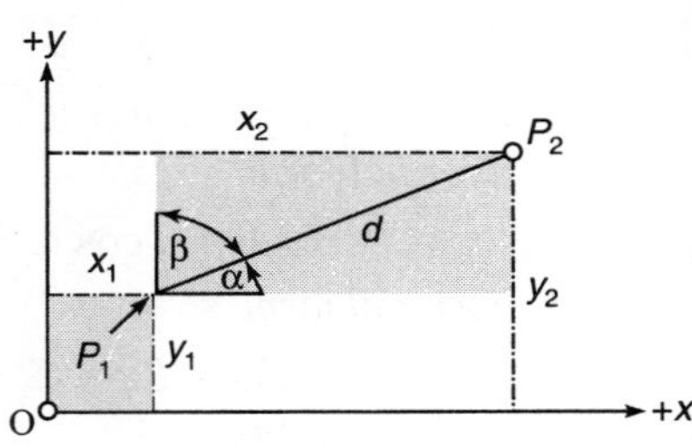

Fig. 6.40

b. *Skew coordinates* $P_1(u_1, v_1)$; $P_2(u_2, v_2)$

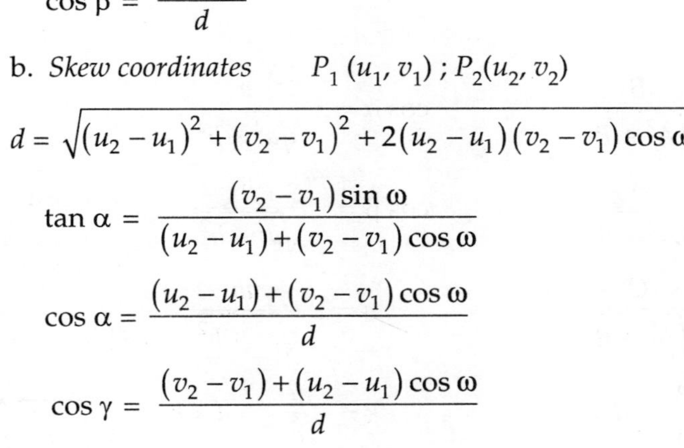

$$d = \sqrt{(u_2 - u_1)^2 + (v_2 - v_1)^2 + 2(u_2 - u_1)(v_2 - v_1)\cos\omega}$$

$$\tan\alpha = \frac{(v_2 - v_1)\sin\omega}{(u_2 - u_1) + (v_2 - v_1)\cos\omega}$$

$$\cos\alpha = \frac{(u_2 - u_1) + (v_2 - v_1)\cos\omega}{d}$$

$$\cos\gamma = \frac{(v_2 - v_1) + (u_2 - u_1)\cos\omega}{d}$$

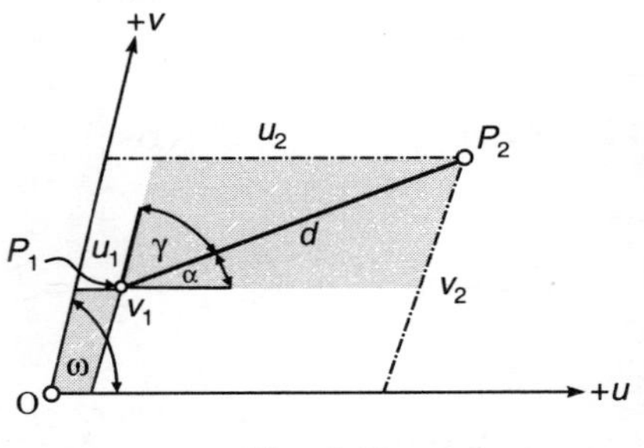

Fig. 6.41

c. *Polar coordinates* $P_1(r_1, \theta_1)$; $P_2(r_2, \theta_2)$

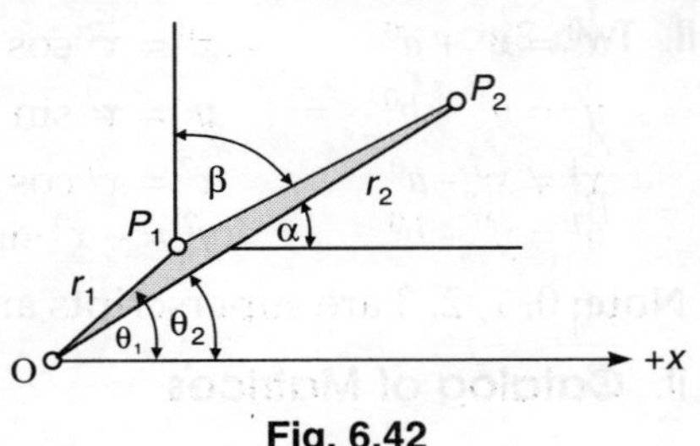

Fig. 6.42

$$d = \sqrt{r_1^2 + r_2^2 - 2r_1r_2\cos(\theta_1 - \theta_2)}$$

$$\tan\alpha = \frac{r_2\sin\theta_2 - r_1\sin\theta_1}{r_2\cos\theta_2 - r_1\cos\theta_1}$$

$$\cos\alpha = \frac{r_2\cos\theta_2 - r_1\cos\theta_1}{d}$$

$$\cos\beta = \frac{r_2\sin\theta_2 - r_1\sin\theta_1}{d}$$

ii. Three Points $P_1(x_1, y_1)$; $P_2(x_2, y_2)$: $P_3(x_3, y_3)$

a. *Area of triangle* $(P_1 P_2 P_3)$

$$A = \frac{1}{2}\begin{vmatrix} x_1 & y_1 & 1 \\ x_2 & y_2 & 1 \\ x_3 & y_3 & 1 \end{vmatrix}$$

b. $P_1 P_2 P_3$, *on a straight line*

$$0 = \begin{vmatrix} x_1 & y_1 & 1 \\ x_2 & y_2 & 1 \\ x_3 & y_3 & 1 \end{vmatrix}$$

6.5 STRAIGHT LINE IN A PLANE

i. Forms of a Straight Line

a. *Direction form*

$$y = kx + 1$$

b. *Intercept form*

$$\frac{x}{a} + \frac{y}{b} = 1$$

c. *Normal form*

$$x\cos\beta + y\cos\alpha = n$$

d. *General form*

$$Ax + By + C = 0$$

Polar equation

$$r\cos(\theta - \beta) = n$$

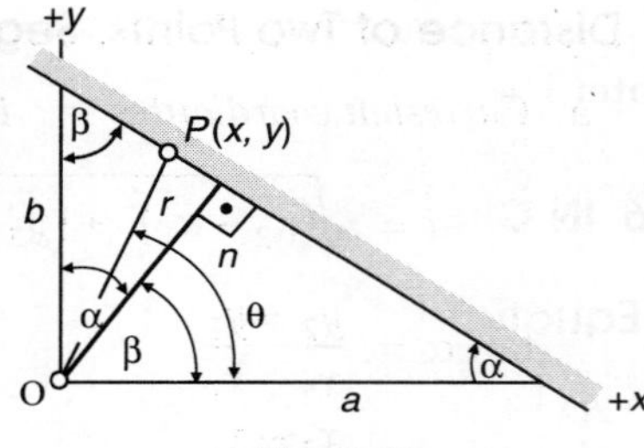

Fig. 6.43

ii. Parameters of Straight Line

$A = b$	$\cos\alpha = \pm\frac{B}{\sqrt{A^2+B^2}}$	$\cos\alpha = \pm\frac{a}{\sqrt{a^2+b^2}}$
$B = b$	$\cos\beta = \pm\frac{A}{\sqrt{A^2+B^2}}$	$\cos\beta = \pm\frac{b}{\sqrt{a^2+b^2}}$
$C = -ab$	$n = \pm\frac{C}{\sqrt{A^2+B^2}}$	$n = \pm\frac{ab}{\sqrt{a^2+b^2}}$
$k = \tan\alpha$	$k = -\frac{A}{B}$ $l = -\frac{C}{B}$	$k = -\frac{b}{a}$ $l = b$

iii. Two Straight Lines in Plane $(A_1x + B_1y + C_1 = 0; A_2x + B_2y + C_2 = 0)$

$\frac{A_1}{B_1} \neq \frac{A_2}{B_2}$: lines intersect at a point $\frac{A_1}{A_2} = \frac{B_1}{B_2} = \frac{C_1}{C_2}$: lines coincide

$\frac{A_1}{B_1} = -\frac{B_2}{A_2}$: lines are normal $\frac{A_1}{B_1} = \frac{A_2}{B_2}$: lines are parallel

The angle ω between two lines = $\tan\omega = \frac{A_1B_2 - B_1A_2}{A_1A_2 + B_1B_2}$ is the clockwise rotation required to transform line 2 into line 1.

iv. Distances $P(x_0, y_0)$; $(A_1x + B_1y + C_1 = 0; A_2x + B_2y + C_2 = 0)$

a. *From a line to a point*

$$d = \frac{A_1x_0 + B_1y_0 + C_1}{\pm\sqrt{A_1^2 + B_1^2}}$$

b. *Between two parallel lines*

$$d = \frac{C_2}{\pm\sqrt{A_2^2 + B_2^2}} - \frac{C_1}{\pm\sqrt{A_1^2 + B_2^2}}$$

Note: The sign of the denominator is opposite to the sign of C.

6.6 IN CIRCLE

i. Equation of Circle

$Ax^2 + Ay^2 + 2Dx + 2Ey + F = 0$

Centre: $x_M = -\frac{D}{A}$; $y_M = -\frac{E}{A}$

Radius: $R = \frac{\sqrt{D^2 + E^2 - AF}}{A}$

$$(x - x_M)^2 + (y - y_M)^2 = R^2$$

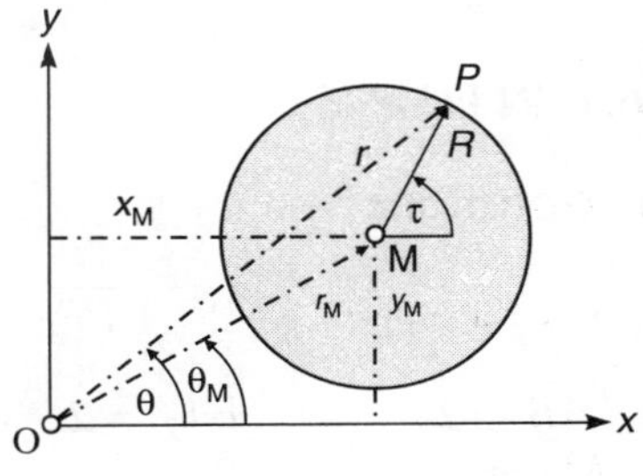

Fig. 6.44

ii. Parametric Equation

$$x = x_M + R\cos\tau \qquad y = y_M + R\sin\tau$$

iii. Polar Equation

$$r^2 - 2r_M r\cos(\theta - \theta_M) + r_M^2 = R^2$$

iv. Special positions

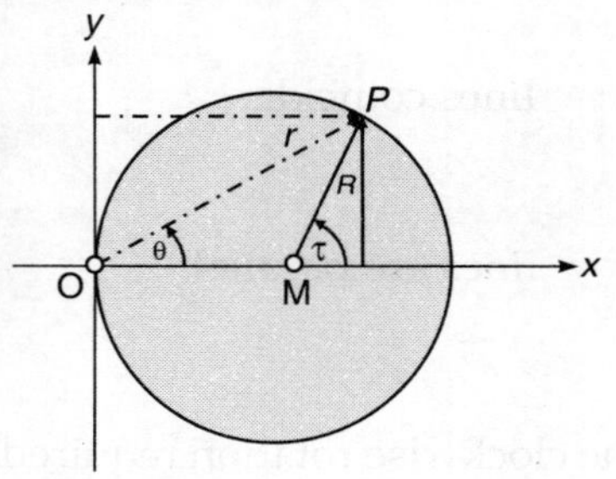

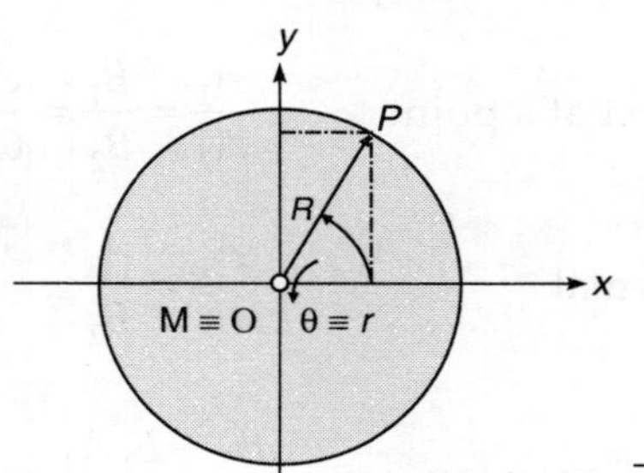

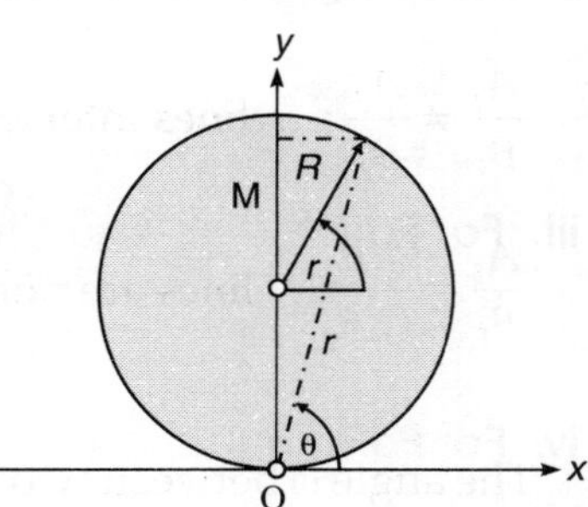

Fig. 6.45

$x^2 - 2Rx + y^2 = 0$
$x = R(1 + \cos \tau)$
$y = R \sin \tau$
$r = 2R \cos \theta$

$x^2 + y^2 = R^2$
$x = R \cos \tau$
$y = R \sin \tau$
$r = R \quad \theta = \tau$

$x^2 - 2Ry + y^2 = 0$
$x = R \cos \tau$
$y = R(1 + \sin \tau)$
$r = 2R \sin \theta$

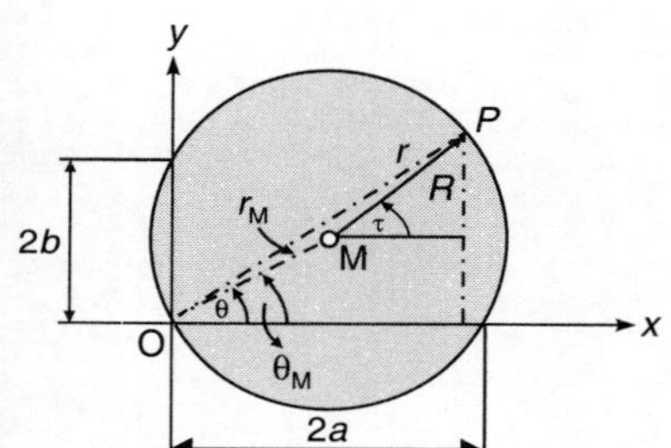

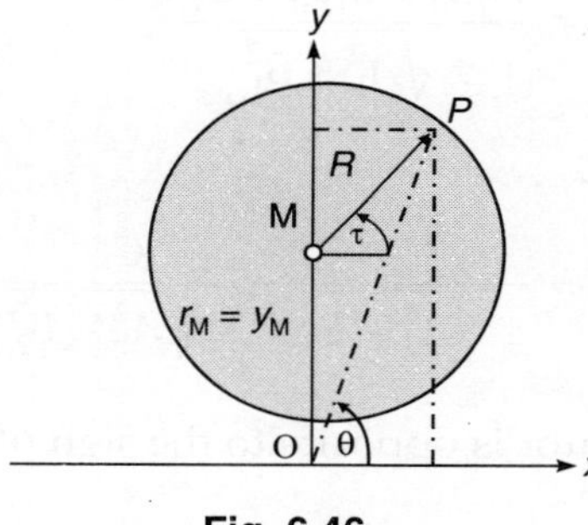

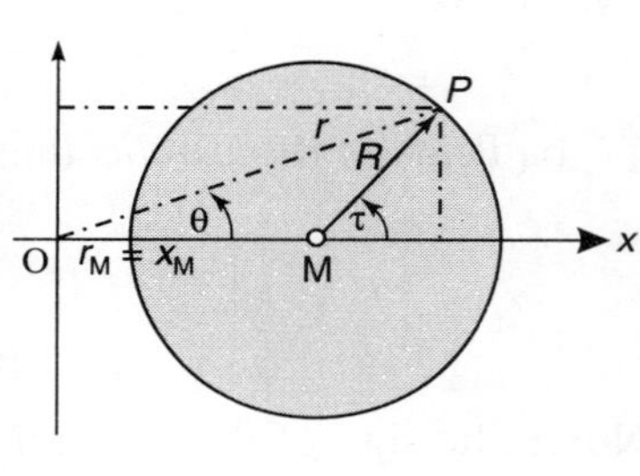

Fig. 6.46

$x(x - 2a) + y(y - 2b) = 0$
$x = a + R \cos \tau$
$y = b + R \sin \tau$
$r = 2a \cos \theta + 2b \sin \theta$

$x^2 + (y - y_M)^2 = R^2$
$x = R \cos \tau$
$y = y_M + R \sin \tau$
$r^2 - 2rr_M \sin \theta + {r_M}^2 = R^2$

$(x - x_M)^2 + y^2 = R^2$
$x = x_M + R \cos \tau$
$y = R \sin \tau$
$r^2 - 2rr_M \cos \theta + r^2_M = R^2$

6.7 IN ELLIPSE

i. Notations

F = focus $\quad \overline{PF_1} + \overline{PF_2} = 2a \quad$ M = center

$\overline{AB} = 2a \quad \overline{F_1F_2} = 2e \quad \overline{CD} = 2b$

Major axis $\qquad$ *Minor axis*

$e = \sqrt{a^2 - b^2}$; linear eccentricity

$\frac{e}{a} = \varepsilon < 1;$ numerical eccentricity

$2p = \frac{2b^2}{a} \qquad 2q = \frac{2a^2}{b}$

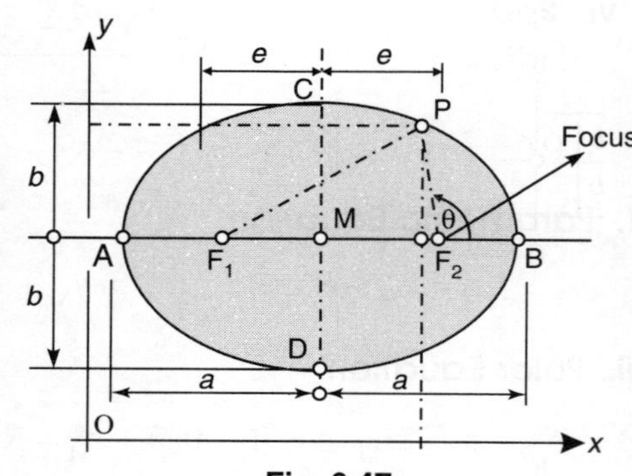

Fig. 6.47

ii. For Cartesian Equation (M ≡ 0)

$$\frac{x_2}{a^2}+\frac{y^2}{b^2}=1$$

iii. For Parametric Equation (M ≡ 0)

$$x = a\cos\tau \qquad\qquad y = b\sin\tau$$

iv. For Polar Equations

Pole at 0; $$r^2=\frac{b^2}{1-\varepsilon^2\cos^2\theta}$$

Pole at F_2; $$r=\frac{p}{1+\varepsilon\cos\theta}$$

v. For Normal Position

$Ax^2 + Cy^2 + 2Dx + 2Ey + F = 0$

Centre:

$$x_M=-\frac{D}{A} \qquad y_M=-\frac{E}{C}$$

$$a=\sqrt{\frac{CD^2+AE^2-ACF}{A^2C}}$$

$$b=\sqrt{\frac{CD^2+AE^2-ACF}{AC^2}}$$

$$\frac{(x-x_M)^2}{a^2}+\frac{(y-y_M)^2}{b^2}=1$$

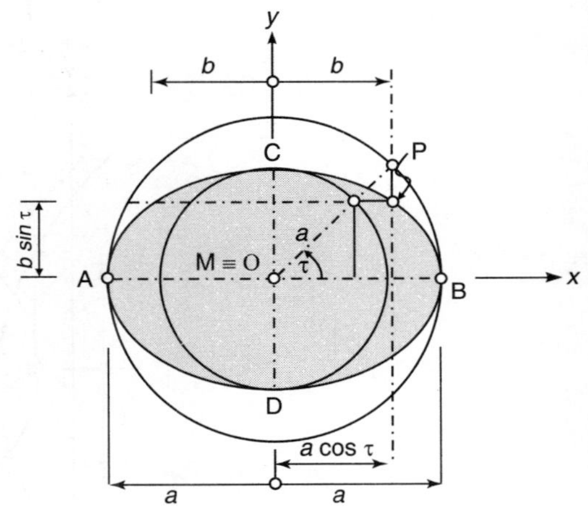

Fig. 6.48

vi. Special Positions

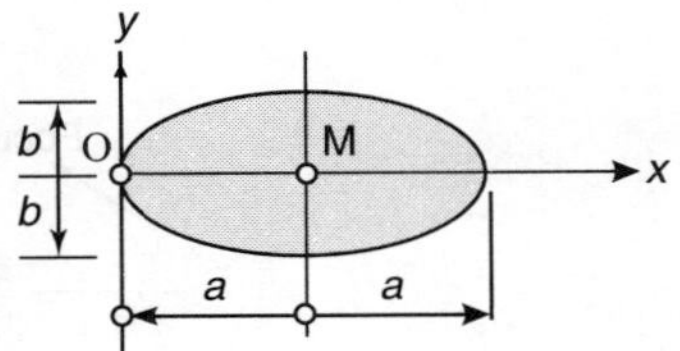

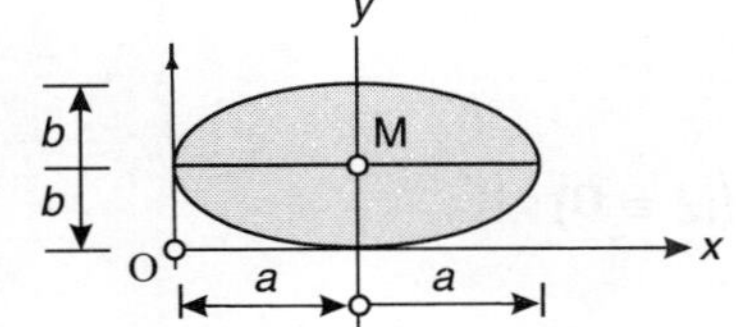

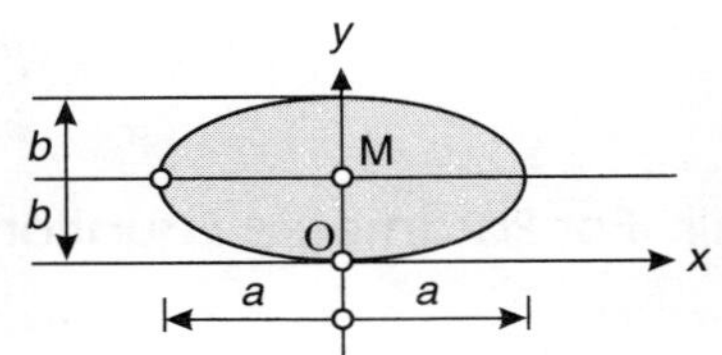

Fig. 6.49

$$y^2=2px-\frac{p}{a}\,x^2 \qquad \frac{x^2}{a^2}-2\left(\frac{x}{a}+\frac{y}{b}\right)+\frac{y^2}{b^2}=-1 \qquad x^2=2qy-\frac{q}{b}\,y^2$$

6.8 IN HYPERBOLA

i. Notations

F = focus $\overline{PF_1} - \overline{PF_2} = 2a$ M = center

$\overline{AB} = 2a$ $\overline{F_1F_2} = 2e$ $\overline{CD} = 2b$

Major axis *Minor axis*

$e = \sqrt{a^2 + b^2}$ linear eccentricity

$\frac{e}{a} = e > 1$ numerical eccentricity

$2p = \frac{2b^2}{a}$ $2q = \frac{2a^2}{b}$

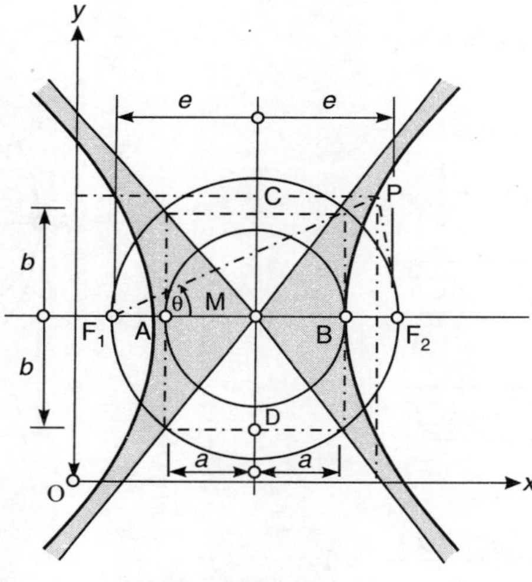

Fig. 6.50

ii. For Cartesian Equation (M ≡ 0)

$$\frac{x^2}{a^2} - \frac{y^2}{b^2} = 1$$

iii. For Parametric Equation (M ≡ 0)

$$x = \frac{a}{\cos\tau} \qquad y = \pm b \tan\tau$$

iv. For Polar Equations

Pole at 0; $r^2 = \frac{b^2}{\varepsilon^2 \cos^2\theta - 1}$ Pole at F; $r = \frac{p}{1 + \varepsilon\cos\theta}$

v. For Normal Position

$$Ax^2 - Cy^2 + 2Dx + 2Ey + F = 0$$

Center: $x_M = -\dfrac{D}{A}; \quad y_M = \dfrac{E}{C}$

$$a = \sqrt{\frac{CD^2 - AE^2 - ACF}{A^2C}}$$

$$b = \sqrt{\frac{CD^2 - AE^2 - ACF}{AC^2}}$$

$$\frac{(x - x_M)^2}{a^2} - \frac{(y - y_M)^2}{b^2} = 1$$

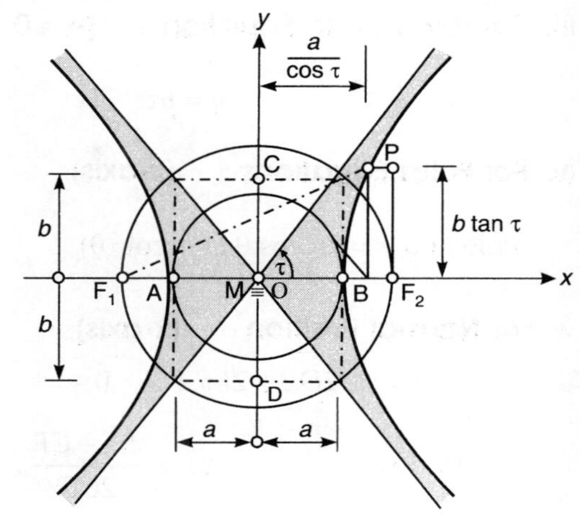

Fig. 6.51

vi. For Special Positions

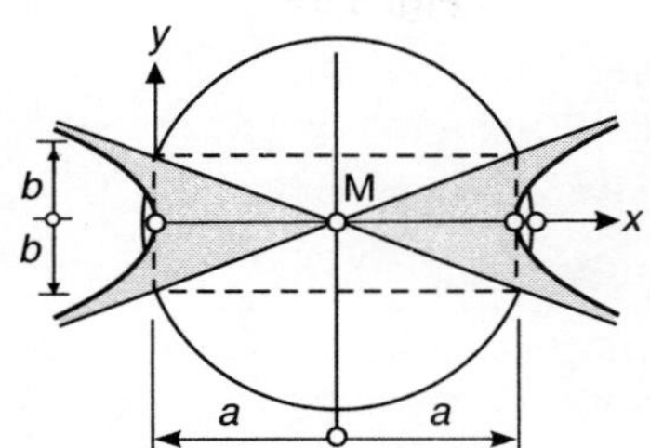

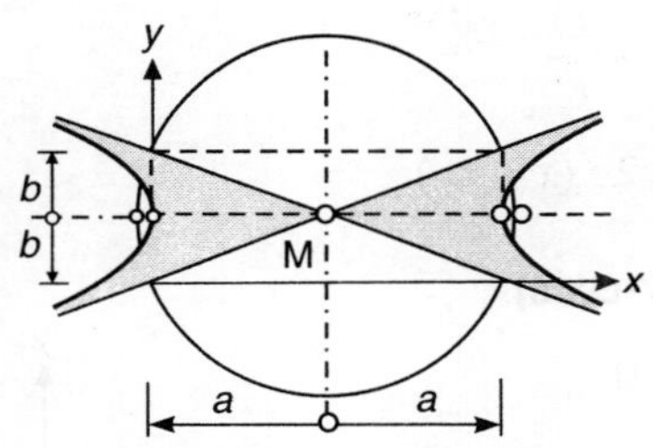

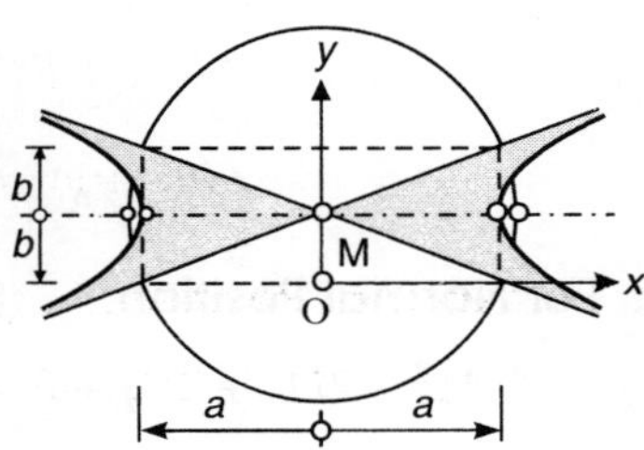

Fig. 6.52

$$y^2 = -2px + \frac{p}{a}x^2 \qquad \frac{x^2}{a^2} - 2\left(\frac{x}{a} - \frac{y}{b}\right) - \frac{y^2}{b^2} = 1 \qquad x^2 = -2qy + \frac{2}{b}y^2 + 2a^2$$

6.9 IN PARABOLA

i. Notations

F = focus $\qquad \overline{PF} = \overline{PE} \qquad$ 0 = vertex

$\overline{AF} = \dfrac{p}{2} \qquad \overline{BF} = \overline{FC} = 0 \qquad \overline{BA} = \dfrac{p}{2}$

$2p$ = parameter = latus rectum

$\varepsilon = 1$ (numerical eccentricity)

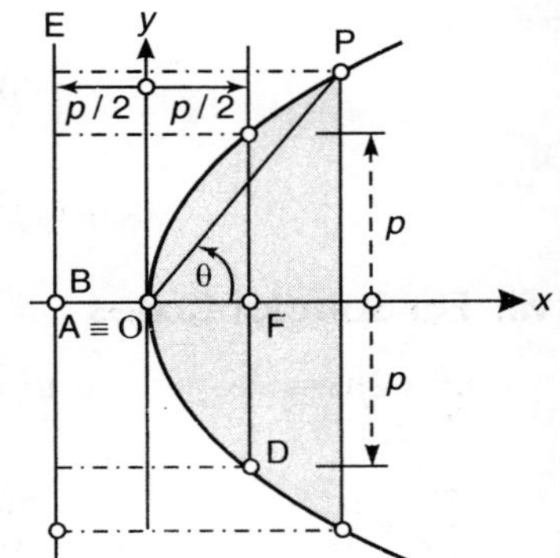

Fig. 6.53

ii. Cartesian Equation (A ≡ 0, x-axis)

$$y^2 = 2px \qquad \begin{cases} p > 0 & \text{Open right} \\ p < 0 & \text{Open left} \end{cases}$$

iii. For Parametric Equation (A ≡ 0, x-axis)

$$x = \frac{p}{2}\tau^2 \qquad y = p\tau$$

iv. For Polar Equations (x-axis)

Pole at 0, $r = 2p \cos\theta\,(1 + \cot^2\theta)$ Pole at F, $r = \dfrac{p}{1 - \cos\theta}$

v. For Normal Position (x-axis)

$$Cy^2 + 2Dx + 2Ey + F = 0$$

Vertex: $x_A = \dfrac{E^2 - CF}{2CD}$

$$y_A = -\frac{E}{C}$$

$$p = -\frac{D}{C}$$

$$(y - y_A)^2 = 2p\,(x - x_A)$$

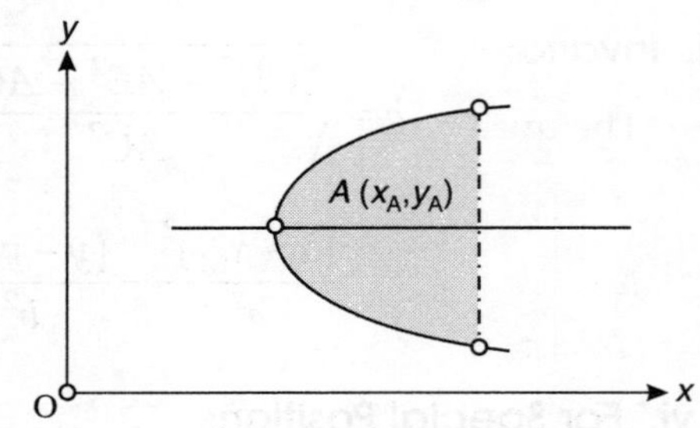

Fig. 6.54

vi. For Normal Position (y-axis)

$$Ax^2 + 2Dx + 2Ey + F = 0$$

Vertex: $x_A = -\dfrac{D}{A}$

$$y_A = \frac{D^2 - AF}{2AE}$$

$$p = -\frac{E}{A}$$

$$(x - x_A)^2 = 2p\,(y - y_A)$$

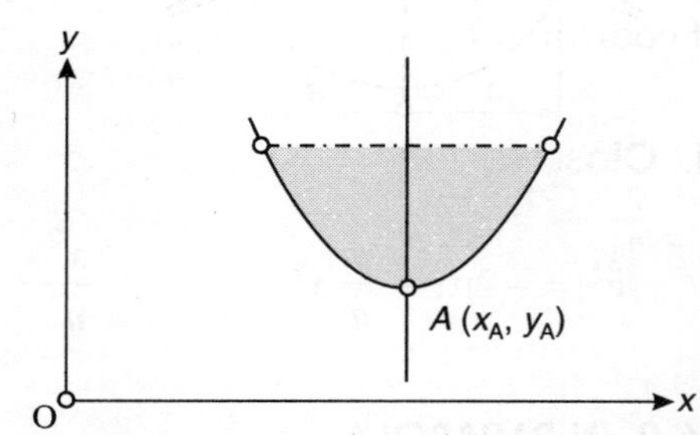

Fig. 6.55

vii. For Special Cases

$y^2 = 2px$ $y^2 = -2px$ $x^2 = 2py$ $x^2 = -2py$

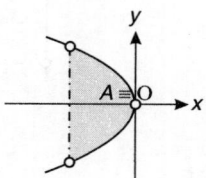

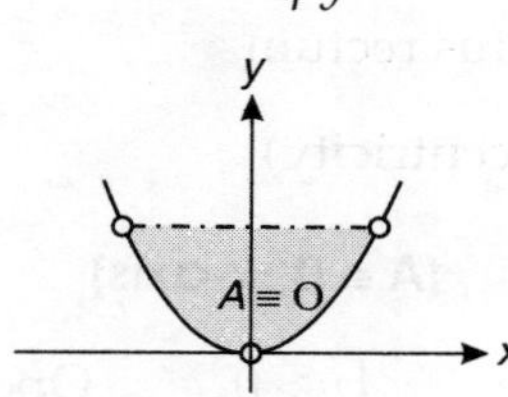

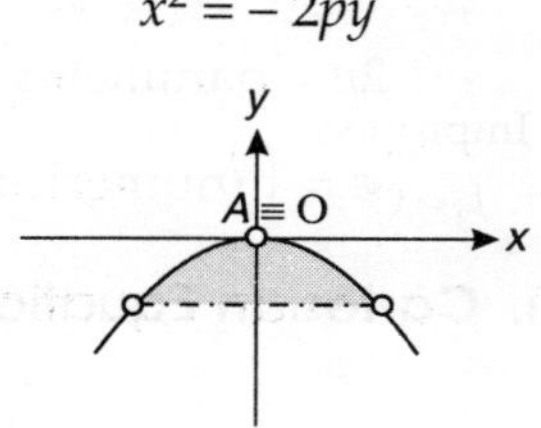

Fig. 6.56

6.10 GENERAL CURVES OF THE SECOND DEGREE

i. General Equation

The general equation of the second degree,

$a_{11}x^2 + 2a_{12}xy + a_{22}y^2 + 2a_{13}x + 2a_{23}y + a_{33} = 0$, defines the following conic sections: a circle, an ellipse, a hyperbola, a parabola, a pair of stright lines, a straight line, or a point (the last three cases are degenerate curves of the second degree).

ii. Invariants

The quantities (note that $a_{ik} = a_{ki}$)

$$I_3 = \begin{vmatrix} a_{11} & a_{12} & a_{13} \\ a_{21} & a_{22} & a_{33} \\ a_{31} & a_{32} & a_{33} \end{vmatrix} \qquad I_2 = \begin{vmatrix} a_{11} & a_{12} \\ a_{21} & a_{22} \end{vmatrix} \qquad I_1 = a_{11} + a_{22}$$

and the sign of the quantity

$$A = \begin{vmatrix} a_{22} & a_{23} \\ a_{32} & a_{33} \end{vmatrix} + \begin{vmatrix} a_{11} & a_{13} \\ a_{31} & a_{33} \end{vmatrix}$$

are invariants of transformation (they are not affected by the translation or the rotation of coordinate axes) and define properties of conics.

iii. Classification (CS = conic section)

<table>
<tr><td>Type</td><td colspan="2">$I_2 \pm O$</td><td colspan="3">$I_2 = O$</td></tr>
<tr><td></td><td colspan="2">Central CS</td><td colspan="3">Noncentral CS</td></tr>
<tr><td rowspan="3">Proper
$I_3 \neq O$</td><td>$I_2 > 0$</td><td>$I_2 < 0$</td><td rowspan="3" colspan="3">Parabola</td></tr>
<tr><td>Real
ellipse
$I_1 I_3 > 0$</td><td rowspan="2">Hyperbola</td></tr>
<tr><td>Imaginary
ellipse
$I_1 I_3 > 0$</td></tr>
<tr><td rowspan="2">Improper
$I_3 = O$</td><td colspan="2">Two nonparallel straight lines</td><td colspan="2">Two parallel straight lines</td><td>One straight line</td></tr>
<tr><td>Imaginary</td><td>Real</td><td>Real
$A < 0$</td><td>Imaginary
$A > 0$</td><td>Real
$A = 0$</td></tr>
</table>

PLANE CURVES

Exponential curve

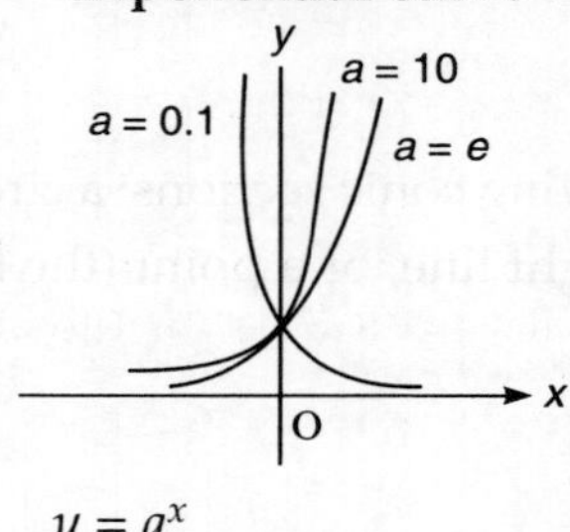

$y = a^x$

Logarithmic curve

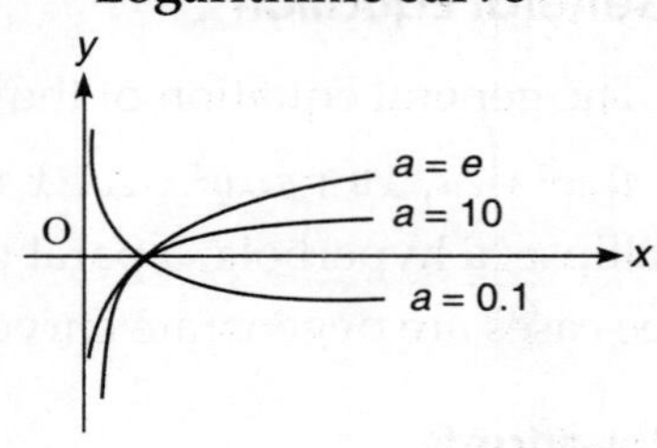

$y = \log_a x$

Fig. 6.57

Catenary

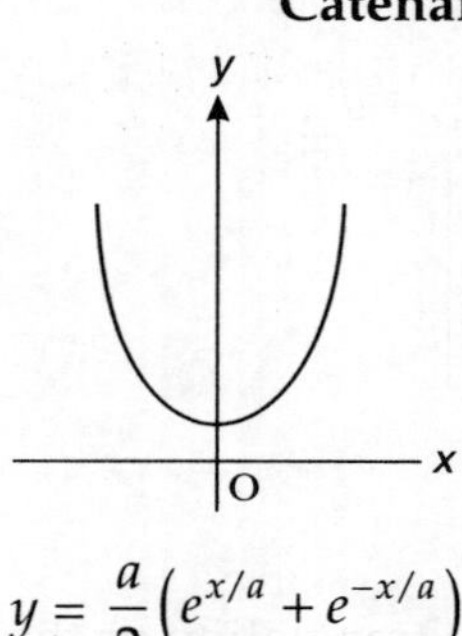

$y = \frac{a}{2}\left(e^{x/a} + e^{-x/a}\right)$

Probability curve

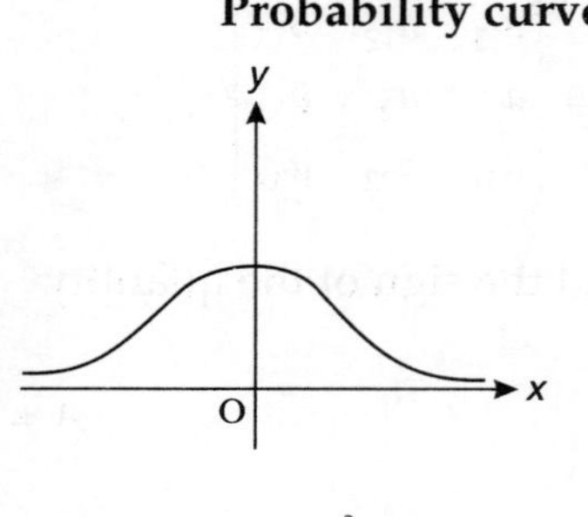

$y = e^{-x^2}$

Fig. 6.58

Exponential curve with end point

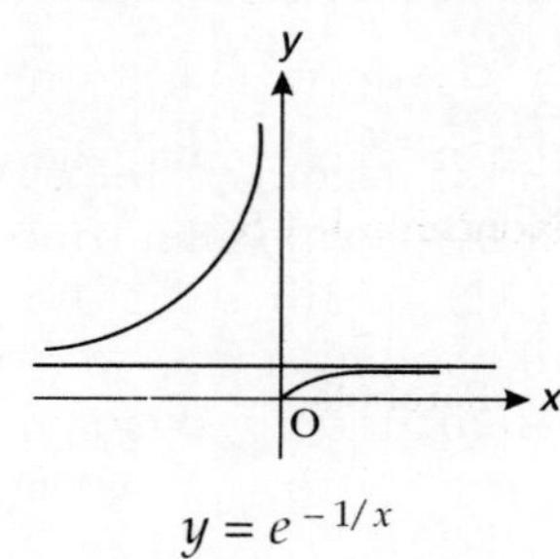

$y = e^{-1/x}$

Logarithmic curve with end point

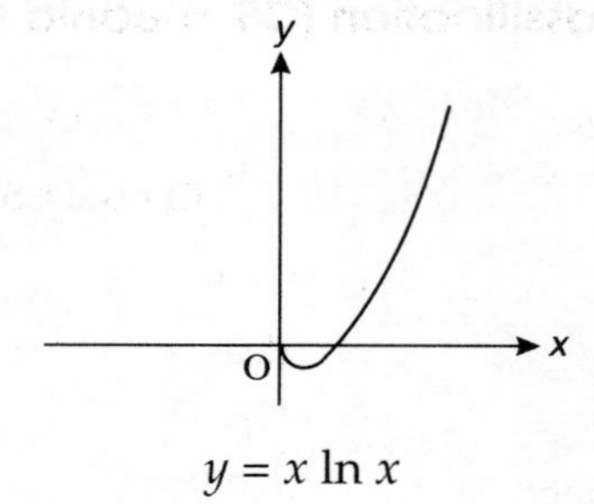

$y = x \ln x$

Fig. 6.59

Cubic parabola

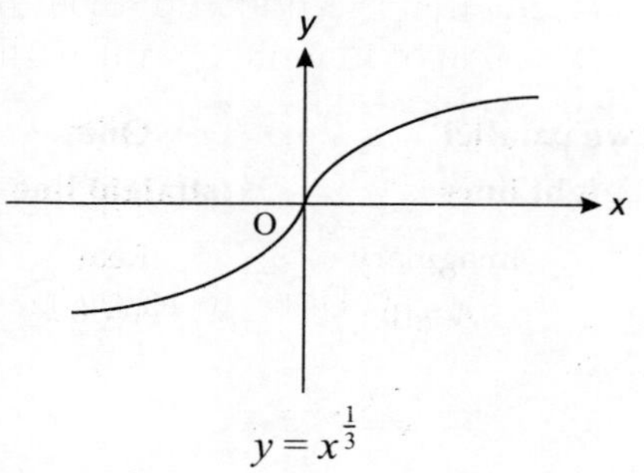

$y = x^{\frac{1}{3}}$

Semicubic parabola

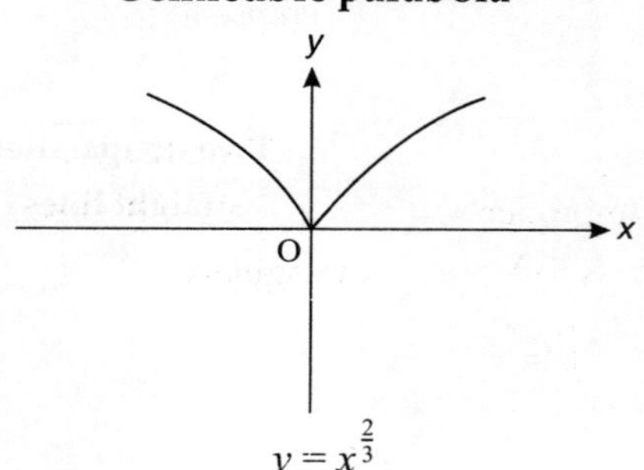

$y = x^{\frac{2}{3}}$

Fig. 6.60

Parabola

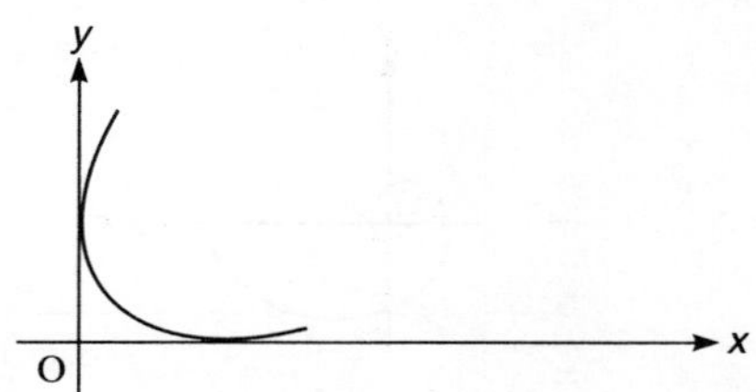

$x^{1/2} + y^{1/2} = a^{1/2}$

Equilateral hyperbola

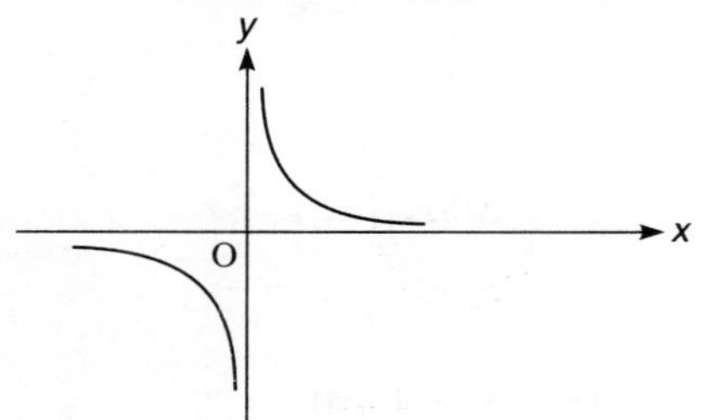

$xy = a$

Fig. 6.61

Cusp of first kind

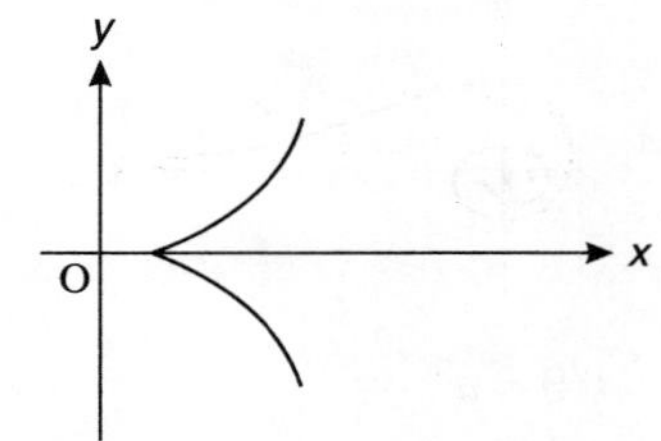

$y^2 = (x-1)^5$

Cusp of second kind

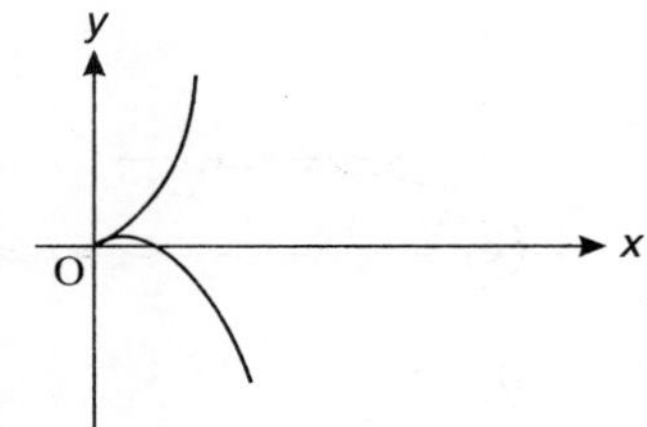

$(y - x^2)^2 = x^5$

Fig. 6.62

Power curve

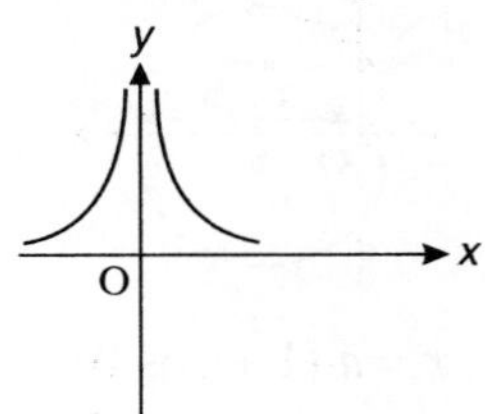

$y = x^{-2}$

Cissiod

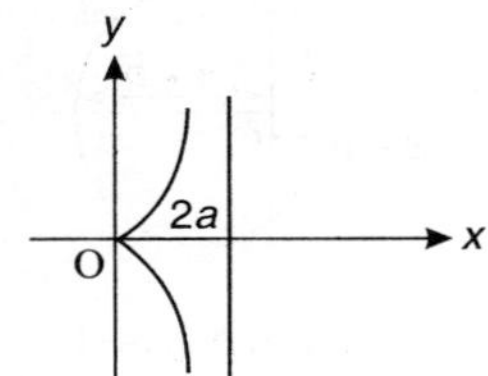

$y^2(2a - x) = x^3$

Fig. 6.63

Strophoid

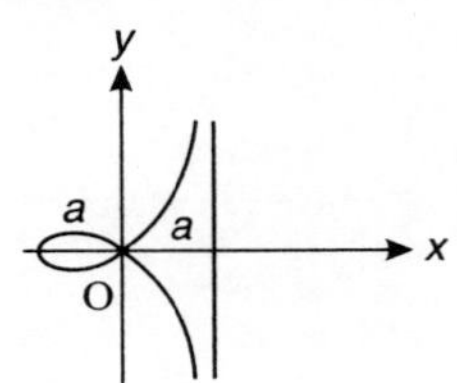

$y = x^2 \dfrac{a+x}{a-x}$

Folium of descartes

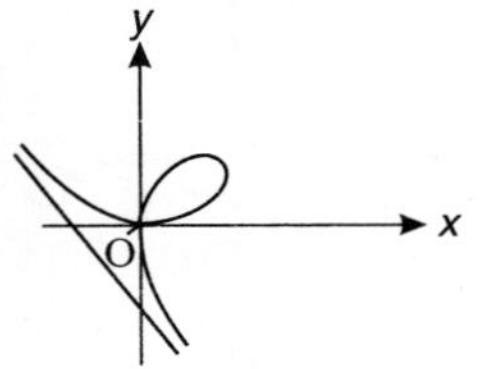

$x^3 + y^3 - 3axy = 0$

Fig. 6.64

Parabolic spiral

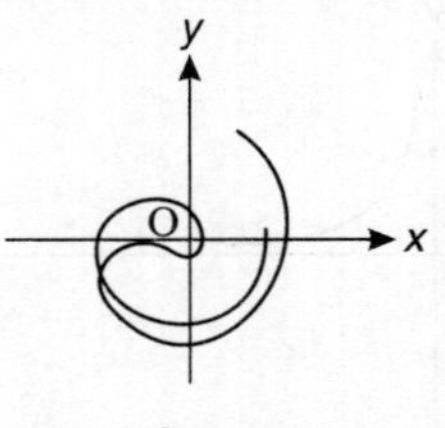

$(r - a)^2 = 4\, ak\theta$

Logarithmic spiral

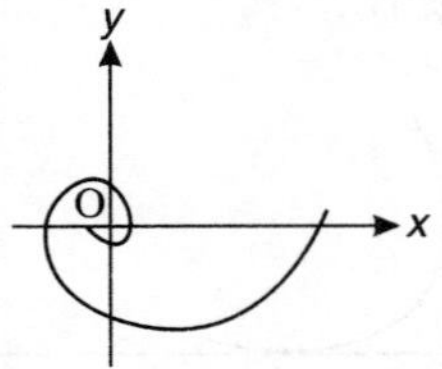

$\ln r = a\,\theta$

Fig. 6.65

Hyperbolic spiral

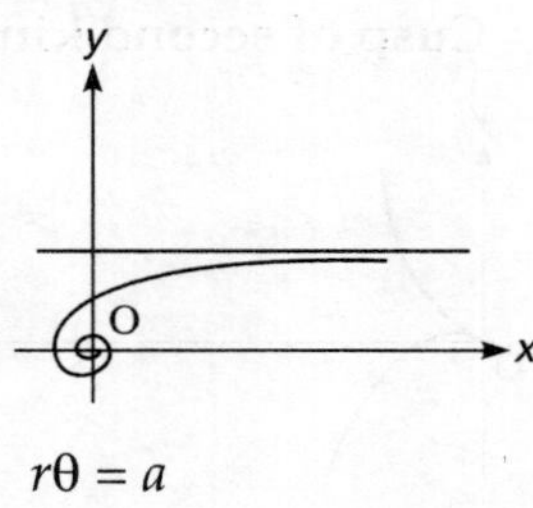

$r\theta = a$

Lituus

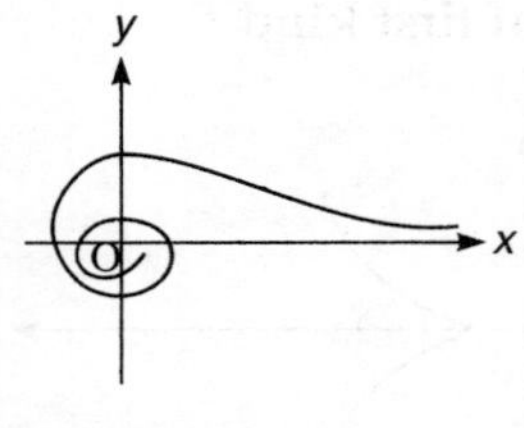

$r^2\theta = a^2$

Fig. 6.66

Limacon $(a > b > 0)$

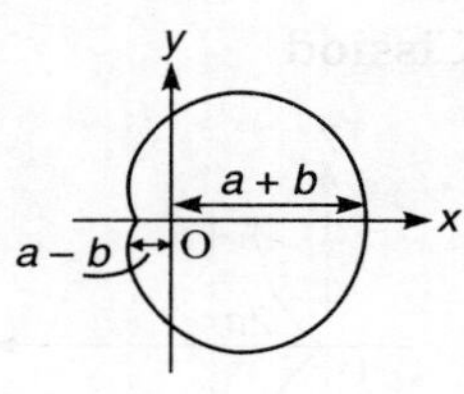

$r = a + b \cos \theta$

Cardioid $(a = b)$

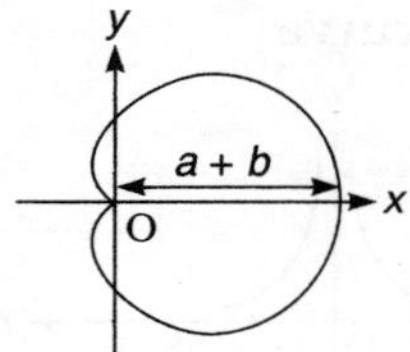

$r = a\,(1 + \cos \theta)$

Fig. 6.67

Limacon $(b > a > 0)$

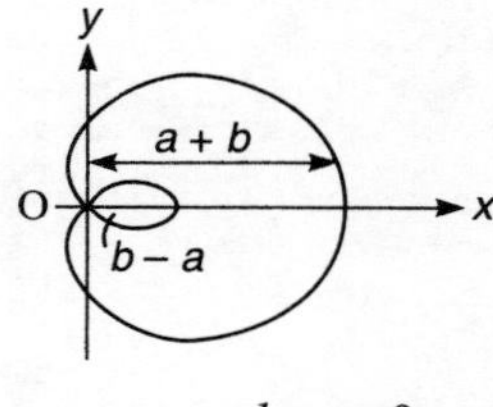

$r = a + b \cos \theta$

Fig. 6.68

Ordinary Cycloids

Cusp at origin

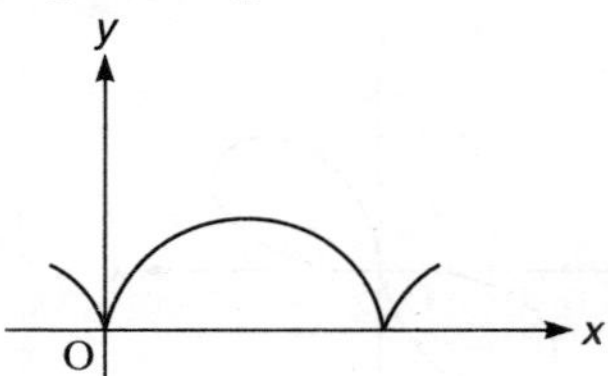

Vertex at origin

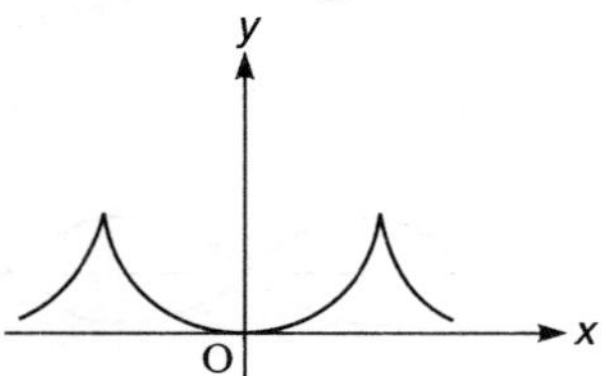

Fig. 6.69

$x = a \cos^{-1}\dfrac{a-y}{a} - \sqrt{2ay - y^2}$

$x = a(\theta - \sin\theta) \qquad y = a(1 - \cos\theta)$

$x = a \cos^{-1}\dfrac{a-y}{a} + \sqrt{2ay - y^2}$

$x = a(\theta + \sin\theta)\; y = a(1 - \cos\theta)$

Prolate cycloid $(a < b)$

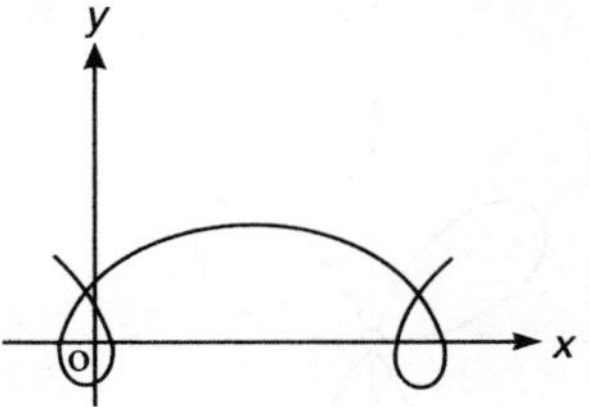

Curtate cycloid $(a > b)$

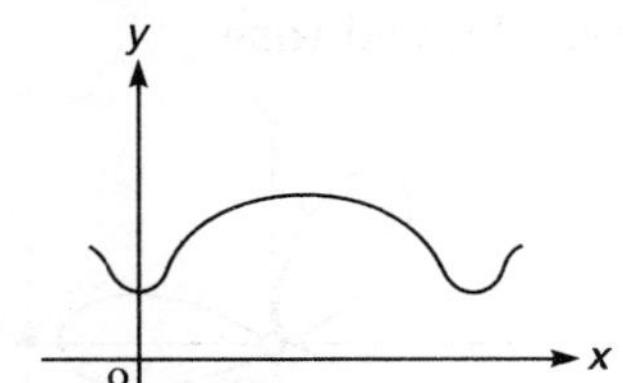

Fig. 6.70

$x = a\theta - b\sin\theta$

$y = a - b\cos\theta$

$x = a\theta - b\sin\theta$

$y = a - b\cos\theta$

Hypocycloid

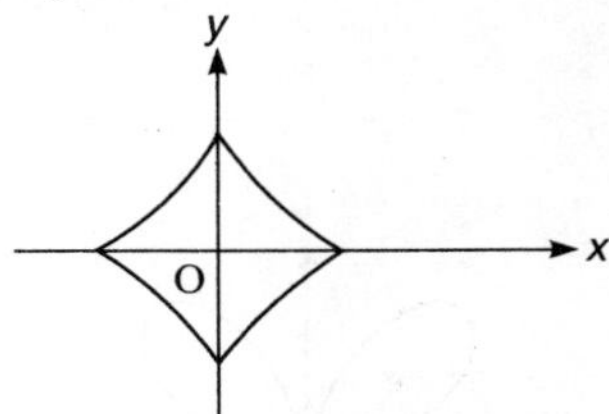

Evolute of ellipse

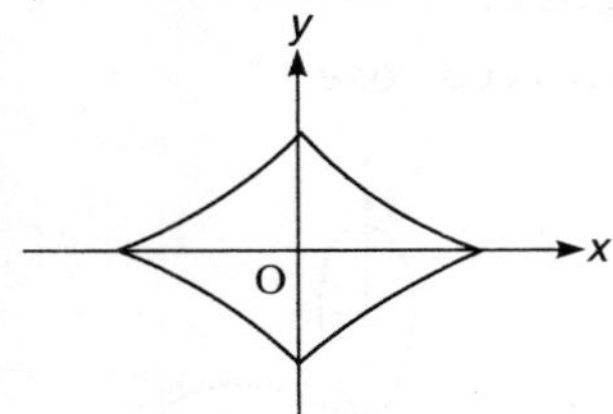

Fig. 6.71

$x^{2/3} + y^{2/3} = a^{2/3}$

$(ax)^{2/3} + (by)^{2/3} = (a^2 - b^2)^{2/3}$

Conchoids

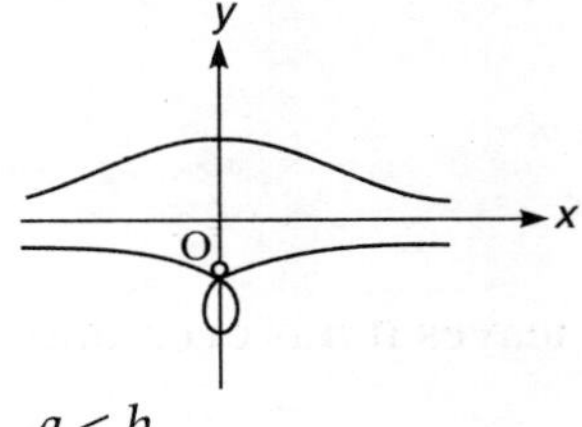

$a < b$

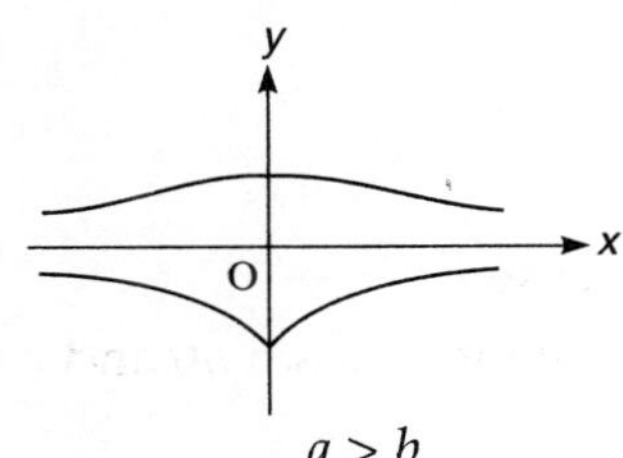

$a > b$

Fig. 6.72

$(y - a)^2 (x^2 + y^2) = b^2y^2$

$r = a \csc\theta \pm b$

Lemniscates of Bernouilli

Two-leaved rose

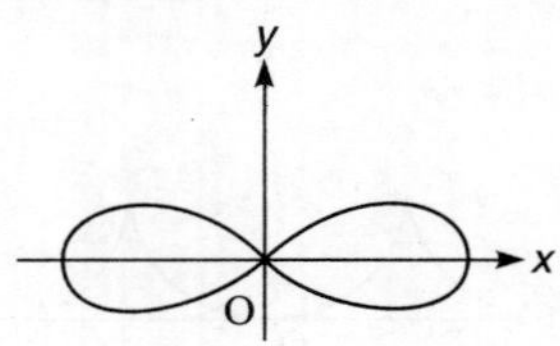

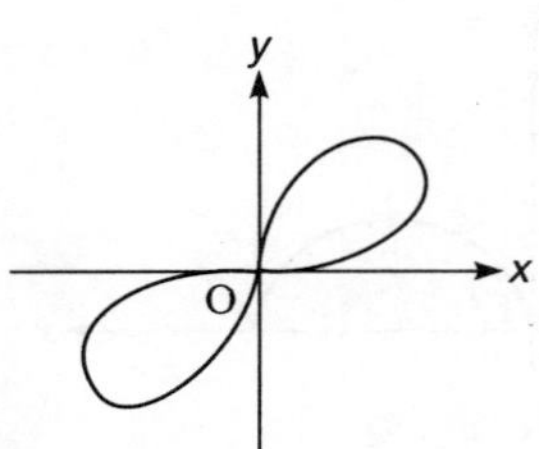

Fig. 6.73

$(x^2 + y^2)^2 = a^2(x^2 - y^2)$

$r^2 = a^2 \cos 2\theta$

$(x^2 + y^2)^2 = 2a^2xy$

$r^2 = a^2 \sin 2\theta$

Three-leaved rose

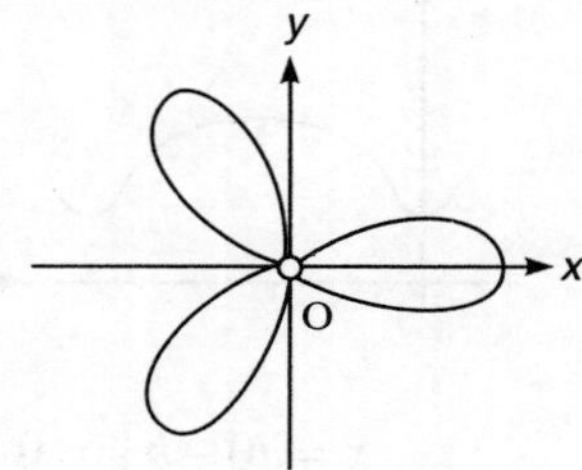

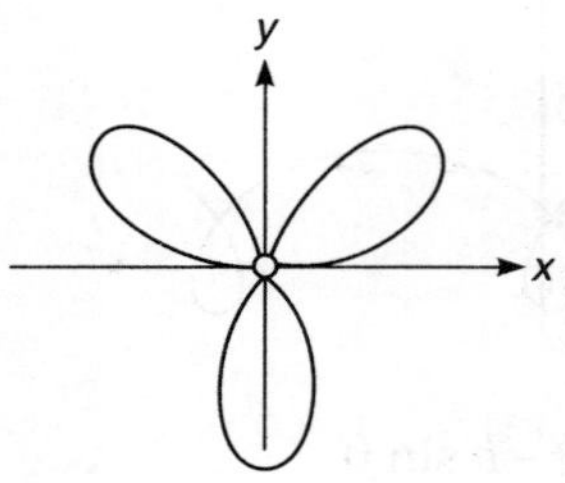

Fig. 6.74

$r = a \cos 3\theta$

$r = a \sin 3\theta$

Four-leaved rose

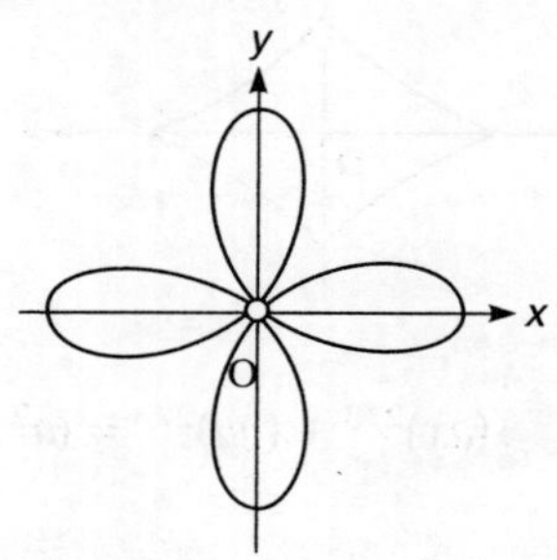

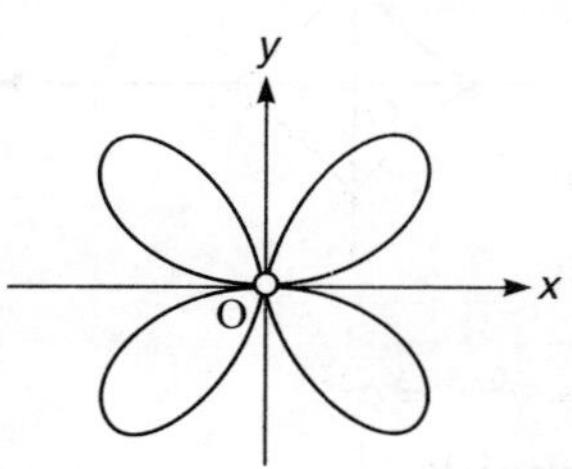

Fig. 6.75

$r = a \cos 2\theta$

$r = a \sin 2\theta$

***n*-leaved rose**

The roses given by $r = a \sin n\theta$ and $r = a \cos n\theta$, have $2n$ leaves if n is even and n leaves if n is odd.

The roses given by $r^2 = a^2 \sin n\theta$ and $r^2 = a^2 \cos n\theta$, have n leaves if n is even and $2n$ leaves if n is odd.

CHAPTER

7

Space Coordinate Geometry

7.1 COORDINATES

i. Systems of Coordinates

a. *Cartesian coordinates*

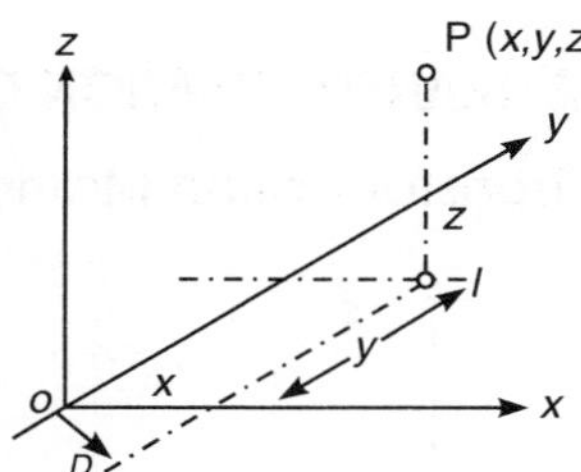

Fig. 7.1

A point P is given by three mutually perpendicular distances x, y, z (coordinates measured from three mutually perpendicular yz, zx, xy planes, respectively. The lines of intersection of these planes are the coordinate axes x, y, z, and the point of their intersection is the origin O. The right-hand system is shown in Fig. 7.1.

b. *Cylindrical coordinates*

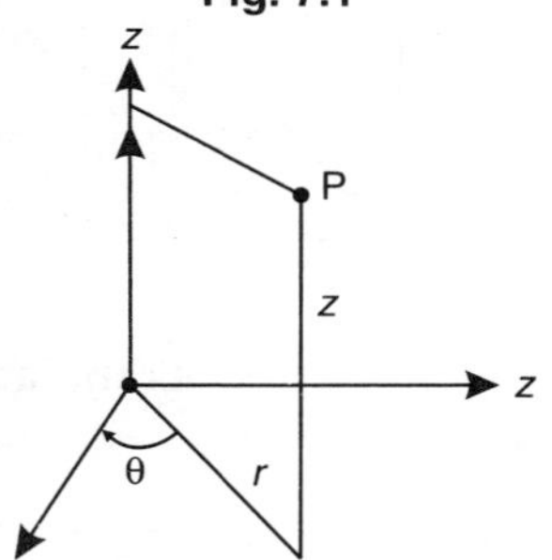

Fig. 7.2

A point P is given by its polar coordinates r and θ, in the xy plane, and the cartesian coordinate z.

c. *Spherical coordinates*

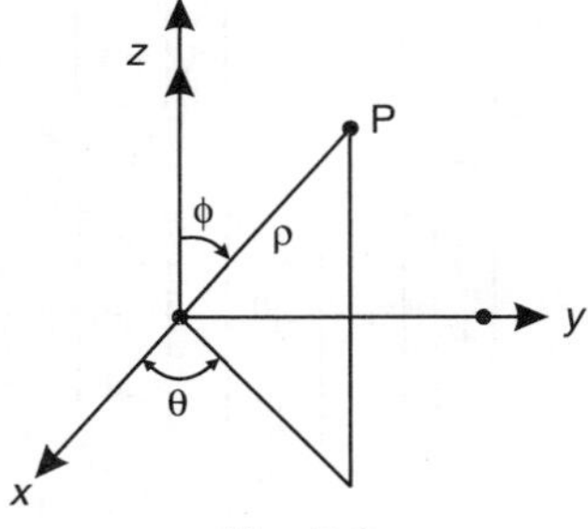

Fig. 7.3

A point P is given by a position angle θ measured from a fixed axis x, a position angle ϕ measured from another fixed axis z, normal to x, and the position radius ρ, measured from the point of intersection of x and y, designated as the pole O.

ii. Relationship between Coordinate Systems

Coordinate	*Cartesian*	*Cylindrical*	*Spherical*
Cartesian	$x = x$ $y = y$ $z = z$	$x = r\cos\theta$ $y = r\sin\theta$ $z = z$	$x = \rho\cos\theta\sin\phi$ $y = \rho\sin\theta\sin\phi$ $z = \rho\cos\phi$
Cylindrical	$r = \sqrt{x^2+y^2}$ $\theta = \tan^{-1}\dfrac{y}{x}$ $z = z$	$r = r$ $\theta = \theta$ $z = z$	$r = \rho\sin\phi$ $\theta = \theta$ $z = \rho\cos\phi$
Spherical	$\rho = \sqrt{x^2+y^2+z^2}$ $\theta = \tan^{-1}\dfrac{y}{x}$ $\phi = \cos^{-1}\dfrac{z}{\sqrt{x^2+y^2+z^2}}$	$\rho = \sqrt{r^2+z^2}$ $\theta = \theta$ $\phi = \cos^{-1}\dfrac{z}{\sqrt{r^2+z^2}}$	$\rho = \rho$ $\theta = \theta$ $\phi = \phi$

7.2 TRANSFORMATION OF CARTESIAN COORDINATES

i. Transformation Matrices

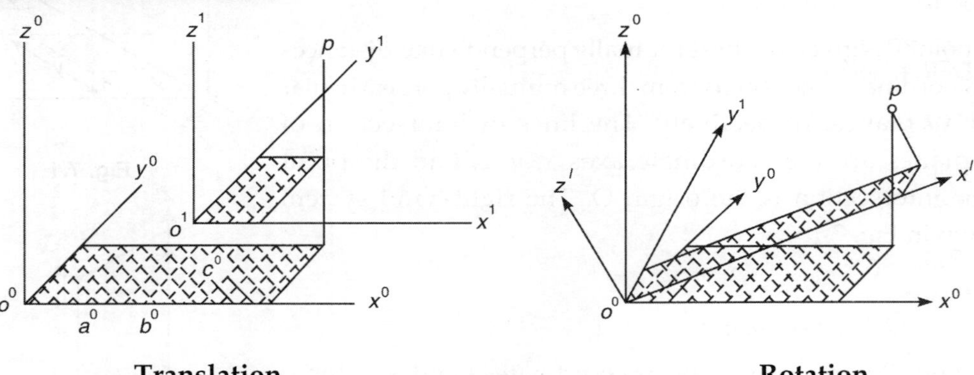

Fig. 7.4

$$\begin{bmatrix} x^0 \\ y^0 \\ z^0 \end{bmatrix} = \begin{bmatrix} x^1 \\ y^1 \\ z^1 \end{bmatrix} + \begin{bmatrix} a^0 \\ b^0 \\ c^0 \end{bmatrix}$$
$$s^0 \qquad s^1 \qquad d^0$$

$$\begin{bmatrix} x^0 \\ y^0 \\ z^0 \end{bmatrix} = \begin{bmatrix} \alpha_x & \alpha_y & \alpha_z \\ \beta_x & \beta_y & \beta_z \\ \gamma_x & \gamma_y & \gamma_z \end{bmatrix} \begin{bmatrix} x^l \\ y^l \\ z^l \end{bmatrix}$$
$$s^0 \qquad \omega^{01} \qquad s^l$$

$$\begin{bmatrix} x^1 \\ y^1 \\ z^1 \end{bmatrix} = \begin{bmatrix} x^0 \\ y^0 \\ z^0 \end{bmatrix} - \begin{bmatrix} a^0 \\ b^0 \\ c^0 \end{bmatrix}$$
$$s^1 \qquad s^0 \qquad d^0$$

$$\begin{bmatrix} x^1 \\ y^1 \\ z^1 \end{bmatrix} = \begin{bmatrix} \alpha_x & \beta_x & \gamma_x \\ \alpha_y & \beta_y & \gamma_y \\ \alpha_z & \beta_z & \gamma_z \end{bmatrix} \begin{bmatrix} x^0 \\ y^0 \\ z^0 \end{bmatrix}$$
$$s^l \qquad \omega^{10} \qquad s^0$$

Note: 0, 1 and l are superscripts designating the system.

ii. Direction Cosines

$\bullet = \cos()$

$\alpha_x = \bullet(x^0 x^l) = \bullet(x^l x^0)$

$\beta_x = \bullet(y^0 x^l) = \bullet(x^l y^0)$

$\gamma_x = \bullet(z^0 x^l) = \bullet(x^l z^0)$

$\bullet = \cos()$

$\alpha_y = \bullet(x^0 y^l) = \bullet(y^l x^0)$

$\beta_y = \bullet(y^0 y^l) = \bullet(y^l y^0)$

$\gamma_y = \bullet(z^0 u^l) = \bullet(y^l z^0)$

$\bullet = \cos()$

$\alpha_z = \bullet(x^0 z^l) = \bullet(z^l x^0)$

$\beta_z = \bullet(y^n z^l) = \bullet(z^l y^0)$

$\gamma_z = \bullet(z^0 z^l) = \bullet(z^l z^0)$

iii. Properties of ω Matrices

$s^0 = \omega^{0l} s^l$

$s^l = \omega^{l0} s^0$

$\omega^{0l} \omega^{l0} = I$

$\omega^{l0} \omega^{0l} = I$

$\omega^{01} = (\omega^{l0})^T = (\omega^{l0})^{-1}$

$\omega^{10} = (\omega^{0l})^T = (\omega^{0l})^{-1}$

Note: ω matrices are orthogonal.

iv. Properties of Direction Cosines

$\alpha_x^2 + \alpha_y^2 + \alpha_z^2 = 1$

$\beta_x^2 + \beta_y^2 + \beta_z^2 = 1$

$\gamma_x^2 + \gamma_y^2 + \gamma_z^2 = 1$

Diagonal terms of matrix product $\omega^{0l}\,\omega^{l0}$

$\alpha_x\beta_x + \alpha_y\beta_y + \alpha_z\beta_z = 0$

$\beta_x\gamma_x + \beta_y\gamma_y + \beta_z\gamma_z = 0$

$\gamma_x\alpha_x + \gamma_y\alpha_y + \gamma_z\alpha_z = 0$

Off-diagonal terms of matrix product $\omega^{0l}\,\omega^{l0}$

$\alpha_x\alpha_y + \beta_x\beta_y + \gamma_x\gamma_y = 0$

$\alpha_y\alpha_z + \beta_y\beta_z + \gamma_y\gamma_z = 0$

$\alpha_z\alpha_x + \beta_z\beta_x + \gamma_z\gamma_x = 0$

Off-diagonal terms of matrix product $\omega^{0l}\,\omega^{l0}$

7.3 DERIVATION OF ω MATRICES

i. Successive Rotation

Every space rotation can be resolved into three components, and every component rotation can be computed independently.

Rotational ω^j about Z^0

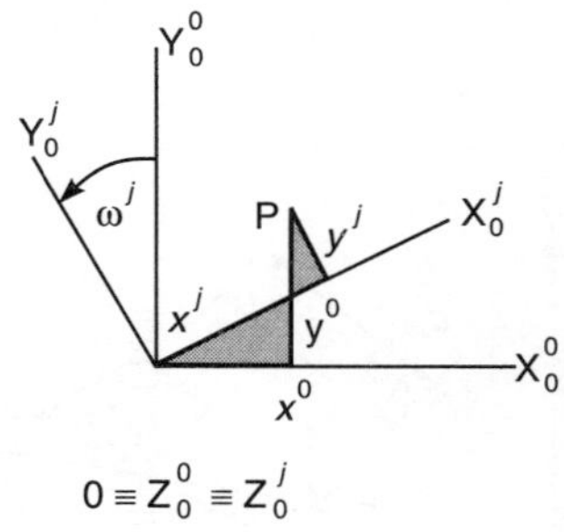

Rotational ω^k about $Y^{j'}$

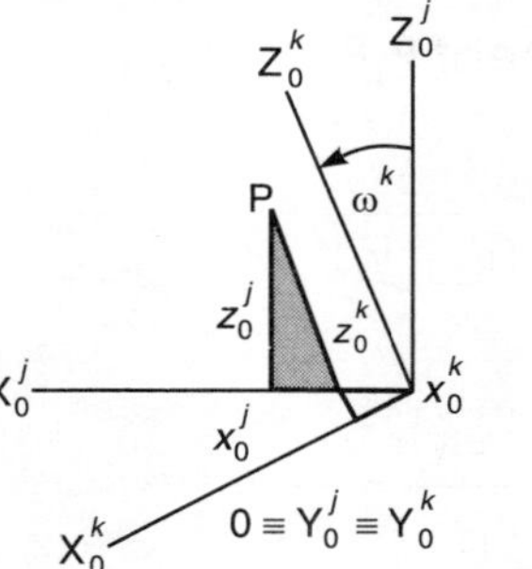

Rotational ω^l about X^k

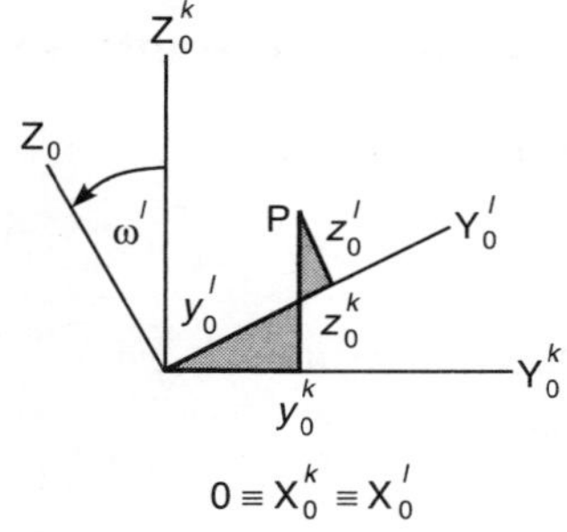

Fig. 7.5

$$[\omega^{0j}] = \begin{bmatrix} \cos\omega^j & -\sin\omega^j & 0 \\ \sin\omega^j & \cos\omega^j & 0 \\ 0 & 0 & 1 \end{bmatrix} \quad [\omega^{jk}] = \begin{bmatrix} \cos\omega^k & 0 & \sin\omega^k \\ 0 & 1 & 0 \\ -\sin\omega^k & 0 & \cos\omega^k \end{bmatrix} \quad [\omega^{kl}] = \begin{bmatrix} 1 & 0 & 0 \\ 0 & \cos\omega^l & -\sin\omega^l \\ 0 & \sin\omega^l & \cos\omega^l \end{bmatrix}$$

$$\omega^{jo} = \begin{bmatrix} \cos\omega^j & \sin\omega^j & 0 \\ -\sin\omega^j & \cos\omega^j & 0 \\ 0 & 0 & 1 \end{bmatrix} \quad \omega^{kj} = \begin{bmatrix} \cos\omega^k & 0 & -\sin\omega^k \\ 0 & 1 & 0 \\ \sin\omega^k & 0 & \cos\omega^k \end{bmatrix} \quad [\omega^{lk}] = \begin{bmatrix} 1 & 0 & 0 \\ 0 & \cos\omega^l & \sin\omega^l \\ 0 & -\sin\omega^l & \cos\omega^l \end{bmatrix}$$

ii. Successive Matrix Multiplication

The resulting rotational transformation matrix (ω^{0l} or ω^{l0}) is equal to the chain-matrix product of the component matrix executed in the order of rotation.

$$\omega^{0l} = \omega^{0j}\,\omega^{jk}\,\omega^{kl}$$

$$\begin{bmatrix} \alpha_x & \alpha_y & \alpha_z \\ \beta_x & \beta_y & \beta_z \\ \gamma_x & \gamma_y & \gamma_z \end{bmatrix} = \left[\begin{array}{c|c|c} +\cos\omega^j\cos\omega^k & \begin{array}{c}-\sin\omega^l\cos\omega^l \\ +\cos\omega^j\sin\omega^k\sin\omega^l\end{array} & \begin{array}{c}+\sin\omega^j\sin\omega^l \\ +\cos\omega^j\sin\omega^k\cos\omega^l\end{array} \\ \hline \begin{array}{c}+\sin\omega^j\cos\omega^k \\ 1\end{array} & \begin{array}{c}+\cos\omega^j\cos\omega^l \\ +\sin\omega^l\sin\omega^k\sin\omega^l\end{array} & \begin{array}{c}-\cos\omega^l\sin\omega^l \\ +\sin\omega^l\sin\omega^k\cos\omega^j\end{array} \\ \hline -\sin\omega^k & +\cos\omega^k\sin\omega^l & +\cos\omega^k\cos\omega^l \end{array}\right]$$

$$\omega^{l0} = \omega^{lk}\,\omega^{kj}\,\omega^{j0}$$

$$\begin{bmatrix} \alpha_x & \beta_x & \gamma_x \\ \alpha_y & \beta_y & \gamma_y \\ \alpha_z & \beta_z & \gamma_z \end{bmatrix} = \left[\begin{array}{c|c|c} +\cos\omega^j\cos\omega^k & +\sin\omega^j\cos\omega^k & -\sin\omega^k \\ \hline \begin{array}{c}-\sin\omega^j\cos\omega^l \\ +\cos\omega^j\sin\omega^k\sin\omega^l\end{array} & \begin{array}{c}+\cos\omega^j\cos\omega^l \\ +\sin\omega^j\sin\omega^k\sin\omega^l\end{array} & +\cos\omega^k\sin\omega^l \\ \hline \begin{array}{c}+\sin\omega^j\sin\omega^l \\ +\cos\omega^j\sin\omega^k\cos\omega^l\end{array} & \begin{array}{c}-\cos\omega^j\sin\omega^l \\ +\sin\omega^j\sin\omega^k\cos\omega^l\end{array} & +\cos\omega^k\cos\omega^l \end{array}\right]$$

7.4 POINTS IN SPACE

i. Distance of Two Points, Segment $P_1P_2 = d_{12}$

Segment in space

$$d_{12} = \sqrt{(x_2 - x_1)^2 + (y_2 - y_1)^2 + (z_2 - z_1)^2}$$

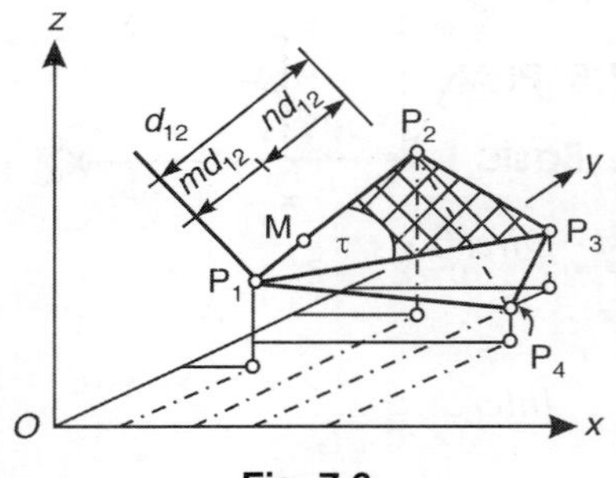

Fig. 7.6

Direction cosines in space

$$\alpha_{12} = \cos Xd_{12} = \frac{x_2 - x_1}{d_{12}} = \frac{d_{12x}}{d_{12}}$$

$$\beta_{12} = \cos Yd_{12} = \frac{y_2 - y_1}{d_{12}} = \frac{d_{12y}}{d_{12}}$$

$$\gamma_{12} = \cos Zd_{12} = \frac{z_2 - z_1}{d_{12}} = \frac{d_{12z}}{d_{12}}$$

Relationship in space

$$\alpha^2 + \beta^2 + \gamma^2 = +1$$

ii. Components of Segment d_{12}

$$d_{12x} = d_{12}\alpha_{12} \qquad d_{12y} = d_{12}\beta_{12} \qquad d_{12z} = d_{12}\gamma_{12}$$

$$d_{12} = d_{12x}\alpha_{12} + d_{12y}\beta_{12} + d_{12z}\gamma_{12}$$

iii. Coordinates of M Dividing P_1, P_2 in Ratio $m : n$

$$x_M = \frac{nx_1 + mx_2}{m+n} \qquad y_M = \frac{ny_1 + my_2}{m+n} \qquad z_M = \frac{nz_1 + mz_2}{m+n}$$

iv. Area and Centroid of Triangle with Vertices P_1, P_2, P_3

$$A = \sqrt{A_1^2 + A_2^2 + A_3^2}$$

$$A_1 = \frac{1}{2}\begin{vmatrix} y_1 & z_1 & 1 \\ y_2 & z_2 & 1 \\ y_3 & z_3 & 1 \end{vmatrix} \qquad A_2 = \frac{1}{2}\begin{vmatrix} z_1 & x_1 & 1 \\ z_2 & x_2 & 1 \\ z_3 & x_3 & 1 \end{vmatrix} \qquad A_3 = \frac{1}{2}\begin{vmatrix} x_1 & y_1 & 1 \\ x_2 & y_2 & 1 \\ x_3 & y_3 & 1 \end{vmatrix}$$

$$x_c = \frac{x_1 + x_2 + x_3}{3} \qquad y_c = \frac{y_1 + y_2 + y_3}{3} \qquad z_c = \frac{z_1 + z_2 + z_3}{3}$$

v. Volume of Tetrahedron with Vertices P_1, P_2, P_3, P_4

$$V = \frac{1}{6}\begin{vmatrix} x_1 & y_1 & z_1 & 1 \\ x_2 & y_2 & z_2 & 1 \\ x_3 & y_3 & z_3 & 1 \\ x_4 & y_4 & z_4 & 1 \end{vmatrix}$$

vi. Angle τ between d_{12} and d_{13}

$$\cos\tau = \alpha_{12}\alpha_{13} + \beta_{12}\beta_{13} + \gamma_{12}\gamma_{13}$$

$$= \frac{d_{12x}d_{13x} + d_{12y}d_{13y} + d_{12z}d_{13z}}{d_{12}d_{13}}$$

7.5 PLANE IN SPACE

i. Basic Equations of a Plane

Direction form

$$z = k_1x + k_2y + l$$

Intercept form

$$\frac{x}{a} + \frac{y}{b} + \frac{z}{c} = 1$$

Normal form

$$\alpha x + \beta y + \gamma z = n$$

General form

$$Ax + By + Cz + D = 0$$

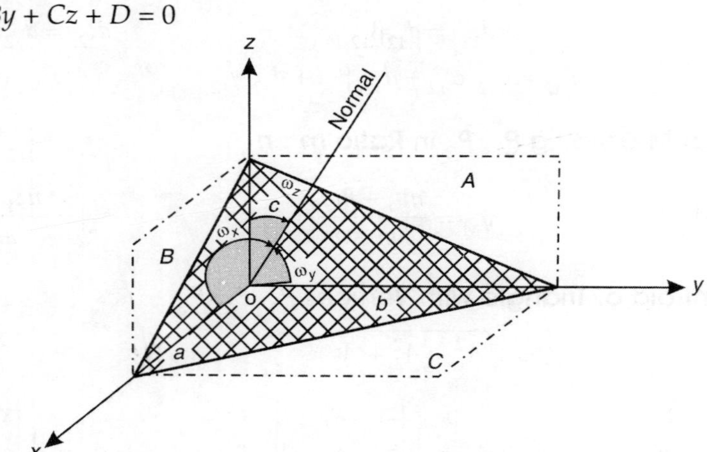

Fig. 7.7

Direction cosines of planes in space

$\alpha = \cos \omega_x$ $\beta = \cos \omega_y$ $\gamma = \cos \omega_z$

ii. Relationship between Different Parameters

$A = bc$	$\alpha = \dfrac{A}{\pm\sqrt{A^2 + B^2 + C^2}}$	$\alpha = \dfrac{bc}{\pm\sqrt{(ab)^2 + (bc)^2 + (ca)^2}}$
$B = ca$	$\beta = \dfrac{B}{\pm\sqrt{A^2 + B^2 + C^2}}$	$\beta = \dfrac{ca}{\pm\sqrt{(ab)^2 + (bc)^2 + (ca)^2}}$
$C = ab$	$\gamma = \dfrac{C}{\pm\sqrt{A^2 + B^2 + C^2}}$	$\gamma = \dfrac{ab}{\pm\sqrt{(ab)^2 + (bc)^2 + (ca)^2}}$
$D = -abc$	$n = \dfrac{-D}{\pm\sqrt{A^2 + B^2 + C^2}}$	$n = \dfrac{abc}{\pm\sqrt{(ab)^2 + (bc)^2 + (ca)^2}}$
$a = -\dfrac{D}{A}$	$k_1 = -\dfrac{A}{C}$	$k_1 = -\dfrac{c}{a}$
$b = -\dfrac{D}{B}$	$k_2 = -\dfrac{B}{C}$	$k_2 = -\dfrac{c}{b}$
$c = -\dfrac{D}{C}$	$l = -\dfrac{D}{C}$	$l = c$

iii. Plane Passing through a Point P_i in a Given Direction

$$A(x - x_i) + B(y - y_i) + C(z - z_i) = 0$$

iv. Plane Passing through Three Points P_i, P_j, P_k

$$\begin{vmatrix} y_i & z_i & 1 \\ y_j & z_j & 1 \\ y_k & z_k & 1 \end{vmatrix} x + \begin{vmatrix} z_i & x_i & 1 \\ z_j & x_j & 1 \\ z_k & x_k & 1 \end{vmatrix} y + \begin{vmatrix} x_i & y_i & 1 \\ x_j & y_j & 1 \\ x_k & y_k & 1 \end{vmatrix} z = \begin{vmatrix} x_i & y_i & z_i \\ x_j & y_j & z_j \\ x_k & y_k & z_k \end{vmatrix}$$

v. Distance between the Point P_i and the Plane $Ax + By + Cz + D = 0$

$$d = \frac{Ax_i + By_i + Cz_i + D}{\pm\sqrt{A^2 + B^2 + C^2}}$$

Note: The sign of the denominator is opposite to the sign of D.

7.6 TWO AND THREE PLANES IN SPACE

i. Relationships of Two Planes

If two planes are given by their equations as

$$A_1x + B_1y + C_1z + D_1 = 0; \qquad A_2x + B_2y + C_2z + D_2 = 0$$

then the following are their relationships:

$A_1 : B_1 : C_1 \neq A_2 : B_2 : C_2$	Planes intersect
$A_1 : B_1 : C_1 = A_2 : B_2 : C_2$	Planes are parallel
$A_1A_2 : B_1B_2 : C_1C_2 = 0$	Planes are normal
$A_1 : A_2 = B_1 : B_2 = C_1 : C_2 = D_1 : D_2$	Planes coincide

ii. Angle between the Normals of Two Planes in Space

$$\cos\tau = \frac{A_1A_2 + B_1B_2 + C_1C_2}{\sqrt{A_1^2 + B_1^2 + C_1^2}\,\sqrt{A_2^2 + B_2^2 + C_2^2}}$$

iii. Distance between Two parallel Planes in Space

$$d = \frac{D_1 - D_2}{\pm\sqrt{A^2 + B^2 + C^2}}; \qquad \begin{aligned} A_1 &= A_2 = A \\ B_1 &= B_2 = B \\ C_1 &= C_2 = C \end{aligned}$$

iv. Intersection of Two Planes in Space

$$\begin{vmatrix} C_1 & C_2 \\ A_1 & A_2 \end{vmatrix} x + \begin{vmatrix} C_1 & C_2 \\ B_1 & B_2 \end{vmatrix} y + \begin{vmatrix} C_1 & C_2 \\ D_1 & D_2 \end{vmatrix} = 0$$

$$\begin{vmatrix} A_1 & A_2 \\ B_1 & B_2 \end{vmatrix} y + \begin{vmatrix} A_1 & A_2 \\ C_1 & C_2 \end{vmatrix} z + \begin{vmatrix} A_1 & A_2 \\ D_1 & D_2 \end{vmatrix} = 0$$

$$\begin{vmatrix} B_1 & B_2 \\ C_1 & C_2 \end{vmatrix} z + \begin{vmatrix} B_1 & B_2 \\ A_1 & A_2 \end{vmatrix} x + \begin{vmatrix} B_1 & B_2 \\ D_1 & D_2 \end{vmatrix} = 0$$

Each one of these equations represents the projected line of intersection in the respective coordinate plane.

v. Point of Intersection of Three Planes in Space

$$x = -\frac{\begin{vmatrix} D_1 & B_1 & C_1 \\ D_2 & B_2 & C_2 \\ D_3 & B_3 & C_3 \end{vmatrix}}{\begin{vmatrix} A_1 & B_1 & C_1 \\ A_2 & B_2 & C_2 \\ A_3 & B_3 & C_3 \end{vmatrix}} \qquad y = -\frac{\begin{vmatrix} A_1 & D_1 & C_1 \\ A_2 & D_2 & C_2 \\ A_3 & D_3 & C_3 \end{vmatrix}}{\begin{vmatrix} A_1 & B_1 & C_1 \\ A_2 & B_2 & C_2 \\ A_3 & B_3 & C_3 \end{vmatrix}} \qquad z = -\frac{\begin{vmatrix} A_1 & B_1 & D_1 \\ A_2 & B_2 & D_2 \\ A_3 & B_3 & D_3 \end{vmatrix}}{\begin{vmatrix} A_1 & B_1 & C_1 \\ A_2 & B_2 & C_2 \\ A_3 & B_3 & C_3 \end{vmatrix}}$$

vi. Plane of Symmetry of Two Planes in Space

$$\frac{A_1x + B_1y + C_1z + D_1}{\pm\sqrt{A_1^2 + B_1^2 + C_1^2}} \pm \frac{A_2x + B_2y + C_2z + D_2}{\pm\sqrt{A_2^2 + B_2^2 + C_2^2}} = 0$$

7.7 STRAIGHT LINE IN SPACE

i. General Form of A Straight Line

Two linearly independent equations,

$$A_1x + B_1y + C_1z + D_1 = 0 \qquad A_2x + B_2y + C_2z + D_2 = 0$$

represent a straight line in space. The projections of this line in the coordinate planes and their constants are as follows:

$$\overline{B}x + \overline{A}y + \overline{D}_{xy} = 0 \qquad \overline{A}z + \overline{C}x + \overline{D}_{zx} = 0 \qquad \overline{C}y + \overline{B}z + \overline{D}_{yz} = 0$$

$$\overline{A} = \begin{vmatrix} B_1 & C_1 \\ B_2 & C_2 \end{vmatrix} \qquad \overline{D}_{xy} = \begin{vmatrix} C_1 & D_1 \\ C_2 & D_2 \end{vmatrix} = \overline{D}_{yx} \qquad \overline{\alpha} = \frac{\overline{A}}{\sqrt{\overline{A}^2 + \overline{B}^2 + \overline{C}^2}}$$

$$\overline{B} = \begin{vmatrix} C_1 & A_1 \\ C_2 & A_2 \end{vmatrix} \qquad \overline{D}_{zx} = \begin{vmatrix} A_1 & D_1 \\ A_2 & D_2 \end{vmatrix} = \overline{D}_{xz} \qquad \overline{\beta} = \frac{\overline{B}}{\sqrt{\overline{A}^2 + \overline{B}^2 + \overline{C}^2}}$$

$$\overline{C} = \begin{vmatrix} A_1 & B_1 \\ A_2 & B_2 \end{vmatrix} \qquad \overline{D}_{yz} = \begin{vmatrix} B_1 & D_1 \\ B_2 & D_2 \end{vmatrix} = \overline{D}_{zy} \qquad \overline{\gamma} = \frac{\overline{C}}{\sqrt{\overline{A}^2 + \overline{B}^2 + \overline{C}^2}}$$

ii. Direction Form

$y = k_{yx}x + l_{yx}$ $z = k_{zx}x + l_{zx}$	$x = k_{xy}y + l_{xy}$ $z = k_{zy}y + l_{zy}$	$x = k_{xz}z + l_{xz}$ $y = k_{yz}z + l_{yz}$
$k_{yz} = -\frac{\overline{B}}{\overline{A}}$	$k_{xy} = -\frac{\overline{A}}{\overline{B}}$	$k_{xz} = -\frac{\overline{A}}{\overline{C}}$
$k_{zx} = -\frac{\overline{C}}{\overline{A}}$	$k_{zy} = -\frac{\overline{C}}{\overline{B}}$	$k_{yz} = -\frac{\overline{B}}{\overline{C}}$

$l_{yx} = -\dfrac{\overline{D}_{xy}}{\overline{A}}$	$l_{xy} = -\dfrac{\overline{D}_{yx}}{\overline{B}}$	$l_{xz} = -\dfrac{\overline{D}_{xz}}{\overline{C}}$
$l_{zx} = -\dfrac{\overline{D}_{zx}}{A}$	$l_{zy} = -\dfrac{\overline{D}_{zy}}{\overline{B}}$	$l_{yz} = -\dfrac{\overline{D}_{yz}}{\overline{C}}$

iii. Straight Line Passing through a Point P_i in a Given Direction

$$\frac{x - x_i}{\overline{\alpha}} = \frac{y - y_i}{\overline{\beta}} = \frac{z - z_i}{z_j - \overline{y}_j}$$

iv. Straight Line Passing through Points P_i and P_j

$$\frac{x - x_i}{x_j - x_i} = \frac{y - y_i}{y_j - y_i} = \frac{z - z_i}{z_j - z_i}$$

v. Parametric Equation of a Straight Line through a Point P_i

$x = x_i + \overline{\alpha}\, t$ $\qquad$ $y = y_i + \overline{\beta}\, t$ $\qquad$ $z = z_i + \overline{\gamma}\, t$

vi. Distance of the Point P_k to a Straight Line through a Point P_i

$$d = \sqrt{\begin{vmatrix} x_k - x_i & y_k - y_i \\ \overline{\alpha} & \overline{\beta} \end{vmatrix} + \begin{vmatrix} y_k - y_i & z_k - z_i \\ \overline{\beta} & \overline{\gamma} \end{vmatrix} + \begin{vmatrix} z_k - z_i & x_k - x_i \\ \overline{\gamma} & \overline{\alpha} \end{vmatrix}}$$

7.8 STRAIGHT LINES AND PLANES IN SPACE

i. Straight Line and Plane

If a plane and straight line are given, respectively as

$Ax + By + Cz + D = 0$ $\qquad$ $\overline{B}x + \overline{A}y + \overline{D}_{xy} = 0$ $\qquad$ $\overline{A}z + \overline{C}x + \overline{D}_{zx} = 0$ $\qquad$ $\overline{C}y + \overline{B}z + \overline{D}_{yz} = 0$

Then the following are their relationship:

$A : B : C \neq \overline{A} : \overline{B} : \overline{C}$	Line and plane intersect
$A : \overline{A} = B : \overline{B} = C : \overline{C}$	Line normal to the plane
$A\overline{A} + B\overline{B} + C\overline{C} = 0$	Line parallel to the plane

Note: The line lies in the plane if they have a common point and $A\overline{A} + B\overline{B} + C\overline{C} = 0$.

ii. Angle between a Straight Line and a Plane

α, β, γ = direction cosines, plane

$\overline{\alpha}, \overline{\beta}, \overline{\gamma}$ = direction cosines, line

$$\sin \tau = \alpha\overline{\alpha} + \beta\overline{\beta} + \gamma\overline{\gamma}$$

iii. Two Straight Lines in a Plane

$$\frac{x-x_i}{\bar{\alpha}_1}=\frac{y-y_i}{\bar{\beta}_1}=\frac{z-z_i}{\bar{\gamma}_1} \qquad \frac{x-x_j}{\bar{\alpha}_2}=\frac{y-y_j}{\bar{\beta}_2}=\frac{z-z_j}{\bar{\gamma}_2}$$

a. The lines are parallel if

$$\bar{\alpha}_1=\bar{\alpha}_2 \qquad \bar{\beta}_1=\bar{\beta}_2 \qquad \bar{\gamma}_1=\bar{\gamma}_2$$

b. The lines are normal if

$$\bar{\alpha}_1\bar{\alpha}_2+\bar{\beta}_1\bar{\beta}_2+\bar{\gamma}_1\bar{\gamma}_2=0$$

c. The lines are coplanar if

$$\begin{vmatrix} x_j-x_i & y_j-y_i & z_j-z_i \\ \bar{\alpha}_1 & \bar{\beta}_1 & \bar{\gamma}_1 \\ \bar{\alpha}_2 & \bar{\beta}_2 & \bar{\gamma}_2 \end{vmatrix}=\Delta=0$$

d. The angle between these lines is given by

$$\cos\tau=\bar{\alpha}_1\bar{\alpha}_2+\bar{\beta}_1\bar{\beta}_2+\bar{\gamma}_1\bar{\gamma}_2$$

e. The distance between these lines if $\Delta\neq 0$ is

$$d=\frac{\begin{vmatrix} x_j-x_i & y_i-y_i & z_j-z_i \\ \bar{\alpha}_1 & \bar{\beta}_1 & \bar{\gamma}_1 \\ \bar{\alpha}_2 & \bar{\beta}_2 & \bar{\gamma}_2 \end{vmatrix}}{\sqrt{\begin{vmatrix} \bar{\alpha}_1 & \bar{\alpha}_2 \\ \bar{\beta}_1 & \bar{\beta}_2 \end{vmatrix}^2+\begin{vmatrix} \bar{\beta}_1 & \bar{\beta}_2 \\ \bar{\gamma}_1 & \bar{\gamma}_2 \end{vmatrix}^2+\begin{vmatrix} \bar{\gamma}_1 & \bar{\gamma}_2 \\ \bar{\alpha}_1 & \bar{\alpha}_2 \end{vmatrix}^2}}$$

7.9 GENERAL EQUATION OF SURFACES OF THE SECOND DEGREE

The general equation for the second degree,

$$a_{11}x^2+a_{22}y^2+a_{33}z^2+2a_{12}xy+2a_{13}xz+2a_{23}yz+2a_{14}x+2a_{24}y+2a_{34}z+a_{44}=0$$

defines the following quadratic surfaces: a sphere, an ellipsoid, a hyperboloid, a paraboloid, a cone, a cylinder, two planes, a line, and a point (the last three are the degenerated surfaces of the second degree).

i. Invariants

The quantities ($aik = aki$)

$$J_4 = \begin{vmatrix} a_{11} & a_{12} & a_{13} & a_{14} \\ a_{21} & a_{22} & a_{23} & a_{24} \\ a_{31} & a_{32} & a_{33} & a_{34} \\ a_{41} & a_{42} & a_{43} & a_{44} \end{vmatrix} \qquad J_3 = \begin{vmatrix} a_{11} & a_{12} & a_{13} \\ a_{21} & a_{22} & a_{23} \\ a_{31} & a_{32} & a_{33} \end{vmatrix}$$

$$J_2 = \begin{vmatrix} a_{11} & a_{12} \\ a_{21} & a_{22} \end{vmatrix} + \begin{vmatrix} a_{11} & a_{13} \\ a_{31} & a_{33} \end{vmatrix} + \begin{vmatrix} a_{22} & a_{23} \\ a_{32} & a_{33} \end{vmatrix} \qquad J_1 = a_{11} + a_{22} + a_{33}$$

are invariants of the general equation; they remain unchanged under transformation of coordinates.

ii. Classification of Conic Section

Type		$J_3 \neq 0$		$J_3 = 0$
		Central CS		*Noncentral conic section*
		$J_3 J_1 > 0$ or $J_2 > 0$	$J_3 J_1$ or $J_2 > 0$	
$J_4 \neq 0$	$J_4 < 0$	Real ellipsoid	Hyperboloid of two sheets	Elliptic Paraboloid
	$J_4 > 0$	Imaginary ellipsoid	Hyperboloid of one sheet	Hyerbolic paraboloid
$J_4 = 0$		Imaginary cone	Real cone	Cylinder or two planes

7.10 TYPICAL SURFACES OF THE SECOND DEGREE

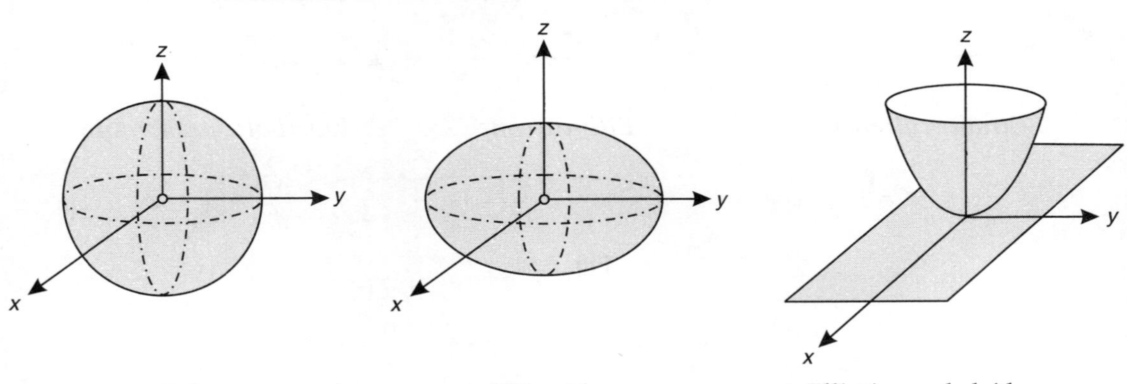

Sphere

$$\frac{x^2}{a^2} + \frac{y^2}{a^2} + \frac{z^2}{a^2} = 1$$

Ellipsoid

$$\frac{x^2}{a^2} + \frac{y^2}{b^2} + \frac{z^2}{c^2} = 1$$

Elliptic paraboloid

$$\frac{x^2}{a^2} + \frac{y^2}{b^2} = z$$

Fig. 7.8

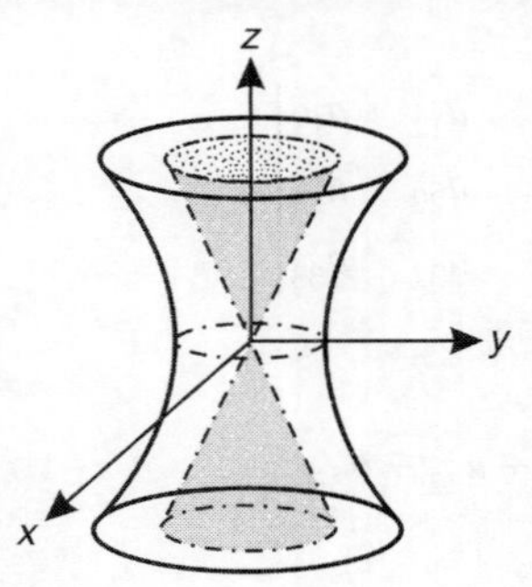

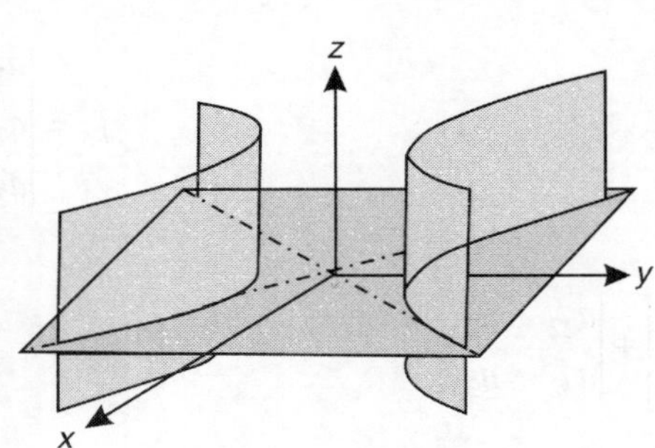

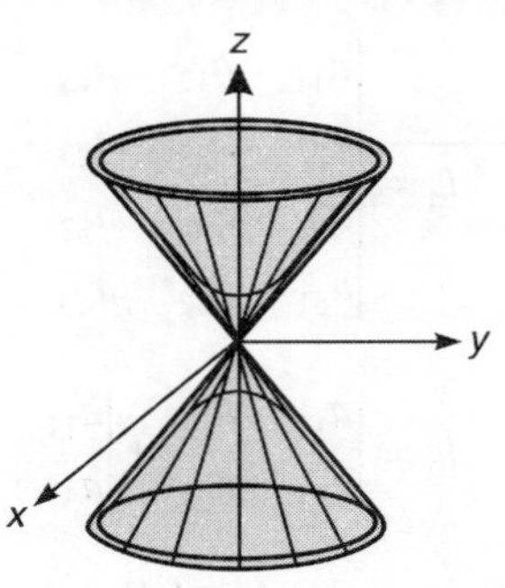

Elliptic hyperboloid *(one sheet)*

$$\frac{x^2}{a^2}+\frac{y^2}{b^2}-\frac{z^2}{c^2}=1$$

Hyperbolic cylinder

$$-\frac{x^2}{a^2}+\frac{y^2}{b^2}=1$$

Elliptic hyperboloid *(two sheets)*

$$\frac{x^2}{a^2}+\frac{y^2}{b^2}-\frac{z^2}{c^2}=-1$$

Fig. 7.9

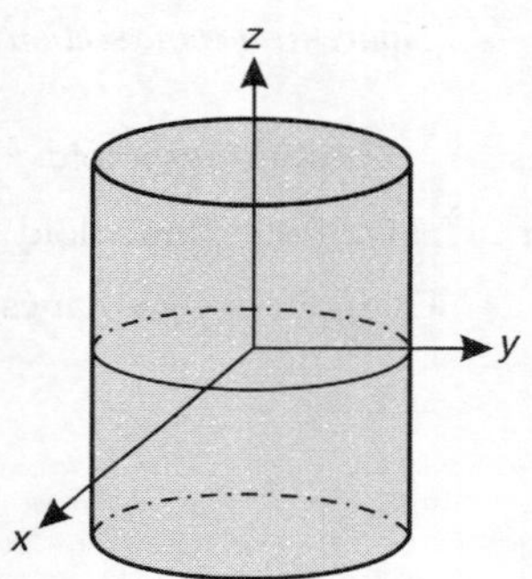

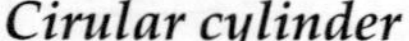

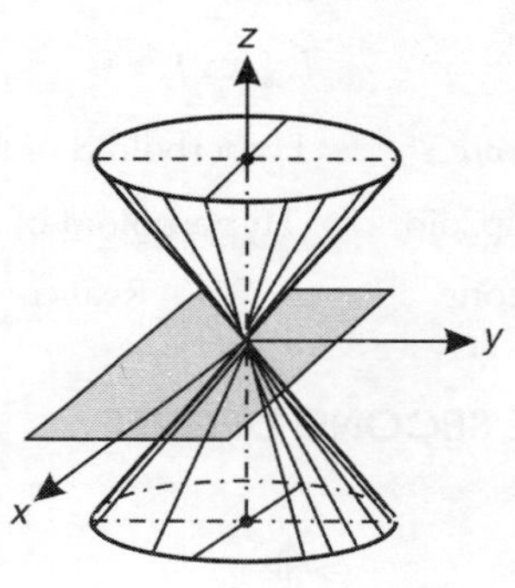

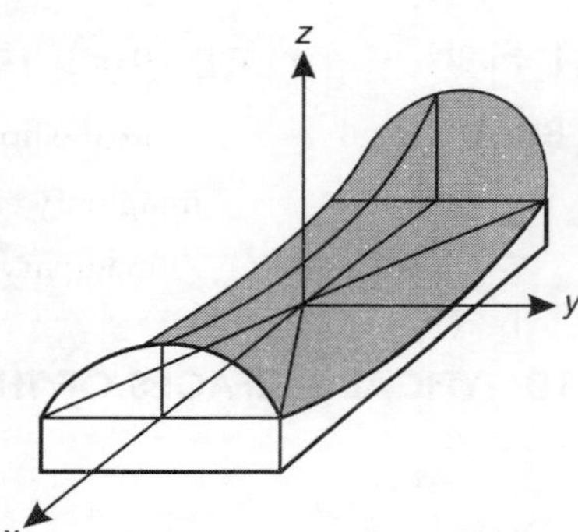

Cirular cylinder

$$\frac{x^2}{a^2}+\frac{y^2}{a^2}=1$$

Elliptic cone

$$\frac{x^2}{a^2}+\frac{y^2}{b^2}-\frac{z^2}{c^2}=0$$

Hyperbolic paraboloid

$$\frac{x^2}{a^2}-\frac{y^2}{b^2}=z$$

Fig. 7.10

CHAPTER

8

Differential Calculus

8.1 FUNCTIONS

i. Basic Terms

a. The quality which changes its value is called a *variable*, and the range of its variation is known as the *interval*.

b. The quality which remains unchanged is called a *constant*, and its range is a *single number*.

c. A *function* is a relationship between two or more *variables* and *constants*.

ii. Definitions

a. A variable y is a function of another variable x, $y = f(x)$;
if for each value in the range of x there are one or more values in the range of y.
x = independent variable or argument
y = dependent variable

b. A variable y is a function of several variables $x_1, x_2,..., x_n$, $y = f(x_1, x_2, ..., x_n)$;
if for each set of values in the range of x's, there is one or more values in the range of y.

iii. Forms

a. The three most common forms of representation of a function are *tabular*, *graphical*, and *analytical*.

b. Analytical representation
Explicit: $y = f(x_1, x_2,..., x_n)$
Implicit: $0 = g(x_1, x_2,..., x_n, y)$
Parametric: $x_1 = h_1(\tau), x_2 = h_2(\tau),..., y = h(\tau)$

iv. Characteristic Properties

a. *Single-valued*

A function is single-valued if it has a single value y for any given value of x.

b. *Multivalued*

A function is multivalued if it has more than one value of y for any given value of x.

c. *Even and odd*

A function is even if $f(-x) = f(x)$ and odd if $f(-x) = -f(x)$.

d. *Periodic*

A function is periodic with the period T if $f(x + T) = f(x)$.

e. *Inverse*

An inverse of a function $y = f(x)$ is another function $x = g(y)$.

v. Interval

$a < x < b$	Bounded open interval	(a, b)
$a < x$	Unbounded open interval	(a, ∞)
$x < b$	Unbounded open interval	$(-\infty, b)$
$a \leq x \leq b$	Bounded dosed interval	$[a, b]$
$a \leq x$	Unbounded closed interval	$[a, \infty]$
$x \leq b$	Unbounded closed interval	$[-\infty, b]$

8.2 LIMIT AND CONTINUITY

i. Definition of Limit

A variable x is said to have a limit a ($\lim x = a$ or $x \to a$) as x takes on consecutively the values $x_1, x_2, x_3, \ldots, x_n$ if for every positive number ε, however small, the numerical value of $|x - a| < \varepsilon$.

A function $f(x)$ is said to have a limit b $\left(\lim\limits_{x \to a} f(x) = b\right)$ as x takes on consecutively the values $x_1, x_2, x_3, \ldots, x_n$ and approaches a, without assuming the value of a, if for every positive number δ, however small, the numerical value of

$$|f(x) - b| < \varepsilon$$

ii. Operations with Limit

$$\lim_{x \to a} [f(x) + g(x)] = \lim_{x \to a} f(x) + \lim_{x \to a} g(x)$$

$$\lim_{x \to a} [f(x)\, g(x)] = \lim_{x \to a} f(x) \lim_{x \to a} g(x)$$

$$\lim_{x \to a} \frac{f(x)}{g(x)} = \frac{\lim\limits_{x \to a} f(x)}{\lim\limits_{x \to a} g(x)} \qquad \left[\lim_{x \to a} g(x) \neq 0\right]$$

iii. Special Cases

$$\lim_{x \to 0} a^x = 1, \quad a > 0 \qquad \lim_{x \to 0} (1+x)^{1/x} = e \qquad \lim_{x \to 0} \frac{e^x - 1}{x} = 1$$

$$\lim_{x \to \infty} \sqrt[x]{x} = 1 \qquad \lim_{x \to \infty} \left(1 + \frac{y}{x}\right)^x = e^y \qquad \lim_{x \to \infty} \frac{a^x - 1}{x} = \ln a, \quad a > 0$$

$$\lim_{m \to \infty} \frac{a^m}{m!} = 0 \qquad \lim_{x \to \infty} \left(1 + \frac{1}{x}\right)^x = e \qquad \lim_{x \to \infty} \frac{x^m}{e^x} = 0$$

$$\lim_{x \to \infty} \frac{(\ln x)^m}{m} = 0 \qquad \lim_{x \to 1} \frac{x - 1}{\ln x} = 1 \qquad \lim_{x \to \infty} \frac{\ln (x+1)}{x} = 1$$

$$\lim_{x \to 0} \frac{\sin x}{x} = 1 \qquad \lim_{x \to 0} \frac{1 - \cos x}{x} = 0 \qquad \lim_{x \to 0} \frac{\tan x}{x} = 1$$

$$\lim_{m \to \infty} \left(1 + \frac{1}{2} + \frac{1}{3} + \ldots + \frac{1}{m} - \ln m\right) = 0.57721\ldots \text{ (Euler's constant)}$$

$$\lim_{m \to \infty} \frac{m!}{m^m e^{-m} \sqrt{m}} = \sqrt{2\pi} \qquad \text{(Stirling's formula)}$$

iv. Continuity of Function

a. A single valued function is *continuous throughout the neighborhood* of $x = a$ if and only if $\lim_{x \to 0} f(x)$ exists and is equal to $f(a)$.

b. A single valued function is *continuous in an interval* (a, b) or $[a, b]$ if and only if it is continuous at each point of this interval.

c. A single valued function has a *discontinuity of the first kind* at the point $x = a$ if $f(a + 0) \neq f(a - 0)$. The difference of these two values is known as the status of $f(x)$.

d. A single valued function is *piecewise-continuous* on a given interval (a, b) or $[a, b]$ if and only if $f(x)$ is continuous throughout this interval except for a finite number of discontinuities of the first kind.

8.3 DERIVATIVE OF A FUNCTION

i. Definition

a. First derivative of $y = f(x)$ with respect to x is defined as

$$\tan \phi = \frac{dy}{dx} = \lim_{\Delta x \to 0} \frac{\Delta y}{\Delta x} = \lim_{\Delta x \to 0} \frac{f(x + \Delta x) - f(x)}{\Delta x}$$

Alternative notations are $f'(x)$, $\frac{df(x)}{dx}$, and y'.

If $\quad y = f(t)$

$$\frac{dy}{dt} = \frac{df(t)}{dt} = y$$

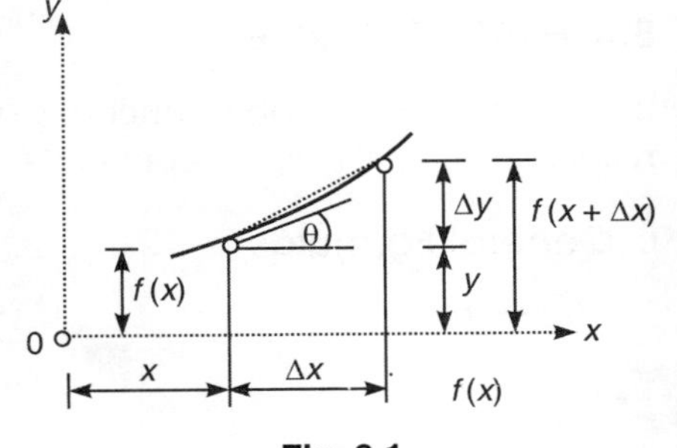

Fig. 8.1

b. *Second and higher derivatives* of the same function are

$$\frac{d^2y}{dx^2} = \frac{d}{dx}\left(\frac{dy}{dx}\right) = \frac{d}{dx}[f'(x)] = y'' \qquad \frac{d^n y}{dx^n} = \frac{d}{dx}\left(\frac{d^{n-1}y}{dx}\right) = \frac{d}{dx}[f^{(n-1)}(x)] = f^{(n)}(x)$$

c. *First partial derivatives of* $y = f(x_1, x_2, ..., x_i, x_j, ...)$ with respect to one of the independent variable x_i or x_j are

$$\frac{\partial y}{\partial x_i} = \frac{\partial f(\ )}{\partial x_i} = F_i \qquad \frac{\partial y}{\partial x_j} = \frac{\partial f(\ )}{\partial x_j} = F_j$$

Thus there are as many possible first partial derivatives as there are independent variables.

d. *Second and higher derivatives are*

$$F_{jj} = \frac{\partial F_j}{\partial x_j} \qquad F_{ij} = \frac{\partial F_j}{\partial x_i} \qquad F_{ji} = \frac{\partial F_i}{\partial x_j} \qquad F_{ii} = \frac{\partial F_i}{\partial x_i}$$

The same process defines derivatives of any order. When the highest derivatives involved are continuous, the result is independent of the order in which the differentiation is performed.

$$F_{ijj} = F_{jij} = F_{jji} \qquad F_{ij} = F_{ji}$$

The number of differentiations performed is the *order of the partial derivatives.*

ii. Rules

$$y = f(x_1) \qquad x_1 = g(x_2) \qquad x_2 = h(x_3) \qquad y = f(x) \qquad x = g(y) \qquad \frac{dx}{dy} \neq 0$$

$$\frac{dy}{dx_3} = \frac{dy}{dx_1}\frac{dx_1}{dx_2}\frac{dx_2}{dx_3} \qquad \frac{dy}{dx} = \frac{1}{dx/dy} \qquad \frac{d^2y}{dx^2} = \frac{-d^2x/dy^2}{(dx/dy)^3}$$

$F(x, y) = 0 \quad F_y \neq 0$

$$\frac{dy}{dx} = -F_x : F_y \qquad \frac{d^2y}{dx^2} = -(F_{xx}F_y^2 - 2F_xF_{xy}F_y + F_x^2 F_{yy}) : F_y^3$$

$$x = x(t); \quad y = y(t) \qquad \underline{\dot{x}}(t) = \frac{dx}{dt} \neq 0; \ \underline{\dot{y}}(t) = \frac{dy}{dt} \neq 0$$

$$\frac{dy}{dx} = \frac{\dot{y}(t)}{\dot{x}(t)} \qquad \frac{d^2y}{dx^2} = \frac{\dot{x}(t)\ddot{y}(t) - \ddot{x}(t)\dot{y}(t)}{[\dot{x}(t)]^2}$$

8.4 FIRST DERIVATIVES

u, v, w are differentiable functions of x; u', v', w' are first derivatives of these functions with respect to x; a, b, c, m are constants.

i. General Formulas

$$(a)' = 0 \qquad (uv)' = u'v + uv'$$

$$(au)' = au' \qquad \left(\frac{u}{v}\right)' = \frac{u'v - uv'}{v^2}$$

$$(u + v + w + ...)' = u' + v' + w' + ...$$

$$(uvw...)' = (uvw...)\left(\frac{u'}{u} + \frac{v'}{v} + \frac{w'}{w} + ...\right)$$

$$\left(\frac{uv}{w}\right)' = \left(\frac{uv}{w}\right)\left(\frac{u'}{u} + \frac{v'}{v} - \frac{w'}{w}\right)$$

$$(u^v)' = vu^{v-1}\, u' + u^v\, v' \ln u$$

ii. Algebraic Functions

$$(au^m)' = amu^{m-1}\, u'$$

$$\left(\frac{a}{u^m}\right)' = -\frac{am}{u^{m+1}}\, u'$$

$$(a\sqrt{u})' = \frac{a}{2\sqrt{u}}\, u'$$

$$\left(\frac{a}{\sqrt{u}}\right)' = \frac{-a}{2\sqrt{u^3}}\, u'$$

$$(a\sqrt[m]{u})' = \frac{a\sqrt[m]{u}}{mu^{m-1}}\, u'$$

$$\left(\frac{a}{\sqrt[m]{u}}\right)' = -\left(\frac{au'}{m\mu\sqrt[m]{u}}\right)$$

iii. Exponential Functions (e = 2.71828...)

$$(e^{mu})' = me^{mu}\, u'$$

$$(e^{-mu})' = -\, me^{-mu}\, u'$$

$$(a^{mu})' = (ma^{mu} \ln a)\, u'$$

$$(e^{-mu})' = -\,(m \ln a)\, a^{-mu}\, u'$$

$$(u^{mu})' = nu^{mu}\,(1 + \ln u)\, u'$$

$$(u^{-mu})' = -\,\frac{m(1+\ln u)\, u^{-mu} u'}{\ln u}$$

iv. Logarithmic Functions

$$(\ln au)' = \frac{u'}{u}$$

$$\left(\ln \frac{a}{u}\right)' = -\frac{u'}{u}$$

$$(\log u)' = \frac{u'}{u} \log e$$

$$\left(\log \frac{1}{u}\right)' = -\frac{u'}{u} \log e$$

$$(\log au)' = \frac{u'}{u} \log e$$

$$\left(\log \frac{a}{u}\right)' = -\frac{u'}{u} \log e$$

8.5 FIRST DERIVATIVES

u = differentiable function of x

u' = first derivative of u with respect to x

i. Trigonometric Functions

$$(\sin u)' = u' \cos u$$

$$\left(\frac{1}{\sin u}\right)' = -\frac{u' \cos u}{\sin^2 u}$$

$$(\cos u)' = -\, u' \sin u$$

$$\left(\frac{1}{\cos u}\right)' = \frac{u' \sin u}{\cos^2 u}$$

$$(\tan u)' = \frac{u'}{\cos^2 u}$$

$$\left(\frac{1}{\tan u}\right)' = -\frac{u'}{\sin^2 u}$$

$$(\cot u)' = -\frac{u'}{\sin^2 u} \qquad \left(\frac{1}{\cot u}\right)' = \frac{u'}{\cos^2 u}$$

$$(\sec u)' = \frac{u' \sin u}{\cos^2 u} \qquad \left(\frac{1}{\sec u}\right)' = -u' \sin u$$

$$(\csc u)' = -\frac{u' \cos u}{\sin^2 u} \qquad \left(\frac{1}{\csc u}\right)' = u' \cos u$$

ii. Inverse Trigonometric Functions (v = principal value)

$$(\sin^{-1} u)' = \frac{u'}{\sqrt{1-u^2}} \qquad (\cos^{-1} u)' = -\frac{u'}{\sqrt{1-u^2}}$$

$$(\tan^{-1} u)' = \frac{u'}{1+u^2} \qquad (\cot^{-1} u)' = \frac{-u'}{1+u^2}$$

iii. Hyperbolic Functions

$$(\sinh u)' = u' \cosh u \qquad (\cosh u)' = u' \sinh u$$

$$(\tanh u)' = \frac{u'}{\cosh^2 u} \qquad (\coth u)' = -\frac{u'}{\sinh^2 u}$$

$$(\text{sech } u)' = -\frac{u' \sinh u}{\cosh^2 u} \qquad (\text{csch } u)' = \frac{-u' \cosh u}{\sinh^2 u}$$

iv. Inverse Hyperbolic Functions (u = principal value)

$$(\sinh^{-1} u)' = \frac{u'}{\sqrt{u^2+1}} \qquad (\cosh^{-1} u)' = \frac{u'}{\sqrt{u^2-1}}$$

$$(\tanh^{-1} u)' = \frac{u'}{1-u^2} \qquad (\coth^{-1} u)' = -\frac{u'}{1-u^2}$$

8.6 TABLE OF HIGHER DERIVATIVES OF FUNCTIONS

(a, k, m = constants; n = order of derivative)

i. Algebraic Functions

y	$y^{(n)} = \dfrac{d^n y}{dx^n}$
ax^m	$am\,(m-1)\,(m-2)\,...\,(m-n+1)\,x^{m-n}$
ax^{-m}	$(-1)^n\,am\,(n+1)\,(m+2)\,...\,(m+n-1)\,\dfrac{1}{x^{m+n}}$
$\sqrt[m]{ax}$	$(-1)^{n-1}\,\dfrac{(m-1)\,(2m-1)\,...\,[(n-1)\,m-1]}{m^n\,\sqrt[m]{x^{mn-1}}}\,\sqrt[m]{a}$

ii. Exponential and Logarithmic Functions

y	$y^{(n)} = \dfrac{d^n y}{dx^n}$	y	$y^{(n)} = \dfrac{d^n y}{dx^n}$
e^x	e^x	a^x	$(\ln a)^n a^x$
a^{kx}	$k^n e^{kx}$	a^{kx}	$(k \ln a)^n a^{kx}$
$\ln x$	$\dfrac{(-1)^{n-1}(n-1)!}{x^n}$	$\log x$	$\dfrac{(-1)^{n-1}(n-1)! \log e}{x^n}$
$\ln kx$	$\dfrac{(-1)^{n-1}(n-1)!}{x^n}$	$\log kx$	$\dfrac{(-1)^{n-1}(n-1)! \log e}{x^n}$

iii. Trigonometric and Hyperbolic Functions

$\sin x$	$\sin\left(x + \dfrac{n\pi}{2}\right)$	$\cos x$	$\cos\left(x + \dfrac{n\pi}{2}\right)$
$\sin kx$	$k^n \sin\left(kx + \dfrac{n\pi}{2}\right)$	$\cos kx$	$k^n \cos\left(kx + \dfrac{n\pi}{2}\right)$
$\sinh x$	$\begin{cases}\sinh x & (n \text{ even}) \\ \cos x & (n \text{ odd})\end{cases}$	$\cosh x$	$\begin{cases}\cosh x & (n \text{ even}) \\ \sinh x & (n \text{ odd})\end{cases}$
$\sinh kx$	$\begin{cases}k^n \sinh kx & (n \text{ even}) \\ k^n \cosh kx & (n \text{ odd})\end{cases}$	$\cosh kx$	$\begin{cases}k^n \cosh kx & (n \text{ even}) \\ k^n \sinh kx & (n \text{ odd})\end{cases}$

iv. Product of Two Functions

If u, v are differentiable functions of x and $\dfrac{d^n u}{dx^n} = u^{(n)}$ $\dfrac{d^n v}{dx^n} = v^{(n)}$

are the nth derivatives of u, v respectively, then

$$\frac{d^n (nv)}{dx^n} = nv^{(n)} + \binom{n}{1} u^{(1)} v^{(n-1)} + \binom{n}{2} u^{(2)} v^{(n-2)} + \ldots + u^{(n)} v$$

8.7 THEOREMS OF DIFFERENTIAL CALCULUS

i. Rolle's Theorem

If a function $f(x)$ is continuous in the closed interval $[a, b]$ and is differentiable in the open interval (a, b) and if $f(a) = f(b)$, then there is at least one point $(x = c)$ in (a, b) in which

$$f'(c) = 0$$

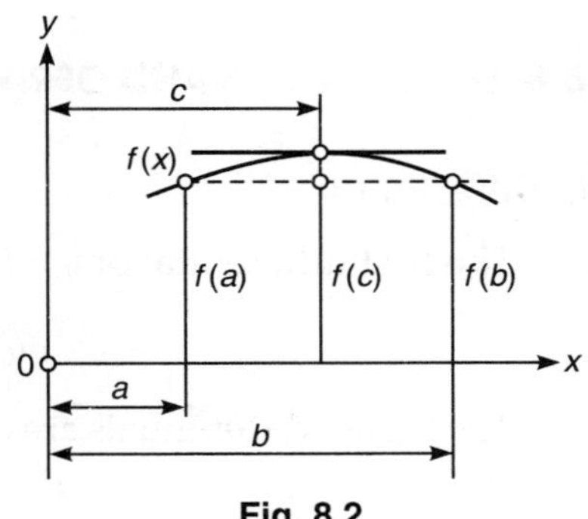

Fig. 8.2

ii. Lagrange's Theorem (First Mean-Value Theorem)

If a function $f(x)$ is continuous in the closed interval $[a, b]$ and is differentiable in the open interval (a, b), then there is at least one point $(x = c)$ in which

$$\frac{f(b) - f(a)}{b - a} = f'(c)$$

iii. Cauchy's Theorem (Second Mean-Value Theorem)

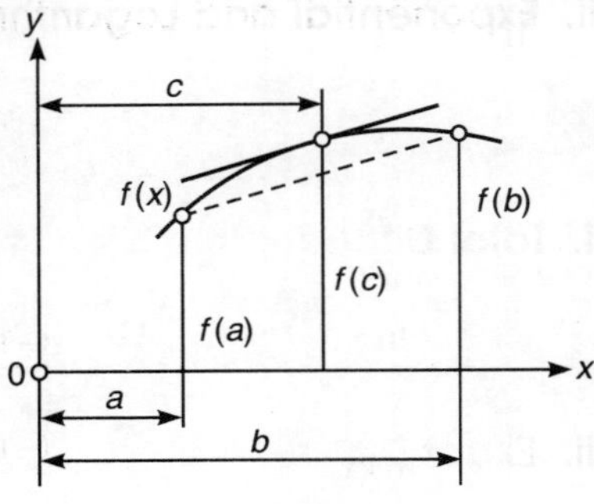

Fig. 8.3

If the function $f(x)$ and $g(x)$ are continuous in the closed interval $[a, b]$ and are differentiable in the open interval (a, b), if $f(x)\, g(x)$ are not simultaneously equal to zero at any point of this open interval, and g if $(a) \neq g(b)$, then there is at least one point $(x = c)$ in (a, b) in which

$$\frac{f(b)-f(a)}{g(b)-g(a)} = \frac{f'(c)}{g'(c)}$$

iv. L'Hospital's Rules

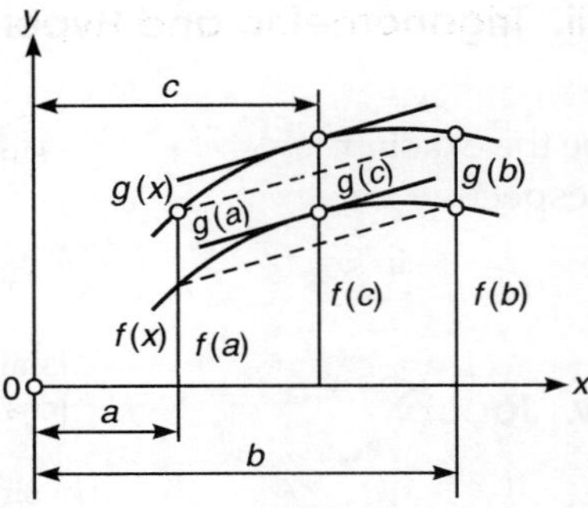

Fig. 8.4

$f(x)$ and $g(x)$ are two continuous functions of x having continuous derivatives at x.

If $f(x) / g(x)$ for $x = a$ is $0/0$ or ∞/∞, then

$$\lim_{x \to a} \frac{f(x)}{g(x)} = \lim_{x \to a} \frac{f'(x)}{g'(x)}$$

If $f(x) - g(x)$ for $x = a$ is $\infty - \infty$, then

$$\lim_{x \to a} [f(x) - g(x)] = \lim_{x \to a} \frac{[1/f(x) - 1/g(x)]'}{[1/f(x)]'\,[1/g(x)]'}$$

If $f(x)\, g(x)$ for $x = a$ is $(0)\,(\infty)$ or $(\infty)\,(0)$, then

$$\lim_{x \to a} [f(x)\, g(x)] = \lim_{x \to a} \frac{f'(x)}{[1/g(x)]'}$$

If $f(x)\, g(x)$ for $x = a$ is 0^0 or ∞^0 or $1^{-\infty}$, then

$$\lim_{x \to a} f(x)^{g(x)} = \lim_{x \to a} e^{g(x)\ln f(x)}$$

v. Leibnitz's Theorem

$$(uv)_n = u_n v + nc_1\, u_{n-1}\, v_1 + nc_2\, u_{n-2}\, v_2 + \ldots + uv_n$$

vi. Euler's Theorem on Homogeneous Functions

If $z = f(x,y)$ is a homogeneous functions of x and y of degree n, then $x\frac{\partial z}{\partial x} + y\frac{\partial z}{\partial y} = nz$

8.8 DIFFERENTIALS AND DERIVATIVES

i. Differential

The first differential of $y = f(x)$ is

$$dy = \frac{dy}{dx}\, dx = df(x) = f'(x)\, dx = y'\, dx$$

The higher differentials are obtained by successive differentiations of the first differential.

$$d^2 y = f''(x)\, dx^2 \qquad d^n y = f^{(n)}(x)\, dx^n$$

If $x = x(t)$ and $y = y(t)$, then

$$dx = \frac{\partial[x(t)]}{\partial t} dt = \dot{x}\, dt \qquad d\dot{y} = \frac{\partial[y(t)]}{\partial t} dt = \dot{y}\, dt$$

ii. Total Differential of $z = f(x, y), x = x(t), y = y(t)$

$$dz = \frac{\partial z}{\partial x} dx + \frac{\partial z}{\partial y} \partial y \qquad dz = \frac{\partial z}{\partial x}\frac{\partial x}{\partial t} dt + \frac{\partial z}{\partial y}\frac{\partial y}{\partial t} dt$$

iii. Exact Differential

In order that

$$dF[x, y] = A(x \cdot y)\, dx + B(x \cdot y)\, dy$$
$$dF[x, y, z] = A(x, y, z)\, dx + B(x, y, z)\, dy + C(x, y, z)\, dz$$

be the exact differentials, the following (necessary and sufficient) conditions must be satisfied, respectively:

$$\frac{\partial A}{\partial y} = \frac{\partial B}{\partial x} \qquad \frac{\partial B}{\partial z} = \frac{\partial C}{\partial y} \qquad \frac{\partial C}{\partial x} = \frac{\partial A}{\partial z}$$

iv. Jacobian Determinant

If $x = x(t)$, $y = y(t)$, and $z = z(t)$,

$$A(x, y, z, t) = 0 \qquad B(x, y, z, t) = 0 \qquad C(x, y, z, t) = 0$$

and if the jacobian determinant

$$J\left[\frac{A, B, C}{x, y, z}\right] = \begin{vmatrix} \frac{\partial A}{\partial x} & \frac{\partial A}{\partial y} & \frac{\partial A}{\partial z} \\ \frac{\partial B}{\partial x} & \frac{\partial B}{\partial y} & \frac{\partial B}{\partial z} \\ \frac{\partial C}{\partial x} & \frac{\partial C}{\partial y} & \frac{\partial C}{\partial z} \end{vmatrix} = \frac{\partial(A, B, C)}{\partial(x, y, z)} \neq 0$$

then the derivatives and the differentials of A, B, C are given by

$$\begin{bmatrix} \frac{\partial A}{\partial t} \\ \frac{\partial B}{\partial t} \\ \frac{\partial C}{\partial t} \end{bmatrix} = \begin{bmatrix} \frac{\partial A}{\partial x} & \frac{\partial A}{\partial y} & \frac{\partial A}{\partial z} \\ \frac{\partial B}{\partial x} & \frac{\partial B}{\partial y} & \frac{\partial B}{\partial z} \\ \frac{\partial C}{\partial x} & \frac{\partial C}{\partial y} & \frac{\partial C}{\partial z} \end{bmatrix} \begin{bmatrix} \frac{\partial x}{\partial t} \\ \frac{\partial y}{\partial t} \\ \frac{\partial z}{\partial t} \end{bmatrix}$$

The higher derivatives and differentials can be found in the same manner.

8.9 ANALYSIS OF A FUNCTION

a. *The rates of change of* $f(x)$ at $x = x$ are

$f'(x) > 0$; $f(x)$ is increasing, $f'(x) < 0$; $f(x)$ is decreasing

$f'(x) = 0$; $f(x)$ has a tangent parallel to the x axis

b. *The curvature of* $f(x)$ at $x = x$ is as follows:

$f''(x) > 0$; $f(x)$ is convex, $f''(x) < 0$; $f(x)$ is concave

$f''(x) = 0$ and changes in sign; $f(x)$ has an inflection point

$f''(x) = 0$ and does not change in sign; $f(x)$ has a flat point

c. *The necessary condition* for a maximum or a minium of $f(x)$ at $x = a$ is $f'(a) = 0$

d. *Sufficient conditions* for extrema of $f(x)$ at $x = a$ are

$f'(a) = 0 \quad f''(a) < 0$ (maximum)

$f'(a) = 0 \quad f'' > 0$ (maximum)

e. Necessary and sufficient conditions for extrema of $f(x)$ at $x = a$ are if $f'(a) = f''(a) = \ldots = f^{(n-1)}(a) = 0$ and $f^{(n)}(a) \neq 0$, then

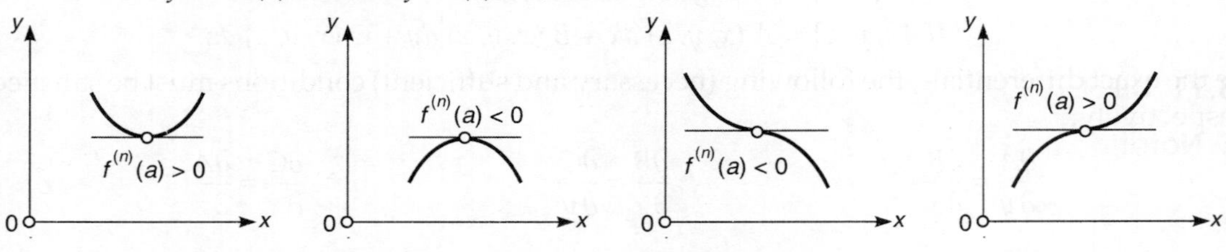

Fig. 8.5

f. The equation of the tangent to a curve at point $P(x,y)$ on it is

$$Y - y = \left(\frac{dy}{dx}\right)(X - x)$$

g. The equation of the normal to a curve at a point $P(x,y)$ on it is

$$Y - y = \left(-\frac{dx}{dy}\right)(X - x)$$

h. The angle ϕ between the radius vector and tangent is given by $\tan\phi = r\dfrac{d\theta}{dr}$

i. The double points of a curve $f(x,y) = 0$ are obtained by solving

$$f(x,y) = 0, \quad \frac{\partial f}{\partial x} = 0, \quad \frac{\partial f}{\partial y} = 0$$

j. If $P(a, b)$ is a double point on the curve $f(x,y) = 0$ then it is a node, cusp, conjugate point accordingly as

$$\left[\left(\frac{\partial^2 f}{\partial x\,\partial y}\right)^2 - \frac{\partial^2 f}{\partial x^2}\frac{\partial^2 f}{\partial y^2}\right]_{P(a,b)} > 0, = 0 \text{ or } < 0$$

k. Radius of curvature of $y = f(x)$ is given by $P = \dfrac{\left(1+(y')^2\right)^{\frac{3}{2}}}{y''}$

l. Radius of curvature of $r = f(\theta)$ is given by

$$\rho = \frac{\left(r^2 + (r')^2\right)^{\frac{3}{2}}}{r^2 + 2(r')^2 - r(r'')} \text{ where } r' = \frac{dr}{d\theta}, r'' = \frac{d^2 r}{d\theta^2}.$$

8.10 DIRECTION DERIVATIVES

a. If $f = f(x, y)$, the derivative of f in the s direction is

$$\frac{\partial f}{\partial s} = \frac{\partial f}{\partial x}\cos\alpha + \frac{\partial f}{\partial y}\cos\beta$$

in which α and β are the angles between the s direction and the positive directions of the X and Y-axes respectively.

b. If $f = f(x, y, z)$, the direction derivative is analogically,

$$\frac{\partial f}{\partial s} = \frac{\partial f}{\partial x}\cos\alpha + \frac{\partial f}{\partial y}\cos\beta + \frac{\partial f}{\partial z}\cos\gamma$$

8.11 DIFFERENTIAL GEOMETRY OF A PLANE CURVE

i. Notation

S = point of contact
ds = element arc
C = center of curvature
ρ = radius of curvature
x, y = coordinates of contact point
X, Y = coordinates of running point
$\overline{AS}$ = tangent
$\overline{DS}$ = normal

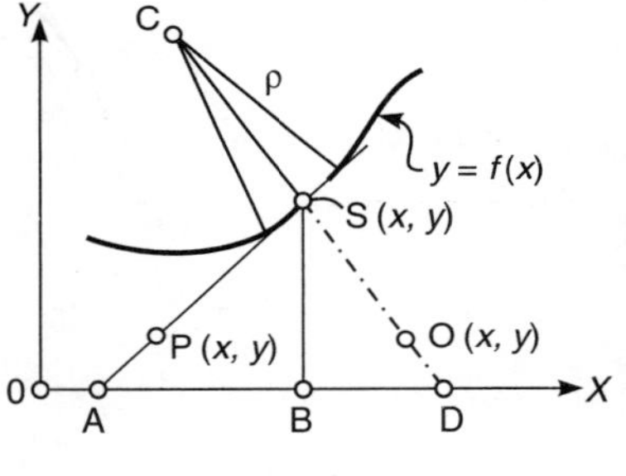

Fig. 8.6

ii. Basic Equations

Curve	$y = f(x)$	$F(x,y) = 0$	$x = x(f) \quad y = y(r)$
Derivatives	$y' = \dfrac{dy}{dx}$	$y' = F_y$	$\dot{x} = \dfrac{dx}{dt} \quad \dot{y} = \dfrac{dy}{dt}$
Arc length ds	$\sqrt{1+(y')^2}\,dx$	$\sqrt{1+\left(\dfrac{F_x}{F_y}\right)^2}\,dx$	$\sqrt{(\dot{x})^2+(\dot{y})^2}\,dt$
Tangent at S	$Y - y = y'(X - x)$	$F_x(X - x) = -F_y(Y - y)$	$(Y - y)\ \dot{x} = (X - x)\,\dot{y}$
Normal at S	$X - x = -y'(Y - y)$	$F_x(Y - y) = -F_y(X - x)$	$(X - x)\ \dot{x} = -(Y - y)\ \dot{y}$
Length of $\overline{AS}$	$\dfrac{y}{y'}\sqrt{1+(y')^2}$	$\dfrac{y}{F_x}\sqrt{F_x^2+F_y^2}$	$y\dfrac{\sqrt{(\dot{x})^2+(\dot{y})^2}}{\dot{y}}$
Length of $\overline{DS}$	$y\sqrt{1+(y')^2}$	$\dfrac{y}{F_y}\sqrt{F_x^2+F_y^2}$	$y\dfrac{\sqrt{(\dot{x})^2+(\dot{y})^2}}{\dot{x}}$

Coordinates of centre of curvature	X_C	$x-\dfrac{y'[1+(y')^2]}{y''}$	$x+F_x\dfrac{F_x^2+F_y^2}{R}$ †	$x-\dot{y}\dfrac{(\dot{x})^2+(\dot{y})^2}{Q}$ ‡
	Y_C	$y+\dfrac{1+(y')^2}{y''}$	$y+F_y\dfrac{F_x^2+F_y^2}{R}$ †	$y+\dot{x}\dfrac{(\dot{x})^2+(\dot{y})^2}{Q}$ ‡
Radius of curvature		$\dfrac{\left[\sqrt{1+(y')^2}\right]^3}{y''}$	$\dfrac{\left[\sqrt{F_x^2+F_y^2}\right]^3}{R}$ †	$\dfrac{\sqrt{(\dot{x})^2+(\dot{y})^2}}{Q}$ ‡

† $R = F_x\begin{vmatrix} F_{xy} & F_x \\ F_{yy} & F_y \end{vmatrix} - F_y\begin{vmatrix} F_{xx} & F_x \\ F_{yx} & F_y \end{vmatrix}$ ‡ $Q = \begin{vmatrix} \dot{x} & \dot{y} \\ \ddot{x} & \ddot{y} \end{vmatrix}$

iii. Shape of a Curve

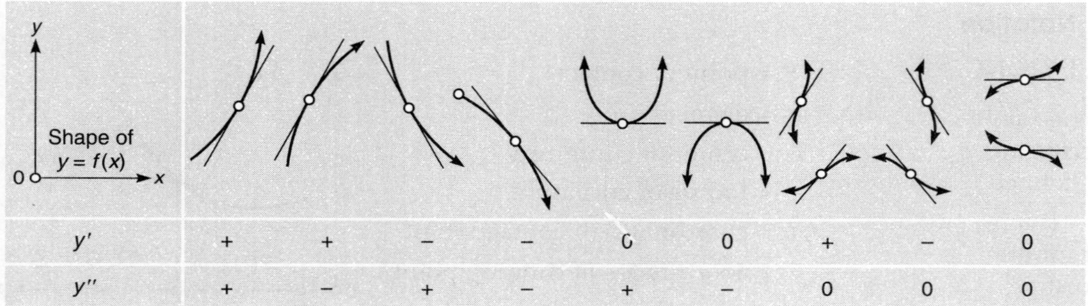

y
Shape of
y = f(x)
0
x
y′ + + − − 0 0 + − 0
y″ + − + − + − 0 0 0

Fig. 8.7

CHAPTER

9

Sequences and Series

9.1 CONCEPT AND DEFINITIONS

A **sequence** is a set of numbers u_1, u_2, u_3,... arranged in a prescribed order and formed according to a definite rule. Each member of the sequence is called a *term*, and the sequence is defined by the number of terms as *finite* or *infinite*.

It can be defined as a function whose domain is either the set of natural numbers (in case of infinite sequences) or the set of first natural numbers (when the sequence in finite). The position of an element in a sequence is its rank or index. It is denoted as $<a_n>$, where a_n is the nth term of the sequences. The number of elements of the sequence determines its length.

i. A sequences $<a_n>$ is said to be *monotonically increasing*, if each term is greater than or equal to the proceeding term, i.e. $a_{n+1} \geq a_n$ for every $n \in N$.

ii. A sequence $<a_n>$ is said to be *strictly monotonically increasing* if $a_{n+1} > a_n$ for every $n \in N$.

iii. A sequence is monotonically decreasing if $a_{n+1} \leq a_n$ for every $n \in N$.

iv. A sequences $<a_n>$ is *striclty monotonically decreasing* if $a_{n+1} < a_n$ for every $n \in N$.

v. A sequence is said to be a *monotonic sequences* if it is either monotonically increasing or decreasing.

vi. A sequence is said to be bounded above if there exist a real number M such that $a_n \leq M \forall n \in N$.

vii. A sequence is bounded below if there exist a real number m such that $a_n \geq m$ $\forall$ $n \in N$.

viii. A sequence is bounded if there exists a real number k such that $|a_n| \leq k \forall n \in N$, equivalently $-k \leq a_n \leq k \forall n \in N$. In other words, a sequence $<a_n>$ is said to be bounded if it is both bounded above and below.

ix. A subsequence of a given sequence is a sequence formed from the given sequence by deleting some of the elements without disturbing the relative portions of the remaining elements.

x. A sequence $<a_n>$ is said to be convergent if there exists a limit L such that the remaining a_n's are arbitrary close to L for some n (large enough).

xi. A sequence $<a_n>$ is said be divergent if it gets arbitrarily large (or small) as $n \to \infty$.

Some Important Results on Sequence

i. Let $<a_n>$ and $<b_n>$ be convergent sequences and c be any scalar. If $<a_n>$ and $<b_n>$ converges to a and b respectively then the sequences $<a_n + b_n>$, $<a_n - b_n>$, $<ca_n>$, $<an.bn>$ are all convergent and they converge to $a + b$, $a - b$, ca, ab respectively.

ii. *Sandwich theorem*: Suppose $<x_n>$, $<y_n>$ and $<z_n>$ are sequences such that $x_n \leq y_n \leq z_n \ \forall\ n$ and that $x_n \to x_o$ and $z_n \to x_o$ then $y_n \to x_o$.

iii. Let $<x_n>$ be a sequence of reals such that $x_n > 0 \ \forall\ n$ and $\lim\limits_{n\to\infty} \dfrac{x_{n+1}}{x_n} = \lambda$.

Then, if $\lambda < 1$ then $\lim\limits_{n\to\infty} x_n = 0$ and if $\lambda > 1$ then $\lim\limits_{n\to\infty} x_n = \infty$.

a. *An infinite series*

$$\sum_{i=1}^{\infty} u_i = u_1 + u_2 + u_3 + \ldots$$

is the sum of an infinite sequence.

If $$S_n = \sum_{i=1}^{n} u_i \quad \text{and} \quad \lim_{n\to\infty} s_n = S$$

exits, the series is called *convergent*, and S is the sum. The infinite series which does not converge is *divergent*. An infinite series is *absolutely convergent* if

$$\sum_{i=1}^{\infty} |u_i| = |S|$$

b. *A power series* in $x - a$ is of the form

$$S(x) = \sum_{i=0}^{\infty} c_i\ (x-a)^i = c_0 + c_1(x-a) + c_2(x-a)^2 + c_3(x-a)^3 \ldots$$

where a and $c_0, c_1, c_2, c_3, \ldots$ are *constants*. For every power series, there exists a value $b \geq 0$, such that $S(x)$ is absolutely convergent for all $|x| < b$ and divergent for all $|x| > b$. Then b is the *radius of convergence*, and the totality $-b < x < b$ is the *interval of convergence*.

c. *A function series* in x is of the form

$$F(x) = \sum_{i=0}^{\infty} a_i\ f_i(x) = a_0 f_0(x) + a_1 f_1(x) + a_2 f_2(x) + \ldots$$

where $a_0, a_1, a_2, \ldots$ are constants. The series converges to $F(x)$ if the sequence of partial sums converges to $F(x)$.

d. *A double series*

$$\sum_{i=0}^{\infty}\sum_{x=0}^{\infty} a_{ix} = \sum_{i=0}^{\infty}\left(\sum_{x=0}^{\infty} a_{ix}\right) = \sum_{x=0}^{\infty}\left(\sum_{i=0}^{\infty} a_{ix}\right)$$

converges to the limit D if

$$\lim_{\substack{m \to \infty \\ n \to \infty}} \sum_{i=0}^{m} \left(\sum_{x=0}^{n} a_{ix} \right) = D$$

e. *A product series*

$$\sum_{i=0}^{\infty} \sum_{x=0}^{\infty} a_i \, b_x = \left(\sum_{i=0}^{\infty} a_i \right) \left(\sum_{x=0}^{\infty} b_x \right)$$

is an *absolutely convergent series* if

$$\sum_{i=0}^{\infty} a_i \quad \text{and} \quad \sum_{x=0}^{\infty} b_x$$

are *absolutely convergent series.*

9.2 TESTS OF CONVERGENCE AND OPERATIONS OF CONSTANT TERMS

i. Tests of Convergence

The following tests are available for the analysis of convergence and divergence of the series

$$\sum_{n=1}^{\infty} u_n = u_1 + u_2 + u_3 + \ldots$$

in which *each term is a constant.*

a. *Comparison test*

If $\qquad |u_n| < a\,|v_n|$

where a is a constant independent of n and v_n is the nth term of another series which is known to be absolutely convergent, the series consisting of u terms is also absolutely convergent.

If $\qquad |u_n| > a\,|v_n|$

and the series consisting of v terms is known to be absolutely divergent, then the series consisting of u terms is also divergent.

b. *Cauchy's test (nth root test)*

If $\qquad \lim_{n \to \infty} \sqrt[n]{|u_n|} = L$

then the series is absolutely convergent for $L < 1$ and divergent for $L > 1$, and the test fails for $L = 1$.

c. *D'Alembert's test (ratio test)*

If $\qquad \lim_{n \to \infty} \left| \frac{u_{n+1}}{u_n} \right| = L$

then the series is absolutely convergent for $L < 1$ and divergent for $L > 1$, and the test fails for $L = 1$.

d. *Raabe's test*

$$\lim_{n \to \infty} n\left(1 - \left|\frac{u_{n+1}}{u_n}\right|\right) = L$$

then the series is absolutely convergent for $L > 1$ and divergent for $L < 1$, and the test fails for $L = 1$.

e. *Integral test*

If each term of the u series is a function of its suffix n and if the function $f(x)$ which represents $f(n)$ when $x = n$ is continuous and monotonic, the given series is convergent if

$$\int_n^{\infty} f(x)\,dx = \lim_{M \to \infty} \int_n^{M} f(x)\,dx$$

is convergent, and it is divergent if this integral is divergent.

ii. Operations with Absolutely Convergent Series

a. The terms of an absolutely convergent series can be rearranged in any order, and the new series will converge to the same sum.
b. The sum, difference, and product of two or more absolutely convergent series is absolutely convergent.

9.3 SERIES OF CONSTANT TERMS

i. Arithmetic Series

$$a + (a + b) + (a + 2b) + \ \dots\ + (a + nb) = \frac{(2a + nb)n}{2}$$

ii. Geometric Series

$$a + ab + ab^2 + ab^3 + \dots + ab^n = \frac{\left(1 - b^{n+1}\right)a}{1 - b}; \qquad b \neq 1$$

$$a + ab + ab^2 + ab^3 + \dots = \frac{a}{1 - b}; \qquad -1 < b < 1$$

$$a^n + a^{n-1}b + a^{n-2}b^2 + \dots + b^n = \frac{a^{n+1} - b^{n+1}}{a - b}; \qquad a \neq b$$

iii. Power Series

$$1 + 2 + 3 + \dots + n = \frac{(n + 1)n}{2}$$

$$1^2 + 2^2 + 3^3 + \dots + n^2 = \frac{(2n + 1)(n + 1)n}{6}$$

$$1^3 + 2^3 + 3^3 + \dots + n^3 = \frac{(n + 1)^2 n^2}{4}$$

$$1^4 + 2^4 + 3^4 + \dots + n^4 = \frac{(3n^2+3n-1)(2n+1)(n+1)n}{30}$$

$$1 + 3 + 5 + \dots + (2n-1) = n^2$$

$$1^2 + 3^2 + 5^2 + \dots + (2n-1)^2 = \frac{n(4n^2-1)}{3}$$

$$1^3 + 3^3 + 5^3 + \dots + (2n-1)^3 = n^2\,(2n^2-1)$$

$$1^4 + 3^4 + 5^4 + \dots + (2n-1)^4 = \frac{n(4n^2-1)(12n^2-7)}{15}$$

iv. Binominal Series

$$\binom{n}{0}+\binom{n}{1}+\binom{n}{2}+\dots+\binom{n}{n} = 2^n \qquad \binom{n}{1}+2\binom{n}{2}+\dots+n\binom{n}{n} = 2^{n-1}\,n$$

$$\binom{n}{1}+2^2\binom{n}{2}+\dots+n^2\binom{n}{n} = 2^{n-2}\,n\,(n+1)$$

v. Trigonometric Series

$$\sum_{1}^{n}\sin k\omega = \frac{\sin(n\omega/2)\sin[(n+1)/2]\omega}{\sin(\omega/2)} \qquad \sum_{0}^{n}\cos k\omega = \frac{\cos n\omega/2\,\sin[(n+1)/2]\omega}{\sin(\omega/2)}$$

$$\sum_{1}^{n-1} k\sin k\omega = \frac{\sin n\omega}{4\sin^2(\omega/2)} - \frac{n\cos[(2n-1)/2]\omega}{2\sin(\omega/2)}$$

$$\sum_{0}^{n-1} k\cos k\omega = \frac{n\sin[(2n-1)/2]\omega}{2\sin(\omega/2)} - \frac{1-\cos n\omega}{4\sin^2(\omega/2)}$$

$$\sum_{1}^{n}\sin^2 k\omega = \frac{n}{2} - \frac{\cos(n+1)\omega\sin n\omega}{2\sin\omega} \qquad \sum_{0}^{n}\cos^2 k\omega = 1 + \frac{n}{2} + \frac{\cos(n+1)\omega\sin n\omega}{2\sin\omega}$$

9.4 SPECIAL SERIES OF CONSTANT TERMS

i. Riemann's Function [Zeta $Z(n)$]

$$Z(n) = 1 + \frac{1}{2^n} + \frac{1}{3^n} + \frac{1}{4^n} + \dots$$

$$\left(1-\frac{1}{2^n}\right)Z(n) = 1 + \frac{1}{3^n} + \frac{1}{5^n} + \frac{1}{7^n} + \dots$$

$$\ln 2 = 1 - \frac{1}{2} + \frac{1}{3} - \frac{1}{4} + \dots$$

$$\left(1-\frac{2}{2^n}\right)Z(n) = 1 - \frac{1}{2^n} + \frac{1}{3^n} - \frac{1}{4^n} + \dots$$

ii. Bernoulli's Number B_n and Euler's Number E_n

$$B_n = \left(1 + \frac{1}{2^{2n}} + \frac{1}{3^{2n}} + \frac{1}{4^{2n}} + \ldots\right)\frac{(2n)!}{2^{2n-1}\,\pi^{2n}}$$

$$B_n = \left(1 - \frac{1}{2^{2n}} + \frac{1}{3^{2n}} - \frac{1}{4^{2n}} + \ldots\right)\frac{(2n)!}{(2^{2n-1}-1)\,\pi^{2n}}$$

$$B_n = \left(1 + \frac{1}{3^{2n}} + \frac{1}{5^{2n}} + \frac{1}{7^{2n}} + \ldots\right)\frac{2(2n)!}{(2^{2n}-1)\,\pi^{2n}}$$

$$E_n = \left(1 - \frac{1}{3^{2n+1}} + \frac{1}{5^{2n+1}} - \frac{1}{7^{2n+1}} + \ldots\right)\frac{2^{2n+2}(2n)!}{\pi^{2n+1}}$$

iii. Table of Constant Terms: Zeta, $Z(n)$, Bernoulli's number (B_n) and Euler's number (E_n)

Zeta [$Z(n)$], Bernoulli's number (b_n) and Euler's number (E_n)

$Z(n)$	B_n	E_n
$Z(1) = \infty$	$B_1 = \frac{1}{6}$	$E_1 = 1$
$Z(2) = 1.64493$	$B_2 = \frac{1}{30}$	$E_2 = 5$
$Z(3) = 1.20206$	$B_3 = \frac{1}{42}$	$E_3 = 61$
$Z(4) = 1.08232$	$B_4 = \frac{1}{30}$	$E_4 = 1{,}385$
$Z(5) = 1.03693$	$B_5 = \frac{5}{66}$	$E_5 = 50{,}521$
$Z(6) = 1.01734$	$B_6 = \frac{691}{2{,}730}$	$E_6 = 2{,}702{,}765$
$Z(7) = 1.00835$	$B_7 = \frac{7}{6}$	$E_7 = 199{,}603{,}981$
$Z(8) = 1.00408$	$B_8 = \frac{3{,}617}{510}$	$E_8 = 19{,}391{,}512{,}145$

iv. Special Numbers

$$e = \sum_{0}^{x} \frac{1}{n!} = 2.718282 \qquad \sum_{1}^{\infty} \frac{n-1}{n!} = 1.000000 \qquad \sum_{1}^{\infty} \frac{1}{4n^2-1} = 0.500000$$

$$1 - \frac{1}{2^2} + \frac{1}{2^4} - \frac{1}{2^6} + \ldots = \frac{4}{5} \qquad 1 - \frac{1}{3^2} + \frac{1}{3^4} - \frac{1}{3^6} + \ldots = \frac{9}{10}$$

$$1 - \frac{1}{4^2} + \frac{1}{4^4} - \frac{1}{4^6} + \ldots = \frac{16}{17} \qquad 1 - \frac{1}{5^2} + \frac{1}{5^4} - \frac{1}{5^6} + \ldots = \frac{25}{26}$$

v. Series Converging to $1, \frac{3}{4}, \frac{1}{2}, \frac{1}{4}$

$$\frac{1}{(1)(2)}+\frac{1}{(2)(3)}+\frac{1}{(3)(4)}+\ldots=1 \qquad \frac{1}{(1)(3)}+\frac{1}{(3)(5)}+\frac{1}{(5)(7)}+\ldots=\frac{1}{2}$$

$$\frac{1}{(1)(2)(3)}+\frac{1}{(2)(3)(4)}+\frac{1}{(3)(4)(5)}+\ldots=\frac{1}{4} \qquad \frac{1}{(1)(3)}+\frac{1}{(2)(4)}+\frac{1}{(3)(5)}+\ldots=\frac{3}{4}$$

9.5 TEST OF CONVERGENCE AND OPERATIONS OF FUNCTION SERIES

i. Tests of Convergence

The following tests are available for the analysis of the convergence of the series

$$F(x) = \sum_{n=1}^{\infty} f_n(x) = f_1(x) + f_2(x) + f_3(x) + \ldots$$

in which each term is a function.

a. *Cauchy's test for uniform convergence*

A series of real (or complex) functions converges uniformly on $F(x)$ in $[a, b]$ if for every real number $\varepsilon > 0$ there exists a real number $N > 0$, independent of x in $[a, b]$, such that

$$[F(x) - f_n(x)| < \varepsilon \qquad \text{for all } n > N$$

This is a necessary and sufficient condition for uniform convergence for a function series.

b. *Weierstrass's test for uniform and absolute convergence*

A series of real (or complex) functions converges uniformly and absolutely on every $F(x)$ in $[a, b]$ if

$$[f_n(x)| \le M_n \qquad \text{for all } n$$

and $M_1 + M_2 + M_3 + \ldots$ is a convergent comparison series of real positive terms. Since this test establishes the absolute (as well as the uniform) convergence, it is applicable only to series which converge absolutely. It must be noted that a function series may converge uniformly but not absolutely, and vice versa.

c. *Dirichlet's test for uniform convergence*

If $\sum_{n=1}^{\infty} a_n = a_1 + a_2 + a_3 + \ldots$ is a monotonic decreasing sequence of real numbers then the infinite function series $\sum_{n=1}^{\infty} a_n f_n(x) = a_1 f_1(x) + a_2 f_2(x) + a_3 f_3(x) + \ldots$ converges uniformly on a set $G(x)$ of values of x if the infinite series

$$\sum_{n=1}^{\infty} f_n(x) = f_1(x) + f_2(x) + f_3(x) + \ldots$$

converges uniformly on the same set $G(x)$ of values of x.

d. *Able's test*

Suppose $\Sigma\, a_n$ is a onvergent series, $<b_n>$ is a monotone and bounded sequence then $\Sigma\, a_n b_n$ is convergent.

e. *Dirichlet's test*

Let $\Sigma\, a_n$ be a_n infinite series such that sequence of partial sums $S_n = a_1 + a_2 + \ldots a_n$ are bounded and let $<b_n>$ be a monotonically decreasing sequence that converges to 0 then $\Sigma\, a_n b_n$ converges to some finite value.

ii. Properties of Uniformly Convergent Function Series

a. *Theorem of continuity*

If any term of a uniformly convergent function series is a continuous function of x in $[a, b]$, then the sum of the series is also a continuous function of x in $[a, b]$.

b. *Theorem of differentiability of a function series*

A uniformly convergent series in (a, b) can be differentiated term by term in (a, b). If each term of the differentiated series is continuous and the differentiated series is uniformly convergent in (a, b), then it will converge to the derivative of the function it represents in (a, b).

iii. Theorem of Integrability of a Function Series

A uniformly convergent series in $[a, b]$ can be integrated term by term in $[a, b]$, and the integrated series will converge uniformly to the integral of the function it represents in $[a, b]$.

9.6 TEST OF CONVERGENCE AND OPERATIONS OF POWER SERIES

i. Interval of Convergence (Ratio Test)

The power series in the real (or complex) variable x

$$S(x) = \sum_{n \to \infty}^{\infty} a_n x^n = a_0 + a_1 x + a_2 x^2 + \ldots$$

where the coefficients $a_0, a_1, a_2, \ldots$ are real or complex numbers independent of x.

convergent if $$\lim_{n \to \infty} \left| \frac{a_{n+1}\, x^{n+1}}{a_n x^n} \right| = \lim_{n \to \infty} \left| \frac{a_{n+1}}{a_n} \right| |x| = r\,|x| < 1$$

and *divergent* if $$\lim_{n \to \infty} \left| \frac{a_{n+1}\, x^{n+1}}{a_n x^n} \right| = \lim_{n \to \infty} \left| \frac{a_{n+1}}{a_n} \right| |x| = r\,|x| > 1$$

The interval of convergence is then

$r\,|x| < 1$ or $-\frac{1}{r} < x < \frac{1}{r}$

and it is symmetrical about the origin of x. The series is convergent in this interval and diverges outside this interval. It may or may not converge at the end points of the interval.

ii. Uniform and Absolute Convergence

The power series which converges in the interval

$$a < x < \beta$$

converges absolutely and uniformly for every value of x within this interval. Since a uniformly convergent series represents a continuous function, a *uniformly convergent series defines a continuous function within the interval of convergence.*

iii. Operations with Power Series

a. *Uniqueness theorem*

If two power series

$$S(x) = \sum_{n=0}^{\infty} a_n x^n \quad \text{and} \quad S(x) = \sum_{n=0}^{\infty} b_n x^n$$

converge to the same sum $S(x)$ for all real values of x, then

$$a_0 = b_0, a_1 = b_1, a_2 = b_2, \ldots$$

b. *Summation theorem*

Two power series can be added or subtracted term by term for each value of x common to their interval of convergence.

c. *Product theorem*

Two power series can be multiplied term by term for each value of x common to their interval of convergence. Thus

$$\left(\sum_{m=0}^{\infty} a_m x^m\right)\left(\sum_{n=0}^{\infty} b_n x^n\right) = \sum_{m=0}^{\infty} \sum_{n=0}^{\infty} a_m b_n x^{m+n}$$

The new series may either converge or diverge.

d. *Theorem of differentiability and integrability*

A power series can be differentiated and integrated term by term in any closed interval if and only if this interval lies entirely within the interval of uniform convergence of the power series.

9.7 INFINITE BINOMIAL SERIES

a. *Binomial series—basic case*

$$(1+x)^n = 1 + \binom{n}{1}x + \binom{n}{2}x^2 + \binom{n}{3}x^3 + \ldots \begin{bmatrix} n = 0,1,2,\ldots, x \gtrless 0; \text{Finite series} \\ n \neq 0,1,2,\ldots, |x|\ |x| < 1; \text{Infinite convergent series} \\ n \neq = 0,1,2,\ldots, |x| > 1; \text{Infinite divergent series} \end{bmatrix}$$

b. *Special cases* $[x^2 < 1]$

$$(1 \pm x)^m = 1 \pm mx + \frac{m(m-1)}{2!}x^2 \pm \frac{m(m-1)(m-2)}{3!}x^3 + \ldots$$

$$(1 \pm x)^{-m} = 1 \mp mx + \frac{m(m+1)}{2!}x^2 \mp \frac{m(m+1)(m+2)}{3!}x^3 + \ldots$$

$$(1 \pm x)^{-1} = 1 \mp x + x^2 \mp x^3 + \ldots$$

$$(1 \pm x)^{-2} = 1 \mp 2x + 3x^2 \mp 4x^3 + \ldots$$

$$(1 \pm x)^{1/4} = 1 \pm \frac{1}{4}x - \frac{(1)(3)}{(4)(8)}x^2 \pm \frac{(1)(3)(7)}{(4)(8)(12)}x^3 - \frac{(1)(3)(7)(11)}{(4)(8)(12)(16)}x^4 \pm \ldots$$

$$(1 \pm x)^{-1/4} = 1 \mp \frac{1}{4}x + \frac{(1)(5)}{(4)(8)}x^2 \mp \frac{(1)(5)(9)}{(4)(8)(12)}x^3 + \frac{(1)(5)(9)(13)}{(4)(8)(12)(16)}x^4 \mp \ldots$$

$$(1 \pm x)^{1/3} = 1 \pm \frac{1}{3}x - \frac{(1)(2)}{(3)(6)}x^2 \pm \frac{(1)(2)(5)}{(3)(6)(9)}x^3 - \frac{(1)(2)(5)(8)}{(3)(6)(9)(12)}x^4 \pm \ldots$$

$$(1 \pm x)^{-1/3} = 1 \mp \frac{1}{3}x + \frac{(1)(4)}{(3)(6)}x^2 \mp \frac{(1)(4)(7)}{(3)(6)(9)}x^3 + \frac{(1)(4)(7)(10)}{(3)(6)(9)(12)}x^4 \mp \ldots$$

$$(1 \pm x)^{1/2} = 1 \pm \frac{1}{2}x - \frac{(1)(1)}{(2)(4)}x^2 \pm \frac{(1)(1)(3)}{(2)(4)(6)}x^3 - \frac{(1)(1)(3)(5)}{(2)(4)(6)(8)}x^4 \pm \ldots$$

$$(1 \pm x)^{-1/2} = 1 \mp \frac{1}{2}x + \frac{(1)(3)}{(2)(4)}x^2 \mp \frac{(1)(3)(5)}{(2)(4)(6)}x^3 + \frac{(1)(3)(5)(7)}{(2)(4)(6)(8)}x^4 \mp \ldots$$

$$(1 \pm x)^{\alpha/\beta} = 1 \pm \frac{\alpha}{\beta}x + \frac{\alpha(\alpha-\beta)}{(\beta)(2\beta)}x^2 \pm \frac{\alpha(\alpha-\beta)(\alpha-2\beta)}{(\beta)(2\beta)(3\beta)}x^3 + \ldots$$

$$(1 \pm x)^{-\alpha/\beta} = 1 \mp \frac{\alpha}{\beta}x + \frac{\alpha(\alpha+\beta)}{(\beta)(2\beta)}x^2 \mp \frac{\alpha(\alpha+\beta)(\alpha+2\beta)}{(\beta)(2\beta)(3\beta)}x^3 + \ldots$$

9.8 POWER SERIES

a. *Product and quotient of two power series*

$$\left(\sum_0^\infty a_n x^n\right)\left(\sum_0^\infty b_n x^n\right) = A_0 + A_1x + A_2x^2 + A_3x^3 + \ldots$$

$A_0 = a_0b_0$ $\quad A_2 = (a_0b_2 + a_1b_1 + a_2b_0)$

$A_1 = (a_0b_1 + a_1b_0)$ $\quad A_3 = (a_0b_3 + a_1b_2 + a_2b_1 + a_3b_0)$

$$\left(\sum_0^\infty a_n x^n\right):\left(\sum_0^\infty b_n x^n\right) = A_0 + A_1x + A_2x^2 + A_3x^3 + \ldots$$

$A_0 = 1$ $\quad A_1 = (a_1 - b_1)$

$A_2 = (a_2 - a_1b_1 - b_2 + b_1^2)$

$A_3 = (a_3 - a_2b_1 - a_1b_2 - b_3 - b_1^3 + a_1b_1^2 + 2b_1b_2)$ $\quad a_0 = b_0 = 1$

b. *Power of power series*

$$(a_0 + a_1x + a_2x^2 + a_3x^3 + \ldots)^2 = A_0 + A_1x + A_2x^2 + A_3x^3 + \ldots$$

$A_0 = 1$ $\quad A_1 = 2a_1$ $\quad A_2 = (a_1^2 + 2a_2)$

$A_3 = 2(a_1a_2 + a_3)$ $\quad A_4 = (a_2^2 + 2a_1a_3 + 2a_4)$ $\quad a_0 = 1$

$$(a_0 + a_1x + a_2x^2 + a_3x^3 + \ldots)^n = A_0 + A_1x + A_2x^2 + A_3x^3 + \ldots$$

$$A_0 = 1 \quad A_1 = \binom{n}{1} a_1$$

$$A_2 = \binom{n}{2} a_1^2 + \binom{n}{1} a_2$$

$$A_3 = \binom{n}{3} a_1^3 + 2\binom{n}{2} a_1 a_2 + \binom{n}{1} a_3$$

$$A_4 = \binom{n}{4} a_1^4 + 3\binom{n}{3} a_1^2 a_2 + \binom{n}{2}(a_2^2 + 2a_1 a_3) + \binom{n}{1} a_4, \qquad a_0 = 1$$

c. *Reversion of power series*

If $y = x + a_2x^2 + a_3x^3 + \ldots$, then its reverse $x = y + b_2y^2 + b_3y^3 + \ldots$ is true, with

$b_2 = -a_2$ $\qquad b_3 = (2a_2^2 - a_3)$

$b_4 = -(5a_2^3 - 5a_2a_3 + a_4)$ $\qquad b_5 = (14a_2^4 - 21a_2^2a_3 + 6a_2a_4 + 3a_3^2 - a_5)$

$b_6 = -(42a_2^5 - 84a_2^3a_3 + 28a_2^2a_4 + 28a_2a_3^2 - 7a_2a_5 - 7a_3a_4 + a_6)$

9.9 REPRESENTATION OF FUNCTIONS BY POWER SERIES

i. Single Variables

a. *MacLaurin's series at $x = 0$*

If a function $f(x)$ is continuous and single-valued and has all derivatives on an interval including $x = 0$, then

$$f(x) = f(0) + \frac{f'(0)}{1!}x + \frac{f''(0)}{2!}x^2 + \ldots + \frac{f^{(n)}(0)}{n!} + R_n$$

in which $R_n = \dfrac{f^{(n+1)}(\theta x)}{(n+1)!}\, x^{n+1} \qquad 0 < \theta < 1$

This series represents $f(x)$ for those values of x for which $R_n \to 0$ as $n \to \infty$.

b. *Taylor's series at $x = a$*

If a function $f(x)$ is continuous and single-valued and has all derivatives on an interval including $x = a$, then

$$f(x) = f(a) + \frac{f'(a)}{1!}(x-a) + \frac{f''(a)}{2!}(x-a)^2 + \ldots + \frac{f^{(n)}(a)}{n!}(x-a)^n + R_n$$

in which $\qquad R_n = \dfrac{f^{(n+1)}(\theta_x)(x-a)^{n+1}}{(n+1)!} \qquad a < \theta_x < x$

The series represents $f(x)$ for those values of x for which $R_n \to 0$ as $n \to \infty$

ii. Two Variables

Taylor's series for a function of two variables is

$$f(x+a, y+b) = f(x, y) + \frac{1}{1!}D_1[f(x,y)] + \frac{1}{2!}D_2[f(x,y)] + \ldots + \frac{1}{n!}D_n[f(x,y)] + R_n$$

in which $D_n = \left(a\frac{\partial}{\partial x} + b\frac{\partial}{\partial y}\right)^n$ and $R_n = \frac{1}{(n+1)!}D_{n+1}[f(x+\theta_1 a, y+\theta_2 b)]$

or at $x = 0, y = 0,$ $0 < \theta_1 < 1;\ 0 < \theta_2 < 1$

$$f(x, y) = f(0, 0) + \frac{1}{1!}D_1[f(0,0)] + \frac{1}{2!}D_2[f(0,0)] + \ldots + \frac{1}{n!}D_n[f(0,0)] + R_n$$

in which $D_n = \left(x\frac{\partial}{\partial x} + y\frac{\partial}{\partial y}\right)^n$ and $R_n = \frac{1}{(n+1)!}D_{n+1}[f(\theta_1 x, \theta_2 y)]$ $0 < \theta_1 < 1, 0 < \theta_2 < 1$

9.10 REPRESENTATION OF TRANSCENDENT FUNCTIONS BY SERIES

$e^x = 1 + \frac{x}{1!} + \frac{x^2}{2!} + \frac{x^3}{3!} + \ldots$	$\lvert x \rvert < \infty$
$\sin x = \frac{x}{1!} - \frac{x^3}{3!} + \frac{x^5}{5!} - \frac{x^7}{7!} + \ldots \quad \lvert x \rvert < \infty$	$\sinh x = \frac{x}{1!} + \frac{x^3}{3!} + \frac{x^5}{5!} + \frac{x^7}{7!} + \ldots \quad \lvert x \rvert < \infty$
$\cos x = 1 - \frac{x^2}{2!} + \frac{x^4}{4!} - \frac{x^6}{6!} + \ldots \quad \lvert x \rvert < \infty$	$\cosh x = 1 + \frac{x^2}{2!} + \frac{x^4}{4!} + \frac{x^6}{6!} + \ldots \quad \lvert x \rvert < \infty$
$\tan x = \frac{x}{1} + \frac{x^3}{3} + \frac{2x^5}{15} + \frac{17x^7}{315} + \frac{62x^9}{2,835} + \ldots \quad \lvert x \rvert < \pi/2$	$\tanh x = \frac{x}{1} - \frac{x^3}{3} + \frac{2x^5}{15} - \frac{17x^7}{315} + \frac{62x^9}{2,835} + \ldots \quad \lvert x \rvert < \pi/2$
$\cot x = \frac{1}{x} - \frac{x}{3} - \frac{x^3}{45} - \frac{2x^5}{945} - \frac{x^7}{4,725} - \ldots \quad 0 < \lvert x \rvert < \pi$	$\coth x = \frac{1}{x} + \frac{x}{3} - \frac{x^3}{45} + \frac{2x^5}{945} - \frac{x^7}{4,725} + \ldots \quad 0 < \lvert x \rvert < \pi$
$\sin^{-1} x = \frac{x}{1} + \frac{x^3}{(2)(3)} + \frac{(1)(3x^5)}{(2)(4)(5)} + \frac{(1)(3)(5x^7)}{(2)(4)(6)(7)} + \ldots$ $\lvert x \rvert < 1$	$\sinh^{-1} x = \frac{x}{1} - \frac{x^3}{(2)(3)} + \frac{(1)(3x^5)}{(2)(4)(5)} - \frac{(1)(3)(5x^7)}{(2)(4)(6)(7)} + \ldots$ $\lvert x \rvert < 1$
$\cos^{-1} x = \frac{\pi}{2} - \left(x + \frac{x^3}{(2)(3)} + \frac{(1)(3x^5)}{(2)(4)(5)} + \frac{(1)(3)(5x^7)}{(2)(4)(6)(7)} + \ldots\right)$ $\lvert x \rvert < 1$	$\cosh^{-1} x =$ $\pm\left[(\text{In } 2x - \frac{1}{(2)(2x^2)} - \frac{(1)(3)}{(2)(4)(4x^4)} - \frac{(1)(3)(5)}{(2)(4)(6)(6x^6)} - \ldots\right]$ $x \geq 1$ (+ if $\cosh^{-1} x > 0$, if $\cosh^{-1} x < 0$)
$\tan^{-1} x = \frac{x}{1} - \frac{x^3}{3} + \frac{x^5}{5} - \frac{x^7}{7} + \ldots \quad \lvert x \rvert < 1$	$\tanh^{-1} x = \frac{x}{1} + \frac{x^3}{3} + \frac{x^5}{5} + \frac{x^7}{7} + \ldots \quad \lvert x \rvert < 1$
$\cot^{-1} x = \frac{\pi}{2} - \frac{x}{1} + \frac{x^3}{3} - \frac{x^5}{5} + \frac{x^7}{7} - \ldots \quad \lvert x \rvert < 1$	$\coth^{-1} x = \frac{1}{x} + \frac{1}{3x^3} + \frac{1}{5x^5} + \frac{1}{7x^7} + \ldots \quad \lvert x \rvert > 1$
$a^x = 1 + \frac{x \text{ in } a}{1!} + \frac{(x \text{ in } a)^2}{2!} + \frac{(x \text{ in } a)^3}{3!} + \ldots$	$\lvert x \rvert < \infty, \quad a > 0$

$\ln x = \frac{(x-1)}{1} - \frac{(x-1)^2}{2} + \frac{(x-1)^3}{3} - \ldots$	$0 < x \le 2$
$\ln x = 2\left[\frac{x-1}{x+1} + \frac{(x-1)^3}{3(x+1)^3} + \frac{(x-1)^5}{5(x+1)^5} + \ldots\right]$	$x > 0$
$\ln(1+x) = \frac{x}{1} - \frac{x^2}{2} + \frac{x^3}{3} - \frac{x^4}{4} + \ldots \quad -1 < x \le 1$	$\ln(1-x) = -\frac{x}{1} - \frac{x^2}{2} - \frac{x^3}{3} - \frac{x^4}{4} - \ldots \quad -1 \le x < 1$
$\ln \frac{x+1}{x-1} = 2\left(\frac{1}{x} + \frac{1}{3x^5} + \frac{1}{5x^5} + \frac{1}{7x^7} + \ldots\right)$ $= 2\,\text{ctnh}^{-1} x \quad \lvert x^2 \rvert > 1$	$\ln \frac{1+x}{1-x} = 2\left(\frac{x}{1} + \frac{x^3}{3} + \frac{x^5}{5} + \frac{x^7}{7} + \ldots\right)$ $= 2\tanh^{-1} x \quad \lvert x^2 \rvert < 1$

9.11 FINITE AND INFINITE PRODUCTS

i. Finite Products

$$1 + x^{2n} = \left(x^2 - 2x\cos\frac{\pi}{2n} + 1\right)\left(x^2 + 2x\cos\frac{3\pi}{2n} + 1\right)\ldots\left(x^2 + 2x\cos\frac{(2n-1)\pi}{2n} + 1\right)$$

$$1 + x^{2n+1} = (1+x)\left[\left(x^2 - 2x\cos\frac{\pi}{2n+1} + 1\right)\left(x^2 - 2x\cos\frac{3\pi}{2n+1} + 1\right)\ldots\left(x^2 - 2x\cos\frac{(2n-1)\pi}{2n+1} + 1\right)\right]$$

$$\sin 2nx = n \sin 2x\,[(1 - a_1 \sin^2 x)(1 - a_2 \sin^2 x)(1 - a_3 \sin^2 x)\ldots(1 - a_{n-1}\sin^2 x)]$$

$$\sin(2n+1)x = n \sin x\,[(1 - b_1 \sin^2 x)(1 - b_2 \sin^2 x)(1 - b_3 \sin^2 x)\ldots(1 - b_n \sin^2 x)]$$

$$a_r = \frac{1}{\sin^2(r\pi/2n)} \qquad b_r = \frac{1}{\sin^2[r\pi/(2n+1)]}; \qquad r = 1, 2, 3, \ldots$$

$$\cos 2nx = (1 - c_1 \sin^2 x)(1 - c_3 \sin^2 x)(1 - c_5 \sin^2 x)\ldots(1 - c_{2n-1}\sin^2 x)$$

$$\cos(2n+1)x = \cos x[(1 - d_1 \sin^2 x)(1 - d_3 \sin^2 x)(1 - d_5 \sin^2 x)\ldots(1 - d_{2n-1}\sin^2 x)]$$

$$c_r = \frac{1}{\sin^2(r\pi/4n)} \qquad d_r = \frac{1}{\sin^2[r\pi/(4n+2)]}; \qquad r = 1, 3, 5\ldots$$

ii. Infinite Products

$$\sin nx = nx\left[1 - \left(\frac{nx}{\pi}\right)^2\right]\left[1 - \left(\frac{nx}{2\pi}\right)^2\right] \times \left[1 - \left(\frac{nx}{3\pi}\right)^2\right]\ldots$$

$$\sinh nx = nx\left[1 + \left(\frac{nx}{\pi}\right)^2\right]\left[1 + \left(\frac{nx}{2\pi}\right)^2\right] \times \left[1 + \left(\frac{nx}{3\pi}\right)^2\right]\ldots$$

$$\cos nx = \left[1-\left(\frac{2nx}{\pi}\right)^2\right]\left[1-\left(\frac{2nx}{3\pi}\right)^2\right] \times\left[1-\left(\frac{2nx}{5\pi}\right)^2\right]\ldots$$

$$\cosh nx = \left[1+\left(\frac{2nx}{\pi}\right)^2\right]\left[1+\left(\frac{2nx}{3\pi}\right)^2\right] \times\left[1+\left(\frac{2nx}{5\pi}\right)^2\right]\ldots$$

$$\sin(x+y) = \left[\left(1+\frac{y}{x}\right)\left(1+\frac{y}{\pi+x}\right)\left(1-\frac{y}{\pi-x}\right)\left(1+\frac{y}{2\pi+x}\right)\left(1-\frac{y}{2\pi-x}\right)\ldots\right]\sin x$$

$$\sin(x-y) = \left[\left(1-\frac{y}{x}\right)\left(1+\frac{y}{\pi-x}\right)\left(1-\frac{y}{\pi+x}\right)\left(1+\frac{y}{2\pi-x}\right)\left(1-\frac{y}{2\pi+x}\right)\ldots\right]\sin x$$

$$\cos(x+y) = \left[\left(1+\frac{2y}{\pi+2x}\right)\left(1-\frac{2y}{\pi-2x}\right)\left(1+\frac{2y}{3\pi+2x}\right)\left(1-\frac{2y}{3\pi-2x}\right)\ldots\right]\cos x$$

$$\cos(x-y) = \left[\left(1+\frac{2y}{\pi-2x}\right)\left(1-\frac{2y}{\pi+2x}\right)\left(1+\frac{2y}{3\pi-2x}\right)\left(1-\frac{2y}{3\pi+2x}\right)\ldots\right]\cos x$$

$$\frac{\pi}{2} = \left(\frac{2}{1}\right)\left(\frac{2}{3}\right)\left(\frac{4}{3}\right)\left(\frac{4}{5}\right)\left(\frac{6}{5}\right)\left(\frac{6}{7}\right)\ldots$$

$$\frac{\sin x}{x} = \cos\frac{x}{2}\cos\frac{x}{4}\cos\frac{x}{8}\cos\frac{x}{16}\ldots$$

CHAPTER

10

Integral Calculus

10.1 INDEFINITE INTEGRAL—CONCEPTS

i. Definition

$F(x)$ is an indefinite integral of $f(x)$ if $\dfrac{dF(x)}{dx} = f(x)$

Since the derivative of $F(x) + C$ is also equal to $f(x)$, all integrals of $f(x)$ are included in the expression.

$$\int f(x)\,dx = F(x) + C$$

in which $f(x)$ is called the integrand and C is an *arbitrary constant*. Because of the indeterminancy of C, there is an infinite number of $F(x) + C$, differing by their relative position to x-axis only. The adjacent graph illustrates the meaning of C for a given function.

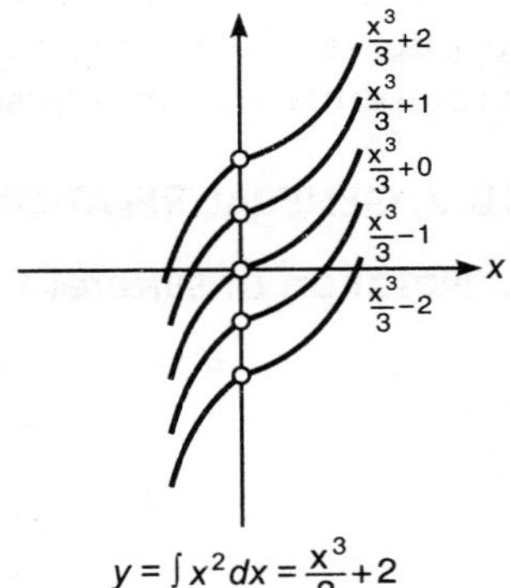

Fig. 10.1

ii. Methods of Integration

a. *Antiderivative method*

If $f(x)$ is a derivative of a known function, the integral is this function plus the constant of integration.

$$\int \frac{dF(x)}{dx}\,dx = F(x) + C$$

b. *Integration by parts*

The integrand can be expressed as a product of two functions

$$f(x) = u(x) \cdot v'(x)$$

then

$$\int u(x)\, v'(x)\, dx = u(x)\, v(x) - \int u'(x)\, v(x)\, dx$$

in which the integral of the right side may be known or can be calculated by one or more repetitions of the same.

c. *Substitution method*

The introduction of a new variable

$$x = \phi(t) \qquad dx = \phi'(t)\, dt$$

yields

$$\int f(x)\, dx = \int f(\phi)(t)\, \phi'(t)\, dt + C$$

The integral of this transformed function may be known, or can be calculated by other methods.

d. *Integration of series (term by term)*

If the integrand can be expressed as a uniformly convergent series of powers of x (within its interval of convergency) and if the result of integration of this series term by term is also a uniformly convergent series, the sum of this series is also the value of the integral.

10.2 GENERAL RELATIONS [$u = f(x)$]

i. Notation of Different Functions

u is differentiable function of x,

$u', u'', \ldots u^{(n)}$ are successive derivatives of u,

$U_1, U_2, \ldots, U_n$ are successive integrals of u; and

a, b, m are constants

The constant of integration C is omitted.

ii. Integral of Different Functions

$\int (0)\, dx = \text{constant}$	$\int (a)\, dx = ax$
$\int f'(x)\, dx = f(x)$	$\int f(a) f'(x)\, dx = f(a) f(x)$
$\int f(x)\, dx = xf(0) + \frac{x^2}{2!} f'(0) + \frac{x^3}{3!} f''(0) + \ldots$	$\int f(x)\, dx = xf(x) - \frac{x^2}{2!} f'(x) + \frac{x^3}{3!} f''(x) - \ldots$
$\int f(0) dx = xf(0)$	$\int f(x)\, dx = xf(x) - \int xf'(x)\, dx$
$\int uu'\, dx = \frac{u^2}{2}$	$\int \frac{u'}{u}\, dx = \ln u$
$\int u^m\, u'\, dx = \frac{u^{m+1}}{m+1}$	$\int \frac{u'}{u^m} dx = -\frac{1}{(m-1)\, u^{m-1}} \qquad m \neq 1$

$\int (a+bu)^m u' dx = \frac{(a+bu)^{m+1}}{b(m+1)}$	$m \neq -1, b \neq 0$
$\int \frac{u' dx}{a+bu} = \frac{\ln(a+bu)}{b}$	$b \neq 0$
$\int \frac{u' dx}{(a+bu)^m} = -\frac{1}{b(m-1)(a+bu)^{m-1}}$	$m \neq 1, b \neq 0$

$$\int ux\, dx = xU_1 - U_2$$

$$\int ux^2\, dx = x^2U_1 - 2xU_2 + U_3$$

$$\int ux^m\, dx = x^mU_1 - mx^{m-1}U_2 + m(m-1)x^{m-2}U_3 - m(m-1)(m-2)x^{m-3}U_4 + \ldots$$

10.3 RELATIONS OF DIFFERENTIABLE FUNCTIONS [$u = f(x)$, $v = g(x)$]

i. Notation of Functions

If u, v are differentiable functions of x,

$\left.\begin{matrix} u', u'', \ldots, u^{(n)} \\ v', v'', \ldots, v^{(n)} \end{matrix}\right]$ are successive derivatives of u and v, respectively

$\left.\begin{matrix} U_1, U_2, \ldots, U_n \\ V_1, V_2, \ldots, V_n \end{matrix}\right]$ are successive integrals of u and v, respectively

and a, b, m are constants

ii. Integrals of Different Functions

$$\int (u+v)\, dx = \int u\, dx + \int v\, dx$$

$$\int (u+v)^m dx = x(u+v) - \frac{x^2}{2!}(u'+v') + \frac{x^3}{3!}(u''+v'') - \ldots$$

$$= \int u(u+v)^{m-1}dx + \int v(u+v)^{m-1}dx$$

$$\int uv\, dx = U_1v - U_2v' + U_3v'' - U_4v''' + \ldots$$

$$= U_1v - \int U_1v'\, dx$$

$$= uV_1 - u'V_2 + u''V_3 - u'''V_4 + \ldots$$

$$= uV_1 - \int u'V_1\, dx$$

$$\int uv'\,dx = uv - \int u'v\,dx$$

$$\int u'v\,dx = uv - \int uv'\,dx$$

$$\int (u'v + uv')\,dx = uv$$

$$\int \frac{u'v - uv'}{(u+v)^2}\,dx = -\frac{v}{u+v}$$

$$\int \frac{u'v - uv'}{(u-v)^2}\,dx = -\frac{v}{u-v}$$

$$\int \frac{uv'}{v^2}\,dx = -\frac{u}{v} + \int \frac{u'}{v}\,dx$$

$$\int \frac{u'v - uv'}{v^2}\,dx = \frac{u}{v}$$

$$\int \frac{u'v - uv'}{uv}\,dx = \ln\frac{u}{v}$$

$$\int \frac{u'v - uv'}{u^2 + v^2}\,dx = \tan^{-1}\frac{u}{v}$$

$$\int \frac{u'v - uv'}{u^2 - v^2}\,dx = \frac{1}{2}\ln\frac{u+v}{u-v}$$

10.4 TABLE: INDEFINITE INTEGRALS—ALGEBRAIC FUNCTIONS

In the following integral formulas, $u = f(x)$, a, m = constants. The constant of integration C, is omitted.

$\int u^m\,du = \frac{u^{m+1}}{m+1}$	$m \neq -1$	$\int \frac{du}{u} = \ln\lvert u\rvert$	$u \neq 0$
$\int \frac{1}{u^m}\,du = \frac{u^{1-m}}{1-m}$	$m \neq 1$	$\int \sqrt[n]{u^m}\cdot du = \frac{nu\sqrt[n]{u^m}}{m+n}$	$m \neq -n$

$\int \frac{du}{a^2+u^2} = \frac{1}{a}\tan^{-1}\frac{u}{a} = -\frac{1}{a}\cot^{-1}\frac{u}{a}$	$a \neq 0$
$\int \frac{du}{a^2-u^2} = \frac{1}{a}\tanh^{-1}\frac{u}{a} = \frac{1}{2a}\ln\frac{a+u}{a-u}$	$u^2 < a^2$
$\int \frac{du}{a^2-u^2} = \frac{1}{a}\coth^{-1}\frac{u}{a} = \frac{1}{2a}\ln\frac{u+a}{u-a}$	$u^2 > a^2$

$$\int \frac{du}{\sqrt{a^2+u^2}} = \sinh^{-1}\frac{u}{a} = \ln\left(u + \sqrt{u^2+a^2}\right)$$

$$\int \frac{du}{\sqrt{a^2-u^2}} = \sin^{-1}\frac{u}{a} = -\cos^{-1}\frac{u}{a}$$

$$\int \frac{du}{\sqrt{u^2-a^2}} = \cosh^{-1}\frac{u}{a} = \ln\left(u + \sqrt{u^2-a^2}\right)$$

$$\int \frac{du}{u\sqrt{a^2+u^2}} = -\frac{1}{a}\operatorname{csch}^{-1}\frac{u}{a} = -\frac{1}{a}\sinh^{-1}\frac{a}{u} = -\frac{1}{a}\ln\frac{a+\sqrt{a^2+u^2}}{u}$$

$$\int \frac{du}{u\sqrt{a^2-u^2}} = -\frac{1}{a}\operatorname{sech}^{-1}\frac{u}{a} = -\frac{1}{a}\cosh^{-1}\frac{a}{u} = -\frac{1}{a}\ln\frac{a+\sqrt{a^2-u^2}}{u}$$

$$\int \frac{du}{u\sqrt{u^2-a^2}} = \frac{1}{a}\sec^{-1}\frac{u}{a} = \cos^{-1}\frac{a}{u}$$

$$\int \frac{du}{\sqrt{(u+a)(u+b)}} = 2\ln\left(\sqrt{u+a}+\sqrt{u+b}\right)$$

$$\int \frac{u\,du}{u^4+a^4} = \frac{1}{2a^2}\tan^{-1}\frac{u^2}{a^2}$$

$$\int \frac{u\,du}{u^4-a^2} = \frac{1}{4a^2}\ln\frac{u^2-a^2}{u^2+a^2}$$

10.5 TABLE OF INDEFINITE INTEGRALS—TRANSCENDENT FUNCTIONS

In the following integral formulas, $u = f(x)$ and a, b, m = constants. The constant of integration C is omitted.

$\int a^u\,du = \frac{a^u}{\ln a}$	$\int \frac{du}{a^u} = -\frac{1}{a^u \ln a}$
$\int e^u\,du = e^u$	$\int \ln u\,du = u\ln u - u$
$\int \sin au\,du = \frac{-1}{a}\cos au$	$\int \sinh au = \frac{1}{a}\cosh au$
$\int \cos au\,du = \frac{1}{a}\sin au$	$\int \cosh au = \frac{1}{a}\sinh au$
$\int \tan au\,du = \frac{-1}{a}\ln\cos au$	$\int \tanh au = \frac{1}{a}\ln\cosh au$
$\int \cot au\,du = \frac{1}{a}\ln\sin au$	$\int \coth au = \frac{1}{a}\ln\sinh au$
$\int \sec au\,du = \frac{1}{2a}\ln\frac{1+\sin au}{1-\sin au}$	$\int \operatorname{sech} au = \frac{1}{a}\tan^{-1}\sinh au$
$\int \csc au\,du = \frac{-1}{2a}\ln\frac{1+\cos au}{1-\cos au}$	$\int \operatorname{csch} au = \frac{1}{a}\ln\tanh\frac{au}{2}$
$\int \frac{du}{\sin au} = \frac{-1}{2a}\ln\frac{1+\cos au}{1-\cos au}$	$\int \frac{du}{\sinh au} = \frac{1}{a}\ln\tanh\frac{au}{2}$
$\int \frac{du}{\cos au} = \frac{1}{2a}\ln\frac{1+\sin au}{1-\sin au}$	$\int \frac{du}{\cosh au} = \frac{1}{a}\tan^{-1}\sinh au$
$\int \frac{du}{\tan au} = \frac{1}{a}\ln\sin au$	$\int \frac{du}{\tanh au} = \frac{1}{a}\ln\sinh au$
$\int \frac{du}{\cot au} = \frac{-1}{a}\ln\cos au$	$\int \frac{du}{\coth au} = \frac{1}{a}\ln\cosh au$

$\int \frac{du}{\sec au} = \frac{1}{a} \sin au$	$\int \frac{du}{\text{sech } au} = \frac{1}{a} \sinh au$
$\int \frac{du}{\csc au} = \frac{-1}{a} \cos au$	$\int \frac{du}{\text{csch } au} = \frac{1}{a} \cosh au$

10.6 INDEFINITE INTEGRALS—TYPICAL ALGEBRAIC SUBSTITUTIONS

Integral	*Substitution*
$\int \frac{dx}{(x-a)^m}$ $\int \frac{f'(x)}{f(x)} dx$	$\frac{1}{x-a} = t \quad dx = -\frac{dt}{t^2}$ $f(x) = t \qquad f'(x)\, dx = dt$
$\int x^m (a + bx^n)^p\, dx$	If $p = \alpha/\beta$ is a fraction and $(m+1)/n$ is an integer, use $t = \sqrt[\beta]{a + bx^n}$ If $p = \alpha/\beta$ is a fraction and $(m+1)/n + p$ is an integer, use $t = \sqrt[\beta]{\frac{a + bx^n}{x^n}}$ If p is an interger, use the binomial expansion.
$\int f(x)^2\, dx$ $\int f\left[x; \left(\frac{ax+b}{cx+d}\right)^m\right] dx$ $\int f\left[x; \left(\frac{ax+b}{cx+d}\right)^m; \left(\frac{ax+b}{cx+d}\right)^n; \ldots\right] dx$	$x^2 = t \qquad 2x\, dx = dt$ $\left(\frac{ax+b}{cx+d}\right)^m = t^m$ $\left(\frac{ax+b}{cx+d}\right)^r = t^r$ r is the least common multiple of m and n.
$\int f(x; \sqrt{ax^2 + bx + c})\, dx$	If $a > 0$, use $x = \frac{t^2 - c}{b - 2t\sqrt{a}}$ $dx = 2\frac{-t^2\sqrt{a} + bt - c\sqrt{a}}{\left(b - 2t\sqrt{a}\right)^2} dt$ If $c > 0$, use $x = \frac{2t\sqrt{c} - b}{a - t^2}$ $dx = \frac{2a\sqrt{c} - 2bt + 2t^2\sqrt{c}}{(a - t^2)^2}\, dt$

	If $ax^2 + bx + c = a(x - \alpha)(x - \beta)$, use $x = \dfrac{t^2 \alpha - a\beta}{t^2 - a}$ $dx = \dfrac{2at(\beta - \alpha)}{(t^2 - a)^2} dt$
$\int f(x); \sqrt[m]{ax+b} \cdot dx$	$\sqrt[m]{ax+b} = t$ $dx = \dfrac{mt^{m-1}\,dt}{a}$

10.7 INDEFINITE INTEGRALS—TYPICAL TRANSCENDENT SUBSTITUTIONS

Integral and substitution		*Transformations*	
$\int f(x; \sqrt{a^2 - x^2})\,dx$		$\sin t = \dfrac{x}{a}$	$\cos t = \dfrac{\sqrt{a^2 - x^2}}{a}$
$x = a \sin t$	$dx = a \cos t\,dt$	$\tan t = \dfrac{x}{\sqrt{a^2 - x^2}}$	$\cot t = \dfrac{\sqrt{a^2 - x^2}}{x}$
$\int f(x; \sqrt{a^2 - x^2})\,dx$		$\sinh t = \dfrac{x}{\sqrt{a^2 - x^2}}$	$\cosh t = \dfrac{a}{\sqrt{a^2 - x^2}}$
$x = a \tanh t$	$dx = \dfrac{a\,dt}{\cosh^2 t}$	$\tanh t = \dfrac{x}{a}$	$\coth t = \dfrac{a}{x}$
$\int f(x; \sqrt{a^2 + x^2})\,dx$		$\sin t = \dfrac{x}{\sqrt{a^2 + x^2}}$	$\cos t = \dfrac{a}{\sqrt{a^2 + x^2}}$
$x = a \tan t$	$dx = \dfrac{a\,dt}{\cos^2 t}$	$\tan t = \dfrac{x}{a}$	$\cot t = \dfrac{a}{x}$
$\int f(x; \sqrt{a^2 + x^2})\,dx$		$\sinh t = \dfrac{x}{a}$	$\cosh t = \dfrac{\sqrt{a^2 + x^2}}{a}$
$x = a \sinh t$	$dx = \cosh t\,dt$	$\tanh t = \dfrac{x}{\sqrt{a^2 + x^2}}$	$\coth t = \dfrac{\sqrt{a^2 + x^2}}{x}$
$\int f(x; \sqrt{x^2 - a^2})\,dx$		$\sin t = \dfrac{\sqrt{x^2 - a^2}}{x}$	$\cos t = \dfrac{a}{x}$
$x = \dfrac{a}{\cos t}$	$dx = \dfrac{a \sin t\,dt}{\cos^2 t}$	$\tan t = \dfrac{\sqrt{x^2 - a^2}}{a}$	$\cot t = \dfrac{a}{\sqrt{x^2 - a^2}}$
$\int f(x; \sqrt{x^2 - a^2})\,dx$		$\sinh t = \dfrac{\sqrt{x^2 - a^2}}{a}$	$\cosh t = \dfrac{x}{a}$
$x = a \cosh t$	$dx = a \sinh t\,dt$	$\tanh t = \dfrac{\sqrt{x^2 - a^2}}{x}$	$\coth t = \dfrac{x}{\sqrt{x^2 - a^2}}$
$\int f(\sin x; \cos x; \tan x; \cot x)\,dx$		$\sin x = \dfrac{2t}{1+t^2}$	$\cos x = \dfrac{1-t^2}{1+t^2}$

$\tan\dfrac{x}{2}=t$	$dx=\dfrac{2dt}{1+t^2}$	$\tan x=\dfrac{2t}{1-t^2}$	$\cot x=\dfrac{1-t^2}{2t}$
$\int f(\sinh x; \cosh x; \tanh x; \coth x)\,dx$		$\sinh x=\dfrac{2t}{1-t^2}$	$\cosh x=\dfrac{1+t^2}{1-t^2}$
$\tanh\dfrac{x}{2}=t$	$dx=\dfrac{2dt}{1-t^2}$	$\tanh x=\dfrac{2t}{1+t^2}$	$\coth x=\dfrac{1+t^2}{2t}$
$\int f(e^x)\,dx$		$\int f(a^x)\,dx$	
$e^x=t$	$dx=\dfrac{dt}{t}$	$a^x=t$	$dx=\dfrac{dt}{t\ln t}$

10.8 DEFINITE INTEGRALS OF FUNCTIONS

i. Definitions

If $f(x)$ is continuous in the closed interval $[a, b]$ and this interval is divided into n equal parts by the points $a, x_1, x_2, \ldots, x_{n-1}, b$ such that $\Delta x=(b-a)/n$, then the *definite integral* of $f(x)$ with respect to x, between the limits $x=a$ to $x=b$ is

$$\int_a^b f(x)\,dx=\lim_{n\to\infty}\sum_1^n f(x_i)\,\Delta x=\left[\int f(x)\,dx\right]_a^b=[F(x)]_a^b=F(b)-F(a)$$

where $F(x)$ is a function, the derivative of which with respect to x is $f(x)$.

The numbers a and b are called, the lower and *upper limits of integration respectively*, and $[a, b]$ is called the *range of integration*.

Geometrically, the definite integral of $f(x)$ with respect to x, between limits $x=a$ to $x=b$, is the *area* bounded by $f(x)$, the x-axis, and the verticals through the end points of a and b.

ii. Rules of Limits

$$\int_a^b=-\int_b^a \qquad \int_a^b+\int_b^c=\int_a^c \qquad \int_a^c-\int_b^c=\int_a^b \qquad \int_a^a=0$$

iii. Fundamental Theorems

$$\int_a^b dx=b-a \qquad \int_a^b \lambda f(x)\,dx=\lambda\int_a^b f(x)dx$$

$$\int_a^b (f(x)+g(x))\,dx=\int_a^b f(x)\,dx+\int_a^b g(x)\,dx$$

$$\int_a^b f(x)\frac{dg(x)}{dx}dx=[f(x)g(x)]_a^b-\int_a^b \frac{df(x)}{dx}g(x)\,dx$$

$$\int_a^b f(x)\,dx=\int_{\phi(a)}^{\phi(b)} f(\phi(t))\,\phi'(t)\,dt$$

$$\int_a^x f(t)\,dt=F(x)-F(a) \qquad \int_a^{\phi(x)} f(t)\,dt=F[\phi(x)]-F(a)$$

$$\frac{d}{dx}\int_a^x f(t)\,dt = f(x) \qquad \frac{d}{dx}\int_a^{\phi(x)} f(t)\,dt = F[\phi(x)]\frac{d\phi(x)}{dx}$$

$$\frac{\partial}{\partial\alpha}\int_{\phi_1(\alpha)}^{\phi_2(\alpha)} f(x,\alpha)\,dx = \int_{\phi_1(\alpha)}^{\phi_2(\alpha)} \frac{\partial f(x,\alpha)}{\partial\alpha}dx + f(\phi_2(\alpha),\alpha)\frac{\partial\phi_2(\alpha)}{\partial\alpha} - f(\phi_1(\alpha),\alpha)\frac{\partial\phi_1(\alpha)}{\partial\alpha}$$

iv. Mean Values

$$\text{Arithmetic mean value} = \frac{\int_a^b f(x)\,dx}{b-a} \qquad \text{Quadratic mean value} = \frac{\int_a^b f(x)^2\,dx}{b-a}$$

10.9 DEFINITE INTEGRALS

$$\int_a^b (x-a)^m (b-x)^n\,dx = (b-a)^{m+n+1}\frac{m!n!}{(m+n+1)!} \qquad m, n = 0, 1, 2,...$$

$$\int_0^a x^{p-1}(a^m - x^m)^n\,dx = \frac{n!m^n a^{mn+p}}{p(p+m)(p+2m)...(p+mn)} \qquad a > 0$$

$p, m = 1, 2,...$
$n = 0, 1, 2,...$

$$\int_0^{\pi/2} \sin x\,dx = 1 \qquad \int_0^{\pi/2} \cos x\,dx = 1$$

$$\int_0^{\pi/2} \sin^2 x\,dx = \frac{\pi}{4} \qquad \int_0^{\pi/2} \cos^2 x\,dx = \frac{\pi}{4}$$

$$\int_0^{\pi/2} \sin^{2n} x\,dx = \int_0^{\pi/2} \cos^{2n} x\,dx = \frac{(1)(3)(5)...(2n-1)}{(2)(4)(6)...(2n)}\frac{\pi}{2}$$

$$\int_0^{\pi/2} \sin^{2n+1} x\,dx = \int_0^{\pi/2} \cos^{2n+1} x\,dx = \frac{(2)(4)(6)...(2n)}{(3)(5)(7)...(2n+1)} \qquad n = 1, 2,...$$

$$\int_0^{\pi/2} \sin^{2m-1} x \cos^{2n-1} x\,dx = \frac{\Gamma(m)\Gamma(n)}{2\Gamma(m+n)} \qquad m, n = \text{positive integers}$$

$\Gamma(m) = (m-1)!$ $\qquad \Gamma(n) = (n-1)!$ $\qquad \Gamma(m+n) = (m+n-1)!$

$$\int_0^{\pi/2} \tan^m x\,dx = \frac{\Gamma[(m+1)/2]\,\Gamma[(1-m)/2]}{2} \qquad |m| < 1$$

$$= \frac{\pi}{2\cos m\pi/2} \qquad 0 < m < 1$$

$$\int_0^{\pi} \sin x dx = 2 \qquad \int_0^{\pi} \cos x dx = 0$$

$$\int_0^{\pi} \sin^2 x dx = \frac{\pi}{2} \qquad \int_0^{\pi} \cos^2 x dx = \frac{\pi}{2}$$

$$\int_0^\pi \cos mx \sin nx\, dx = \begin{cases} 0 & \text{if } m \neq n \\ \dfrac{\pi}{2} & \text{if } m = n \end{cases} \qquad m, n = 1, 2, \ldots$$

$$\int_0^\pi \cos mx \sin nx\, dx = \begin{cases} 0 & \text{if } m \neq n \text{ and } m + b \text{ is even} \\ \dfrac{2n}{n^2 - m^2} & \text{if } m \neq n \text{ and } m + n \text{ is odd} \\ 0 & \text{if } m = n \end{cases} \qquad m, n = 1, 2, \ldots$$

$$\int_0^\pi \sin mx \sin nx\, dx = \begin{cases} 0 & \text{if } m \neq n \\ \dfrac{\pi}{2} & \text{if } m = n \end{cases} \qquad m, n = 1, 2, \ldots$$

$$\int_{-a}^{+a} \cos\frac{m\pi x}{a} \cos\frac{n\pi x}{a}\, dx = \int_{-a}^{+a} \sin\frac{m\pi x}{a} \sin\frac{n\pi x}{a}\, dx = \begin{cases} 0 & m \neq n \\ a & m = n \end{cases} \qquad m, n = 1, 2, \ldots$$

$$\int_{-a}^{+a} \cos\frac{m\pi x}{a} \sin\frac{n\pi x}{a}\, dx = 0 \qquad m, n = 1, 2, \ldots$$

$$\int_0^{\pi/2} \ln \cos x\, dx = \int_0^{\pi/2} \ln \sin x\, dx = -\frac{\pi}{2} \ln 2 \qquad \int_0^{\pi/2} \ln \tan x\, dx = 0$$

10.10 DOUBLE INTEGRALS

The calculation of a double integral is performed by successive evaluation of two definite integrals.

i. Cartesian Coordinates

If P_1, P_2 and Q_1, Q_2 are points on a contour enclosing area A selected, so that they identify extreme coordinates x and y respectively then

$$A = \iint_A f(x, y)\, dx\, dy = \int_a^b \left[\int_{f_1(x)}^{f_2(x)} f(x, y)\, dy \right] dx$$

where the boundary of A consists of two continuous curves $y = f_1(x)$ and $y = f_2(x)$. The boundary of A is met by a line parallel to the y-axis in at most two points, and $x = a$, $x = b$ are the extreme values of x on A; or

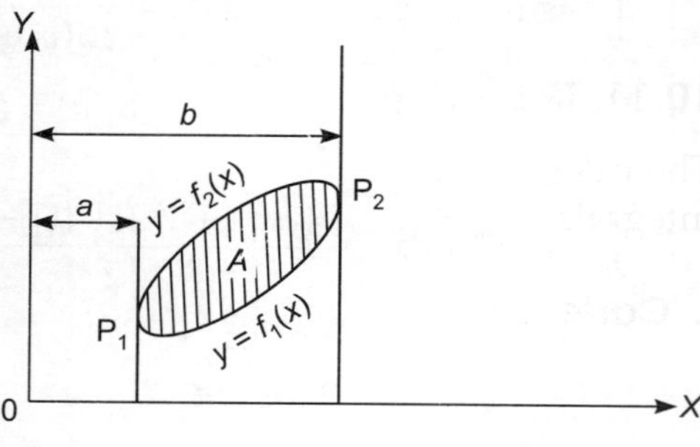

Fig. 10.2

$$A = \iint_A f(x,y)\, dx\, dy = \int_c^d \left[\int_{g_1(y)}^{g_2(y)} f(x,y)\, dx \right] dy$$

where the boundary of A consists of two continuous curves $x = g_1(y)$ and $x = g_2(y)$. The boundary of A is met by a line parallel to the x-axis in at most two points, and $y = c$, $y = d$ are the extreme values of y and A.

If neither of these conditions is satisfied, the area A must be divided in two or more portions. Then

$$A = \iint_A f(x,y)\,dx\,dy$$

$$= \iint_{A_1} f(x,y)\,dx\,dy + \iint_{A_2} f(x,y)\,dx\,dy$$

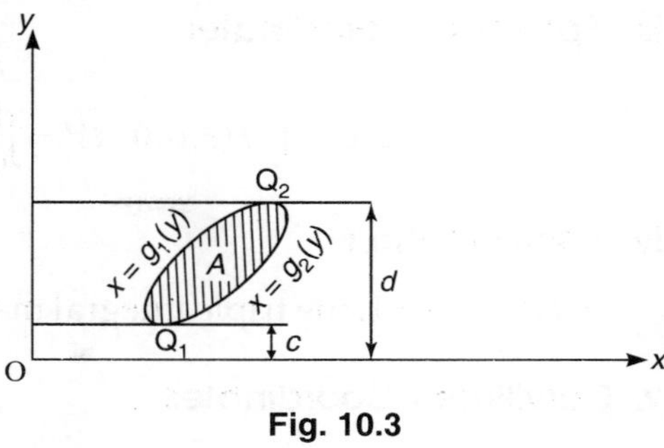

Fig. 10.3

ii. Polar Coordinates

If S_1, S_2 are points on a contour enclosing area A, selected so that they identify extreme polar coordinates, then

$$A = \iint_A f(r,\theta)\,r\,dr\,d\theta = \int_\alpha^\beta \left[\int_{f_1(\theta)}^{f_2(\theta)} f(r,\theta)\,r\,dr\right] d\theta$$

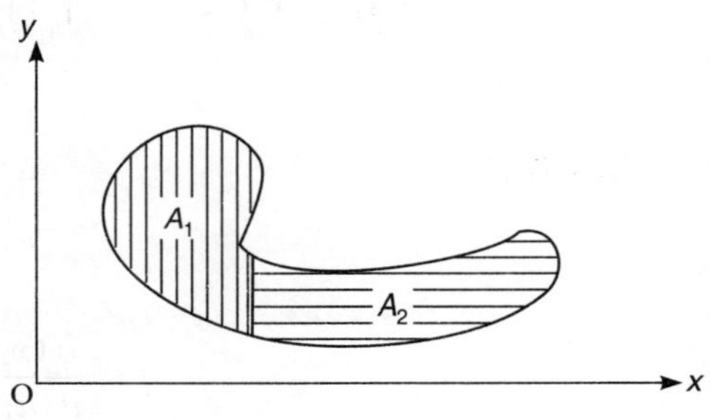

Fig. 10.4

where the boundary of A consists of two continuous curves $r = f_1(\theta)$ and $r = f_2(\theta)$. The boundary of A is met by two tangents from O in at most two points and $\theta = \alpha$, $\theta = \beta$, are the extreme values of θ on A.

iii. Interpretation

If $f(x, y)$ has the same sign over A, the double integral may be interpreted as the *volume of a vertical cylinder* bounded below by the region A projected in the xy plane and above by the surface $z = f(x, y)$.

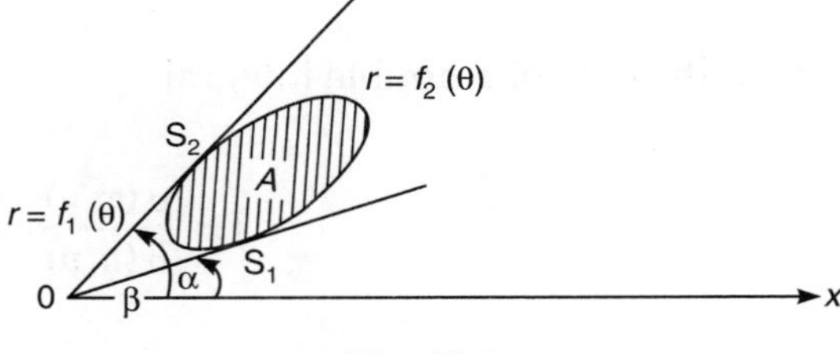

Fig. 10.5

10.11 TRIPLE INTEGRALS

The calculation of a triple integral is performed by successive evaluation of three definite integrals.

i. Cartesian Coordinates

$$V = \iiint_V f(x,y,z)\,dV = \int_a^b \int_{y_1(x)}^{y_2(x)} \int_{z_1(x,y)}^{z_2(x,y)} f(x,y,z)dz\,dy\,dx$$

ii. Cylindrical Coordinates

$$V = \iiint_V f(r,\theta,z)\,dV = \int_\alpha^\beta \int_{r_1(\theta)}^{r_2(\theta)} \int_{z_1(r,\theta)}^{z_2(r,\theta)} f(r,\theta,z)dz\,d\theta\,dr$$

iii. Spherical Coordinates

$$V = \iiint_V f(r,\phi,\theta)\,dV = \int_{\alpha}^{\beta}\int_{\phi_1(\theta)}^{\phi_2(\theta)}\int_{r_1(\phi,\theta)}^{r_2(\phi,\theta)} f(r,\phi,\theta)\,r^2 \sin\phi\,dr\,d\phi\,d\theta$$

iv. Interpretation

If $f(x, y, z) = 1$, the triple integral may be interpreted as the *volume enclosed by the region V*.

v. Curvilinear Coordinates

If $x = x(u, v, w)$, $y = y(u, v, w)$, $z = z(u, v, w)$, and $f(x, y, z) = g(u, v, w)$, then

$$V = \iiint_V f(x,y,z)\,dV = \int_{u_1}^{u_2}\int_{v_1(u)}^{v_2(u)}\int_{w_1(u,v)}^{w_2(u,v)} g(u,v,w)\underbrace{\frac{\partial(x,y,z)}{\partial(u,v,w)}}_{J}\,du\,dv\,dw$$

in which

$$J = \frac{\partial(x,y,z)}{\partial(u,v,w)} = \begin{vmatrix} \frac{\partial x}{\partial u} & \frac{\partial y}{\partial u} & \frac{\partial z}{\partial u} \\ \frac{\partial x}{\partial v} & \frac{\partial y}{\partial v} & \frac{\partial z}{\partial v} \\ \frac{\partial x}{\partial w} & \frac{\partial y}{\partial w} & \frac{\partial z}{\partial w} \end{vmatrix}$$

In the case of a double integral,

$$J = \frac{\partial(x,y)}{\partial(u,v)} = \begin{vmatrix} \frac{\partial x}{\partial u} & \frac{\partial y}{\partial u} \\ \frac{\partial x}{\partial v} & \frac{\partial y}{\partial v} \end{vmatrix}$$

Note: The order of integration is arbitrary; thus a double integral can be evaluated in two ways, and a triple integral can be evaluated in six ways.

10.12 DEFINITE INTEGRALS OF PLANE GEOMETRY

a. *Area by single integration*

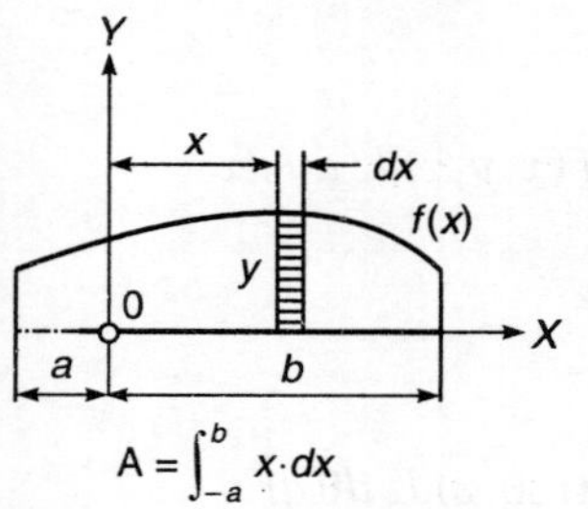

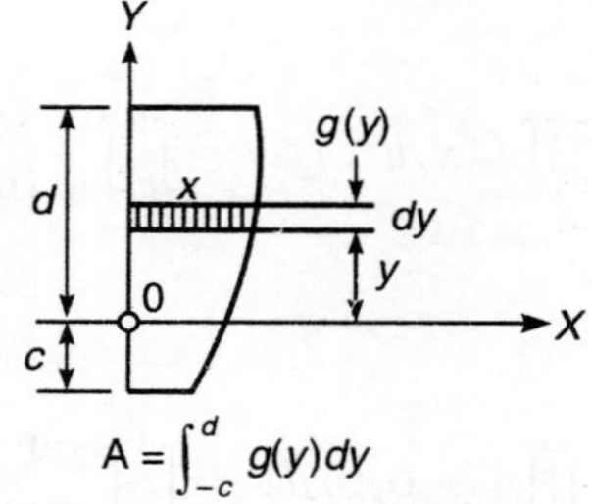

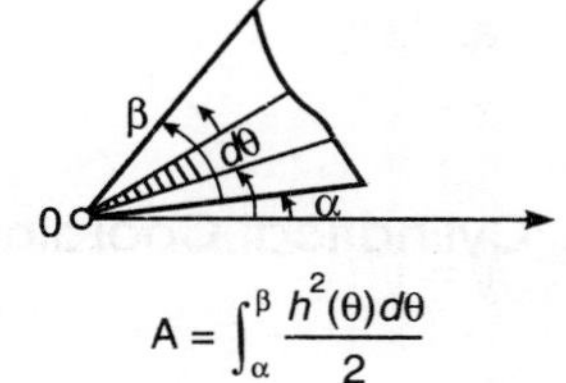

Fig. 10.6

b. *Area by double integration*

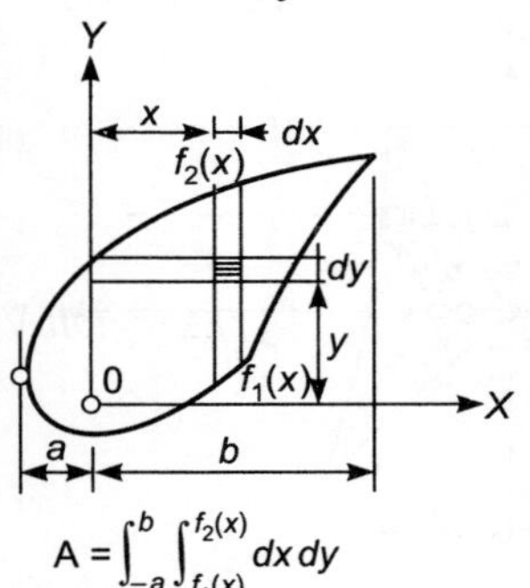

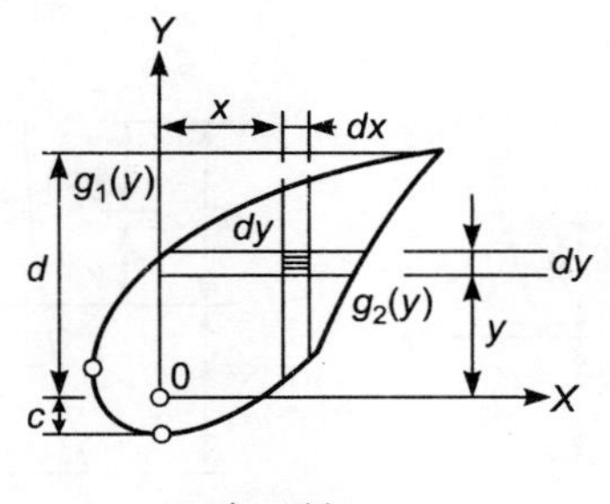

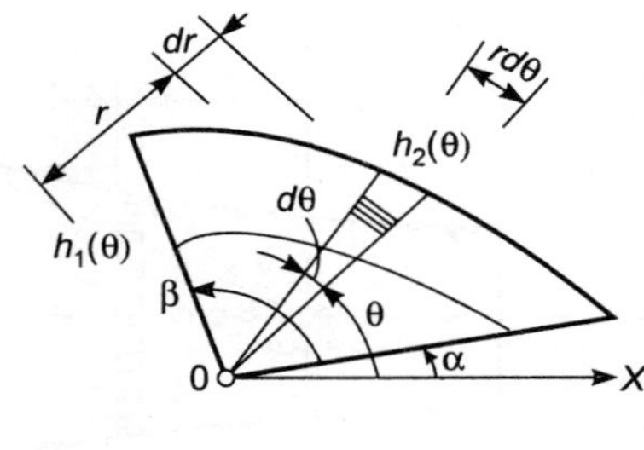

Fig. 10.7

c. *Length of plane curve*

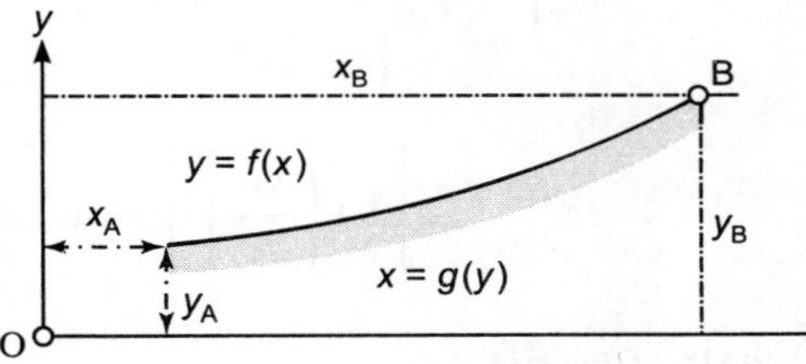

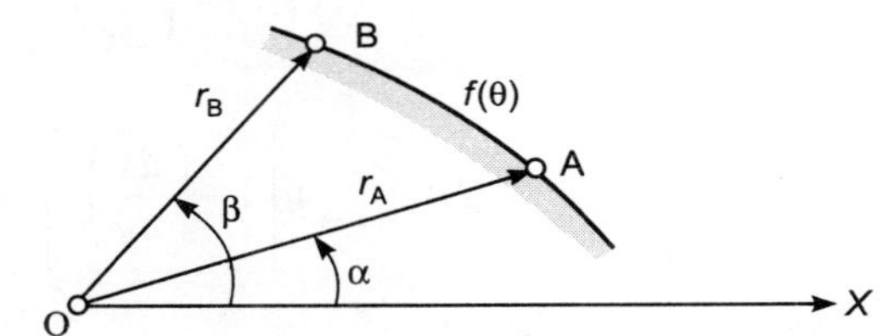

Fig. 10.8

$$L_{AB} = \int_{x_A}^{x_B} \sqrt{1+\left(\frac{dy}{dx}\right)^2}\, dx \qquad\qquad L_{AB} = \int_{\alpha}^{\beta} \sqrt{r^2+\left(\frac{dr}{d\theta}\right)^2}\, d\theta$$

$$= \int_{y_A}^{y_B} \sqrt{1+\left(\frac{dx}{dy}\right)^2}\, dy \qquad\qquad = \int_{r_A}^{r_B} \sqrt{1+r^2\left(\frac{d\theta}{dr}\right)^2}\, dr$$

$$x = x(t) \qquad y = y(t) \qquad L_{AB} = \int_{t_A}^{t_B} \sqrt{[\dot{x}(t)]^2+[\dot{y}(t)]^2}\, dt$$

$$\dot{x} = \frac{dx(t)}{dt} \qquad \dot{y} = \frac{dy(t)}{dt}$$

10.13 DEFINITE INTEGRALS OF SPACE GEOMETRY

a. *Length of space curve*

$x = x(t)$

$y = y(t)$ $$L_{AB} = \int_{t_A}^{t_B} \sqrt{\left(\frac{dx}{dt}\right)^2+\left(\frac{dy}{dt}\right)^2+\left(\frac{dz}{dt}\right)^2}\, dt$$

$z = z(t)$

b. *Area of surface of revolution generated by rotation of plane curve*

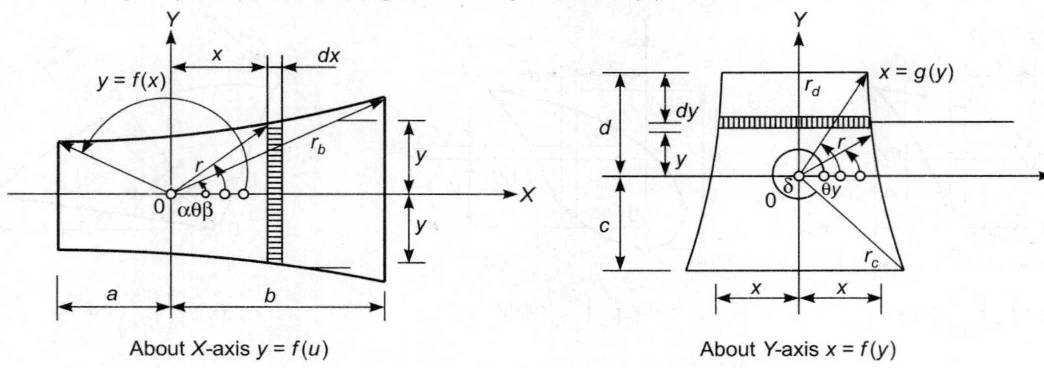

Fig. 10.9

$$S = 2\pi\int_{-a}^{+b} f(x)\left[1+\left(\frac{dy}{dx}\right)^2\right]^{1/2} dx \qquad S = 2\pi\int_{-c}^{d} g(y)\left[1+\left(\frac{dx}{dy}\right)^2\right]^{1/2} dy$$

$$= 2\pi\int_{f(-a)}^{f(+b)} y\left[1+\left(\frac{dx}{dy}\right)^2\right]^{1/2} dy \qquad = 2\pi\int_{g(-c)}^{g(+d)} x\left[1+\left(\frac{dy}{dx}\right)^2\right]^{1/2} dx$$

About X-axis $r = f(\theta)$ About Y-axis $r = f(\theta)$

$$S = 2\pi\int_{\alpha}^{\beta} r\sin\theta\left[r^2+\left(\frac{dr}{d\theta}\right)^2\right]^{1/2} d\theta \qquad S = 2\pi\int_{\gamma}^{\delta} r\cos\theta\left[r^2+\left(\frac{dr}{d\theta}\right)^2\right]^{1/2} d\theta$$

$$= 2\pi\int_{r_a}^{r_b} r\sin\theta\left[1+r^2\left(\frac{d\theta}{dr}\right)^2\right]^{1/2} dr \qquad = 2\pi\int_{r_c}^{r_d} r\cos\theta\left[1+r^2\left(\frac{d\theta}{dr}\right)^2\right]^{1/2} dr$$

c. *Area of surface* $z = f(x, y)$

$$S = \iint_A \left[1+\left(\frac{\partial z}{\partial x}\right)^2+\left(\frac{\partial z}{\partial y}\right)^2\right]^{1/2} dx\,dy \quad \text{integration over the projection A on } xy \text{ plane}$$

d. *Volume of body of revolution generated by rotation of plane curve*

About X-axis $y = f(x)$ About Y-axis $x = g(y)$

$$V = \pi\int_{-a}^{+b} [f(x)]^2 dx \qquad V = \pi\int_{-c}^{+d} [g(y)]^2 dy$$

e. *Volume of body with known parallel cross-section*

$$V = \int_{-a}^{+b} A_x\, dx \qquad V = \int_{-c}^{d} A_y\, dy$$

f. *Volume of body of general shape*

$$V = \iiint_V dx\,dy\,dz = \iiint_V r\,dr\,d\theta\,dz = \iiint_V r^2 \sin\phi\, d\theta\, d\phi\, dr$$

10.14 PROPERTIES OF PLANE AREAS USING DOUBLE INTEGRALS

i. System $\overline{XY}$ (O = origin, C = centroid)

Area $dA = dx\,dy$ $A = \iint_A dA$

Static moments $M_x = \iint_A y\,dA$ $M_y = \iint_A x\,dA$

Coordinates of centroid

$$x_C = \frac{M_y}{A} \qquad y_C = \frac{M_x}{A}$$

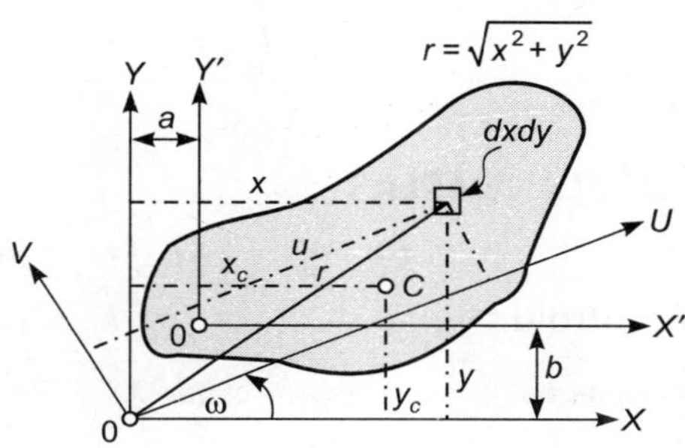

Fig. 10.10

Moments of inertia	Products of inertia	Polar moment of inertia
$I_{xx} = \iint_A y^2\,dA$	$I_{xy} = \iint_A xy\,dA$	$J_0 = \iint_A r^2\,dA$
$I_{yy} = \iint_A x^2\,dA$	$I_{yx} = \iint_A yx\,dA$	$= I_{xx} + I_{yy}$
Radii of gyration		
$k_x = \sqrt{\frac{I_{xx}}{A}}$	$k_y = \sqrt{\frac{I_{yy}}{A}}$	$k_0 = \sqrt{\frac{J_0}{A}}$

ii. System $\overline{X'Y'}$ (O′ origin, C = centroid)

Static moments $M'_x = M_x - bA$ $M'_y = M_y - aA$

Coordinates of centroid $x'_C = x_C - a$ $y'_C = y_C - b$

Moments of inertia	Products of inertia	Polar moment of inertia
$I'_{xx} = I_{xx} - 2bM_x + b^2A$ $I'_{yy} = I_{yy} - 2aM_y + a^2A$	$I'_{xy} = I'_{yx}$ $= I_{xy} - bM_x - aM_y + abA$	$J'_0 = J_0 + r^2A - 2(aM_x + bM_y)$
Radii of gyration		
$k'_x = \sqrt{k_x^2 - 2by_C + b^2}$	$k'_y = \sqrt{k_y^2 - 2ax_c + a^2}$	$k'_0 = \sqrt{k_0^2 - 2[ax_C + 2(ax_C + by_C) + r^2]}$

iii. System UV (O = origin, C = centroid)

Static moments

$$M_u = M_x \cos\omega + M_y \sin\omega$$
$$M_v = -M_x \sin\omega + M_y \cos\omega$$

Moments and products of inertia

$$I_{uu} = I_{xx}\cos^2\omega + I_{yy}\sin^2\omega - I_{xy}\sin 2\omega$$
$$I_{vv} = I_{xx}\sin^2\omega + I_{yy}\cos^2\omega + I_{xy}\sin 2\omega$$
$$I_{uv} = I_{vu} = \frac{I_{xx} - I_{yy}}{2}\sin 2\omega + I_{xy}\cos 2\omega$$

Principal moments of inertia

Position of principal axes

$$\tan 2\omega = -\frac{2I_{xy}}{I_{xx} - I_{yy}}$$

Principal values

$$I_{1,2} = \frac{I_{xx} + I_{yy}}{2} \pm \frac{1}{2}\sqrt{\left(I_{xx} - I_{yy}\right)^2 + 4I_{xy}^2}$$

10.15 TABLE: PROPERTIES OF PLANE AREAS

A = area

x_C, y_C = coordinates of centroid

$I_{AA}, I_{BB}, I_{CC}, I_{xx}, I_{yy}$ = moments of inertia

I_{AB}, I_{xy} = products of inertia

Area and centroid	*Moments of inertia*	*(Products of inertia)*	*Plane area*
Square $A = a^2$ $x_C = \frac{a}{2}$ $y_C = \frac{a}{2}$	$I_{AA} = \frac{a^4}{12}$ $I_{BB} = \frac{a^4}{12}\ I_{AB} = 0$	$I_{xx} = \frac{a^4}{3}$ $I_{yy} = \frac{a^4}{3}$ $I_{xy} = \frac{a^4}{4}$	
Rectangle $A = ab$ $x_C = \frac{a}{2}$ $y_C = \frac{b}{2}$	$I_{AA} = \frac{ab^3}{12}$ $I_{BB} = \frac{a^3b}{12}$ $I_{AB} = 0$	$I_{xx} = \frac{ab^3}{3}$ $I_{yy} = \frac{a^3b}{3}$ $I_{xy} = \frac{a^2b^2}{4}$	
Triangle $A = \frac{ah}{2}$ $y_C = \frac{h}{3}$	$I_{AA} = \frac{ah^3}{36}$	$I_{xx} = \frac{ah^3}{12}$ $I_{TT} = \frac{ah^3}{4}$	

Equal rectangle $A = a(h-b)$ $x_C = \frac{a}{2}$ $y_C = \frac{h}{2}$	$I_{AA} = \frac{a(h^3-b^3)}{12}$ $I_{BB} = \frac{ua^3}{6}$ $I_{AB} = 0$	$I_{xx} = \frac{a(4h^3-3bu^2-b^3)}{12}$ $I_{yy} = \frac{2ua^3}{3}$ $I_{xy} = \frac{a^2h}{4}(h-b)$ 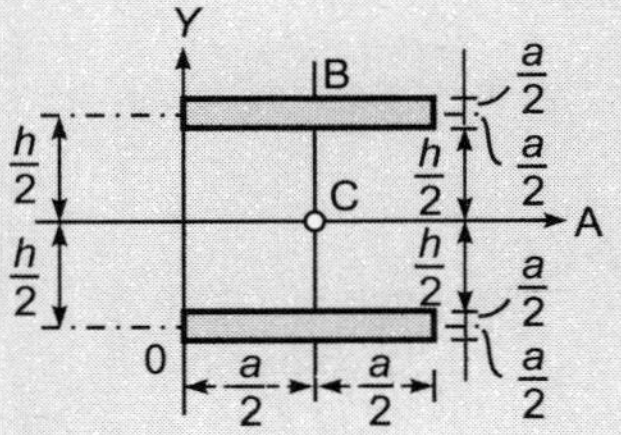
Trapezoid $A = (a+b)h$ $e = \frac{h}{3}\frac{a+2b}{a+b}$ $f = \frac{h}{3}\frac{2a+b}{a+b}$	$I_{AA} = \frac{h^3,}{18}\frac{a^2+4ab+b^2}{(a+b)}$ $I_{BB} = \frac{h}{6}\frac{a^4-b^4}{a-b}$ $I_{AB} = 0$	$I_{xx} = \frac{h^3}{6}(a+3b)$ $I_{TT} = \frac{h^3}{6}(3a+b)$ 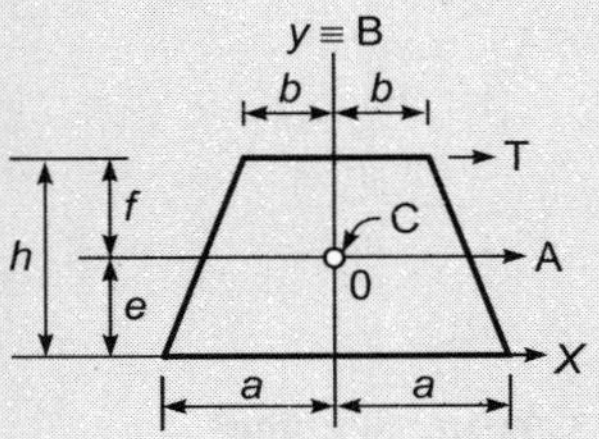
Regular polygon $A = na^2 \cot\alpha$ $r = a\cot\alpha$ $R = \frac{a}{\sin\alpha}$	$I_{AA} = \frac{A}{12}[3r^2+a^2]$ $= \frac{A}{12}[3R^2-2a^2]$ $= I_{BB} = I_{CC}$	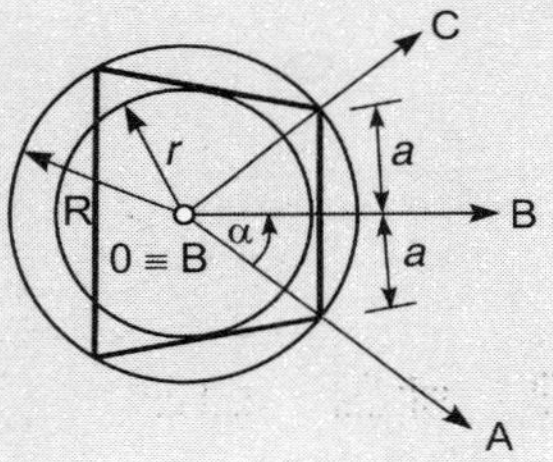

10.16 TABLE: PROPERTIES OF PLANE AREAS

A = area

x_C, y_C = coordinates of centroid

$I_{AA}, I_{BB}, I_{CC}, I_{xx}, I_{yy}$ = moments of inertia

I_{AB}, I_{xy} = products of inertia

Area and centroid	*Moments of inertia*	*Products inertia*
Circle $A = \pi a^2$ $x_C = a$ $y_C = a$	$I_{AA} = \frac{\pi a^4}{4}$ $I_{BB} = \frac{\pi a^4}{4}$ $I_{AB} = 0$	$I_{xx} = \frac{5\pi a^4}{4}$ $I_{yy} = \frac{5\pi a^4}{4}$ $I_{xy} = \pi a^4$ 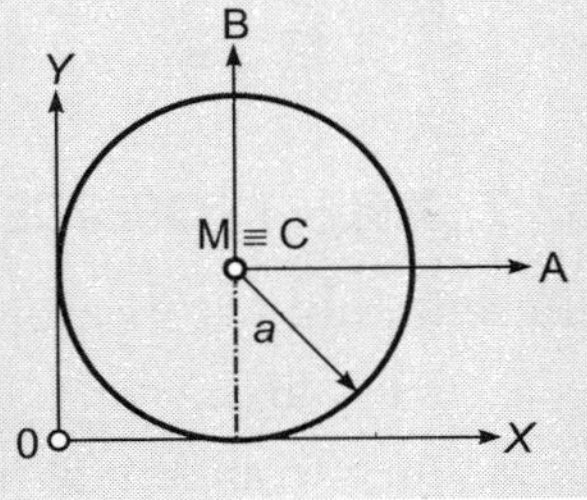
Hollow circle $A = \pi(b^2-a^2)$ $x_C = b$ $y_C = b$	$I_{AA} = \frac{\pi(b^4-a^4)}{4}$ $I_{BB} = \frac{\pi(b^4-a^4)}{4}$ $I_{AB} = 0$	$I_{xx} = I_{AA} + b^2A$ $I_{yy} = I_{xx}$ $I_{xy} = \pi b^2(b^2-a^2)$ 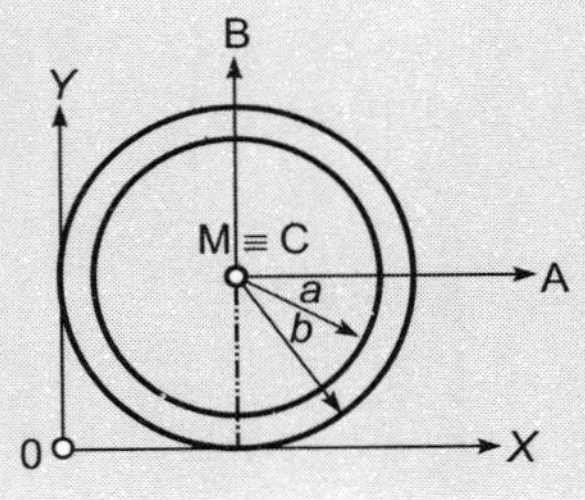

Half hollow circle $A = \frac{\pi(b^2 - a^2)}{2}$ $x_C = b$ $y_C = \frac{4(b^3 - a^3)}{3\pi(b^2 - a^2)}$	$I_{AA} = I_{xx} - y_C^2 A$ $I_{BB} = \frac{\pi(b^4 - a^4)}{8}$ $I_{AB} = 0$	$I_{xx} = \frac{\pi(b^4 - a^4)}{8}$ $I_{yy} = I_{xx} + b^2 A$ $I_{xy} = x_C\, y_C\, A$	
Circular sector $A = \alpha a^2$ $x_C = \frac{2a \sin\alpha}{3\alpha}$ $y_C = 0$	$I_{xx} = \frac{a^2}{4}(\alpha - \sin\alpha\cos\alpha)$ $I_{yy} = \frac{a^4}{4}(\sin\alpha\cos\alpha)$ $I_{xy} = 0$		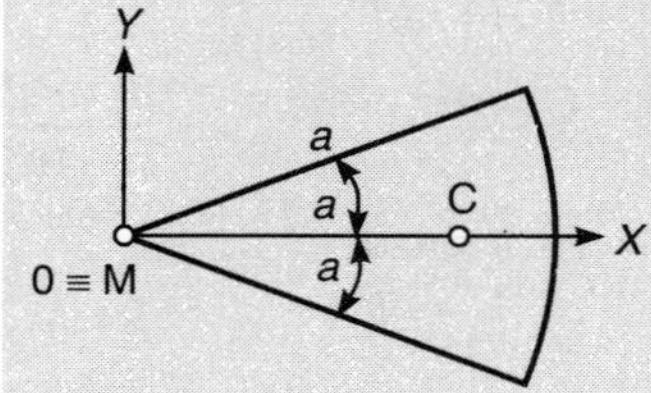
Circular segment $A = a^2(\alpha - \sin\alpha\cos\alpha)$ $x_C = \frac{2a}{3}\frac{\sin^3\alpha}{\alpha - \sin\alpha\cos\alpha}$ $y_C = 0$	$I_{xx} = \frac{Aa^2}{4}\left(1 - \frac{2\Delta}{3}\right)$ $I_{yy} = \frac{Aa^2}{4}(1 + 2\Delta)$ $\Delta = \frac{\sin^3\alpha\cos\alpha}{\alpha - \sin\alpha\cos\alpha}$		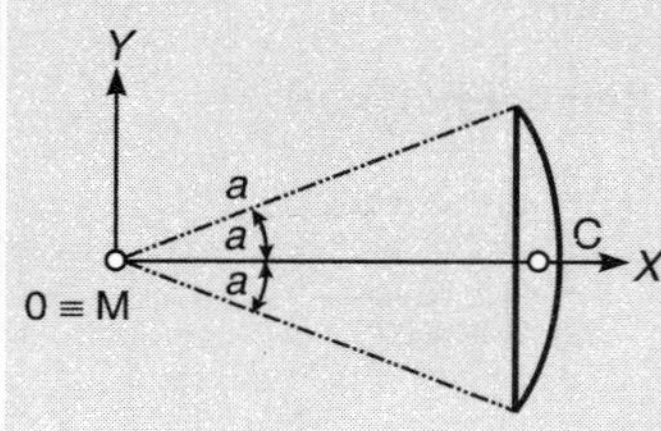
Half circle $A = \frac{\pi b^2}{2}$ $x_C = b$ $y_C = \frac{4b}{3\pi}$	$I_{AA} = 0.1098\, b^4$ $I_{BB} = \frac{\pi b^4}{8}$ $I_{AB} = 0$	$I_{xx} = \frac{\pi b^4}{8}$ $I_{yy} = \frac{5\pi b^4}{8}$ $I_{xy} = \frac{2b^4}{3}$	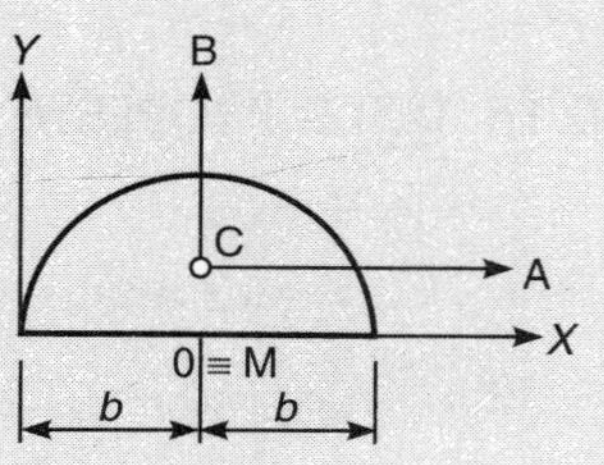

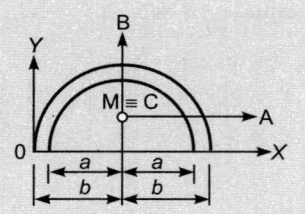

10.17 TABLE: PROPERTIES OF PLANE AREAS

Different parameters are:

A = area

x_C, y_C = coordinates of centroid

$I_{AA}, I_{BB}, I_{CC}, I_{xx}, I_{yy}$ = moments of inertia

I_{AB}, I_{xy} = products of inertia

Area and centroid	*Moments of inertia*	*Products of inertia*	
Ellipse $A = pab$ **Centroid** $xC = a$ $yC = b$	$I_{AA} = \frac{\pi ab^3}{4}$ $I_{BB} = \frac{\pi a^3 b}{4}$ $I_{AB} = 0$	$I_{xx} = \frac{5\pi ab^3}{4}$ $I_{yy} = \frac{5\pi a^3 b}{4}$ $I_{xy} = \pi a^2 b^2$	

Half ellipse $A = \frac{\pi ab}{2}$ **Centroid** $x_C = a$ $y_C = \frac{4b}{3\pi}$	$I_{AA} = \frac{\pi ab^3}{8}\left(1 - \frac{64}{9\pi^2}\right)$ $I_{BB} = \frac{\pi a^3 b}{8}$ $I_{AB} = 0$	$I_{xx} = \frac{\pi ab^3}{8}$ $I_{yy} = \frac{5\pi a^3 b}{8}$ $I_{xy} = \frac{2a^2b^2}{3}$	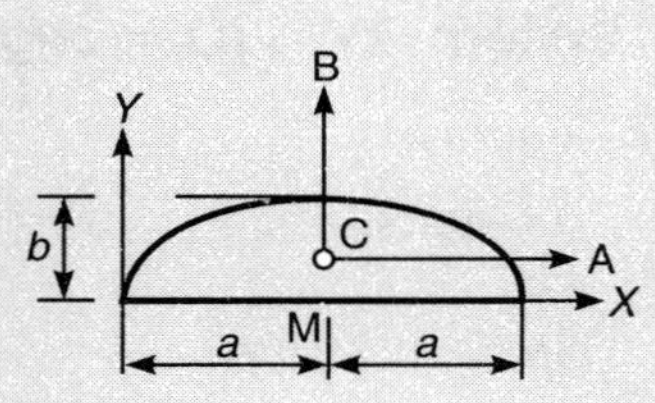
Quarter ellipse $A = \frac{\pi ab}{4}$ **Centroid** $e = \frac{4\alpha}{3\pi}$ $f = \frac{4b}{3\pi}$	$I_{AA} = \frac{\pi a^3 b}{16}\left(1 - \frac{64}{9\pi^2}\right)$ $I_{BB} = \frac{\pi ab^3}{16}\left(1 - \frac{64}{9\pi^2}\right)$ $I_{AB} = \frac{a^2b^2}{8}\left(1 - \frac{32}{9\pi^2}\right)$	$I_{xx} = \frac{\pi ab^3}{16}$ $I_{yy} = \frac{\pi a^3 b}{16}$ $I_{xy} = \frac{a^2b^2}{8}$	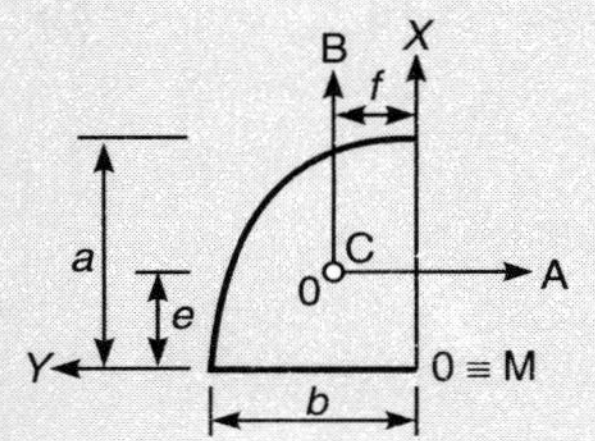
Parabola $A = \frac{4ab}{3} = \frac{2bl}{3}$ **Centroid** $x_C = a$ $y_C = \frac{2b}{5}$	$I_{AA} = \frac{16ab^3}{175}$ $I_{BB} = \frac{4a^3b}{15}$ $I_{AB} = 0$ $I_{TT} = \frac{4ab^3}{7}$	$I_{xx} = \frac{32ab^3}{105}$ $I_{yy} = \frac{8a^3b}{5}$ $I_{xy} = \frac{8a^2b^2}{15}$	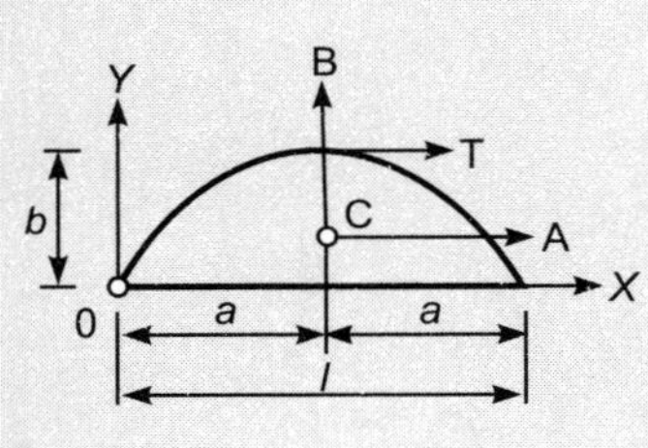
Half parabola $A = \frac{2ab}{3}$ **Centroid** $x_C = \frac{3a}{5}$ $y_C = \frac{3b}{8}$	$I_{AA} = \frac{19ab^3}{480}$ $I_{BB} = \frac{8a^3b}{175}$ $I_{AB} = \frac{a^2b^2}{60}$	$I_{xx} = \frac{2ab^3}{15}$ $I_{yy} = \frac{2a^3b}{7}$ $I_{xy} = \frac{a^2b^2}{6}$	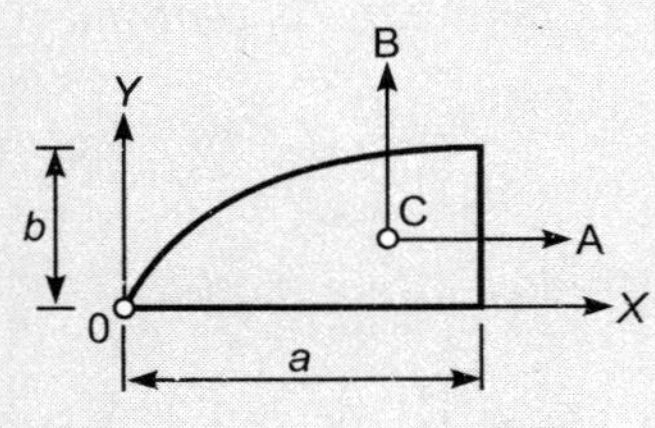
Parabolic complement $A = \frac{ab}{3}$ $x_C = \frac{3a}{5}$ $y_C = \frac{3b}{10}$	$I_{AA} = \frac{37ab^3}{2,100}$ $I_{BB} = \frac{a^3b}{80}$ $I_{AB} = \frac{a^2b^2}{120}$	$I_{xx} = \frac{ab^3}{21}$ $I_{yy} = \frac{a^3b}{5}$ $i_{xy} = \frac{a^2b^2}{12}$	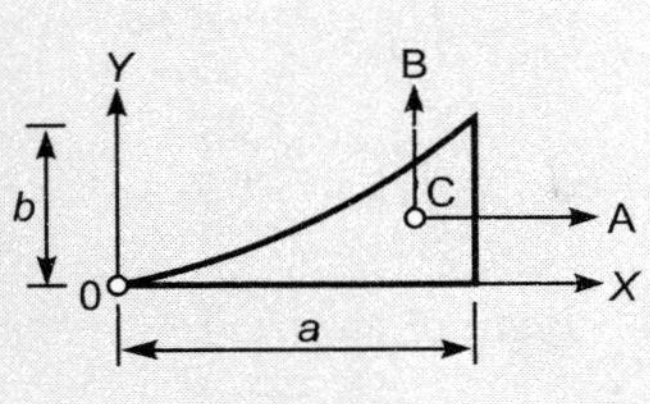

10.18 PROPERTIES OF SOLIDS USING TRIPLE INTEGRALS

i. System XY (O = origin, C = centroid)

Volume

δ = density $\quad dV = dx\,dy\,dz\,\delta \quad V = \iiint_V dV$

Static moments

$$M_{xy} = \iiint_V z\,dV \qquad M_{yz} = \iiint_V x\,dV$$

$$M_{zx} = \iiint_V y\,dV$$

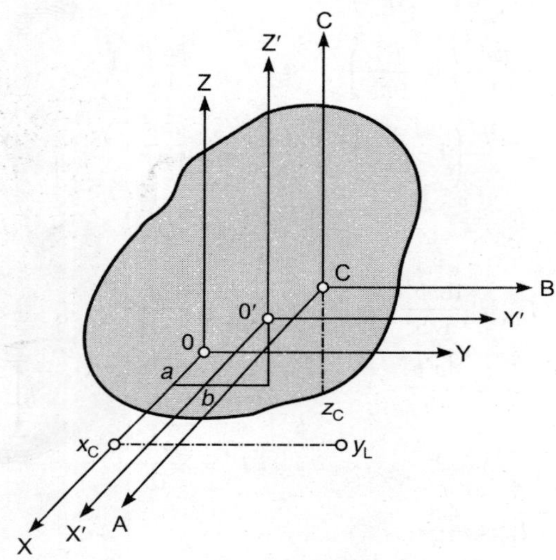

Fig. 10.11

Coordinates of centroid

$$x_C = \frac{M_{yz}}{V} \qquad z_C = \frac{M_{xy}}{V} \qquad y_C = \frac{M_{zx}}{V}$$

Moments of inertia	*Products of inertia*	*Polar moment of inertia*
$I_{xx} = \iiint_V (y^2 + z^2)\,dV$	$I_{yz} = I_{zy} = \iiint_V yz\,dV$	$J_0 = \iiint_V (x^2 + y^2 + z^2)\,dV$
$I_{yy} = \iiint_V (x^2 + z^2)\,dV$	$I_{xz} = I_{zx} = \iiint_V xz\,dV$	$= \dfrac{I_{xx} + I_{yy} + I_{zz}}{2}$
$I_{zz} = \iiint_V (x^2 + y^2)\,dV$	$I_{xy} = I_{yx} = \iiint_V xy\,dV$	

Radii of gyration

$$k_x = \sqrt{\frac{I_{xx}}{V}} \qquad k_y = \sqrt{\frac{I_{yy}}{V}} \qquad k_z = \sqrt{\frac{I_{zz}}{V}} \qquad k_0 = \sqrt{\frac{J_0}{V}}$$

ii. System X′ Y′ (O′ = origin, C = centroid)

Static moments $M'_{yz} = M_{yz} - aV$ $M'_{zx} = M_{zx} - bV$ $M'_{xy} = M_{xy} - cV$

Coordinates of centroid $x'_C = x_C - a$ $y'_C = y_C - b$ $z'_C = z_C - c$

Moments of inertia

$I'_{xx} = I_{xx} - 2bM_{zx} - 2cM_{xy} + (b^2 + c^2)\,V$

$I'_{yy} = I_{yy} - 2aM_{yz} - 2cM_{xy} + (a^2 + c^2)\,V$

$I'_{zz} = I_{zz} - 2aM_{yz} - 2bM_{zx} + (a^2 + b^2)\,V$

Products of inertia

$I'_{yz} = I_{yz} - bM_{zx} - cM_{xy} + bcV$

$I'_{xz} = I_{xz} - aM_{yz} - cM_{xy} + acV$

$I'_{xy} = I_{xy} - aM_{yz} - bM_{zx} + abV$

Polar moment of inertia $(r^2 = a^2 + b^2 + c^2)$

$$J'_0 = \frac{I_{xx} + I_{yy} + I_{zz}}{2} - 2aM_{yz} - 2bM_{zx} - 2cM_{xy} + r^2 V$$

Radii of gyration

$k'_x = \sqrt{k_x^2 - 2by_C - 2cz_C + (b^2 + c^2)}$

$k'_z = \sqrt{k_z^2 - 2ax_C - 2by_C + (a^2 + b^2)}$

$k'_y = \sqrt{k_y^2 - 2ax_C - 2cz_C + (a^2 + c^2)}$

$k'_0 = \sqrt{k_0^2 - 2(ax_C + by_C + cz_C) + r^2}$

10.19 TABLE: PROPERTIES OF SOLIDS

V = volume

δ = density

m = mass

$m = V\delta$

$I_{xx}, I_{yy}, I_{zz}, I_{AA}, I_{BB}, I_{CC}$ = moments of inertia

$I_{xy}, I_{yz}, I_{zx}, I_{AB}, I_{BC}, I_{CA}$ = products of inertia

x_C, y_C, z_C = coordinates of centroid

Mass, coordinates of centroid	*Moments of inertia*	*Products of inertia*	
Straight bar $m = a\delta$ **Centroid** $x_C = \frac{a}{2}$ $y_C = 0$	$I_{AA} = 0$ $I_{BB} = \frac{ma^2}{12}$ $I_{AB} = 0$	$I_{xx} = 0$ $I_{yy} = \frac{ma^2}{3}$ $I_{xy} = 0$	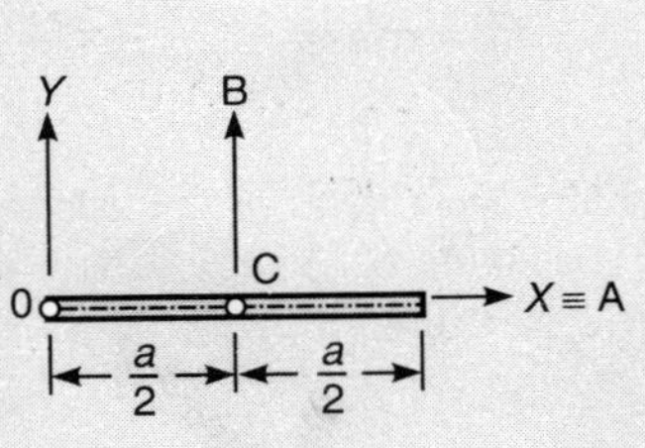
Straight bar $m = c\delta$ $x_C = \frac{a}{2}$ $y_C = \frac{b}{2}$	$I_{AA} = \frac{mb^2}{12}$ $I_{BB} = \frac{ma^2}{12}$ $I_{AB} = \frac{mab}{12}$	$I_{xx} = \frac{mb^2}{3}$ $I_{yy} = \frac{ma^2}{3}$ $I_{xy} = \frac{mab}{3}$	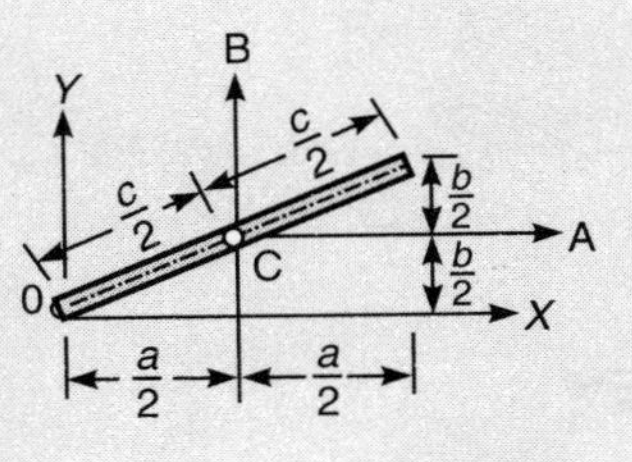

Bent bar $m = 2c\delta$ **Centroid** $x_C = a$ $y_C = \frac{b}{2}$	$I_{AA} = \frac{mb^2}{12}$ $I_{BB} = \frac{ma^2}{12}$ $I_{AB} = 0$	$I_{xx} = \frac{mb^2}{3}$ $I_{yy} = \frac{ma^2}{3}$ $I_{xy} = \frac{mab}{2}$	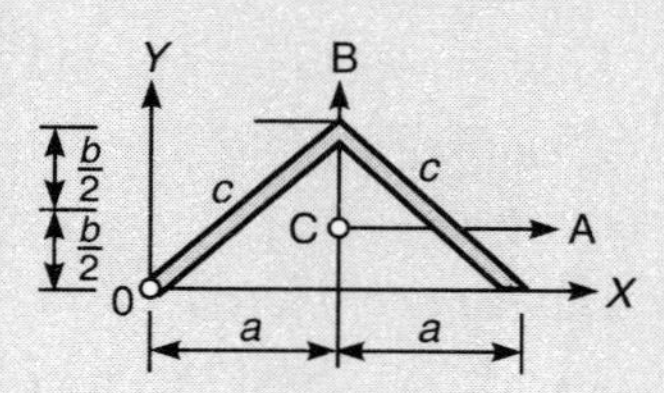
Circular bar $m = 2a\alpha\delta$ **Centroid** $x_C = c = a \sin \alpha$ $y_C = a\frac{\sin \alpha - \cos \alpha}{\alpha}$	$I_{AA} = \frac{ma^2}{2\alpha^2}[\alpha(\alpha + \sin \alpha \cos \alpha) - 2 \sin^2 \alpha]$ $I_{BB} = \frac{ma^2}{2\alpha}(\alpha - \sin \alpha \cos \alpha)$ $I_{AB} = 0$		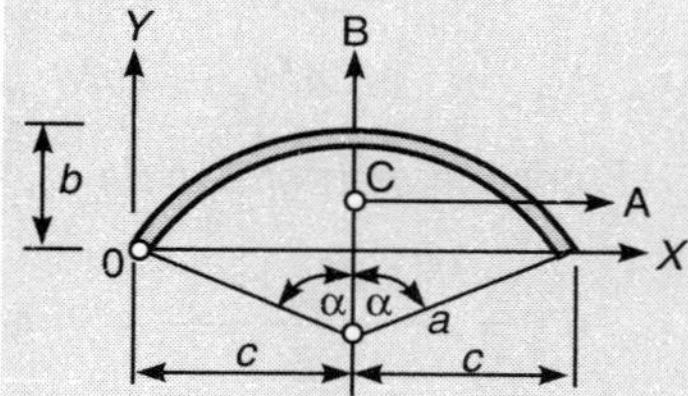
2° parabolic bar $\delta(\phi) = \frac{\delta V}{\cos \phi}$ δ_V = density at V $m = 2a\delta_V$ **Centroid** $x_C = a$ $y_C = \frac{2b}{3}$	$I_{AA} = \frac{4mb^2}{45}$ $I_{BB} = \frac{ma^2}{3}$ $I_{AB} = 0$	$I_{xx} = \frac{8mb^2}{15}$ $I_{yy} = \frac{4ma^2}{3}$ $I_{xy} = \frac{2abm}{3}$	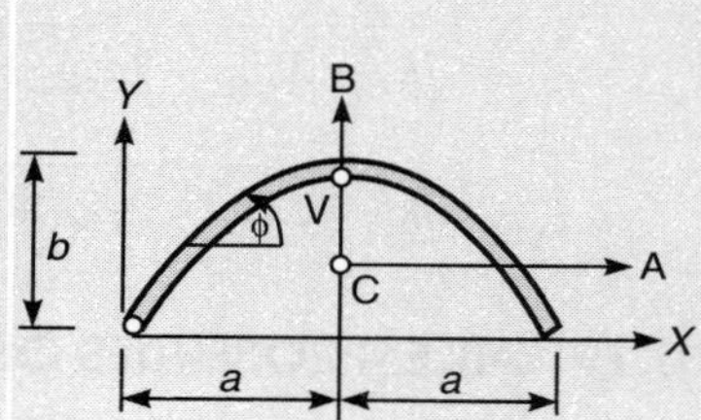

10.20 TABLE: PROPERTIES OF SOLIDS

V = volume
δ = density
m = mass
$I = V\delta$

$I_{xx}, I_{yy}, I_{zz}, I_{AA}, I_{BB}, I_{CC}$ = moments of inertia
$I_{xy}, I_{yz}, I_{zx}, I_{AB}, I_{BC}, I_{CA}$ = products of inertia
x_C, y_C, z_C = coordinates of centroid

Mass, coordinates of centroid	*Moments of inertia*	*Products of inertia*	
Cube $m = a^3\delta$ **Centroid** $x_C = \frac{a}{2}$ $y_C = \frac{a}{2}$ $z_C = \frac{a}{2}$	$I_{AA} = \frac{ma^2}{6}$ $I_{BB} = I_{CC} = I_{AA}$ $I_{TT} = \frac{2ma^2}{3}$	$I_{xx} = \frac{2ma^2}{3}$ $I_{yy} = I_{xx}$ $I_{zz} = I_{xx}$	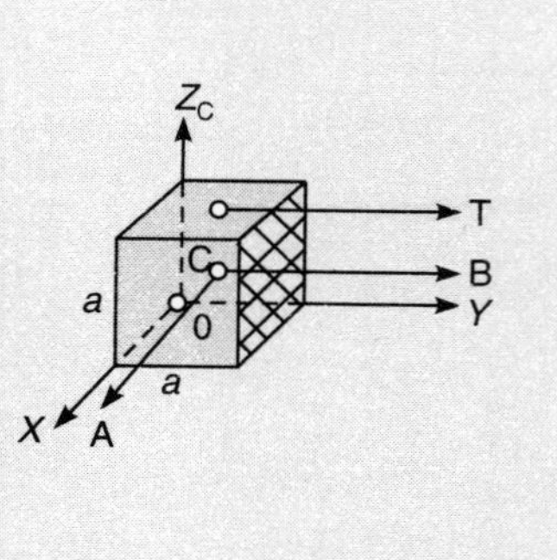
	$I_{AB} = 0$	$I_{xy} = I_{yz} = I_{zx} = \frac{ma^2}{4}$	

Prism $m = abc\,\delta$ **Centroid** $x_C = \frac{a}{2}$ $y_C = \frac{b}{2}$ $z_C = \frac{c}{2}$	$I_{AA} = m\frac{b^2+c^2}{12}$ $I_{BB} = m\frac{a^2+c^2}{12}$ $I_{TT} = m\frac{a^2+4c^2}{12}$	$I_{xx} = m\frac{b^2+c^2}{3}$ $I_{yy} = m\frac{a^2+c^2}{3}$ $I_{zz} = m\frac{a^2+b^2}{3}$	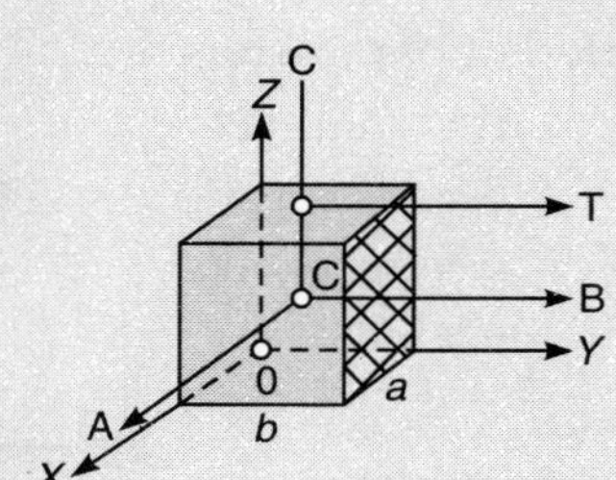
	$I_{AB} = 0$	$I_{xy} = \frac{mab}{4}$ ...	
Cylinder $m = \pi a^2 h\,\delta$ **Centroid** $x_C = 0$ $y_C = 0$ $z_C = \frac{h}{2}$	$I_{AA} = m\frac{3a^2+h^2}{12}$ $I_{BB} = m\frac{3a^2+h^2}{12}$ $I_{CC} = \frac{ma^2}{2}$	$I_{xx} = m\frac{3a^2+4h^2}{12}$ $I_{yy} = m\frac{3a^2+4h^2}{12}$ $I_{zz} = \frac{ma^2}{2}$	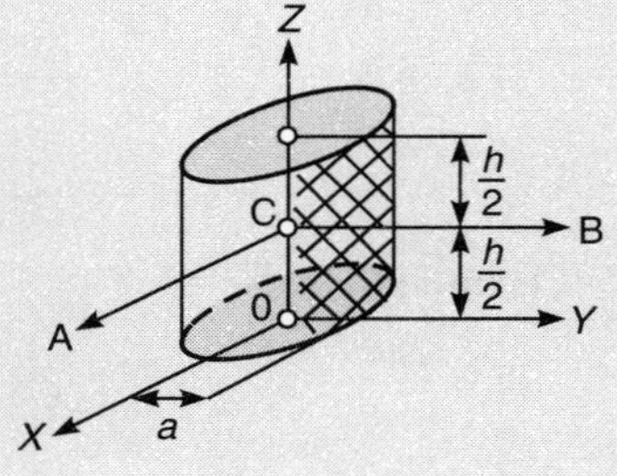
	$I_{AB} = 0$	$I_{xy} = I_{yz} = I_{zx} = 0$	
Cone $m = \frac{\pi a^2 h\delta}{3}$ **Centroid** $x_C = 0$ $y_C = 0$ $z_C = \frac{h}{a}$	$I_{AA} = \frac{3m}{80}(4a^2+h^2)$ $I_{BB} = \frac{3m}{80}(4a^2+h^2)$ $I_{TT} = \frac{3m}{20}(a^2+4h^2)$	$I_{xx} = \frac{m}{20}(3a^2+h^2)$ $I_{yy} = \frac{m}{20}(3a^2+h^2)$ $I_{zz} = \frac{3ma^2}{10}$	
	$I_{AB} = 0$	$I_{xy} = I_{yz} = I_{zx} = 0$	
Rectangular pyramid $m = \frac{abh\delta}{3}$ **Centroid** $x_C = 0$ $y_C = 0$ $z_C = \frac{h}{4}$	$I_{AA} = \frac{m}{80}(4b^2+3h^2)$ $I_{BB} = \frac{m}{80}(4a^2+3h^2)$ $I_{TT} = \frac{m}{20}(b^2+12h^2)$	$I_{xx} = \frac{m}{20}(b^2+2h^2)$ $I_{yy} = \frac{m}{20}(a^2+2h^2)$ $I_{xy} = \frac{m}{20}(a^2+b^2)$	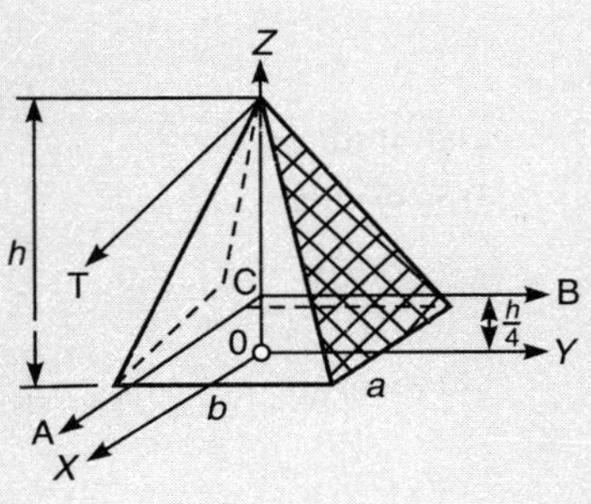
	$I_{AB} = 0$	$I_{xy} = I_{yz} = I_{zx} = 0$	

10.21 TABLE: PROPERTIES OF SOLIDS

Different parameters are:

V = volume $I_{xx}, I_{yy}, I_{zz}, I_{AA}, I_{BB}, I_{CC}$ = moments of inertia

δ = density $I_{xy}, I_{yz}, I_{zx}, I_{AB}, I_{BC}, I_{CA}$ = products of inertia

m = mass x_C, y_C, z_C = coordinates of centroid

$m = V\delta$

Mass, coordinates of centroid	*Moments of inertia*	*Products of inertia*
Sphere $m = \frac{4\pi a^3\delta}{3}$ **Centroid** $x_C = 0$ $y_C = 0$ $z_C = 0$	$I_{TT} = \frac{7ma^2}{5}$ $I_{NN} = I_{TT}$	$I_{xx} = \frac{2ma^2}{5}$ $I_{zz} = I_{yy} = I_{zz}$ 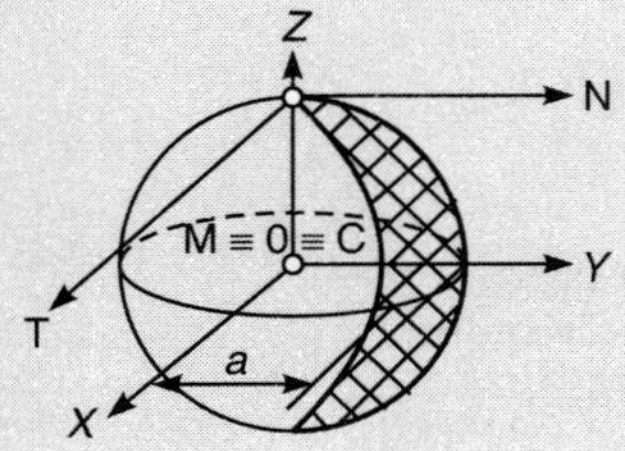
	$I_{TN} = 0$	$I_{xy} = I_{yz} = I_{zx} = 0$
Hemisphere $m = \frac{2\pi a^3\delta}{3}$ **Centroid** $x_C = 0$ $y_C = 0$ $z_C = \frac{3a}{8}$	$I_{AA} = \frac{83ma^2}{320}$ $I_{TT} = \frac{208ma^2}{320}$	$I_{xx} = \frac{2ma^2}{5}$ $.I_{xx} = I_{yy} = I_{zz}$ 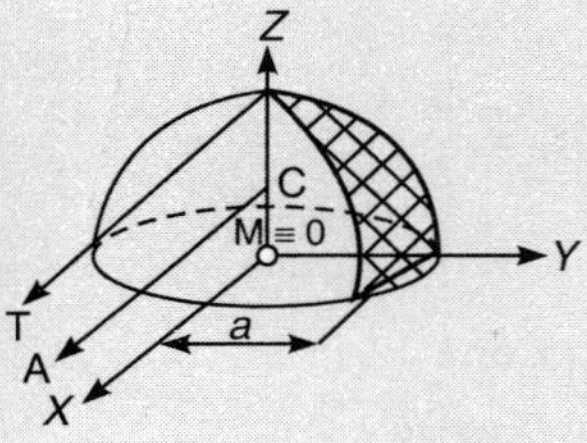
	$I_{TN} = 0$	$I_{xy} = I_{yz} = I_{zx} = 0$
Ellipsoid $m = \frac{4\pi abc\delta}{3}$ **Centroid** $x_C = 0$ $y_C = 0$ $z_C = 0$	$I_{TT} = m\frac{b^2+6c^2}{5}$ $I_{NN} = m\frac{a^2+6c^2}{5}$	$I_{xx} = m\frac{b^2+c^2}{5}$ $I_{yy} = m\frac{a^2+c^2}{5}$ $I_{zz} = m\frac{a^2+b^2}{5}$ 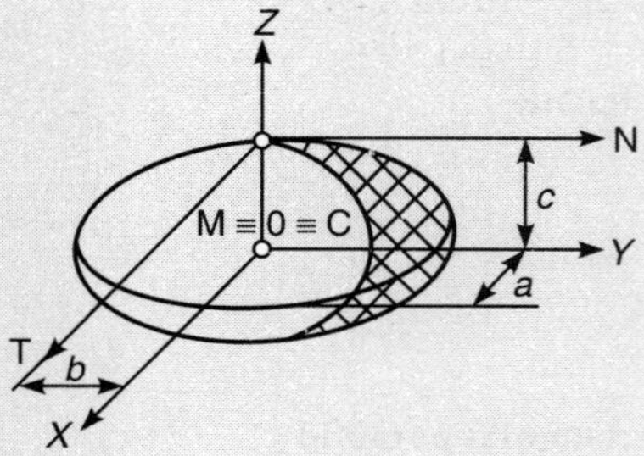
	$I_{TN} = 0$	$I_{xy} = I_{yz} = I_{zx} = 0$
Paraboloid of revolution $m = \frac{\pi a^2 h\delta}{2}$ **Centroid** $x_C = 0$ $y_C = 0$ $z_C = \frac{2h}{3}$	$I_{AA} = \frac{m}{18}(3a^2 + h^2)$ $I_{BB} = \frac{m}{18}(3a^2 + h^2)$	$I_{xx} = \frac{m}{6}(a^2 + 3h^2)$ $I_{yy} = \frac{m}{6}(a^2 + 3h^2)$ $I_{zz} = \frac{ma^2}{3}$ 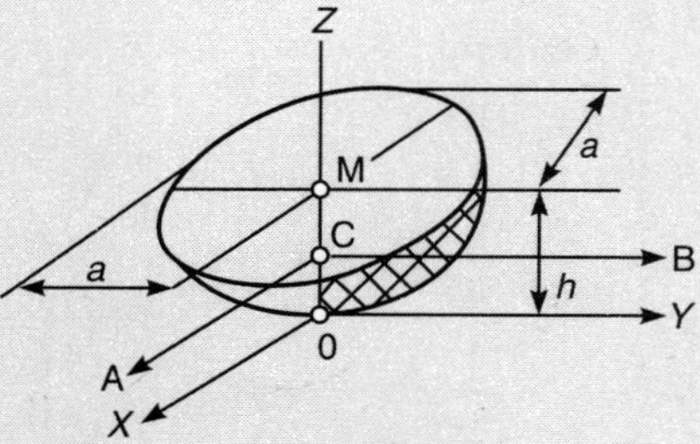

	$I_{AB} = 0$	$I_{xy} = I_{yz} = I_{zx} = 0$	
Elliptic paraboloid $m = \dfrac{\pi abh\delta}{2}$ $x_C = 0$ $y_C = 0$ $z_C = \dfrac{2h}{3}$	$I_{AA} = \dfrac{m}{18}(3b^2 + h^2)$ $I_{BB} = \dfrac{m}{18}(3a^2 + h^2)$	$I_{xx} = \dfrac{m}{6}(b^2 + 3h^2)$ $I_{yy} = \dfrac{m}{6}(a^2 + 3h^2)$ $I_{zz} = \dfrac{m}{6}(a^2 + b^2)$	Z M a h b C B Y 0 A X
	$I_{AB} = 0$	$I_{xy} = I_{yz} = I_{zx} = 0$	

10.22 IMPROPER INTEGRALS

i. Definition

If either or both of the *limits* of a definite integral are *infinitely large* or if the *integrand becomes infinite* in the interval of integration, the integral is called an *improper integral.*

$$\int_a^{+\infty} f(x)\,dx = \lim_{b \to +\infty} \int_a^b f(x)\,(dx) \qquad \int_{-\infty}^b f(x)\,dx = \lim_{a \to -\infty} \int_a^b f(x)\,dx$$

$$\int_{-\infty}^{+\infty} f(x)\,dx = \lim_{\substack{a \to -\infty \\ b \to +\infty}} \int_a^b f(x)\,dx$$

If $\lim_{x \to c} f(x) = \infty$, then

$$\int_a^b f(x)\,dx = \int_a^c f(x)\,dx + \int_c^b f(x)\,dx = \lim_{\varepsilon_1 \to 0} \int_a^{c-\varepsilon_1} f(x)\,dx + \lim_{\varepsilon_2 \to 0} \int_{c+\varepsilon_2}^b f(x)dx \qquad \begin{matrix} \varepsilon_1 > 0 \\ \varepsilon_2 > 0 \end{matrix}$$

ii. Table

$\int_0^{+\infty} \dfrac{dx}{a + bx^2} = \dfrac{\pi}{2\sqrt{ab}}$	$a > 0$ $b > 0$	$\int_0^{+\infty} \dfrac{x^{n-1}}{x+1} dx = \dfrac{\pi}{\sin n\pi}$	$0 < n < 1$
$\int_0^{+\infty} \dfrac{dx}{(1+x)\sqrt{x}} = \pi$		$\int_0^{+\infty} \dfrac{dx}{(1-x)\sqrt{x}} = 0$	
$\int_0^1 \dfrac{dx}{x^n} dx = \dfrac{1}{1-n}$	$n < 1$	$\int_0^1 \dfrac{dx}{\sqrt{1-x^2}} = \dfrac{\pi}{2}$	
$\int_0^a \dfrac{dx}{\sqrt{a^2 - x^2}} = \dfrac{\pi}{2}$	$a > 0$	$\int_0^a \dfrac{dx}{\sqrt{ax - x^2}} = \dfrac{3\pi a^2}{8}$	$a > 0$
$\int_a^b \dfrac{dx}{(x-a)(b-x)} = \pi$		$a > 0$ $b > 0$	

$\int_0^{+\infty} e^{-\lambda x} dx = \frac{1}{\lambda}$ $\quad \lambda > 0$	$\int_0^{+\infty} e^{-x^2} dx = \sqrt{\frac{\pi}{4}}$
$\int_0^{+\infty} \frac{x\,dx}{e^x + 1} = \frac{\pi^2}{12}$	$\int_0^{+\infty} \frac{x\,dx}{e^x - 1} = \frac{\pi^2}{6}$
$\int_0^{+\infty} x^n e^{-\lambda x} dx = \frac{n!}{\lambda^{n+1}}$	$\lambda > 0, n = 1, 2, \ldots$
$\int_0^1 \frac{\ln x}{x+1} dx = -\frac{\pi^2}{12}$	$\int_0^1 \frac{\ln x}{x-1} dx = \frac{\pi^2}{6}$
$\int_0^{\infty} \frac{\sin mx \sin nx\,dx}{x} = \frac{1}{2} \ln \frac{m+n}{m-n}$ $\quad m > n \geq 0$	
$\int_0^{\infty} \frac{\sin mx \sin nx\,dx}{x^2} = \begin{cases} \frac{n\pi}{2} & \text{if } m \geq n > 0 \\ \frac{m\pi}{2} & \text{if } n \geq m > 0 \end{cases}$	

CHAPTER

11

Vector Analysis

11.1 CONCEPT AND DEFINITIONS

i. Definitions

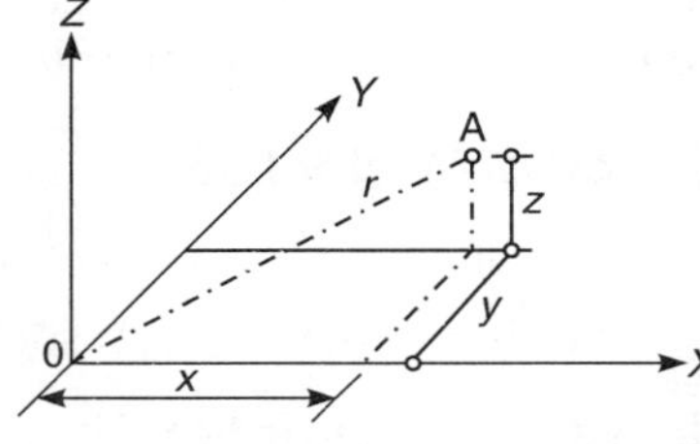

Fig. 11.1

A *scalar* is a quantity defined by magnitude only (mass, length, time, temperature).

If to each point (x, y, z) of a region R in space there is a scalar $r = \overline{\text{OA}}$, then $\overline{\text{OA}}$ is called a *scalar point function*, and the region R is denoted as *scalr field*.

A **vector** is a quantity defined by magnitude and direction (force, moment, displacement, velocity, acceleration).

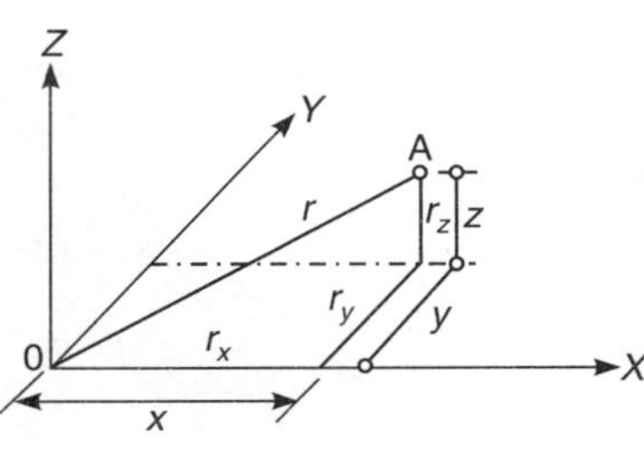

Fig. 11.2

If to each point (x, y, z) of a region R in space there is vector $r = \overline{\text{OA}}$, then $\overline{\text{OA}}$ is called a *vector point function*, and the region R is denoted as a *vector field*.

ii. Components and Magnitudes

a. *The vector* $\boldsymbol{r}$ may be resolved into any number of components. In the cartesian-coordinate system, $\boldsymbol{r}$ is resolved into three mutually perpendicular components, each parallel to the respective coordinate axis.

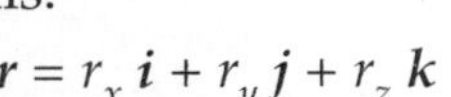

$$\boldsymbol{r} = r_x\,\boldsymbol{i} + r_y\,\boldsymbol{j} + r_z\,\boldsymbol{k}$$

b. *The magnitude* r is given by the magnitudes of components.

$$r = |r| = \sqrt{r_x^2 + r_y^2 + r_z^2}$$

c. The unit vector is the ratio of the vector to its magnitude. $\boldsymbol{i}, \boldsymbol{j}, \boldsymbol{k}$ are the unit vectors in the X-, Y-, Z-axes and directions, respectively $\boldsymbol{e}$ is the unit vector in the $\boldsymbol{r}$ direction.

$$\boldsymbol{i} = \frac{\boldsymbol{r}_x}{r_x} \quad \boldsymbol{j} = \frac{\boldsymbol{r}_y}{r_y} \quad \boldsymbol{k} = \frac{\boldsymbol{r}_z}{r_z} \quad \boldsymbol{e} = \frac{\boldsymbol{r}}{r}$$

(d) *The direction cosines* of vector **r** are

$$\alpha = \frac{r_x}{r};\quad \beta = \frac{r_y}{r};\quad \gamma = \frac{r_z}{r}$$

and relate **e** to **i, j, k** and

$$e = \alpha\, i + \beta\, j + \gamma\, k$$

11.2 VECTOR SUMMATION

i. Vector Addition and Subtraction

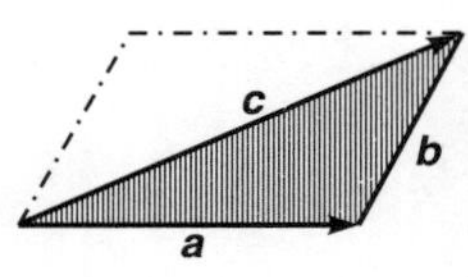

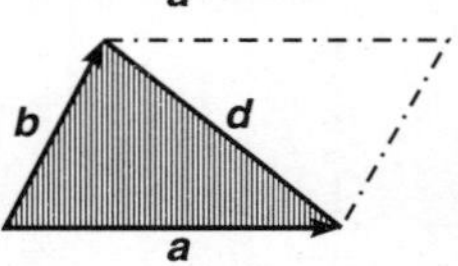

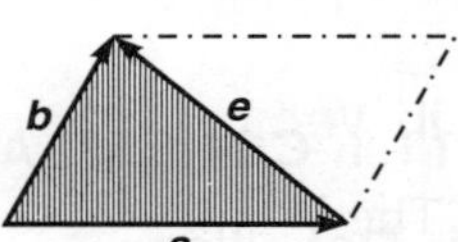

Fig. 11.3

A sum of vectors **a** and **b** is a vector **c** formed by placing the initial point of **b** on the terminal point of **a** and joining the initial point of **a** to the terminal point of **b**.

A difference of vectors **a** *and* **b** is a vector **d** formed by placing the initial point of **b** on the initial point of **a** and joining the terminal point of **b** with the terminal point of **a**.

A difference of vectors **b** *and* **a** is a vector **e** formed by placing the initial point of **b** on the initial point of **a** and joining the terminal point of **a** with terminal point of **b**.

ii. Scalar-Vector Laws (*a*, *b* = vectors; *m*, *n* = scalars)

$ma = a\,m$	Commutative law	$(m + n)\,a = ma + na$	Distributive law
$m(na) = (mn)a$	Associative law	$m\,(a + b) = ma + mb$	Distributive law

iii. Vector Summation Laws

a. *Commutative law*

$$a + b + c = b + c + a = c + a + b = f$$

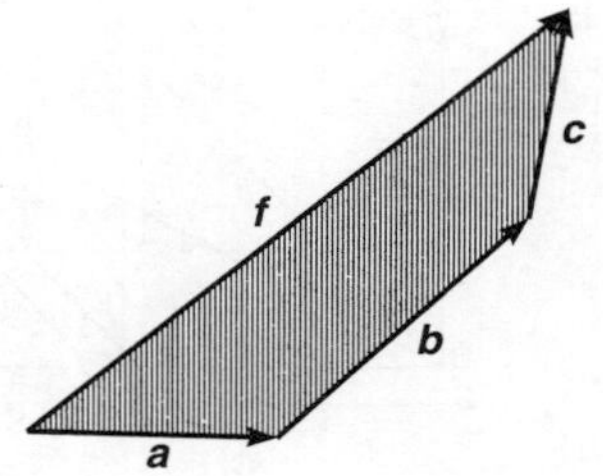

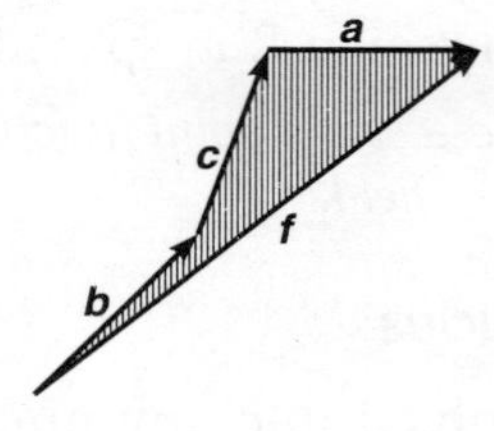

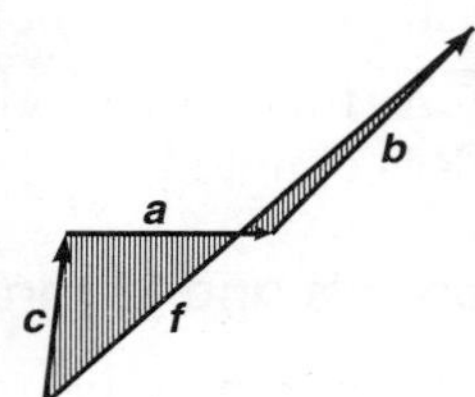

Fig. 11.4

b. *Associative law*

$$(a + b) + c = a + (b + c) = f$$

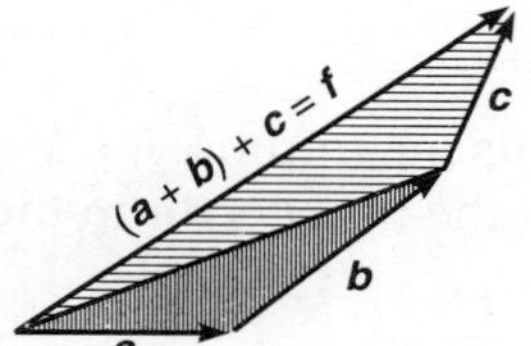

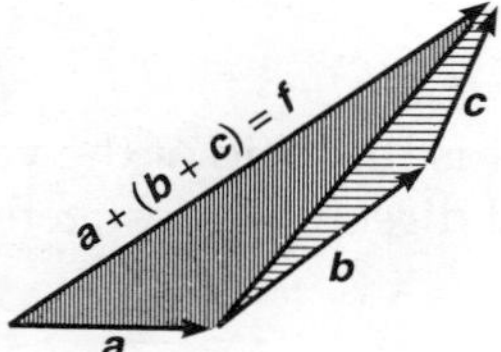

Fig. 11.5

11.3 SCALAR AND VECTOR PRODUCTS

i. Scalar Product

The scalar product (or a dot product) of two vectors $\boldsymbol{a}$ and $\boldsymbol{b}$ is defined as the product of their magnitudes and the cosine of the angle between them. *The result is a scalar.*

$$a \,.\, b = ab \cos \omega = a_x b_x + a_y b_y + a_z b_z$$

The multiplication is *commutative*, and *distributive*.

$\boldsymbol{a.b} = \boldsymbol{b.a}$ $\qquad$ $\boldsymbol{a. (b + c) = a.b + a.c}$

Two vectors $a \neq 0, b \neq 0$ are *normal* ($\omega = 90°$) if	Two vectors $\boldsymbol{a} \neq 0, \boldsymbol{b} \neq 0$ are parallel ($\omega = 0°$) if
$\boldsymbol{a.b} = 0$	$\boldsymbol{a.b} = ab$
From this,	
$\boldsymbol{i.j} = \boldsymbol{j.k} = \boldsymbol{k.i} = 0$	$\boldsymbol{i.i} = \boldsymbol{j.j} = \boldsymbol{k.k} = 1$

ii. Vector Product

The vector product (or a cross product) of two vectors $\boldsymbol{a}$ and $\boldsymbol{b}$ is defined as a product of their magnitudes, the sine of the angle between them, and the unit vector $\boldsymbol{n}$ normal to their plane. *The result is a vector.*

$$\boldsymbol{a} \times \boldsymbol{b} = ab \sin \omega \boldsymbol{n} = (a_y b_z - a_z b_y)\boldsymbol{i} - (a_x b_z - a_z b_x)\boldsymbol{j} + (a_x b_y - a_y b_x)\boldsymbol{k}$$

The multiplication is *noncommutative*, but it is *distributive*.

$\boldsymbol{a} \times \boldsymbol{b} = -\boldsymbol{b} \times \boldsymbol{a}$ $\qquad$ $\boldsymbol{a} \times (b + c) = \boldsymbol{a} \times \boldsymbol{b} + a \times c$

Two vectors $\boldsymbol{a} \neq 0, \boldsymbol{b} \neq 0$ are normal ($\omega = 90°$).	Two vectors $\boldsymbol{a} \neq 0, \boldsymbol{b} \neq 0$ are *parallel* ($\omega = 0°$)
if	if
$\boldsymbol{a} \times \boldsymbol{b} = ab\boldsymbol{n}$	$\boldsymbol{a} \times \boldsymbol{b} = 0$
From this,	
$\boldsymbol{i} \times \boldsymbol{j} = \boldsymbol{k}, \boldsymbol{j} \times \boldsymbol{k} = \boldsymbol{i}, \boldsymbol{k} \times \boldsymbol{i} = \boldsymbol{j}$	$\boldsymbol{i} \times \boldsymbol{i} = \boldsymbol{j} \times \boldsymbol{j} = \boldsymbol{k} \times \boldsymbol{k} = 0$

The vector product in *determinant form* is

$$\boldsymbol{a} \times \boldsymbol{b} = \begin{vmatrix} \boldsymbol{i} & \boldsymbol{j} & \boldsymbol{k} \\ a_x & a_y & a_z \\ b_x & b_y & b_z \end{vmatrix} = \begin{vmatrix} \mathrm{i} & a_x & b_x \\ \mathrm{j} & a_y & b_y \\ \mathrm{k} & a_z & b_z \end{vmatrix} = \boldsymbol{i}\begin{vmatrix} a_y & a_z \\ b_y & b_z \end{vmatrix} - \boldsymbol{j}\begin{vmatrix} a_x & a_z \\ b_x & b_z \end{vmatrix} + \boldsymbol{k}\begin{vmatrix} a_x & a_y \\ b_x & b_y \end{vmatrix}$$

11.4 TRIPLE PRODUCTS

i. Scalar Triple Product (Results a Scalar)

$$\boldsymbol{a.(b \times c) = b.(c \times a) = c.(a \times b) = abc}$$

$$= \begin{vmatrix} a_x & a_y & a_z \\ b_x & b_y & b_z \\ c_x & c_y & c_z \end{vmatrix} = \begin{vmatrix} a_x & b_x & c_x \\ a_y & b_y & c_y \\ a_z & b_z & c_z \end{vmatrix}$$

ii. Vector Triple Product (Results a Vector)

$$a \times (b \times c) = (a.c)b - (a.b)c = \begin{vmatrix} i & j & k \\ a_x & a_y & a_z \\ \begin{vmatrix} b_y & b_z \\ c_y & c_z \end{vmatrix} & \begin{vmatrix} b_z & b_x \\ c_z & c_x \end{vmatrix} & \begin{vmatrix} b_x & b_y \\ c_x & c_y \end{vmatrix} \end{vmatrix}$$

$$(a \times b) \times c = (a.c)b - (b.c)a = \begin{vmatrix} i & j & k \\ \begin{vmatrix} a_y & a_z \\ b_y & b_z \end{vmatrix} & \begin{vmatrix} a_z & a_x \\ b_z & b_x \end{vmatrix} & \begin{vmatrix} a_x & a_y \\ b_x & b_y \end{vmatrix} \\ c_x & c_y & c_z \end{vmatrix}$$

iii. Special Products

$$(c \times b) . (c \times d) = \begin{vmatrix} a.c & b.c \\ a.d & b.d \end{vmatrix} \qquad (a \times b)^2 = \begin{vmatrix} a.a & a.b \\ a.b & b.b \end{vmatrix}$$

$$(a \times b) \times (c \times d) = \begin{vmatrix} i & j & k \\ \begin{vmatrix} a_y & a_z \\ b_y & b_z \\ c_y & c_z \\ d_y & d_z \end{vmatrix} & \begin{vmatrix} a_z & a_x \\ b_z & b_x \\ c_z & c_x \\ d_z & d_x \end{vmatrix} & \begin{vmatrix} a_x & a_y \\ b_x & b_y \\ c_x & c_y \\ d_x & d_y \end{vmatrix} \end{vmatrix}$$

If $a \times (b \times c) = b \times (c \times a) = c \times (a \times b)$ then a, b, c are orthogonal.

If $a \times b = c, b \times c = a, c \times a = b$ then a, b, c are orthogonal unit vectors, and

$a \times (b \times c) = b \times (c \times a) = c \times (a \times b) = 0$

11.5 VECTOR DIFFERENTIAL CALCULUS

i. Ordinary Derivatives

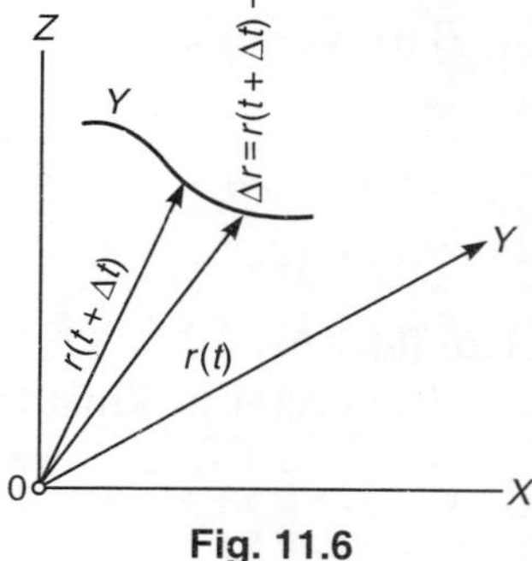

Fig. 11.6

a. *Definitions* of limit, continuity, and vector function are formally identical with the definitions of scalar functions.

If

$$r = r(t) = x(t)\, i + y(t)\, j + z(t)\, k$$

= vector function of scalar variable t

then
$$\frac{dr}{dt} = \frac{dx(t)}{dt} i + \frac{dy(t)}{dt} j + \frac{dz(t)}{dt} k$$

b. *Basic formulas* $[\alpha = \alpha(t) = \text{scalar function of } t]$

$$\frac{d}{dt}(r_1 + r_2) = \frac{dr_1}{dt} + \frac{dr_2}{dt} \qquad \frac{d}{dt}(r_1 . r_2) = r_1 . \frac{dr_2}{dt} + \frac{dr_1}{dt} . r_2$$

$$\frac{d}{dt}(\alpha r) = \alpha \frac{dr}{dt} + \frac{d\alpha}{dt} r \qquad \frac{d}{dt}(r_1 \times r_2) = r_1 \times \frac{dr_2}{dt} + \frac{dr_1}{dt} \times r_2$$

$$\frac{d}{dt}(r_1 . r_2 \times r_3) = r_1 . r_2 \times \frac{dr_3}{dt} + r_1 . \frac{dr_3}{dt} \times r_3 + \frac{dr_1}{dt} . r_2 \times r_3$$

$$\frac{d}{dt}[r_1 \times (r_2 \times r_2)] = r_1 \times \left(r_2 \times \frac{dr_3}{dt}\right) + r_1 \times \left(\frac{dr_2}{dt} \times r_3\right) + \frac{dr_1}{dt} \times (r_2 \times r_3)$$

ii. Partial Derivatives

If $\boldsymbol{r}(x, y, z)$ is a *vector function of several scalar variables* x, y, z, then partial derivatives are obtained by differentiating the magnitude of each component as a scalar function.

$$\frac{\partial \boldsymbol{r}}{\partial x} = \frac{\partial \boldsymbol{r}(x, y, z)}{\partial x} \qquad \frac{\partial \boldsymbol{r}}{\partial y} = \frac{\partial \boldsymbol{r}(x, y, z)}{\partial y} \qquad \frac{\partial \boldsymbol{r}}{\partial z} = \frac{\partial \boldsymbol{r}(x, y, z)}{\partial z}$$

$$\frac{\partial^2 \boldsymbol{r}}{\partial x^2} = \frac{\partial}{\partial x}\left(\frac{\partial \boldsymbol{r}}{\partial x}\right) \qquad \frac{\partial^2 \boldsymbol{r}}{dy^2} = \frac{\partial}{\partial y}\left(\frac{\partial \boldsymbol{r}}{\partial y}\right) \qquad \frac{\partial^2 \boldsymbol{r}}{dz^2} = \frac{\partial}{\partial z}\left(\frac{\partial \boldsymbol{r}}{\partial z}\right)$$

$$\frac{\partial^2 \boldsymbol{r}}{\partial x \partial y} = \frac{\partial}{\partial x}\left(\frac{\partial \boldsymbol{r}}{\partial y}\right) \qquad \frac{\partial^2 \boldsymbol{r}}{\partial y \partial z} = \frac{\partial}{\partial y}\left(\frac{\partial \boldsymbol{r}}{\partial z}\right) \qquad \frac{\partial^2 \boldsymbol{r}}{\partial z \partial x} = \frac{\partial}{\partial z}\left(\frac{\partial \boldsymbol{r}}{\partial x}\right)$$

$$\frac{\partial(\boldsymbol{r}_1 \cdot \boldsymbol{r}_2)}{\partial x} = \boldsymbol{r}_1 \cdot \frac{\partial \boldsymbol{r}_2}{\partial x} + \frac{\partial \boldsymbol{r}_1}{\partial x} \cdot \boldsymbol{r}_2 \qquad \frac{\partial^2(\boldsymbol{r}_1 \cdot \boldsymbol{r}_2)}{\partial x \partial y} = \boldsymbol{r}_1 \cdot \frac{\partial^2 \boldsymbol{r}_2}{\partial x \partial y} + \frac{\partial \boldsymbol{r}_1}{\partial x} \cdot \frac{\partial \boldsymbol{r}_2}{\partial y} + \frac{\partial \boldsymbol{r}_1}{\partial y} \cdot \frac{\partial \boldsymbol{r}_2}{\partial x} + \frac{\partial^2 \boldsymbol{r}_1}{\partial x \partial y} \cdot \boldsymbol{r}_2$$

$$\frac{\partial(\boldsymbol{r}_1 \times \boldsymbol{r}_2)}{\partial x} = \boldsymbol{r}_1 \times \frac{\partial \boldsymbol{r}_2}{\partial x} + \frac{\partial \boldsymbol{r}_1}{\partial x} \times \boldsymbol{r}_2 \qquad \frac{\partial^2(\boldsymbol{r}_1 \times \boldsymbol{r}_2)}{\partial x \partial y} = \boldsymbol{r}_1 \times \frac{\partial^2 \boldsymbol{r}_2}{\partial x \partial y} + \frac{\partial \boldsymbol{r}_1}{\partial x} \times \frac{\partial \boldsymbol{r}_2}{\partial y} + \frac{\partial \boldsymbol{r}_1}{\partial y} \times \frac{\partial \boldsymbol{r}_2}{\partial x} + \frac{\partial^2 \boldsymbol{r}_1}{\partial x \partial y} \times \boldsymbol{r}_2$$

iii. Partial Differentials

$$d\,(\boldsymbol{r}_1 \cdot \boldsymbol{r}_2) = \boldsymbol{r}_1 \cdot d\boldsymbol{r}_2 + d\boldsymbol{r}_1 \cdot \boldsymbol{r}_2$$
$$d\,(\boldsymbol{r}_1 \times \boldsymbol{r}_2) = \boldsymbol{r}_1 \times d\boldsymbol{r}_2 + d\boldsymbol{r}_1 \times \boldsymbol{r}_2$$

iv. Total Differentials

$$d\boldsymbol{r} = \frac{\partial \boldsymbol{r}}{\partial x} dx + \frac{\partial \boldsymbol{r}}{\partial y} dy + \frac{d\boldsymbol{r}}{\partial z} dz$$

$$d\boldsymbol{r} = \left(\frac{\partial \boldsymbol{r}}{\partial x}\frac{\partial x}{\partial t} + \frac{\partial \boldsymbol{r}}{\partial y}\frac{\partial y}{\partial t} + \frac{\partial \boldsymbol{r}}{\partial z}\frac{\partial z}{\partial t}\right) dt$$

11.6 TRANSFORMATION OF UNIT VECTORS

Cartesian

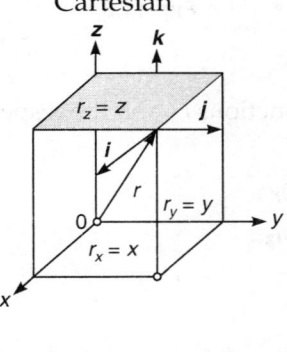

$$\boldsymbol{r} = r_x\,\boldsymbol{i} + r_y\,\boldsymbol{j} + r_z\,\boldsymbol{k}$$

Cylindrical coordinates

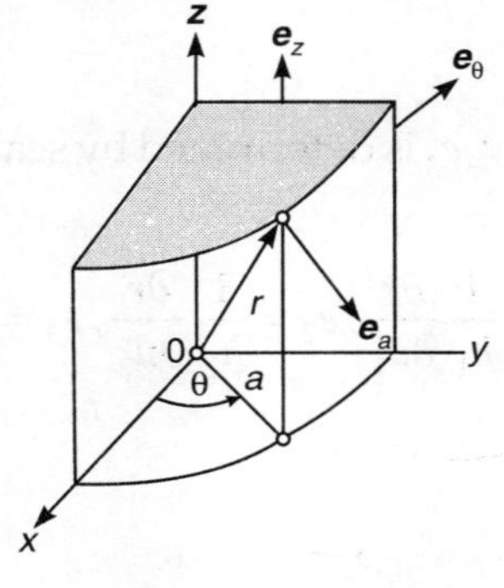

$$r = r_a\,\boldsymbol{e}_\mathbf{a} + \boldsymbol{r}_\theta\,\boldsymbol{e}_\theta + \boldsymbol{r}_z\,\boldsymbol{e}_\mathbf{z}$$

Spherical coordinates

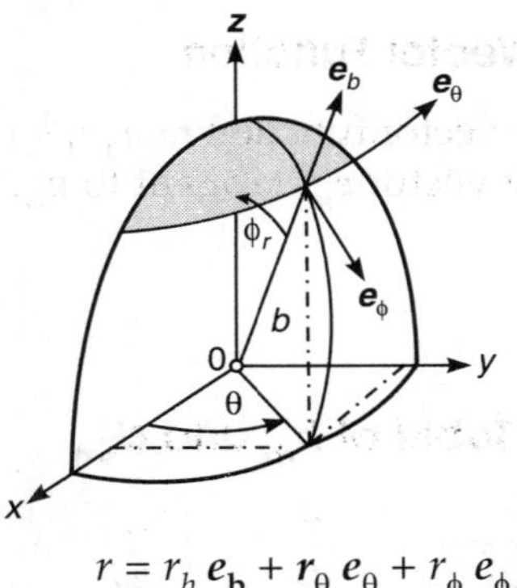

$$r = r_b\,\boldsymbol{e}_\mathbf{b} + \boldsymbol{r}_\theta\,\boldsymbol{e}_\theta + \boldsymbol{r}_\phi\,\boldsymbol{e}_\phi$$

Fig. 11.7

i. Cartesian-Cylindrical Transformation

$$\begin{bmatrix} i \\ j \\ k \end{bmatrix} = \begin{bmatrix} \cos\theta & -\sin\theta & 0 \\ \sin\theta & \cos\theta & 0 \\ 0 & 0 & 1 \end{bmatrix} \begin{bmatrix} e_a \\ e_\theta \\ e_z \end{bmatrix} \qquad \begin{bmatrix} e_a \\ e_\theta \\ e_z \end{bmatrix} = \begin{bmatrix} \cos\theta & \sin\theta & 0 \\ -\sin\theta & \cos\theta & 0 \\ 0 & 0 & 1 \end{bmatrix} \begin{bmatrix} i \\ j \\ k \end{bmatrix}$$

ii. Cylindrical-Spherical Transformation

$$\begin{bmatrix} e_a \\ e_\theta \\ e_z \end{bmatrix} = \begin{bmatrix} \sin\phi & 0 & \cos\phi \\ 0 & 1 & 0 \\ \cos\phi & 0 & -\sin\phi \end{bmatrix} \begin{bmatrix} e_b \\ e_\theta \\ e_\phi \end{bmatrix} \qquad \begin{bmatrix} e_b \\ e_\theta \\ e_\phi \end{bmatrix} = \begin{bmatrix} \sin\phi & 0 & \cos\phi \\ 0 & 1 & 0 \\ \cos\phi & 0 & -\sin\phi \end{bmatrix} \begin{bmatrix} e_a \\ e_\theta \\ e_z \end{bmatrix}$$

iii. Spherical-Cartesian Transformation

$$\begin{bmatrix} e_b \\ e_\theta \\ e_\phi \end{bmatrix} = \begin{bmatrix} \sin\phi\ \cos\theta & \sin\phi\sin\theta & \cos\phi \\ -\sin\phi\sin\theta & \sin\phi\cos\theta & 0 \\ \cos\phi\cos\theta & \cos\phi\sin\theta & -\sin\theta \end{bmatrix} \begin{bmatrix} i \\ j \\ k \end{bmatrix}$$

$$\begin{bmatrix} i \\ j \\ k \end{bmatrix} = \begin{bmatrix} \sin\phi\ \sin\theta & -\sin\phi\sin\theta & \cos\phi\cos\theta \\ \sin\phi\sin\theta & \sin\phi\cos\theta & \cos\phi\sin\theta \\ \cos\phi & o & -\sin\phi \end{bmatrix} \begin{bmatrix} e_b \\ e_\theta \\ e_\phi \end{bmatrix}$$

11.7 ORTHOGONAL CURVILINEAR COORDINATES

i. Differential Elements

The *curvilinear coordinates* u_1, u_2, u_3 are said to be *orthogonal* if the coordinate curves are mutually perpendicular at every point. The differential element of length $ds_m = h_m du_m$ $(m = 1, 2, 3,)$ is defined by the differential element of the respective coordinate and the corresponding *scaling factor*.

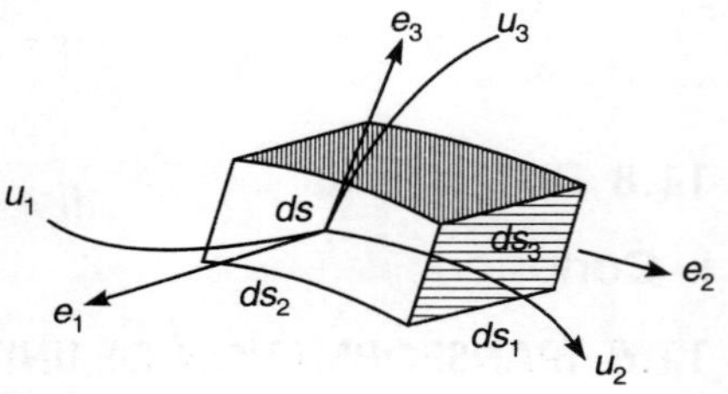

Fig. 11.8

ii. Vector Function

The vector function $\boldsymbol{r} = r_1 e_1 + r_2 e_2 + r_3 e_3$ is determined by scalar functions r_m and the respective unit vector e_m, tangent to u_m.

$$e_1 = \frac{1}{h_1}\frac{\partial \boldsymbol{r}}{\partial u_1},\ e_2 = \frac{1}{h_2}\frac{\partial \boldsymbol{r}}{\partial u_2},\ e_3 = \frac{1}{h_3}\frac{\partial \boldsymbol{r}}{\partial u_3}$$

iii. Table of h_m and d_{sm}

	x	y	z	a	θ	z	b	θ	ϕ
ds_m	dx	dy	dz	da	$a\,d\theta$	dz	db	$b\sin\phi\,d\theta$	$bd\phi$
h_m	1	1	1	1	a	1	1	$b\sin\phi$	b

iv. Differential Operators—General Formulas

a. *Gradient of scalar function f* = Grad *f*

$$\nabla f = \frac{1}{h_1}\frac{\partial f}{\partial u_1}e_1 + \frac{1}{h_2}\frac{\partial f}{\partial u_2}e_2 + \frac{1}{h_3}\frac{\partial f}{\partial u_3}e_3$$

b. *Divergence of vector function* $\mathbf{V}$ = Div *V*

$$\nabla . \mathbf{V} = \frac{1}{h_1h_2h_3}\left[\frac{\partial}{\partial \mu_1}(V_1h_2h_3) + \frac{\partial}{\partial \mu_2}(h_1V_2h_3) + \frac{\partial}{\partial \mu_3}(h_1h_2V_3)\right]$$

c. *Curl of vector function* $\mathbf{V}$ = Curl *V* = Rot *V*

$$\nabla \times \mathbf{V} = \frac{1}{h_1h_2h_3}\begin{vmatrix} h_1\, e_1 & \dfrac{\partial}{\partial\, u_1} & h_1\, V_1 \\ h_2\, e_2 & \dfrac{\partial}{\partial\, u_2} & h_2\, V_2 \\ h_3\, e_3 & \dfrac{\partial}{\partial\, u_3} & h_3\, V_3 \end{vmatrix}$$

d. *Laplacian of scalar function f*

$$\nabla^2 f = \frac{1}{h_1h_2h_3}\left[\frac{\partial}{\partial \mu_1}\left(\frac{h_2h_3}{h_1}\frac{\partial f}{\partial \mu_1}\right) + \frac{\partial}{\partial \mu_2}\left(\frac{h_1h_3}{h_2}\frac{\partial f}{u_2}\right) + \frac{\partial}{\partial \mu_3}\left(\frac{h_1h_2}{h_3}\frac{\partial f}{\partial \mu_3}\right)\right]$$

e. *General operator's formulas*

$\nabla(f_1 + f_2) = \nabla f_1 + \nabla f_2$

$\nabla . (V_1 + V_2) = \nabla . V_1 + \nabla . V_2$

$\nabla \times (V_1 + V_2) = \nabla \times V_1 + \nabla \times V_2$

$\nabla . (fV) = f(\nabla . \nabla) + (\nabla f). V$

$\nabla \times (fV) = f(\nabla \times \nabla) + (\nabla f) \times V$

$\nabla . (\nabla f) = \nabla^2 f$

11.8 DIFFERENTIAL OPERATORS–SPECIAL CASES

i. Cartesian Operators

a. *Gradient*

$$\nabla f = \frac{\partial f}{\partial x}\boldsymbol{i} + \frac{\partial f}{\partial y}\boldsymbol{j} + \frac{\partial f}{\partial z}\boldsymbol{k}$$

b. *Divergence*

$$\nabla . V = \frac{\partial\, V_x}{\partial x} + \frac{\partial V_y}{\partial y} + \frac{\partial V_z}{\partial z}$$

c. *Curl*

$$\nabla \times V = \begin{vmatrix} i & \dfrac{\partial}{\partial x} & V_x \\ j & \dfrac{\partial}{\partial y} & V_y \\ k & \dfrac{\partial}{\partial z} & V_z \end{vmatrix}$$

d. *Laplacian*

$$\nabla^2 f = \frac{\partial^2 f}{\partial x^2} + \frac{\partial^2 f}{\partial y^2} + \frac{\partial^2 f}{\partial z^2}$$

ii. Cylindrical Operators

a. *Gradient*

$$\nabla f = \frac{\partial f}{\partial a} e_a + \frac{1}{a}\frac{\partial f}{\partial \theta} e_\theta + \frac{\partial f}{\partial z} e_z$$

b. *Divergence*

$$\nabla \cdot V = \frac{1}{a}\left[\frac{\partial (aV_a)}{\partial a} + \frac{\partial V_\theta}{\partial \theta} + a\frac{\partial V_z}{\partial z}\right]$$

c. *Curl*

$$\nabla \times V = \frac{1}{a}\begin{vmatrix} e_a & \dfrac{\partial}{\partial a} & V_a \\ ae_\theta & \dfrac{\partial}{\partial \theta} & a V_\theta \\ e_z & \dfrac{\partial}{\partial z} & V_z \end{vmatrix}$$

d. *Laplacian*

$$\nabla^2 f = \frac{1}{a}\left[\frac{\partial}{\partial a}\left(a\frac{\partial f}{\partial a}\right) + \frac{1}{a}\frac{\partial}{\partial \theta}\left(\frac{\partial f}{\partial \theta}\right) + \frac{\partial}{\partial z}\left(a\frac{\partial f}{\partial z}\right)\right]$$

iii. Spherical Operators

a. *Gradient*

$$\nabla f = \frac{\partial f}{\partial b} e_b + \frac{1}{b\sin\phi}\frac{\partial f}{\partial \theta} e_\theta + \frac{1}{b}\frac{\partial f}{\partial \phi} e_\phi$$

b. *Divergence*

$$\nabla \cdot V = \frac{1}{b^2 \sin\phi}\left[\frac{\partial}{\partial b}(b^2 \sin\phi V_b) + \frac{\partial}{\partial \theta}(bV_\theta) + \frac{\partial}{\partial \phi}(b\sin\phi V_\phi)\right]$$

c. *Curl*

$$\nabla \times V = \frac{1}{b^2 \sin\phi}\begin{vmatrix} e_b & \dfrac{\partial}{\partial b} & V_b \\ b\sin\phi\, e_\theta & \dfrac{\partial}{\partial \theta} & b\sin\phi\, V_\theta \\ be_\phi & \dfrac{\partial}{\partial \phi} & b\, V_\phi \end{vmatrix}$$

d. *Laplacian*

$$\nabla^2 f = \frac{1}{b^2 \sin\phi}\left[\frac{\partial}{\partial b}\left(b^2 \sin\phi \frac{\partial f}{\partial b}\right) + \frac{\partial}{\partial \theta}\left(\frac{1}{\sin\phi\, \partial\phi} + \frac{\partial f}{\partial \theta}\right) + \frac{\partial}{\partial \phi}\left(\sin\phi \frac{\partial f}{\partial \phi}\right)\right]$$

11.9 VECTOR INTEGRAL CALCULUS

i. Indefinite Integral

a. *Definition*

The *indefinite integral of a vector function* $f(t)$ of a single scalar variable t is the anti-derivative of that vector function.

$$f(t) = \dot{F}(t) = \frac{dF(t)}{dt}$$

$$\int f(t)\, dt = \int \dot{F}(t)\, dt = F(t) + c$$

c = vector constant of integration

b. *General formulas*

E = constant vector

m = constant scalar

t = variable scalar

$F(t)$ = vector function of t

$\dot{F}(t)$ = first derivative of $F(t)$ *with respect to t*

$\ddot{F}(t)$ = second derivative of $F(t)$ *with respect to t*

$$\int m\dot{F}(t)\, dt = mF(t) + c$$

$$\int E \,.\, \dot{F}(t)\, dt = E \,.\, F(t) + c$$

$$\int E \times \dot{F}(t)\, dt = E \times F(t) + c$$

$$\int F(t) \,.\, \dot{F}(t)\, dt = \frac{F^2(t)}{2} + c$$

$$\int F(t) \times \ddot{F}(t)\, dt = F(t) \times \dot{F}(t) + c$$

$$\int\int \ddot{F}(t)\, dt\, dt = F(t) + c_1 t + c_2$$

ii. Definite Integral

a. *Definition*

The *definite integral of a vector function* $f(t)$ of a single scalar variable t is the limit of a sum between the lower and the upper unit.

$$\int_{t=a}^{t=b} f(t)\, dt = \int_{t=a}^{t=b} \dot{\boldsymbol{F}}(t)\, dt = \boldsymbol{F}(b) - \boldsymbol{F}(a)$$

a, b = scalar constants

b. *Interchange of limits*

$$\int_{t=a}^{t=b} f(t)\, dt = -\int_{t=b}^{t=a} f(t)\, dt = \mathrm{F}(b) - \mathrm{F}(a)$$

c. *Decomposition of limits*

$$\int_{t=a}^{t=b} f(t)\, dt = \int_{t=a}^{t=c} f(t)\, dt + \int_{t=c}^{t=b} f(t)\, dt$$

$$= \boldsymbol{F}(b) - \boldsymbol{F}(a)$$

d. *Sum of several functions*

$$\int_{t=a}^{t=b} [f_1(t) + f_2(t) + f_3(t)]\, dt = \int_{t=a}^{t=b} f_1(t)\, dt + \int_{t=a}^{t=b} f_2(t)\, dt + \int_{t=a}^{t=b} f_3(t)\, dt$$

e. *Constant factors E, m*

$$\int_{t=a}^{t=b} \boldsymbol{E} \times f(t)\, dt = \boldsymbol{E} \times \int_{t=a}^{t=b} f(t)\, dt \qquad \int_{t=a}^{t=b} mf(t) = m\int_{t=a}^{t=b} f(t)\, dt$$

$$\int_{t=a}^{t=b} \boldsymbol{E} \,.\, f(t)\, dt = \boldsymbol{E} \,.\int_{t=a}^{t=b} f(t)\, dt$$

11.10 LINE AND SURFACE INTEGRALS

i. Line Integral

a. *Definition*

The *line integral of a vector* **F** over a path C is defined as the integral of the scalar product of $\boldsymbol{F}$ and the elemental path $\boldsymbol{dr}$ along C. $\boldsymbol{r}$ defines a curve which has continuous derivative, and $\boldsymbol{F}$ is continuous along C.

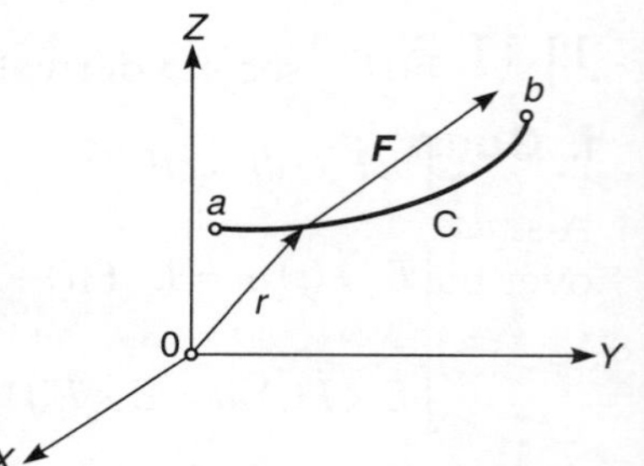

Fig. 11.9

$$\int_a^b \boldsymbol{F} \,.\, \boldsymbol{dr} = \int_a^b \boldsymbol{F}_x\, dx + \int_a^b \boldsymbol{F}_y\, dy + \int_a^b \boldsymbol{F}_z\, dz$$

b. *Theorems*

If $\boldsymbol{F} = \nabla\phi = \dfrac{\partial\phi}{\partial x}\boldsymbol{i} + \dfrac{\partial\phi}{\partial y}\boldsymbol{j} + \dfrac{\partial\phi}{\partial z}\boldsymbol{k}$ and C is *an open curve*, then

$\int_a^b \nabla\phi \,.\, \boldsymbol{dr}$ is *independent of the path* between a and b.

If $\boldsymbol{F} = \nabla\phi = \dfrac{\partial\phi}{\partial x}\boldsymbol{i} + \dfrac{\partial\phi}{\partial y}\boldsymbol{j} + \dfrac{\partial\phi}{\partial z}\boldsymbol{k}$ and C is *a closed curve*, then

$\oint \nabla\phi \,.\, \boldsymbol{dr}$ is *equal to zero.*

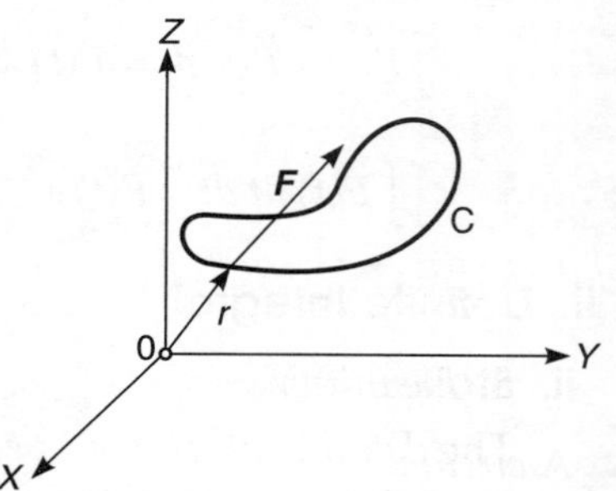

Fig. 11.10

ii. Surface Integral

a. *A normal surface vector* is a vector whose length is equal to the area bounded by a closed curve C and whose direction is perpendicular to the plane of the area.

b. *Flow of a scalar field*

$$\mathbf{PS} = \iint_{s} \phi\, d\mathbf{S} = \boldsymbol{i} \iint_{\Sigma yz} \phi\, dy\, dz + \boldsymbol{j} \iint_{\Sigma xz} \phi\, dx\, dz + \boldsymbol{k} \iint_{\Sigma xy} \phi\, dx\, dy$$

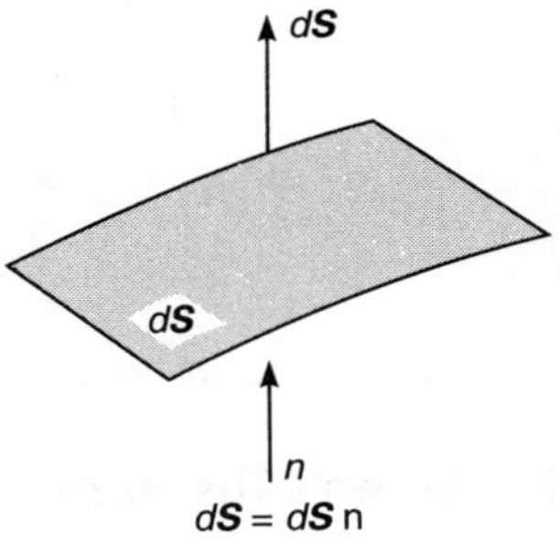

c. *Scalar flow of a vector field*

$$\mathbf{Q} = \iint_{s} \boldsymbol{F} . d\mathbf{S} = \iint_{\Sigma yz} F_x\, dy\, dz + \iint_{\Sigma xz} F_y dx\, dz + \iint_{\Sigma xy} F_z dx\, dy$$

d. *Vector flow of a vector field*

$$\mathbf{R} = \iint_{s} \boldsymbol{F} \times d\mathbf{S} = \iint_{\Sigma yz} (F_z \boldsymbol{j} - F_y \boldsymbol{k})\, dydz$$

$$+ \iint_{\Sigma xz} (F_x \boldsymbol{k} - F_z \boldsymbol{i})\, dxdz + \iint_{\Sigma xy} (F_y \boldsymbol{i} - F_x \boldsymbol{j})\, dx\, dy$$

Fig. 11.11

e. *Closed surface integral notation*

$$\mathbf{P} = \oint_{s} \phi\, d\mathbf{S} \qquad \mathbf{Q} = \oint_{s} F . d\mathbf{S} \qquad \mathbf{R} = \oint_{s} F \times d\mathbf{S}$$

Note: In integrals **P, Q, R** each integral is taken over projection of the surface on the respective coordinate plane.

11.11 INTEGRAL THEOREMS

i. Gauss' Theorem

A *scalar flow of a vector function* **F** through a closed surface S is equal to the integral of ∇ . **F** over the volume V bounded by S.

$$\underbrace{\oint_{s} \boldsymbol{F} . d\mathbf{S}}_{\text{Closed surface integral}} = \underbrace{\int_{V} \nabla . \boldsymbol{F}\, dV}_{\text{Closed volume integral}}$$

In Cartesian coordinates

$$\oint_{s} \boldsymbol{F} . d\mathbf{S} = \iint_{\Sigma yz} F_x dy\, dz + \iint_{\Sigma zx} F_y\, dz\, dx + \iint_{\Sigma xy} F_z\, dx\, dy = \iiint_{\Sigma xyz} \left(\frac{\partial F_x}{\partial x} + \frac{\partial F_y}{\partial y} + \frac{\partial F_z}{\partial z} \right) dx\, dy\, dz$$

Thus a closed volume integral can be reduced to a closed surface integral.

ii. Stokes' Theorem

A *circulation of a vector function* **F** about a closed path C is equal to a vector flow of the same vector function over an arbitrary surface bounded by C.

$$\underbrace{\oint_C \boldsymbol{F} \cdot d\boldsymbol{r}}_{\substack{\text{closed path} \\ \text{integral}}} = \underbrace{\oint_S (\nabla \times \boldsymbol{F}) \cdot d\mathbf{S}}_{\substack{\text{surface} \\ \text{integral}}}$$

In Cartesian coordinates

$$\oint_c F \cdot dr = \oint_c (F_x \, dx + F_y \, dy + F_z \, dz)$$

$$= \iint_{\Sigma yz} \left(\frac{\partial F_z}{\partial y} - \frac{\partial F_y}{\partial z} \right) dydz + \iint_{\Sigma zx} \left(\frac{\partial F_x}{\partial z} - \frac{\partial F_z}{\partial x} \right) dz\, dx + \iint_{\Sigma xy} \left(\frac{\partial F_y}{\partial x} - \frac{\partial F_x}{\partial y} \right) dx\, dy$$

Thus a *surface integral* can be reduced to a *line integral.*

iii. Green's Theorem

If in Gauss' theorem $\boldsymbol{F} = \alpha \nabla \beta$, where α, β are scalar functions of x, y, z then

$$\underbrace{\oint_s (\alpha \nabla \beta) \cdot d\mathbf{S}}_{\substack{\text{Closed surface} \\ \text{integral}}} = \underbrace{\int_v \nabla \cdot (\alpha \nabla \beta)\, dV}_{\substack{\text{Closed volume} \\ \text{integral}}} = \underbrace{\int_v (\alpha \nabla^2 \beta) + \nabla \alpha \cdot \nabla \beta)\, dV}_{\substack{\text{From operator's} \\ \text{formula}}}$$

If $\alpha = +1$,

$$\oint_s \nabla \beta \cdot d\mathbf{S} = \int_V \nabla^2 \beta \, dV$$

and in *cartesian coordinates*

$$\iint_{\Sigma yz} \frac{\partial \beta}{\partial x} dy\, dz + \iint_{\Sigma zx} \frac{\partial \beta}{\partial y} dz\, dx + \iint_{\Sigma xy} \frac{\partial \beta}{\partial z} dx\, dy = \iiint_{\Sigma xyz} \left(\frac{\partial^2 \beta}{\partial x^2} + \frac{\partial^2 \beta}{\partial y^2} + \frac{\partial^2 \beta}{\partial z^2} \right) dx\, dy\, dz$$

11.12 VECTOR ALGEBRA

A *scalar* is a quantity that is determined by its magnitude. For instance length, temperature and voltage are scalars.

A *vector* is a quantity that is determined by both its magnitude and its direction, thus it is directed live segment. For instance, a force is a vector and so it is a velocity giving the *speed and direction* of motion.

i. Components of a Vector

In cartisian coordinate system in space, if a given vector a has initial point P: $(x, y, z,)$ and terminal point Q: (x_2, y_2, z_2) the three numbers

$$a_1 = x_2 - x_1, \qquad a_2 = y_2 - y_1, \qquad \text{and} \qquad a_3 = z_2 - z_1$$

are called the components of the vector Q with respect to that coordinate system; can be written as

$$Q = [a_1, a_2, a_3],$$

length $\qquad |\, a \,| = \sqrt{a_1^2 + a_2^2 + a_3^2}$

ii. Length in terms of Component

The length | a | of a vector a is the distance between the initial point P and terminal point Q is given by

$$|a| = \sqrt{a_1^2 + a_2^2 + a_3^3}$$

iii. Position Vector

In cartisian coordinate system the position vector r of a point A $(x, y, z,)$ is the vector with oregin (000) as the initial point A as the terminal point thus $r = [x, y, z]$.

iv. Vector Calculus

Convergence: An infinite sequence of vectors a_n, $n = 1, 2, \ldots$ is said to be converge if there is a vector a such that $\lim_{n\to\infty} |a_n - a| = 0$.

a is called *unit vector* of that sequence $\lim_{n\to\infty} a_n = a$.

Thus sequence of vector converges to a if and only if the three sequences of components of the vectors converges to the corresponding components of a.

Continuity: A vector function $v(t)$ is said to be continuous at $t = t_o$ if it is defined in some neighborhood of t_o and $\lim_{t\to t_o} v(t) = v(t_o)$.

For a Cartesian Coordinate System

$v(t) = [v_1(t), v_2(t), v_3(t)] = v_1(t)i + v_2(t)j + v_3(t)k.$

Then $v(t)$ is a continuous at t_o if and only if its three components are continuous at t_o.

Definition: Derivative of a vector function.

A vector function $v(t)$ is said to be differentiable at a point t if the following limit exists

$$v'(t) = \lim_{\Delta t\to 0} \frac{v(t+\Delta t) - v(t)}{\Delta t}$$

The vector $v'(t)$ is called the derivative of $v(t)$.

In terms of components with respect to a given Cartesian coordinate system, $v(t)$ is differentiable at a point t if and only if its three components are differentiable at t and then the derivative $v'(t)$ is obtained by differentiating each component separately

$$v'(t) = [v_2'(t)\, v_2'(t), v_1'(t)]$$

v. Green's Theorem in the Plane

(Transformation between double integrals and line integrals)

Let R be a closed bounded region in the xy-plane whose boundary C consists of finitely many smooth curves. Let $F_1(x,y)$ and $F_2(x,y)$ be the functions that are continuous and have continuous partial derivatives of $\partial \frac{F_1}{\partial y}$ and $\partial \frac{2F_2}{\partial x}$ every where in some domain containing R,

Then

$$\int_R \int \left(\frac{\partial F_2}{\partial x} - \frac{\partial F_1}{\partial y} \right) \partial x \partial y = \oint_c (F_1\, \partial x + F_2\, \partial y)$$

This can be written in vectorial form

$$\int_R\int (\text{curl } F).k\, dxdy = \oint_c F.dr \qquad F = [F_1, F_2] = F_1\, i + F_2\, j$$

The formula in divergence theorem becomes
'Green's first formula'

$$\iiint_T \left(f\nabla^2 g + gradf \cdot gradg\right) dv = \iint_S f\frac{dg}{dn}\, dA.$$

'Green's second formula'

$$\iiint_T \left(f\nabla^2 g - g\Delta^2 f\right) dV = \iint_S \left(f\frac{\partial g}{\partial n} - g\frac{\partial f}{\partial n}\right) dA$$

vi. Divergence Theorem of Gauss

Transformation between volume integrals and surface integrals:

Let T be a closed bounded region in space whose boundary is a piecewise smooth orient table surface S. Let $F(x, y, z)$ be a vector function that is continuous and has continuous first partial derivatives in some domain containg T. Then

$$\iiint_T \text{div } F dv = \iint_S F.\boldsymbol{n}\, dA$$

Where $\boldsymbol{n}$ is the outer unit normal vector of S.

Above formula in components can be given by:

$$\iiint_T \left(\frac{\partial F_1}{\partial x} + \frac{\partial F_2}{\partial y} + \frac{\partial F_3}{\partial z}\right) dx\, dy\, dz = \iint \left(F_1 dydz + F_2 dzdx + F_3 dxdy\right).$$

vii. Stroke's Theorem

Transformation between surface integrals and line integrals:

Let S be a piecewise smooth oriented surface in space and let the boundary of S be a piecewise smooth simple closed curve C. Let $F(x, y, z)$ be a continuous vector function that has continuous first partial derivatives in a domain in space containg S. Then

$$\iint_S (\text{curl } F).n\, dA = \oint_a F.r'(s)\, ds$$

Where n is a unit normal vector of S and depending on n the integration around C. Further $r' = \frac{dr}{ds}$ is the unit tangent vector and S, the are length of C.

The above formula in Components is given as:

$$\iiint_R \left[\left(\frac{\partial F_3}{\partial y} - \frac{\partial F_2}{\partial z}\right) N_1 + \left(\frac{\partial F_1}{\partial z} - \frac{\partial F_3}{\partial x}\right) N_2 + \left(\frac{\partial F_2}{\partial x} - \frac{\partial F_1}{\partial y}\right) \cdot N_3\right] du\, dv = \oint_{\overline{C}} \left(F_1\, dx + F_2 dy + F_3 dz\right)$$

where R is the region with boundary curve $\overline{C}$ in the uv-plane corresponding to S.

CHAPTER

12

Functions of Complex Variables

12.1 COMPLEX NUMBERS

i. Definitions

The second root of a negative number

$$\sqrt{-b^2} = b\sqrt{-1} = bi$$

is called the *imaginary number*. The basic of imaginary numbers is the *imaginary unity i*. The *complex expression* consists of a real and an imaginary part. The *conjugate-complex expression* is a pair of binomic complex terms differing in sign only.

$$\sqrt{-1} = i \qquad i^{4k+1} = i$$

$$i^2 = -1 \qquad i^{4k+2} = -1$$

$$i^3 = -i \qquad i^{4k+3} = -i$$

$$i^4 = 1 \qquad i^{4k+4} = 1$$

$$k = 0, 1, 2, \ldots$$

ii. Operations

$$p = a + bi \qquad q = a - bi$$

$$p + q = 2a \qquad p - q = 2bi$$

$$pq = a^2 + b^2$$

$$\frac{p}{q} = \frac{a^2 + 2abi - b^2}{a^2 + b^2}$$

$$p = a + bi \qquad q = c + di$$

$$p \pm q = (a \pm c) + (b \pm d)\, i$$

$$pq = (ac - bd) + (ad + bc)\, i$$

$$\frac{p}{q} = \frac{(ac + bd) + (bc - ad)\, i}{c^2 + d^2}$$

iii. Complex Surds

$$\sqrt{\pm bi} = \sqrt{\frac{b}{2}} \pm i\sqrt{\frac{b}{2}} = \frac{\sqrt{2b}}{2}(1 \pm i)$$

$$\sqrt{a + bi} \pm \sqrt{a - bi} = \sqrt{2\left(a \pm \sqrt{a^2 + b^2}\right)}$$

$$\sqrt{\pm i} = \sqrt{\frac{1}{2}} \pm i\sqrt{\frac{1}{2}} = \frac{\sqrt{2}}{2}(1 \pm i) \qquad \sqrt{a \pm bi} = \sqrt{\frac{\sqrt{a^2+b^2}+a}{2}} \pm i\sqrt{\frac{\sqrt{a^2+b^2}-a}{2}}$$

iv. Representation of Complex Number

A *complex number* $z = a + bi$ can be represented as a point in the complex plane (Gauss plane).

$$z_1 = r_1(\cos\phi_1 + i\sin\phi_1)$$
$$z_2 = r_2(\cos\phi_2 + i\sin\phi_2)$$
$$z_1 z_2 = r_1 r_2[\cos(\phi_1+\phi_2) + i\sin(\phi_1+\phi_2)]$$
$$\frac{z_1}{z_2} = \frac{r_1}{r_2}[\cos(\phi_1-\phi_2) + i\sin(\phi_1-\phi_2)]$$
$$z^p = r^p(\cos\phi + i\sin\phi)^p$$
$$= r^p[\cos p\phi + i\sin p\phi]$$
$$x = r\cos\phi$$
$$y = r\sin\phi$$
$$z = a + bi = re^{i\phi} = r(\cos\phi + i\sin\phi)$$
$$\phi = \tan^{-1}\frac{b}{a}$$

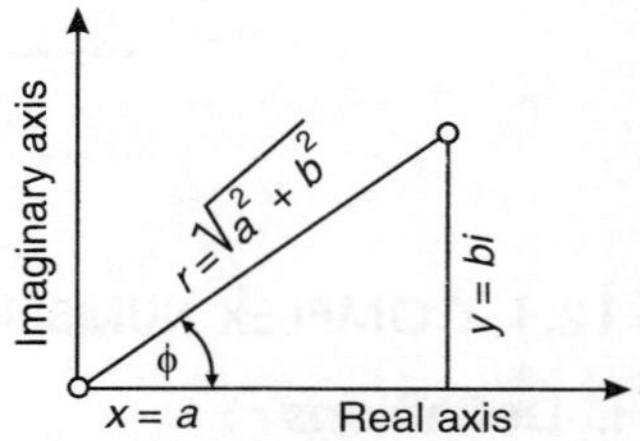

Fig. 12.1

$$\sqrt[s]{1} = \cos\frac{2k\pi}{s} + i\sin\frac{2k\pi}{s}$$
$$\sqrt[s]{-1} = \cos\frac{(2k+1)\pi}{s} + i\sin\frac{(2k+1)\pi}{s} \qquad s = 1, 2, 3, \ldots \quad k = 0, 1, 2, 3, \ldots, s-1$$

12.2 EXPONENTIAL AND TRIGONOMETRIC FUNCTIONS ($z = x + iy$)

i. Exponential Functions (k = 0, 1, 2, ...) (e = 2.71828 ...) $\left(r = \sqrt{x^2+y^2}\right)$

$$e^{i2k\pi} = 1 \qquad e^{i(2k+1)\pi} = -1 \qquad e^{\phi + i2k\pi} = e^{\phi} \qquad e^{\phi + i(2k+i)\pi} = -e^{\phi}$$

$$z^{m/n} = r^{m/n}\left(\cos\frac{m\phi}{n} + i\sin\frac{m\phi}{n}\right) = r^{m/n}e^{im\phi/n} \qquad \phi = \tan^{-1}\frac{y}{x} \qquad i^i = (0.20788\ldots)\,e^{2k\pi}$$

$$e^{ix} = 1 + \frac{ix}{1!} + \frac{(ix)^2}{2!} + \frac{(ix)^3}{3!} + \ldots + \frac{(ix)^n}{n!} + \ldots$$
$$= \left(1 - \frac{x^2}{2!} + \frac{x^4}{2!} - \ldots\right) + i\left(\frac{x}{1!} - \frac{x^3}{3!} + \frac{x^5}{5!} - \ldots\right) = \cos x + i\sin x$$

$$e^{-ix} = 1 - \frac{ix}{1!} + \frac{(ix)^2}{2!} - \frac{(ix)^3}{3!} + \ldots \pm \frac{(ix)^n}{n!} + \ldots + \ldots$$
$$= \left(1 + \frac{x^2}{2!} + \frac{x^4}{4!} + \ldots\right) - i\left(\frac{x}{1!} + \frac{x^3}{3!} + \frac{x^5}{5!}\ldots\right) = \cos x - i\sin x$$

ii. Derivative of Exponential Function

$$\frac{d(e^{(e+bix)})}{dx} = (a + bi)\, e^{(e + bi)x} = (a + bi)\, e^{ax} (\cos bx + i \sin x)$$

iii. Integral of Exponential Function

$$\int e^{(a+bi)x}\, dx = \frac{e^{(a+bi)x}}{a+bi} = \frac{e^{ax}}{a^2+b^2} [(a \cos bx + b \sin bx) + i\, (a \sin bx - b \cos bx)]$$

iv. Trigonometric Functions (*A*, *B*, *C*, *D*, *b* = constants)

$e^{ibx} = \cos bx + i \sin bx$ $\qquad \cos bx = \dfrac{e^{ibx} + e^{-ibx}}{2}$

$e^{-ibx} = \cos bx - i \sin bx$ $\qquad \sin bx = \dfrac{e^{ibx} - e^{-ibx}}{2i}$

$$(\cos bx \pm i \sin bx)^n = \cos nbx \pm i \sin nbx \qquad n = 1, 2, 3, \ldots$$

$$\sqrt[n]{(\cos bx \pm i \sin bx)} = \cos\left[\frac{b}{n}(x + 2\pi k)\right] \pm i \sin\left[\frac{b}{n}(x + 2\pi k)\right]$$

$$n = 1, 2, 3, \ldots ; K = 0, 1, 2, \ldots , n-1$$

$\sin ibx = i \sinh bx$ $\qquad \cos ibx = \cosh bx$ $\qquad \tan ibx = i \tanh bx$

$\mathrm{A}e^{ibx} + \mathrm{B}e^{-ibx} = \bar{\mathrm{A}} \cos bx + \bar{\mathrm{B}} \sin bx$ $\qquad \mathrm{C}\,(e^{ibx} + e^{-ibx}) = \bar{\mathrm{C}} \cos bx$

$\mathrm{D}\,(e^{ibx} - e^{-ibx}) = \bar{\mathrm{D}} \sin bx$

$\bar{\mathrm{A}} = (\mathrm{A} + \mathrm{B})$ $\qquad \bar{\mathrm{B}} = (\mathrm{A} - \mathrm{B})i$ $\qquad \bar{\mathrm{C}} = 2\,\mathrm{C}$ $\qquad \bar{\mathrm{D}} = 2\,\mathrm{D}i$

$$\sin (ax \pm iby) = \sin ax \cosh by \pm i \cos ax \sinh by$$

$$\cos (ax \pm iby) = \cos ax \cosh by \mp i \sin ax \sinh by$$

$$\tan (ax \pm iby) = \frac{\sin 2ax \pm i \sinh 2by}{\cos 2ax + \cosh 2by}$$

12.3 LOGARITHMIC AND HYPERBOLIC FUNCTIONS ($z = x + iy$)

i. Logarithmic Functions ($k = 0, \pm 1, \pm 2, \ldots$) ($\pi = 3.14159\ldots$) ($r = \sqrt{x^2 + y^2}$)

$\ln z = \ln re^{i\phi} = \ln re^{\,i(\phi + 2k\pi)}$ $\qquad \ln 1 = 2k\pi i$

$= \ln r + i\phi = \ln r + i\,(\phi + 2k\pi)$ $\qquad \phi = \tan^{-1}\dfrac{y}{x}$ $\qquad \ln (-1) = (2k + 1)\pi i$

$\ln a = \ln a + 2\,k\pi i$ $\qquad \ln bi = \ln b + i\left(\dfrac{\pi}{2} + 2k\pi\right)$ $\qquad \ln i = (4k + 1)\,\dfrac{\pi i}{2}$

$\ln (-a) = \ln a + i\,(2k + 1)\,\pi$ $\qquad \ln (-bi) = \ln b + i\left(\dfrac{3\pi}{2} + 2k\pi\right)$ $\qquad \ln (-i) = (4k + 3)\dfrac{\pi i}{2}$

ii. Hyperbolic Functions (*A, B, C, D, b* = constants)

$\sinh ibx = i \sin bx$ $\qquad \tanh ibx = i \tan bx$ $\qquad \text{sech}\, ibx = \sec ibx$

$\cosh ibx = \cos bx$ $\qquad \coth ibx = i \cot bx$ $\qquad \text{csch}\, ibx = i \csc bx$

$$(\cosh bx \pm \sinh bx)^n = \cosh nbx \pm \sinh nbx \qquad n = 1, 2, 3,...$$

$$\sqrt[n]{(\cosh bx \pm \sinh bx)} = \cosh \frac{b(x + 2\pi ki)}{n} \pm \sinh \frac{b(x + 2\pi ki)}{n}$$

$$n = 1, 2, 3, ...; k = 0, 1, 2, ..., n - 1$$

$$Ae^{bx} + Be^{-bx} + Ce^{ibx} + De^{-ibx} = \bar{A} \cosh bx + \bar{B} \sinh bx + \bar{C} \cos bx + \bar{D} \sin bx$$

$$= A \underbrace{\frac{\cosh bx + \cos bx}{2}}_{\phi_1} + B \underbrace{\frac{\sinh bx + \sin bx}{2}}_{\phi_2} + C \underbrace{\frac{\cosh bx - \cos bx}{2}}_{\phi_3} + D \underbrace{\frac{\sinh bx - \sin bx}{2}}_{\phi_4}$$

$m = 1, 2, 3, 4$	ϕ_1	ϕ_2	ϕ_3	ϕ_4
$d\phi_m/dx$	$b\,\phi_4$	$b\,\phi_1$	$b\,\phi_2$	$b\,\phi_3$
$d^2\phi_m/dx^2$	$b^2\,\phi_3$	$b^2\,\phi_4$	$b^2\,\phi_1$	$b^2\,\phi_2$
$d^3\phi_m/dx^3$	$b^3\,\phi_2$	$b^3\,\phi_3$	$b^3\,\phi_4$	$b^3\,\phi_1$
$d^4\phi_m/dx^4$	$b^4\,\phi_1$	$b^4\,\phi_2$	$b^4\,\phi_3$	$b^4\,\phi_4$

$\bar{A} = (A + B)$ $\qquad \bar{B} = (A - B)$ $\qquad \bar{C} = (C + D)$ $\qquad \bar{D} = i\,(C - D)$

$A^* = (A + B) + (C + D)$ $\qquad C^* = (A + B) - (C + D)$

$B^* = (A - B) + i\,(C - D)$ $\qquad D^* = (A - B) - i\,(C - D)$

$$\sinh (ax \pm iby) = \sinh ax \cos bx \pm i \cosh ax \sin bx$$

$$\cosh (ax \pm iby) = \cosh ax \cos bx \pm i \sinh ax \sin bx$$

$$\tanh (ax \pm iby) = \frac{\sinh 2ax \pm i \sin 2bx}{\cosh 2ax + \cos 2bx}$$

12.4 INVERSE FUNCTIONS OF TRIGONOMETRIC FUNCTIONS

i. Inverse Trigonometric Functions (*z* = *x* + *iy*)

$$\sin^{-1} z = -i \sinh^{-1} iz = -i \ln \left(iz + \sqrt{1 - z^2}\right)$$

$$\cos^{-1} z = -i \cosh^{-1} z = -i \ln \left(z + i\sqrt{1 - z^2}\right)$$

$$\tan^{-1} z = -i \tanh^{-1} iz = \frac{i}{2} \ln \frac{1 + iz}{1 - iz}$$

$$\sin^{-1}(ax \pm iby) = \sin^{-1}\left[\frac{\sqrt{b^2y^2 + (1+ax)^2} - \sqrt{b^2y^2 + (1-ax)^2}}{2}\right]$$

$$\pm i\cosh^{-1}\left[\frac{\sqrt{b^2y^2 + (1+ax)^2} + \sqrt{b^2y^2 + (1-ax)^2}}{2}\right]$$

$$\cos^{-1}(ax \pm iby) = \cos^{-1}\left[\frac{\sqrt{b^2y^2 + (1+ax)^2} - \sqrt{b^2y^2 + (1-ax)^2}}{2}\right]$$

$$\mp \cosh^{-1}\left[\frac{\sqrt{b^2y^2 + (1+ax)^2} + \sqrt{b^2y^2 + (1-ax)^2}}{2}\right]$$

$$\tan^{-1}(ax \pm iby) = \frac{\pi - \tan^{-1}\left[\dfrac{ax}{(\pm by - 1)}\right] + \tan^{-1}\left[\dfrac{ax}{(\pm by + 1)}\right]}{2} \pm \frac{i}{4}\ln\frac{a^2x^2 + (1 \pm by)^2}{a^2x^2 + (1 \mp by)^2}$$

ii. Inverse Hyperbolic Functions ($z = x + iy$)

$$\sinh^{-i} z = -i \sin^{-1} iz = \ln\left(z + \sqrt{z^2 + 1}\right)$$

$$\cosh^{-i} z = \quad i \cos^{-1} z = \ln\left(z + \sqrt{z^2 - 1}\right)$$

$$\tanh^{-i} z = -i \tan^{-1} iz = \frac{1}{2}\ln\frac{1+z}{1-z}$$

$$\sinh^{-1}(ax \pm iby) = \cosh^{-1}\frac{\sqrt{a^2x^2 + (1+by)^2} + \sqrt{a^2x^2 + (1-by)^2}}{2}$$

$$\pm i\sin^{-1}\frac{\sqrt{a^2x^2 + (1+by)^2} - \sqrt{a^2x^2 + (1-by)^2}}{2}$$

$$\cosh^{-1}(ax \pm iby) = \cosh^{-1}\frac{\sqrt{b^2y^2 + (1+ax)^2} - \sqrt{b^2y^2 + (1-ax)^2}}{2}$$

$$\pm i\cosh^{-1}\frac{\sqrt{b^2y^2 + (1+ax)^2} - \sqrt{b^2y^2 + (1-ax)^2}}{2}$$

$$\tanh^{-1}(ax \pm iby) = \frac{1}{2}\tanh^{-1}\frac{2ax}{1 + a^2x^2 + b^2x^2} + \frac{i}{2}\tan^{-1}\frac{\pm 2by}{1 - a^2x^2 - b^2y^2}$$

iii. De Moivre's Theorem for Integers

De Moivre's Theorem for Integers. A very important application of De Moivre's theorem is computing nth roots of complex numbers, where n is a positive integer.

Let n be any integer, then

$(\cos\theta + i\sin\theta)^n = \cos n\theta + i\sin n\theta$

iv. De Moivre's Theorem for Rationals

Let $\frac{p}{q}$ be a rational number, where $q \neq 0$, then

$$(\cos\theta + i\sin\theta)^{\frac{p}{q}} = \left(\cos p\frac{\theta}{q} + i\sin p\frac{\theta}{q}\right)$$

v. The roots of the complex polynomial $z^n - 1 = 0$ are $\frac{\cos 2k\pi}{n} + i\sin\frac{2k\pi}{n}$, where $k = 0, 1, 2, \ldots n-1$.

These are called the nth roots of unity in particular

a. If $z^2 - 1 = 0$, then square roots of unity are $\cos 0 + i\sin 0$ and $\cos\pi + i\sin\pi$, i.e. 1 and -1.

b. If $z^3 - 1 = 0$, then cube roots of unity are $\cos 0 + i\sin 0$, $\cos\frac{2\pi}{3} + i\sin\frac{2\pi}{3}$, $\cos\frac{4\pi}{3} + i\sin\frac{4\pi}{3}$ i.e. $1, -\frac{1}{2} + \frac{i\sqrt{3}}{2}; -\frac{1}{2} - \frac{i\sqrt{3}}{2}$.

vi. The roots of the complex polynomial $z^n + 1 = 0$ are $\cos\frac{(2k+1)\pi}{n} + i\sin\frac{(2k+1)\pi}{n}$, where $k = 0, 1, 2, \ldots n-1$.

In particular,

a. If $z^2 + 1 = 0$, then roots are $\cos\frac{\pi}{2} + i\sin\frac{\pi}{2}$ and $\cos\frac{3\pi}{2} + i\sin\frac{3\pi}{2}$, i.e. i and $-i$

b. If $z^3 + 1 = 0$ then roots are $\cos\frac{\pi}{3} + i\sin\frac{\pi}{3}$, $\cos\pi + i\sin\pi$ and $\cos\frac{5\pi}{3} + i\sin\frac{5\pi}{3}$, i.e. $\frac{1}{2} + \frac{i\sqrt{3}}{2}, -1, \frac{1}{2} - \frac{i\sqrt{3}}{2}$.

CHAPTER

13

Fourier Series

13.1 CONCEPT OF A FOURIER SERIES

i. Fourier Series

A Fourier series is an accurate representation of a periodic signal which consists of the sum of sinosoids at the fundamental and harmonic frequencies. The expression for a finite of harmonically related sinosoids called a Fourier series.

Let $F(x)$ satisfy the following condition:

a. $F(x)$ is defined in the interval $c < x < c + 2l$.

b. $F(x)$ and $F'(x)$ are sectionally continuous in $c < x < c + 2l$.

c. $F(x + 2l) = F(x)$, *i.e.* $F(x)$ is periodic with period $2l$.

Then at every point of continuity:

$$F(x) = \frac{a_0}{2} + \sum_{n=1}^{\infty}\left(a_n \cos\frac{n\pi x}{l} + b_n \sin\frac{n\pi x}{l}\right) \tag{13.1}$$

where

$$\left.\begin{aligned} a_n &= \frac{1}{l}\int_c^{c+2l} F(x)\cos\frac{n\pi x}{l}\,dx \\ b_n &= \frac{1}{l}\int_c^{c+2l} F(x)\sin\frac{n\pi x}{l}\,dx \end{aligned}\right\} \tag{13.2}$$

At a point of discontinuity, the left side of [Eq. (13.1)] is replaced by $\frac{1}{2}\{F(x + 0) + F(x - 0)\}$, *i.e.* the mean value at the discontinuity.

The series [Eq. (13.1)] with coefficients [Eq. (13.2)] is called the *Fourier series* of $F(x)$. For many problems, $c = 0$ or $-l$. In case $l = \pi$, $F(x)$ has period 2π and Eqs (13.1) and (13.2) are simplified.

The above conditions are often called *Dirichlet conditions* and are sufficient (but not necessary) conditions for convergence of Fourier series.

ii. Odd and Even Functions

A function $F(x)$ is called *odd* if $F(-x) = -F(x)$. Thus x^3, $x^5 - 3x^3 + 2x$, $\sin x$, $\tan 3x$ are odd functions.

A function $F(x)$ is called *even* if $F(-x) = -F(x)$. Thus x^4, $2x^6 - 4x^1 + 5$, $\cos x$, $e^x + e^{-x}$ are even functions.

In the Fourier series corresponding to an odd function, only sine terms can be present. In the Fourier series corresponding to an even function, only cosine terms can be present.

iii. Half Range Fourier Sine and Cosine Series

A half range Fourier sine or cosine series is a series in which only sine terms or only cosine terms are present respectively. When a half range series corresponding to a given function is desired, the function is generally defined in the interval $(0, l)$ [which is half of the interval $(-l, l)$, thus accounting for the name half range] and then the function is specified as odd or even, so that it is clearly defined in the other half of the interval, namely $(-l, 0)$. In such case:

$$\left.\begin{aligned} a_n = 0, \quad b_n &= \frac{2}{l}\int_c^1 F(x)\sin\frac{n\pi x}{l}dx \quad \text{for half range sine series} \\ b_n = 0, \quad a_n &= \frac{2}{l}\int_c^1 F(x)\cos\frac{n\pi x}{l}dx \quad \text{for half range cosine series} \end{aligned}\right\} \tag{13.3}$$

iv. Complex Form of Fourier Series

In complex notation, the Fourier series [Eq. (13.1)] and coefficients [Eq. (13.2)] can be written as

$$F(x) = \sum_{n=-\infty}^{\infty} c_n \, e^{in\pi x/l}$$

where, taking $c = -l$,

$$c_n = \frac{1}{2l}\int_{-1}^{1} F(x)\, e^{in\pi x/l}\, dx$$

v. Parseval's Identity for Fourier Series

Parseval's identity states that

$$\frac{1}{l}\int_{-1}^{1}\{F(x)\}^2 dx = \frac{a_0^2}{2} + \sum_{n=1}^{\infty}(a_n^2 + b_n^2)$$

where a_n and b_n are given by Eq. (13.2).

An important consequence is that

$$\left.\begin{aligned} \lim_{n\to\infty}\int_{-1}^{1} F(x)\sin\frac{n\pi x}{l}dx &= 0 \\ \lim_{n\to\infty}\int_{-1}^{1} F(x)\cos\frac{n\pi x}{l}dx &= 0 \end{aligned}\right\} \tag{13.4}$$

This is called *Riemann's theorem.*

vi. Finite Fourier Transforms

The *finite Fourier sine transform* of $F(x)$, $0 < x < l$, is defined as

$$f_s(n) = \int_0^l F(x) \sin\frac{n\pi x}{l}\, dx$$

where n is an integer. The function $F(x)$ is then called the *inverse finite Fourier sine transform* of $f_s(n)$ and is given by

$$F(x) = \frac{2}{l}\sum_{n=1}^{\infty} f_s(n)\sin\frac{n\pi x}{l}$$

The *finite Fourier cosine transform* of $F(x)$, $0 < x < l$, is defined as

$$f_s(n) = \int_0^l F(x) \cos\frac{n\pi x}{l}\, dx$$

where n is an integer. The function $F(x)$ is then called the *inverse finite Fourier cosine transform* of $f_c(n)$ and is given by

$$F(x) = \frac{1}{l} f_c(0) + \frac{2}{l}\sum_{n=1}^{\infty} f_s(n)\cos\frac{n\pi x}{l}$$

vii. The Fourier Integral

Let $F(x)$ satisfy the following conditions:

1. $F(x)$ satisfies the Dirichlet conditions in every finite interval $-l \le x \le l$.
2. $\int_{-\infty}^{\infty} |F(x)|\, dx$ converges, *i.e.* $F(x)$ is absolutely integrable in $-\infty < x < \infty$.

Then *Fourier's integral theorem* states that

$$\mathbf{F(x)} = \int_0^{\infty} \{A(\lambda) \cos \lambda x + B(\lambda) \sin \lambda x\}\, d\lambda$$

where

$$A(\lambda) = \frac{1}{\pi}\int_{-\infty}^{\infty} F(x) \cos \lambda x\, dx$$

$$B(\lambda) = \frac{1}{\pi}\int_{-\infty}^{\infty} F(x) \sin \lambda x\, dx$$

This can be written equivalently as

$$F(x) = \frac{1}{2\lambda}\int_{\infty=-\infty}^{\infty} \int_{u=-\infty}^{\infty} F(u) \cos \lambda\, (x-u) du\, d\lambda$$

viii. Complex Form of Fourier Integrals

In complex notation, the Fourier integral with coefficients can be written as

$$F(x) = \frac{1}{2\pi}\int_{-\infty}^{\infty} e^{i\lambda x}\, d\lambda \int_{-\infty}^{\infty} F(u)\, e^{-i\lambda u}\, du$$

$$= \frac{1}{2\pi}\int_{-\infty}^{\infty}\int_{-\infty}^{\infty} F(u)\, e^{-i\lambda(x-u)}\, du\, d\lambda$$

ix. Fourier Transforms

The Fourier integral, if

$$f(\lambda) = \int_{-\infty}^{\infty} e^{-i\lambda u} F(u)\, du$$

then
$$F(u) = \frac{1}{2\pi}\int_{-\infty}^{\infty} e^{i\lambda u} f(\lambda)\, d\lambda$$

which gives $F(x)$ on replacing u by x.

The function $f(\lambda)$ is called the Fourier transform of $F(x)$ and is sometimes written $F(\lambda) = \mathcal{F}[\{F(x)\}$. The function $F(x)$ is the inverse Fourier transform of $f(\lambda)$ and is written $F(x) = \mathcal{F}^{-1}\{f(\lambda)\}$.

x. Fourier Sine and Cosine Transforms

The *(infinite) Fourier sine transform* of $F(x)$, $0 < x < \infty$, is defined as

$$f_s(\lambda) = \int_0^{\infty} F(u) \sin \lambda u\, du$$

The function $F(x)$ is then called the *inverse Fourier sine transform* of $f_s(\lambda)$ and is given by

$$F(x) = \frac{2}{\pi}\int_0^{\infty} f_s(\lambda) \sin \lambda x\, d\lambda$$

The *(infinite) Fourier cosine transform* of $F(x)$, $0 < x < \infty$, is defined as

$$f_c(\lambda) = \int_0^{\infty} F(u) \cos \lambda u\, du$$

The function $F(x)$ is then called the *inverse Fourier cosine transform* of $f_c(\lambda)$

xi. The Convolution Theorem

The *convolution* of two functions $F(x)$ and $G(x)$, where $-\infty < x < \infty$, is defined as

$$F * G = \int_{-\infty}^{\infty} F(u)\, G(x-u)\, du = H(x)$$

An important result, known as the *convolution theorem for Fourier transforms.*

Theorem: If $H(x)$ is the convolution of $F(x)$ and $G(x)$, then

$$\int_{-\infty}^{\infty} H(x)\, e^{-i\lambda x}\, dx = \left\{\int_{-\infty}^{\infty} F(x)\, e^{-i\lambda x}\, dx\right\}\left\{\int_{-\infty}^{\infty} G(x)\, e^{-i\lambda x}\, dx\right\}$$

or $$'F\,\{F * G\} = 'F\{F\}\;'F^{(G)}$$

i.e. the Fourier transform of the convolution of F and G is the product of the Fourier transforms of F and G.

xii. Parseval's Identity for Fourier Integrals

If the Fourier transform of $F(x)$ is $f(\lambda)$, then

$$\int_{-\infty}^{\infty} |F(x)|^2\, dx = \frac{1}{2\pi} |f(\lambda)|^2\, d\lambda$$

This is called *Parseval's identity for Fourier integrals.*

xiii. Relationship of Fourier and Laplace Transforms

Consider the function

$$f(t) = \begin{cases} e^{-xt}\Phi(t) & t > 0 \\ 0 & t < 0 \end{cases}$$

$$'F\{F(t)\} = \int_0^{\infty} e^{-(x+iy)t}\, \Phi(t)\, dt = \int_0^{\infty} e^{-st}\Phi(t)\, dt$$

where we have written $s = x + iy$. The right side of above equation is the Laplace transform of $\Phi(t)$ and the result indicates *a* relationship of Fourier and Laplace transforms. It also indicates a need for considering s as a complex variable $x + iy$.

If $F(t)$ and $G(t)$ are zero for $t < 0$, the convolution of F and G is given by

$$F * G = \int_0^{t} F(u)\, G(t-u)\, du,$$

if corresponds to

$$\mathcal{L}\,\{F * G\} = \mathcal{L}\{F\}\, \mathcal{L}\{G\}$$

Table of Inverse Laplace Transforms

	$F(s)$	$\mathcal{L}^{-1}\{F(s)\} = f(t)$
1.	$\frac{1}{s}$	1
2.	$\frac{1}{s^2}$	t
3.	$\frac{1}{s^{n+1}}$ $n = 0, 1, 2, \ldots$	$\frac{t^n}{n!}$

	$F(s)$	$\mathcal{L}^{-1}\{F(s)\} = f(t)$
4.	$\frac{1}{s-a}$	e^{at}
5.	$\frac{1}{s^2+a^2}$	$\frac{\sin at}{a}$
6.	$\frac{8}{s^2+a^2}$	$\cos at$
7.	$\frac{1}{s^2-a^2}$	$\frac{8}{s^2-a^2}$
8.	$\frac{\sinh at}{a}$	$\cosh at$

13.2 SOME IMPORTANT PROPERTIES OF INVERSE LAPLACE TRANSFORMS

i. *Linearity Property*

Theorem: If c_1 and c_2 are any constants while $F_1(s)$ and $F_2(s)$ are the Laplace transforms of $f_1(t)$ and $f_2(t)$ respectively, then

$$\mathcal{L}^{-1}\{c_1 F_1(s) + c_2 F_2(s)\} = c_1\mathcal{L}^{-1}\{F_1(s)\} + c_2\mathcal{L}^{-1}\{F_2(s)\} \quad (13.5)$$

$$= c_1 f_1(t) + c_2 f_2(t)$$

ii. *First translation or shifting property*

Theorem: If $\mathcal{L}^{-1}\{F(s)\} = f(t)$, then

$$\mathcal{L}^{-1}\{F_1(s-a)\} = e^{at} f(t) \quad (13.6)$$

i. Definations

a. *Definition*

Any *single-valued function* $f(\theta)$ that is *continuous* except for a *finite number of discontinuities* in an interval $-\pi < \theta < +\pi$, and has a finite number of maxima and minima in this interval may be represented by a *Convergent Fourier Series.*

$$f(\theta) = \frac{a_0}{2} + a_1 \cos\theta + a_2 \cos 2\theta + a_3 \cos 3\theta + \ldots, + b_1 \sin\theta + b_2 \sin 2\theta + b_3 \sin 3\theta + \ldots$$

$$= \frac{a_0}{2} + \sum_{n=1}^{\infty} (a_n \cos n\theta + b_n \sin n\theta)$$

If $f(\theta)$ is a *periodic function* of θ with *period* 2π,

$$a_n = \frac{1}{\pi}\int_{-\pi}^{+\pi} f(\theta)\cos n\theta\, d\theta \qquad n = 0, 1, 2, \ldots b_n = \frac{1}{\pi}\int_{-\pi}^{+\pi} f(\theta)\sin n\theta\, d\theta \qquad n = 1, 2, \ldots$$

b. *Phase angle α and β*

The cosine and sine terms in the Fourier series may be combined in a single cosine or sine series with phase angles α or β, respectively.

$$f(\theta) = \frac{A_0}{2} + \sum_{n=1}^{\infty} A_n \cos(n\theta + \alpha) \qquad f(\theta) = \frac{B_0}{2} + \sum_{n=1}^{\infty} B_n \sin(n\theta + \beta)$$

$$A_n = \sqrt{a_n^2 + b_n^2} \qquad B_n = \sqrt{a_n^2 + b_n^2}$$

$$\tan\alpha = \frac{a_n}{b_n} \qquad \tan\beta = \frac{b_n}{a_n}$$

ii. Special Cases

a. *Change in variable* $\left(\theta = \frac{nx}{1} \text{ and } -l < x < +l\right)$

$$\bar{a}_n = \frac{1}{l}\int_{-l}^{+l} f(x)\cos\frac{n\pi x}{l}\,dx$$

$$f(x) = \frac{\bar{a}_0}{2} + \sum_{n=1}^{\infty}\left(\bar{a}_n \cos\frac{n\pi x}{l} + \bar{b}_n \sin\frac{n\pi x}{l}\right)$$

$$\bar{b}_n = \frac{1}{l}\int_{-l}^{+l} f(x)\sin\frac{n\pi x}{l}\,dx$$

b. *Change in variable* $\left(\theta = \frac{2\pi t}{T} \text{ and } -\frac{T}{2} < t < +\frac{T}{2}\right)$

$$a_n^* = \frac{2}{T}\int_{-T/2}^{+T/2} f(t)\cos\frac{2n\pi t}{T}\,dt$$

$$f(t) = \frac{a_0^*}{2} + \sum_{n=1}^{\infty}\left(a_n^* \cos\frac{2n\pi t}{T} + b_n^* \sin\frac{2n\pi t}{T}\right)$$

$$b_n^* = \frac{2}{T}\int_{-T/2}^{+T/2} f(t)\sin\frac{2n\pi t}{T}\,dt$$

13.3 DEVELOPMENT OF SERIES

i. Change in Limits

In the development of Fourier series the *limits of integral may be changed* (shifting of interval) as shown in the table that follows.

	$-2\pi < \theta < 0$	$\phi < \theta < \phi + 2\pi$	$0 < \theta < 2\pi$
a_n	$\frac{1}{\pi}\int_{-2\pi}^{0} f(\theta) \cos n\theta \, d\theta$	$\frac{1}{\pi}\int_{\phi}^{\phi+2\pi} f(\theta) \cos n\theta \, d\theta$	$\frac{1}{\pi}\int_{0}^{2\pi} f(0) \cos n\theta \, d\theta$
b_n	$\frac{1}{\pi}\int_{-2\pi}^{0} f(\theta) \sin n\theta \, d\theta$	$\frac{1}{\pi}\int_{\phi}^{\phi+2\pi} f(\theta) \sin n\theta \, d\theta$	$\frac{1}{\pi}\int_{0}^{2\pi} f(0) \sin n\theta \, d\theta$
	$-2l < x < 0$	$a < x < a + 2l$	$0 < x < 2l$
$\overline{a}_n$	$\frac{1}{l}\int_{-2l}^{0} f(x)\cos\frac{n\pi x}{l}dx$	$\frac{1}{l}\int_{a}^{a+2l} f(x)\cos\frac{n\pi x}{l}dx$	$\frac{1}{l}\int_{0}^{2l} f(x)\cos\frac{n\pi x}{l}dx$
$\overline{b}_n$	$\frac{1}{l}\int_{-2l}^{0} f(x)\sin\frac{n\pi x}{l}dx$	$\frac{1}{l}\int_{a}^{a+2l} f(x)\sin\frac{n\pi x}{l}dx$	$\frac{1}{l}\int_{0}^{2l} f(x)\cos\frac{n\pi x}{l}dx$
	$-T < t < 0$	$C < t < C + T$	$0 < t < T$
a^*_n	$\frac{2}{T}\int_{-T}^{0} f(t)\cos\frac{2n\pi t}{T}dt$	$\frac{2}{T}\int_{C}^{C+T} f(t)\cos\frac{2n\pi t}{T}dt$	$\frac{2}{T}\int_{0}^{T} f(t)\cos\frac{2n\pi t}{T}dt$
b^*_n	$\frac{2}{T}\int_{-T}^{0} f(t)\sin\frac{2n\pi t}{T}dt$	$\frac{2}{T}\int_{C}^{C+T} f(t)\sin\frac{2n\pi t}{T}dt$	$\frac{2}{T}\int_{0}^{T} f(t)\sin\frac{2n\pi t}{T}dt$

ii. Identities

In the Fourier series expansion the following identities are useful:

	n	n *even*	n *odd*	$\frac{n}{2}$ *odd*	$\frac{n}{2}$ *even*
$\sin n\pi$	0	0	0	0	0
$\cos n\pi$	$(-1)^n$	+1	−1	+1	+1
$\sin\frac{n\pi}{2}$		0	$(-1)^{n-1/2}$	0	0
$\cos\frac{n\pi}{2}$		$(-1)^{n/2}$	0	−1	+1

If all derivatives are finite and the series is convergent, then

$$\frac{f'(2\pi) - f'(0)}{n^2} - \frac{f'''(2\pi) - f'''(0)}{n^4} + \ldots = \int_0^{2\pi} f(\theta) \cos n\theta \, d\theta$$

$1 - \frac{1}{3} + \frac{1}{5} - \frac{1}{7} + \ldots = \frac{\pi}{4}$	$1 + \frac{1}{2^2} + \frac{1}{3^2} + \frac{1}{4^2} + \ldots = \frac{\pi^2}{6}$
$1 - \frac{1}{3^3} + \frac{1}{5^3} - \frac{1}{7^3} + \ldots = \frac{\pi^3}{32}$	$1 - \frac{1}{2^2} + \frac{1}{3^2} - \frac{1}{4^2} + \ldots = \frac{\pi^2}{12}$
$1 + \frac{1}{3^4} + \frac{1}{5^4} + \frac{1}{7^4} + \ldots = \frac{\pi^4}{96}$	$1 + \frac{1}{2^4} + \frac{1}{3^4} + \frac{1}{4^4} + \ldots = \frac{\pi^4}{90}$

13.4 SPECIAL FORMS

i. Closed Form (s = constant; n = 1, 2 ...)

$$\sum_{n=1}^{\infty} s^n \sin nx = \frac{s \sin x}{1 - 2s\cos x + s^2}, \qquad s^2 < 1$$

$$\sum_{n=0}^{\infty} s^n \sin nx = \frac{1 - s\cos x}{1 - 2s\cos x + s^2}, \qquad s^2 < 1$$

$$\sum_{n=1}^{\infty} \frac{s^n}{n} \sin nx = \tan^{-1}\frac{s \sin x}{1 - s\cos x}, \qquad s^2 \le 1$$

$$\sum_{n=1}^{\infty} \frac{s^n}{n} \cos nx = \ln\frac{1}{\sqrt{1 - 2s\cos x + s^2}}, \qquad s^2 \le 1$$

$$\sum_{n=1}^{\infty} \frac{\sin nx}{n} = \frac{\pi - x}{2}$$

$$\sum_{n=1}^{\infty} \frac{\cos nx}{n} = \frac{1}{2}\ln\frac{1}{2(1-\cos x)}$$

$$\sum_{n=1}^{\infty} \frac{\sin nx}{n^3} = \frac{\pi^2 x}{6} - \frac{\pi x^2}{4} + \frac{x^3}{12}, \qquad 0 < x < 2\pi$$

$$\sum_{n=1}^{\infty} \frac{\cos nx}{n^2} = \frac{\pi^2}{6} - \frac{\pi x}{2} + \frac{x^2}{4}, \qquad 0 < x < 2\pi$$

$$\sum_{n=1}^{\infty} \frac{\sin nx}{n^5} = \frac{\pi^4 x}{90} - \frac{\pi^2 x^3}{36} + \frac{\pi x^4}{48} - \frac{x^5}{240}$$

$$\sum_{n=1}^{\infty} \frac{\cos nx}{n^4} = \frac{\pi^4}{90} - \frac{\pi^2 x^2}{12} + \frac{\pi x^3}{12} - \frac{x^4}{48}$$

ii. Complex Form

Since the *exponential* and *trigonometric functions* are given by

$$\cos\theta = \frac{e^{i\theta} + e^{-i\theta}}{2} \qquad \sin\theta = \frac{e^{i\theta} - e^{-i\theta}}{2i}$$

with $\omega_n = \dfrac{n\pi}{l}$ and $n = 0, \pm 1, \pm 2 \ldots$ then

$$f(x) = \frac{1}{2}\left(C_0 + \sum_{n=1}^{\infty} C_n e^{i\omega_n x} + \sum_{n=1}^{\infty} D_n e^{-j\omega_n x} \right)$$

in which

$$C_n = \frac{1}{l}\int_{-l}^{l} f(x)\, e^{-i\omega_n x}\, dx \qquad D_n = \frac{1}{l}\int_{-l}^{l} f(x)\, e^{-j\omega_n x}\, dx$$

and

$$C_n = \bar{a}_n - i\bar{b}_n \qquad D_n = \bar{a}_n + i\bar{b}_n$$

or in simpler form

$$f(x) = \frac{1}{2} \sum_{n=-x}^{+\infty} C_n e^{-i\omega_n x}$$

The set of coefficients $|C_n|$ is called the *spectrum* of $f(x)$.

13.5 EVEN AND ODD FUNCTIONS

i. Even Functions

$$\bar{a}_n = \frac{2}{l}\int_0^l f(x)\cos\frac{n\pi x}{l}dx$$

$$\bar{b}_n = 0 \qquad n = 0, 1, 2, 3, \ldots$$

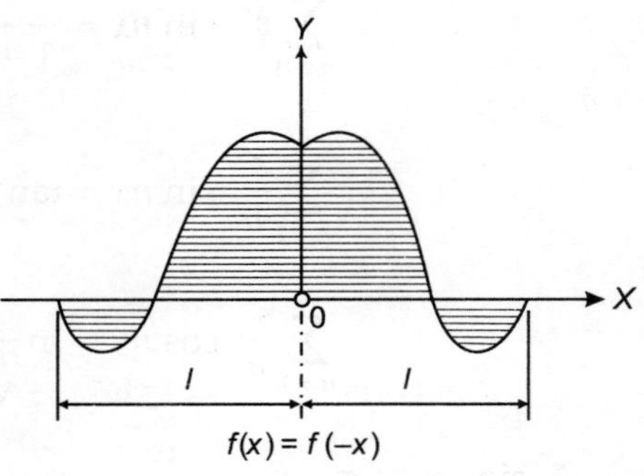

Fig. 13.1

$$\bar{a}_{2n} = \frac{2}{l}\int_0^l f(x)\cos\frac{2n\pi x}{l}dx$$

$$\bar{b}_{2n} = \frac{2}{l}\int_0^l f(x)\sin\frac{2n\pi x}{l}dx$$

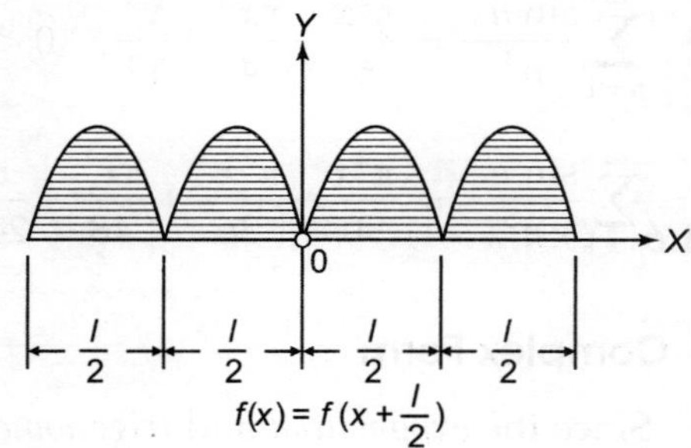

Fig. 13.2

$$\bar{a}_{2n+1} = 0 \qquad \bar{b}_{2n+1} = 0 \qquad n = 0, 1, 2, \ldots$$

ii. Odd Functions

$$\bar{b}_n = \frac{2}{l}\int_0^1 f(x)\sin\frac{n\pi x}{l}dx$$

$$\bar{a}_n = 0 \qquad n = 1, 2, 3, \ldots$$

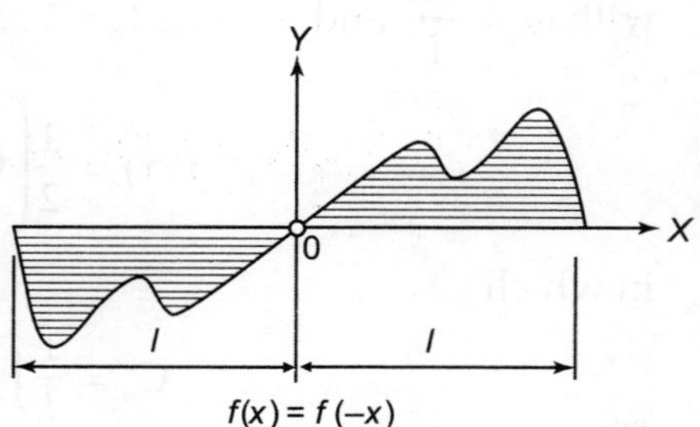

Fig. 13.3

$$\overline{a}_{2n+1} = \frac{2}{l}\int_0^1 f(x)\cos\frac{(2n+1)\pi x}{l}dx$$

$$\overline{b}_{2n+1} = \frac{2}{l}\int_0^1 f(x)\sin\frac{(2n+1)\pi x}{l}dx$$

$\overline{a}_{2n} = 0 \qquad \overline{b}_{2n} = 0 \qquad n = 0, 1, 2, \ldots$

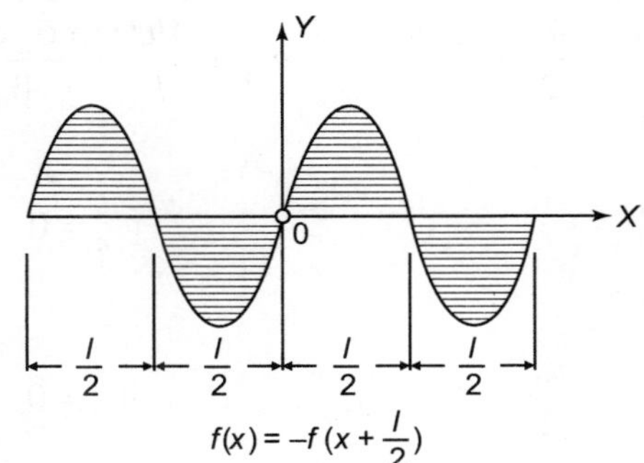

Fig. 13.4

$$\overline{b}_{2n+1} = \frac{2}{l}\int_0^{1/2} f(x)\sin\frac{(2n+1)\pi x}{l}dx$$

$\overline{a}_{2n} = 0 \qquad \overline{b}_{n} = 0 \qquad n = 0, 1, 2, \ldots$

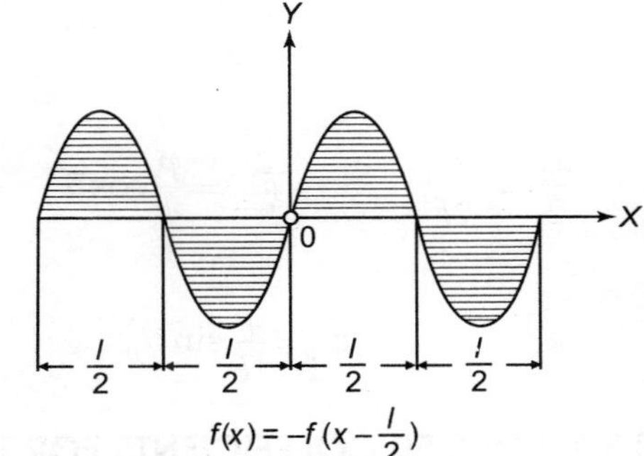

Fig. 13.5

13.6 FOURIER COEFFICIENTS FOR RECTANGULAR PERIODIC FUNCTIONS

$$f(x) = \frac{\overline{a}_0}{2} + \sum_{n=1}^{\infty}\overline{a}_n\cos\frac{n\pi x}{l} + \sum_{n=1}^{\infty}\overline{b}_n\sin\frac{n\pi x}{l} \qquad \overline{\alpha}_n = \frac{n\pi s}{l}$$

$$\overline{\beta}_n = \frac{n\pi r}{l}$$

$\frac{\overline{a}_0}{2} = \frac{hs}{l} \qquad \overline{a}_n = \frac{2h}{n\pi}\sin\overline{\alpha}_n$

$\overline{b}\, n = 0$

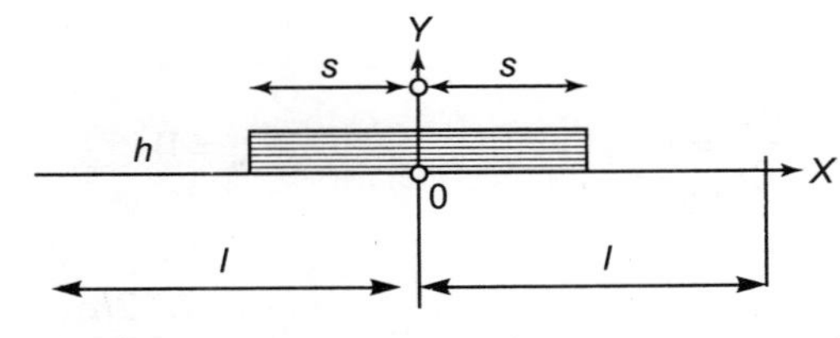

Fig. 13.6

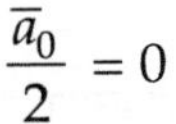

$\frac{\overline{a}_0}{2} = 0 \qquad \overline{a}_n = 0$

$n = 1, 3, 5, \ldots \qquad \overline{b}_n = \frac{4h}{n\pi}$

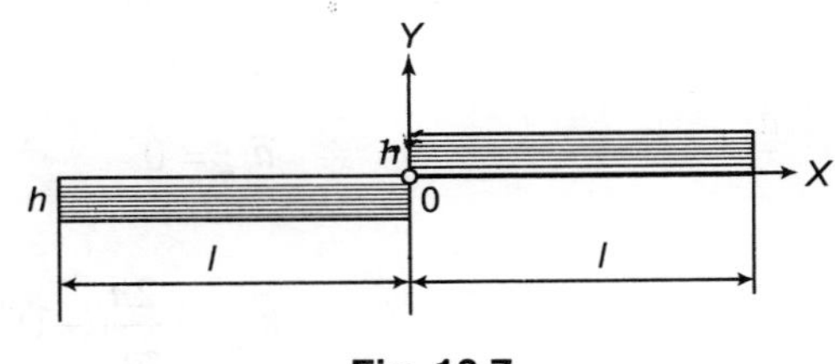

Fig. 13.7

$$\frac{\bar{a}_0}{2} = \frac{2hs}{l} \qquad \bar{a}_n = \frac{4hr}{l} \frac{\sin \bar{\alpha}_n \cos \bar{\beta}_n}{\beta_n}$$

$$\bar{b}_n = 0$$

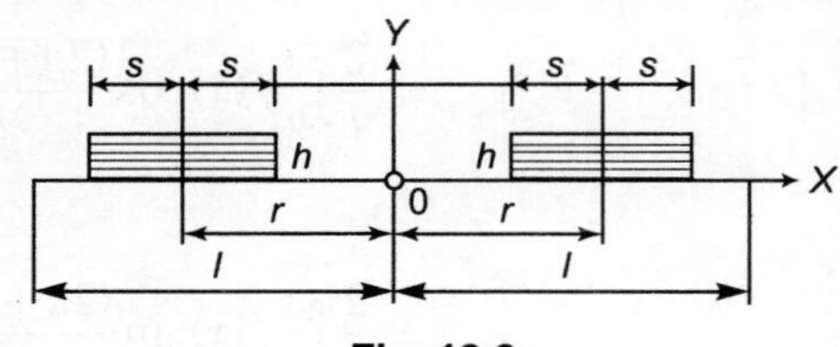

Fig. 13.8

$$\frac{\bar{a}_0}{2} = 0 \qquad \bar{a}_n = 0$$

$$\bar{b}_n = \frac{4hr}{l} \frac{\sin \bar{\alpha}_n \sin \bar{\beta}_n}{\bar{\beta}_n}$$

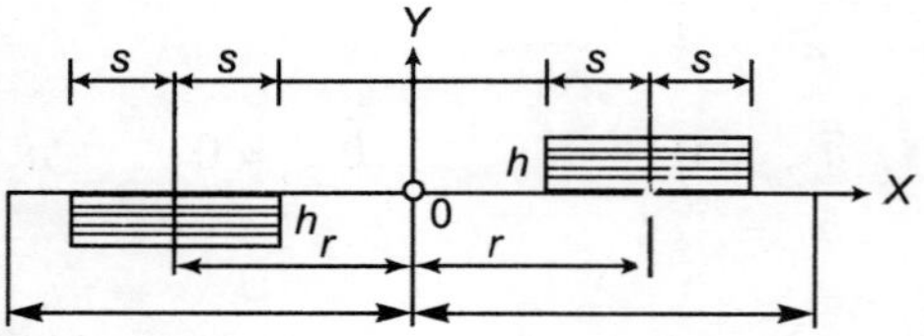

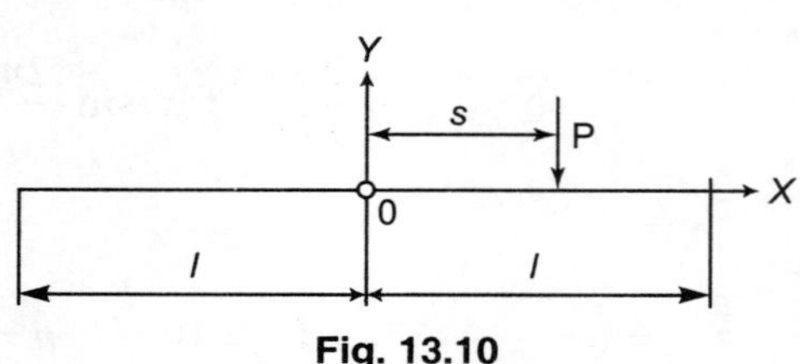

Fig. 13.9

$$\frac{\bar{a}_0}{2} = \frac{p}{2l} \qquad \bar{a}_n = \frac{p}{l} \cos \bar{\alpha}_n$$

$$\bar{b}_n = \frac{p}{l} \sin \bar{\alpha}_n$$

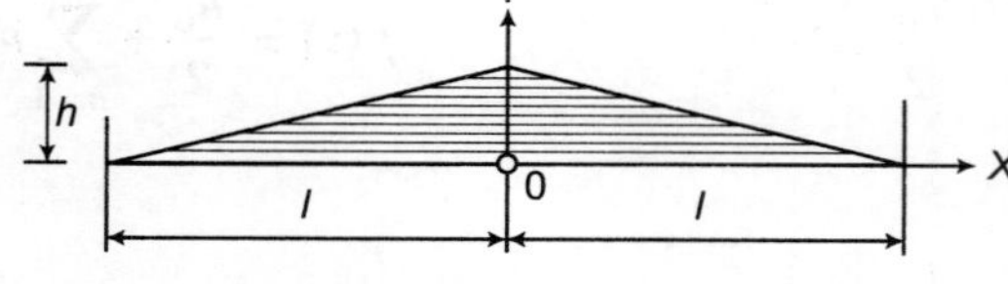

Fig. 13.10

13.7 FOURIER COEFFICIENTS FOR TRIANGULAR PERIODIC FUNCTIONS

$$f(x) = \frac{\bar{a}_0}{2} + \sum_{n=1}^{\infty} \bar{a}_n \cos \frac{n\pi x}{l} + \sum_{n=l}^{\infty} \bar{b}_n \sin \frac{n\pi x}{l} \qquad \bar{\alpha}_n = \frac{n\pi s}{l}$$

$$\frac{\bar{a}_0}{2} = \frac{h}{2} \qquad \bar{a}_n = \frac{4h}{\pi^2 n^2}$$

$$n = 1, 3, 5, \ldots \qquad \bar{b}_n = 0$$

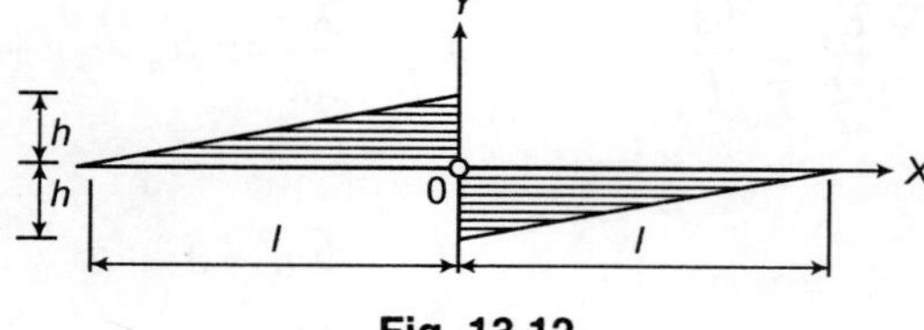

Fig. 13.11

$$\frac{\bar{a}_0}{2} = 0 \qquad \bar{a}_n = 0$$

$$\bar{b}_n = -\frac{2h}{\pi n}$$

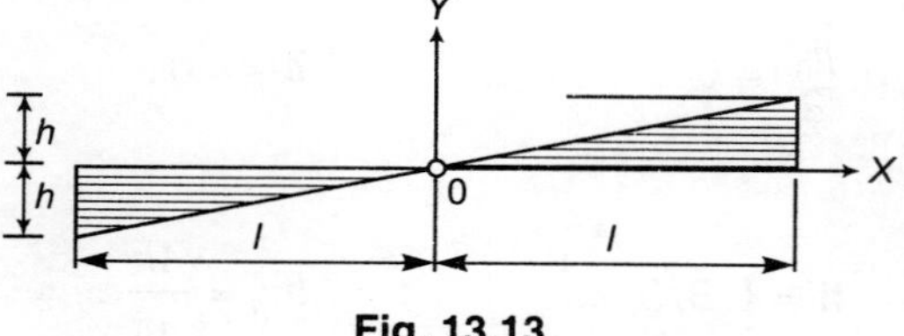

Fig. 13.12

$$\frac{\bar{a}_0}{2} = 0 \qquad \bar{a}_n = 0$$

$$\bar{b}_n = \frac{2h}{\pi n} (-1)^{n-1}$$

Fig. 13.13

$$\frac{\bar{a}_0}{2} = \frac{hs}{2l} \qquad \bar{a}_n = \frac{2h(1-\cos\bar{\alpha}_n)}{n^2\pi\bar{\alpha}_n}$$

$$\bar{b}_n = 0$$

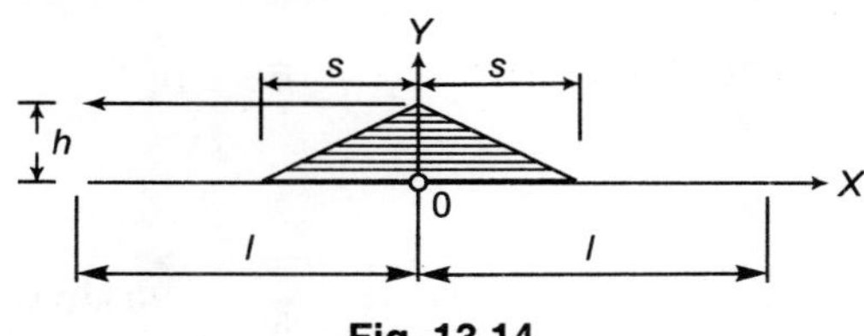

Fig. 13.14

$$\frac{\bar{a}_0}{2} = \frac{h}{4} \qquad \bar{a}_n = -\frac{2h}{\pi^2 n^2} \qquad n = 1, 3, 5, \ldots$$

$$\bar{b}_n = \frac{h}{\pi n} \qquad n = 1, 2, 3, \ldots$$

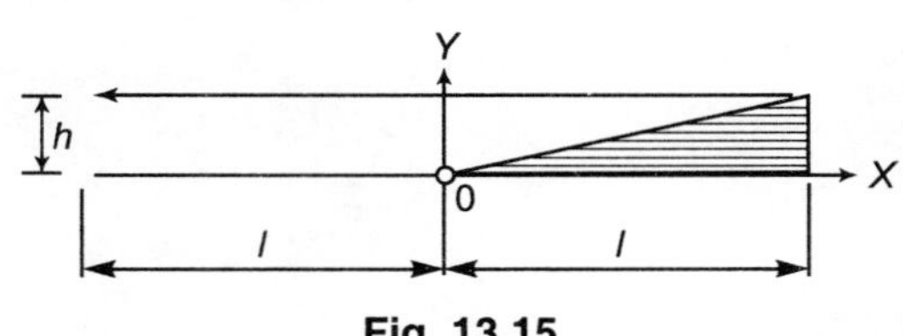

Fig. 13.15

13.8 FOURIER COEFFICIENTS FOR CURVILINEAR PERIODIC FUNCTIONS

$$f(x) = \frac{\bar{a}_0}{2} + \sum_{n=1}^{\infty} \bar{a}_n \cos\frac{n\pi x}{l} - \sum_{n=1}^{\infty} \bar{b}_n \sin\frac{n\pi x}{l} \qquad \bar{\alpha}_n = \frac{n\pi s}{l}$$

$$\frac{\bar{a}_0}{2} = \frac{2h}{\pi} \qquad \bar{a}_n = \frac{4h}{(n-1)(n+1)\pi}$$

$$n = 2, 4, \ldots \qquad \bar{b}_n = 0$$

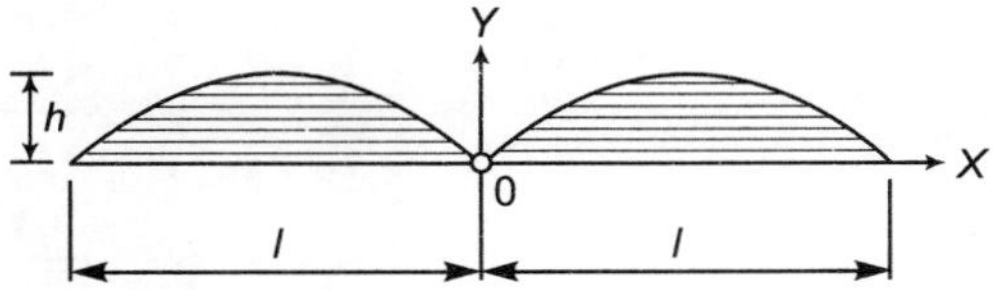

Fig. 13.16

$$\frac{\bar{a}_0}{2} = \frac{2h}{\pi} \qquad \bar{a}_n = \frac{4(-1)^{n/2+1}h}{(n-1)(n+1)\pi}$$

$$n = 2, 4, \ldots \qquad \bar{b}_n = 0$$

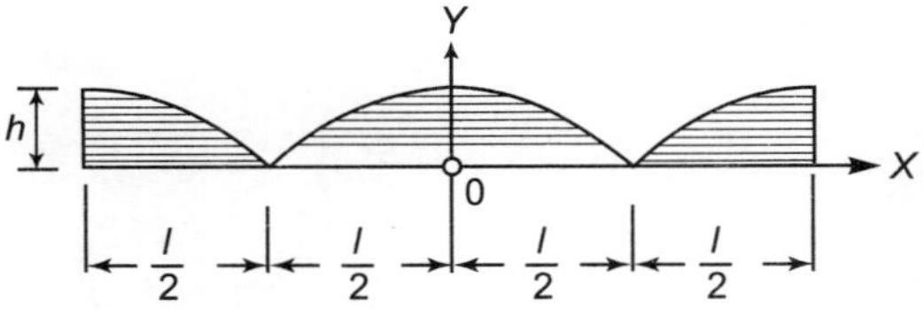

Fig. 13.17

$$\frac{\bar{a}_0}{2} = \frac{h}{\pi} + \frac{h}{2}\cos\frac{\pi x}{l} \qquad \bar{a}_n = \frac{2h(-1)^{n/2+1}}{(n-1)(n+1)\pi}$$

$$n = 2, 4, \ldots \qquad \bar{b}_n = 0$$

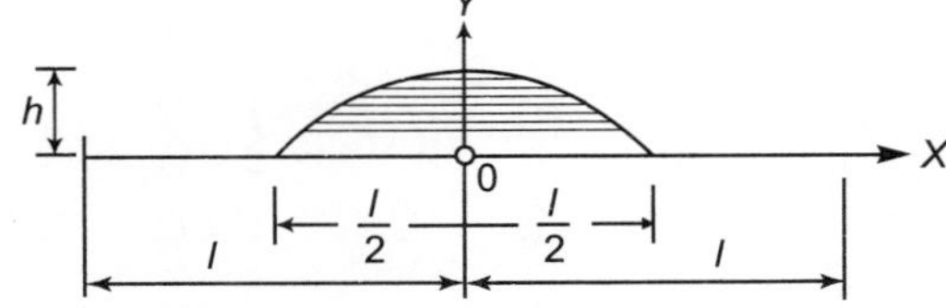

Fig. 13.18

$$\frac{\bar{a}_0}{2} = \frac{h}{\pi} + \frac{h}{2}\sin\frac{\pi x}{l} \qquad \bar{a}_n = \frac{-2h}{(n-1)(n+1)\pi}$$

$$n = 2, 4, \ldots \qquad \bar{b}_n = 0$$

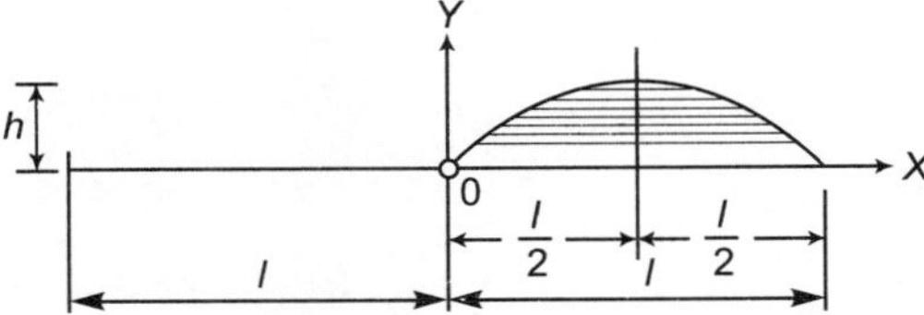

Fig. 13.19

$$\frac{\bar{a}_0}{2} = 0 \qquad \bar{a}_n = 0$$

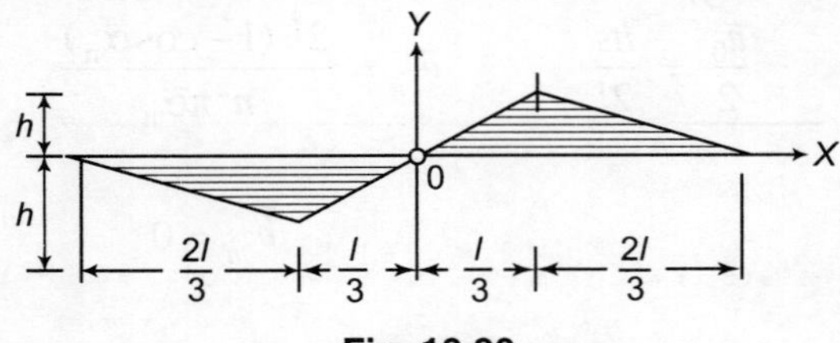

Fig. 13.20

$$\bar{b}_n = \frac{9h \sin(n\pi/3)}{n^2\pi^2}$$

CHAPTER

14

Higher Transcendent Functions

14.1 INTEGRAL FUNCTIONS—ANALYTICAL EXPRESSIONS

i. Definition

Integrals which cannot be evaluated as finite combinations of elementary functions are called *integral functions*. The most typical functions in this group evaluated by *series expansion* are given below:

Special Functions of Higher Order

Integration in the complex plane is important for two reasons.

1. In applications there occur real integrals that can be evaluated by complex integration, where as the usual methods of real integral calculus fail.
2. some basic properties of analytic functions can be established by complex integration, but would be difficult to prove by other techniques. The existence of higher derivatives of analytic function is a striking property of this type.

The special functions in this group evaluated by series expansion are given below:

a. Integral-sine, cosine and exponential function.
b. fresnel integrals.
c. error function.
d. gamma function Γ.
e. beta function β.
f. pie function.
g. ellpitic functions.

ii. Integrals—Sine, Cosine, and Exponential Functions

$$Si(x) = \int_0^x \frac{\sin x}{x}\,dx = x - \frac{1}{3}\frac{x^3}{3!} + \frac{1}{5}\frac{x^5}{5!} - \frac{1}{7}\frac{x^7}{7!} + \ldots \qquad Si(\infty) = \frac{\pi}{2}$$

$$Ci(x) = \int_{+\infty}^{x} \frac{\cos x}{x} dx = C + \text{In } x - \frac{1}{2}\frac{x^2}{2!} + \frac{1}{4}\frac{x^4}{4!} + \ldots \qquad Ci(\infty) = 0$$

$$Ei(x) = \int_{+\infty}^{x} \frac{e^{-x}}{x} dx = C + \text{In } x - x + \frac{1}{2}\frac{x^2}{2!} - \frac{1}{3}\frac{x^3}{3!} + \ldots \qquad Ei(\infty) = 0$$

$$C = \int_{+\infty}^{0} e^{-x} \text{ In } x dx = 0.57721 = \text{Euler's constant}$$

iii. Fresnel Integrals

$$\int_0^x \frac{\sin x}{\sqrt{x}} dx = 2\sqrt{x}\left(\frac{1}{3}\frac{x}{1!} - \frac{1}{7}\frac{x^3}{3!} + \frac{1}{11}\frac{x^5}{5!} - \ldots\right) \qquad \int_0^{+\infty} \frac{\sin x}{\sqrt{x}} dx = \sqrt{\frac{\pi}{2}}$$

$$\int_0^x \frac{\cos x}{\sqrt{x}} dx = 2\sqrt{x}\left(1 - \frac{1}{5}\frac{x^2}{2!} + \frac{1}{9}\frac{x^4}{4!} - \frac{1}{13}\frac{x^6}{6!} + \ldots\right) \qquad \int_0^{+\infty} \frac{\cos x}{\sqrt{x}} dx = \sqrt{\frac{\pi}{2}}$$

$$S(x) = \sqrt{\frac{2}{\pi}} \int_0^x \sin x^2 \, dx \qquad S(-x) = -S(x)$$

$$S(0) = 0$$

$$= \sqrt{\frac{2}{\pi}}\left(\frac{1}{1!}\frac{x^3}{3} - \frac{1}{3!}\frac{x^7}{7} + \frac{1}{5!}\frac{x^{11}}{11} - \ldots\right) \qquad S(\infty) = \frac{1}{2}$$

$$C(x) = \sqrt{\frac{2}{\pi}} \int_0^x \cos x^2 dx \qquad C(-x) = -C(x)$$

$$C(0) = 0$$

$$= \sqrt{\frac{2}{\pi}}\left(\frac{1}{0!}\frac{x}{1} - \frac{1}{2!}\frac{x^3}{5} + \frac{1}{4!}\frac{x^9}{9} - \ldots.\right) \qquad C(\infty) = \frac{1}{2}$$

iv. Error Function

$$\text{erf}(x) = \frac{2}{\sqrt{\pi}} \int_0^x e^{-x^2} dx \qquad \text{erf}(-x) = -\text{erf}(x)$$

$$\text{erf}(0) = 0$$

$$= \frac{2}{\sqrt{\pi}}\left(\frac{1}{0!}\frac{x}{1} - \frac{1}{1!}\frac{x^3}{3} + \frac{1}{2!}\frac{x^5}{5} - \frac{1}{3!}\frac{x^7}{7} + \ldots\right) \qquad \text{erf}(\infty) = 1$$

14.2 INTEGRAL FUNCTIONS—TABLES

a. *Sine integral*

$$Si(x) = \int_0^x \frac{\sin x}{x} dx$$

x	0	1	2	3	4	5	6	7	8	9	x
0.	0.0000	0.0999	0.1996	0.2985	0.3965	0.4931	0.5881	0.6812	0.7721	0.8605	0.
1.	0.9461	1.0287	0.1080	1.1840	1.2562	1.3247	1.3892	1.4496	1.5058	1.5578	1.

x	0	1	2	3	4	5	6	7	8	9	x
2.	1.6054	1.6487	1.6876	1.7222	1.7525	1.7785	1.8004	1.8182	1.8321	1.8422	2.
3.	1.8487	1.8517	1.8514	1.8481	1.8419	1.8331	1.8219	1.8086	1.7934	1.7765	3.
4.	1.7582	1.7387	1.7184	1.6973	1.6758	1.6541	1.6325	1.6110	1.5900	1.5696	4.

b. *Cosine integral*

$$Ci(x) = \int_{+\infty}^{x} \frac{\cos x}{x} dx$$

x	0	1	2	3	4	5	6	7	8	9	x
0.	$-\infty$	–1.7279	–1.0422	–0.6492	–0.3788	–0.1778	–0.0223	+0.1051	+0.1983	+0.2761	0.
1.	+0.3374	+0.3849	+0.4205	+0.4457	+0.4620	–0.4704	+0.4717	+0.4670	+0.4568	+0.4419	1.
2.	+0.4230	+0.4005	+0.3751	+3472	+3173	+0.2859	+0.2533	+0.2201	+0.1865	+0.1529	2.
3.	+0.1196	+0.0870	+0.553	+0.0247	–0.0045	–0.0321	–0.0580	–0.0819	–0.1038	–0.1235	3.
4.	–0.1410	–0.1562	–0.1690	–0.1795	–0.1877	–0.1935	–0.1970	–0.1984	–0.1976	–0.1948	4.

c. *Exponential integral*

$$Ei(x) = \int_{+\infty}^{x} \frac{e^{-x}}{x} dx$$

x	0	1	2	3	4	5	6	7	8	9	x
0.	$-\infty$	–1.8229	–12227	–0.9057	–0.7024	–0.5598	–0.4544	–0.3738	–0.3106	–0.2602	0.
1.	–0.2194	–0.1860	–0.1584	–0.1355	–0.1162	–0.1000	–0.0863	–0.0747	–0.0647	–0.0562	1.
2.	–0.0489	–0.0426	–0.0372	–0.0325	0.0284	–0.0249	–0.0219	–0.0192	–0.0169	–0.0148	2.

d. *Error integral*

$$\text{erf}\ (x) = \frac{2}{\sqrt{\pi}} \int_{v}^{x} e^{-x^2} dx$$

x	0	1	2	3	4	5	6	7	8	9	x
0.	0.0000	0.1125	0.2227	0.3286	0.4284	0.5205	0.6039	0.6778	0.7421	0.7969	0.
1.	0.8427	0.8802	0.9103	0.9340	0.9523	0.9661	0.9764	0.9838	0.9891	0.9928	1.
2.	0.9953	0.9970	0.9981	0.9989	0.9994	0.9996	0.9998	0.9999	0.9999	1.0000	2.

14.3 GAMMA, PI, AND BETA FUNCTIONS

i. Gamma Function (Γ)

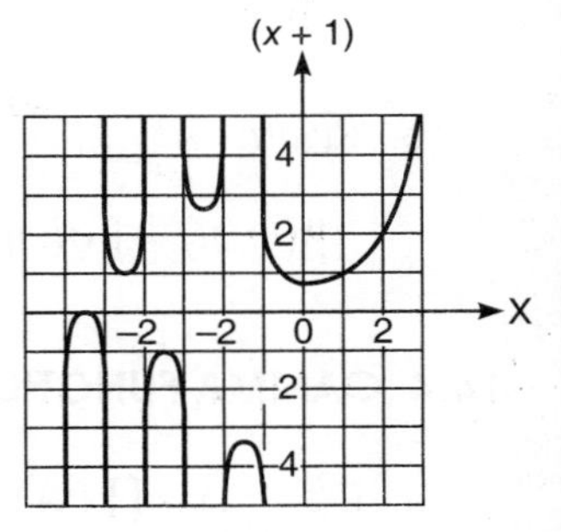

Fig. 14.1

a. *Definition*

$$\Gamma(x) = \int_0^{\infty} t^{x-1} e^{-t}\, dt \qquad x > 0$$

$$\Gamma(x) = \lim_{x \to \infty} \frac{n^x n!}{x(x+1)(x+2)\dots(x+n)}$$

b. *Recursion formulas* $(n = 0, 1, 2 \dots)$

$$\Gamma(n + 1) = n(n - 1)\ (n - 2) \dots 1 = n!$$

$$\Gamma(n) = \frac{1}{n}\,\Gamma(n+1) \qquad\qquad \Gamma(n) = (n-1)\,\Gamma(n-1)$$

c. *Special values*

$$\Gamma\left(\frac{1}{2}+p\right)\Gamma\left(\frac{1}{2}-p\right)=\frac{\pi}{\cos\pi p} \qquad\qquad \Gamma(p)\,\Gamma(1-p) = \frac{\pi}{\sin\pi p}$$

$$\Gamma\left(n+\frac{1}{2}\right)=\frac{(2n)!\sqrt{\pi}}{n!2^{2n}} \qquad\qquad \Gamma\left(-n+\frac{1}{2}\right)=\frac{(-1)^n n!2^{2n}\sqrt{\pi}}{(2n)!}$$

$$n = 0, 1, 2\ldots \qquad\qquad n = 0, 1, 2\ldots$$

d. *Derivatives*

$$\frac{d\Gamma(x)}{dx}=\left[-C+\left(\frac{1}{1}-\frac{1}{x}\right)+\left(\frac{1}{2}-\frac{1}{x+1}\right)+\ldots+\left(\frac{1}{n}-\frac{1}{x+n-1}\right)+\ldots\right]\Gamma(x)$$

$$-C=\int_0^\infty e^{-x}\ln x\,dx=\frac{d\Gamma(1)}{dx}=-0.57721\ldots=\text{Euler's constant}$$

ii. PI Function ($\prod$)

a. *Definition*

$$\prod(x) = \Gamma(x+1) = \int_0^\infty t^x e^{-t}\,dt \qquad\qquad x > 0$$

b. *Recursion formulas*

$$\prod(n) = n\,(n-1)\,(n-2)\ldots 1 = n!$$

$$\prod(n) = \frac{1}{n}\prod(n+1) \qquad \prod(n) = n\prod(n-1) \qquad n = 0, 1, 2\ldots$$

iii. Beta Function (β)

a. *Definition*

$$\beta(x, y) = \int_0^1 t^{x-1}(1-t)^{y-1}\,dt = \frac{\Gamma(x)\Gamma(y)}{\Gamma(x+y)} \qquad\qquad x > 0, y > 0$$

b. *Relations*

$$\beta(m, n) = \beta(n, m) \qquad\qquad \beta(m,n)=\frac{(m-1)!(n-1)!}{(m+n-1)!} \quad m = 0, 1, 2,\ldots;\ n = 0, 1, 2,\ldots,$$

14.4 GAMMA FUNCTION—TABLES

a. $\Gamma(x+1) = \prod(x) = x!$ $\qquad\qquad \Gamma(x+1) = \int_0^\infty t^x e^{-t}\,dt$

x	0	1	2	3	4	5	6	7	8	9	x
0.0	1.0000	0.9943	0.9888	0.9836	0.9784	0.9735	0.9687	0.9642	0.9597	0.9555	0.0

x	0	1	2	3	4	5	6	7	8	9	x
0.1	0.9514	0.9474	0.9436	0.9399	0.9364	0.9330	0.9298	0.9267	0.9237	0.9209	0.1
0.2	0.9182	0.9156	0.9131	0.9108	0.9085	0.9064	0.9044	0.9025	0.9007	0.8990	0.2
0.3	0.8975	0.8960	0.8946	0.8934	0.8922	0.8912	0.8902	0.8893	0.8885	0.8879	0.3
0.4	0.8873	0.8868	0.8864	0.8860	0.8858	0.8857	0.8856	0.8856	0.8857	0.8859	0.4
0.5	0.8862	0.8866	0.8870	0.8876	0.8882	0.8889	0.0096	0.8905	0.8914	0.8924	0.5
0.6	0.8835	0.8947	0.8959	0.8972	0.8986	0.9001	0.9017	0.9033	0.9050	0.9068	0.6
0.7	0.9086	0.9106	0.9126	0.9147	0.9168	0.9191	0.9214	0.9238	0.9262	0.9288	0.7
0.8	0.9314	0.9341	0.9368	0.9397	0.9426	0.9456	0.9487	0.9518	0.9551	0.9584	0.8
0.9	0.9618	0.9652	0.9688	0.9724	0.9761	0.9799	0.9837	0.9877	0.9917	0.9958	0.9
1.0	1.0000	1.0043	1.0086	1.0131	1.0176	1.0222	1.0269	1.0316	1.0365	1.0415	1.0
1.1	1.0465	1.0516	1.0568	1.0621	1.0675	1.0730	1.0786	1.0842	1.0900	1.0959	1.1
1.2	1.1018	1.1078	1.1140	1.1202	1.1266	1.1330	1.1395	1.1462	1.1529	1.1598	1.2
1.3	1.1667	1.1738	1.1809	1.1882	1.1956	1.2031	1.2107	1.2184	1.2262	1.2341	1.3
1.4	1.2422	1.2503	1.2586	1.2670	1.2756	1.2842	1.2930	1.3019	1.3109	1.3201	1.4
1.5	1.3293	1.3388	1.3483	1.3580	1.3678	1.3777	1.3878	1.3981	1.4084	1.4190	1.5
1.6	1.4296	1.4404	1.4514	1.4625	1.4738	1.4852	1.4968	1.5085	1.5204	1.5325	1.6
1.7	1.5447	1.5571	1.5696	1.5824	1.5953	1.6084	1.6216	1.6351	1.6487	1.6625	1.7
1.8	1.6765	1.6907	1.7051	1.7196	1.7344	1.7494	1.7646	1.7799	1.7955	1.8113	1.8
1.9	1.8274	1.8436	1.8600	1.8767	1.8936	1.9108	1.9281	1.9457	1.9636	1.9816	1.9

b. $1/\Gamma(x+1) = 1/\prod(x) = 1/x!$

$$\frac{1}{\Gamma(x+1)} = \frac{1}{\int_0^\infty t^x e^{-t}\, dt^x e^{-t}\, dt}$$

x	0	1	2	3	4	5	6	7	8	9	x
0.0	1.0000	1.0057	1.0113	1.0167	1.0220	1.0272	1.0323	1.0372	1.0420	1.0466	0.0
0.1	1.0511	1.0555	1.0598	1.0639	1.0679	1.0718	1.0755	1.0791	1.0826	1.0859	0.1
0.2	1.0891	1.0922	1.0952	1.0980	1.1007	1.1032	1.1057	1.1080	1.1102	1.1123	0.2
0.3	1.1142	1.1161	1.1178	1.1194	1.1208	1.1222	1.1234	1.1244	1.1254	1.1263	0.3
0.4	1.1270	1.1277	1.1282	1.1286	1.1289	1.1291	1.1292	1.1291	1.1290	1.1287	0.4
0.5	1.1284	1.1279	1.1273	1.1267	1.1259	1.1250	1.1240	1.1230	1.1218	1.1205	0.5
0.6	1.1191	1.1177	1.1161	1.1145	1.1128	1.1109	1.1091	1.1071	1.1049	1.1028	0.6
0.7	1.1005	1.0982	1.0958	1.0933	1.0907	1.0881	1.0854	1.0825	1.0796	1.0767	0.7
0.8	1.0737	1.0706	1.0674	1.0642	1.0609	1.0575	1.0541	1.0506	1.0471	1.0434	0.8
0.9	1.0398	1.0360	1.0322	1.0284	1.0245	1.0206	1.0165	1.0125	1.0083	1.0042	0.9
1.0	1.00000	0.99575	0.99145	0.98711	0.98273	0.97830	0.97383	0.96933	0.96478	0.96020	1.0
1.1	0.95558	0.95092	0.94623	0.94151	0.93676	0.93197	0.92715	0.92231	0.91743	0.91253	1.1
1.2	0.90760	0.90265	0.89767	0.89268	0.88765	0.88261	0.87755	0.87247	0.86737	0.86225	1.2
1.3	0.85711	0.85196	0.84679	0.84161	0.83642	0.83122	0.82600	0.82078	0.81554	0.81030	1.3

x	0	1	2	3	4	5	6	7	8	9	x
0.0	1.0000	1.0057	1.0113	1.0167	1.0220	1.0272	1.0323	1.0372	1.0420	1.0466	0.0
0.1	1.0511	1.0555	1.0598	1.0639	1.0679	1.0718	1.0755	1.0791	1.0826	1.0859	0.1
0.2	1.0891	1.0922	1.0952	1.0980	1.1007	1.1032	1.1057	1.1080	1.1102	1.1123	0.2
0.3	1.1142	1.1161	1.1178	1.1194	1.1208	1.1222	1.1234	1.1244	1.1254	1.1263	0.3
0.4	1.1270	1.1277	1.1282	1.1286	1.1289	1.1291	1.1292	1.1291	1.1290	1.1287	0.4
0.5	1.1284	1.1279	1.1273	1.1267	1.1259	1.1250	1.1240	1.1230	1.1218	1.1205	0.5
0.6	1.1191	1.1177	1.1161	1.1145	1.1128	1.1109	1.1091	1.1071	1.1049	1.1028	0.6
0.7	1.1005	1.0982	1.0958	1.0933	1.0907	1.0881	1.0854	1.0825	1.0796	1.0767	0.7
0.8	1.0737	1.0706	1.0674	1.0642	1.0609	1.0575	1.0541	1.0506	1.0471	1.0434	0.8
0.9	1.0398	1.0360	1.0322	1.0284	1.0245	1.0206	1.0165	1.0125	1.0083	1.0042	0.9
1.0	1.00000	0.99575	0.99145	0.98711	0.98273	0.97830	0.97383	0.96933	0.96478	0.96020	1.0
1.1	0.95558	0.95092	0.94623	0.94151	0.93676	0.93197	0.92715	0.92231	0.91743	0.91253	1.1
1.2	0.90760	0.90265	0.89767	0.89268	0.88765	0.88261	0.87755	0.87247	0.86737	0.86225	1.2
1.4	0.80504	0.79978	0.79452	0.78925	0.78397	0.77868	0.77240	0.76812	0.76283	0.75754	1.4
1.5	0.75225	0.74696	0.74167	0.74167	0.73111	0.72583	0.72055	0.71527	0.71000	0.70474	1.5
1.6	0.69948	0.69423	0.68899	0.68376	0.67853	0.67331	0.66810	0.66291	0.65772	0.65253	1.6
1.7	0.64737	0.64222	0.63708	0.63196	0.62685	0.62175	0.61667	0.61160	0.60654	0.60150	1.7
1.8	0.59648	0.59147	0.58649	0.58151	0.57656	0.57163	0.56671	0.56182	0.55694	0.55208	1.8
1.9	0.54724	0.54242	0.53762	0.53284	0.52808	0.52335	0.51864	0.51394	0.50927	0.50462	1.9

14.5 ELLIPTIC INTEGRALS

i. Elliptic Integrals, Normal Form, Formulas

$$F(k, x) = \int_0^x \frac{dx}{\sqrt{(1-x^2)(1-k^2x^2)}} = F(k, \phi) = \int_0^\phi \frac{d\phi}{\sqrt{1-k^2\sin^2\phi}}$$

$$E(k, x) = \int_0^x \sqrt{\frac{1-k^2x^2}{1-x^2}}\, dx = E(k, \phi) = \int_0^\phi \sqrt{1-k^2\sin^2\phi}\, d\phi$$

where $k = \sin\omega$ = modulus (given constant) in the interval $0 \le k \le +1$, x = independent variable in interval $-1 \le x \le +1$.

ii. Elliptic Integrals, Complete Form, Formulas

$$F\left(k, \frac{\pi}{2}\right) = K = \frac{\pi}{2}\left\{1 + \left(\frac{1}{2}\right)^2 k^2 + \left[\frac{(1)(3)}{(2)(4)}\right]^2 k^4 + \left[\frac{(1)(3)(5)}{(2)(4)(6)}\right]^2 k^6 + \ldots\right\}$$

$$E\left(k, \frac{\pi}{2}\right) = E = \frac{\pi}{2}\left\{1 - \left(\frac{1}{2}\right)^2 \frac{k^2}{1} - \left[\frac{(1)(3)}{(2)(4)}\right]^2 \frac{k^4}{3} - \left[\frac{(1)(3)(5)}{(2)(4)(6)}\right]^2 \frac{k^6}{5} - \ldots\right\}$$

iii. Elliptic Integrals, Degenerated Form

$F(0, x) = \sin^{-1} x$ $\quad K = \frac{\pi}{2}$ $\quad F(1, x) = \tanh^{-1}$ $\quad K = \infty$

$E(0, x) = \sin^{-1} x$ $\quad E = \frac{\pi}{2}$ $\quad E(1, x) = x$ $\quad E = 1$

14.6 ELLIPTIC FUNCTIONS

i. Definition

If $u = F(k, \phi)$, the inverse function is designated as ϕ = am u and is called the *elliptic function of Jacobi.*

$$x = sn\ u = \sin\ (am\ u) \qquad sn^2 + cn^2\ u = 1$$

$$\sqrt{1-x^2} = cn\ u = \cos\ (am\ u) \qquad dn^2\ u + k^2 \sin^2 u = 1$$

$$\sqrt{1-k^2x^2} = dn\ u = \sqrt{1-k^2 sn^2 u} \qquad dn^2\ u - k^2\ cn^2\ u = 1 - k^2$$

$$sn\ u = u - \frac{(1+k^2)u^3}{3!} + \frac{(1+14k^2+k^4)u^5}{5!} - \dots$$

$$cn\ u = u - \frac{u^2}{2!} + \frac{(1+4k^2)u^4}{4!} - \frac{(1+44k^2+16k^4)u^6}{6!} + \dots$$

$$dn\ u = 1 - \frac{k^2u^2}{2!} + \frac{k^2(4+k^2)u^4}{4!} - \frac{k^2(16+44k^2+k^4)u^6}{6!} + \dots$$

$sn\ (k = 0) = \sin u$ $\quad cn\ (k = 0) = \cos u$ $\quad dn\ (k = 0) = 1$

$sn\ (k = 1) = \tanh u$ $\quad cn\ (k = 1) = \frac{1}{\cosh u}$ $\quad dn\ (k = 1) = \frac{1}{\cosh u}$

ii. Derivatives

$$\frac{d}{dx}\ (sn\ u) = cn\ u\ dn\ u$$

$$\frac{d}{dx}\ (cn\ u) = -\ sn\ u\ dn\ u$$

$$\frac{d}{dx}\ (dn\ u) = -\ k^2\ sn\ u\ cn\ u$$

iii. Integrals

$$\int su\ u\ du = \frac{1}{k}\ \ln\ (dn\ u - k\ cn\ u)$$

$$\int cn\ u\ du = \frac{1}{k}\ \cos^{-1}\ (dn\ u)$$

$$\int dn\ u\ du = \sin^{-1}\ (sn\ u)$$

14.7 ELLIPTIC INTEGRALS, NORMAL FORM, TABLES

a. *First kind* $F(k, \phi) = \int_0^{\phi} \frac{d\phi}{\sqrt{1 - k^2 \sin^2 \phi}}$ $k = \sin \omega$

ω	0°	10°	20°	30°	40°	50°	60°	70°	80°	90°	ω
k / φ	0	0.1737	0.3420	0.5000	0.6428	0.7660	0.8660	0.9397	0.9848	1.0000	*k* / φ
0°	0.0000	0.0000	0.0000	0.0000	0.0000	0.0000	0.0000	0.0000	0.0000	0.0000	0°
10°	0.1745	0.1746	0.1746	0.1748	0.1749	0.1751	0.1752	0.1753	0.1754	0.1754	10°
20°	0.3491	0.3493	0.3499	0.3508	0.3520	0.3533	0.3545	0.3555	0.3561	0.3564	20°
30°	0.5236	0.5243	0.5263	0.5294	0.5334	0.5379	0.5422	0.5459	0.5484	0.5493	30°
40°	0.6981	0.6997	0.7043	0.7117	0.7213	0.7323	0.7436	0.7535	0.7604	0.7629	40°
50°	0.8727	0.8756	0.8842	0.8982	0.9173	0.9401	0.9647	0.9876	1.0044	1.0107	50°
60°	1.0472	1.0519	1.0660	1.0896	1.1226	1.1643	1.2126	1.2619	1.3014	1.3170	60°
70°	1.2217	1.2286	1.2495	1.2853	1.3372	1.4068	1.4944	1.5959	1.6918	1.7354	70°
80°	1.3963	1.4057	1.4344	1.4846	1.5597	1.6660	1.8125	2.0119	2.2653	2.4363	80°
90°	1.5708	1.5828	1.6200	1.6858	1.7868	1.9356	2.1565	2.5046	3.1534	∞	90°

b. *Second kind* $E(k, \phi) = \int_0^{\phi} \sqrt{1 - k^2 \sin^2 \phi}\, d\phi$ $k = \sin \omega$

ω	0°	10°	20°	30°	40°	50°	60°	70°	80°	90°	ω
k / φ	0	0.1737	0.3420	0.5000	0.6428	0.7660	0.8660	0.9397	0.9848	1.0000	*x* / φ
0°	0.0000	0.0000	0.0000	0.0000	0.0000	0.0000	0.0000	0.0000	0.0000	0.0000	0°
10°	0.1745	0.1745	0.1744	0.1743	0.1742	0.1740	0.1739	0.1738	0.1737	0.1736	10°
20°	0.3491	0.3489	0.3483	0.3473	0.3462	0.3450	0.3438	0.3429	0.3422	0.3420	20°
30°	0.5236	0.5229	0.5209	0.5179	0.5141	0.5100	0.5061	0.5029	0.5007	0.5000	30°
40°	0.6981	0.6966	0.6921	0.6851	0.6763	0.6667	0.6575	0.6497	0.6446	0.6428	40°
50°	0.8727	0.8698	0.8614	0.8483	0.8317	0.8134	0.7954	0.7801	0.7697	0.7660	50°
60°	1.0472	1.0426	1.0290	1.0076	0.9801	0.9493	0.9184	0.8914	0.8728	0.8660	60°
70°	1.2217	1.2149	1.1949	1.1632	1.1221	1.0750	1.0266	0.9830	0.9514	0.9397	70°
80°	1.3963	1.3870	1.3597	1.3161	1.2590	1.1926	1.1225	1.0565	1.0054	0.9848	80°
90°	1.5708	1.5289	1.5238	1.4675	1.3931	1.3055	1.2111	1.1184	1.0401	1.0000	90°

14.8 OTHER ELLIPTIC INTEGRALS, NORMAL FORM

$$\lambda = \sqrt{1-k^2\sin^2\phi} \qquad k = \sin\omega$$

$$k = \frac{F-E}{k^2} \qquad l = \cos\omega$$

$$\int_0^{\phi}\frac{\sin^2\phi\,d\phi}{\lambda} = k \qquad \int_0^{\phi}\frac{\cos^2\phi\,d\phi}{\lambda} = F-K$$

$$\int_0^{\phi}\frac{\sin^2\phi\,d\phi}{\lambda^2} = \frac{(F-K)\lambda-\sin\phi\cos\phi}{l^2\lambda} \qquad \int_0^{\phi}\frac{\cos^2\phi\,d\phi}{\lambda^2} = \frac{k\lambda+\sin\phi\cos\phi}{\lambda}$$

14.9 ELLIPTIC INTEGRALS, COMPLETE FORM, TABLES

ω	$F\left(k,\frac{\pi}{2}\right)$	$E\left(k,\frac{\pi}{2}\right)$	ω	$F\left(k,\frac{\pi}{2}\right)$	$E\left(k,\frac{\pi}{2}\right)$	ω	$F\left(k,\frac{\pi}{2}\right)$	$E\left(k,\frac{\pi}{2}\right)$
0°	1.5708	1.5708	50°	1.9356	1.3055	82°0′	3.3699	1.0278
1°	1.5709	1.5707	51°	1.9539	1.2963	82°12′	3.3946	1.0267
2°	1.5713	1.5703	52°	1.9729	1.2870	82°24′	3.4199	1.0256
3°	1.5719	1.5697	53°	1.9927	1.2776	82°36′	3.4460	1.0245
4°	1.5727	1.5689	54°	2.0133	1.2682	82°45′	3.4728	1.0234
5°	1.5738	1.5678	55°	2.0347	1.2587	83°0′	3.5004	1.0223
6°	1.5751	1.5665	56°	2.0571	1.2492	83°12′	3.5288	1.0213
7°	1.5767	1.5650	57°	2.0804	1.2397	83°24′	3.5581	1.0202
8°	1.5785	1.5632	58°	2.1047	1.2301	83°36′	3.5884	1.0192
9°	1.5805	1.5611	59°	2.1300	1.2206	83°48′	3.6196	1.0182
10°	1.5828	1.5589	60°	2.1565	1.2111	84°0′	3.6519	1.0172
11°	1.5854	1.5564	61°	2.1842	1.2015	84°12′	3.6853	1.0163
12°	1.5882	1.5537	62°	2.2132	1.1921	84°24′	3.7198	1.0153
13°	1.5913	1.5507	63°	2.2435	1.1826	84°36′	3.7557	1.0144
14°	1.5946	1.5476	64°	2.2754	1.1732	84°48′	3.7930	1.0135
15°	1.5981	1.5442	65°	2.3088	1.1638	85°0′	3.8317	1.0127
16°	1.6020	1.5405	66°	2.3439	1.1546	85°12′	3.8721	1.0118
17°	1.6061	1.5367	67°	2.3809	1.1454	85°24′	3.9142	1.0100
18°	1.6105	1.5326	68°	2.4198	1.1362	85°36′	3.9583	1.0102
19°	1.8151	1.5283	69°	2.4610	1.1273	85°48′	4.0044	1.0094
20°	1.6200	1.5238	70°0′	2.5046	1.1184	86°0′	4.0528	1.0087
21°	1.6252	1.5191	70°30′	2.5273	1.1140	86°12′	4.1037	1.0079

ω	$F\left(k,\frac{\pi}{2}\right)$	$E\left(k,\frac{\pi}{2}\right)$	ω	$F\left(k,\frac{\pi}{2}\right)$	$E\left(k,\frac{\pi}{2}\right)$	ω	$F\left(k,\frac{\pi}{2}\right)$	$E\left(k,\frac{\pi}{2}\right)$
22°	1.6307	1.5142	71°0′	2.5507	1.1096	86°24′	4.1574	1.0072
23°	1.6305	1.5090	71°30′	2.5749	1.1053	86°36′	4.2142	1.0065
24°	1.6426	1.5037	72°0′	2.5998	1.1011	86°48′	4.2746	1.0059
25°	1.6490	1.4981	72°30′	2.6256	1.0968	87°0′	4.3387	1.0053
26°	1.6557	1.4924	73°0′	2.6521	1.0927	87°12′	4.4073	1.0047
27°	1.6627	1.4864	73°30′	2.6796	1.0885	87°24′	4.4812	1.0041
28°	1.6701	1.4803	74°0′	2.7081	1.0844	87°36′	4.5609	1.0036
29°	1.6777	1.4740	74°30′	2.7375	1.0804	87°48′	4.6477	1.0031
30°	1.6858	1.4675	75°0′	2.7681	1.0764	88°0′	4.7427	1.0026
31°	1.6941	1.4608	75°30′	2.7998	1.0725	88°12′	4.8479	1.0022
32°	1.7028	1.4539	76°0′	2.8327	1.0686	88°24′	4.9654	1.0017
33°	1.7119	1.4469	76°30′	2.8669	1.0648	88°36′	5.0988	1.0014
34°	1.7214	1.4397	77°0′	2.9026	1.0611	88°48′	5.2527	1.0010
35°	1.7313	1.4323	77°30′	2.9397	1.0574	89°0′	5.4349	1.0008
36°	1.7415	1.4248	78°0′	2.9786	1.0538	89°6′	5.5402	1.0006
37°	1.7522	1.4171	78°30′	3.0192	1.0502	89°12′	5.6579	1.0005
38°	1.7633	1.4092	79°0′	3.0617	1.0468	89°18′	5.7914	1.0005
39°	1.7748	1.4013	79°30′	3.1064	1.0434	89°24′	5.9455	1.0003
40°	1.7868	1.3931	80°0′	3.1534	1.0401	89°30′	6.1278	1.0002
41°	1.7992	1.3849	80°12′	3.1729	1.0388	89°36′	6.3509	1.0001
42°	1.8122	1.3765	80°24′	3.1928	1.0375	89°42′	6.6385	1.0001
43°	1.8256	1.3680	80°36′	3.2132	1.0363	89°48′	7.0440	1.0000
44°	1.8396	1.3594	80°48′	3.2340	1.0350	89°54′	7.7371	1.0000
45°	1.8541	1.3506	81°0′	3.2553	1.0338	90°	∞	1.0000
46°	1.8692	1.3418	81°12′	3.2771	1.0326			
47°	1.8848	1.3329	81°24′	3.2995	1.0313			
48°	1.9011	1.3238	81°36′	3.3223	1.0302			
49°	1.9180	1.3147	81°48′	3.3458	1.0290			

CHAPTER

15

Ordinary Differential Equations

15.1 GENERAL CONCEPT

i. Definitions

A differential equation is an *algebraic* or *transcendent equality* involving differentials or derivatives. The *order* of differential equation is the order of the highest derivative. The *degree* of a differential equation is the algebraic degree of the highest derivative.

The *number of independent variables* defines the differential equation as an *ordinary differential equation* (single independent variable) or a *partial differential equation* (two or more independent variables). A *homogeneous* differential equation has all terms of the same degree in the independent variable and its derivatives. If one or more terms do not involve the independent variable,the differential equation is *nonhomogeneous*. A homogeneous equation obtained from a non-homogeneous equation by setting the non-homogeneous terms equal to zero is the *reduced equation.* A *linear differential equation* consists of linear terms. A linear term is one which is of the first degree in the independent variables and their derivatives.

Ordinary Differential Equation

If a homogeneous linear differential equation has constant coefficients, it can be solved by algebraic methods. If such equation has variable coefficients, it must be solved by other methods. Legendre's equation, hypergeometric equation and Bessel's equation are very important equations of this type. The power series method which yields solutions in the form of power series, and an extension of it, called Frobeniusm method.

ii. Solution

An ordinary differential equation of order n given in general form as

$F\,[(x, y\,(x), y'\,(x)\,,..., y^n\,(x)] = 0$

has a *general solution* (general integral)

$y = y\,(x, C_1, C_2,..., C_n)$

where $C_1, C_2,..., C_n$ are *arbitrary* and *independent constants*. A general solution of a reduced differential equation is the *complementary solution* (complementary function). A *particular solution* is a special case of the general solution with definite values assigned to the constants. A *singular solution* is a solution which cannot be obtained from the general solution by specifying the values of the arbitrary constants. It is the surface that envelopes all the solutions of the differential equation.

iii. Order and Degree

The order of the highest derivative appearing in a differential equation is called the order of the differential equation. The power of the highest order derivative appearing in the differential equation is called the degree of the differential equation.

15.2 TYPES OF DIFFERENTIAL EQUATIONS

i. A differential equation of the form $F\left(x,y,\frac{dy}{dx}\right)=0$ is of first order and first degree.

ii. A differential equation of the form

$$F\left(x,y,\frac{dy}{dx},\left(\frac{dy}{dx}\right)^2,\ldots\left(\frac{dy}{dx}\right)^n\right)=0$$

is of order one and degree n.

iii. A differential equation of the form

$$\frac{d^n y}{dx^n}+a_1\frac{d^{n-1}}{dx^{n-1}}+\ldots+a_{n-1}\frac{dy}{dx}+a_n y=X$$

where a_i's are constants and X is a function of x is called an nth order linear differential equation with constant coefficients.

iv. A differential equation of the form

$$x^n\frac{d^n y}{dx^n}+a_1 x^{n-1}\frac{d^{n-1}y}{dx^{n-1}}+\ldots+a_{n-1}x\frac{dy}{dx}+a_n y=x$$

is called an nth order homogeneous linear differential equations.

v. A pair of linear differential equations

$$\frac{dx}{dt}+a_1 x+b_1 y=f(t)$$

$$\frac{dy}{dt}+a_2 x+b_2 y=g(t)$$

is called simultaneous linear differential equations with constant coefficients.

vi. Another form of simultaneous equation is

$$\frac{dx}{P}=\frac{dy}{Q}=\frac{dz}{R}$$

where P, Q, R are functions of x, y and z.

vii. A special class of first order differential equation known as total differential equations is $Pdx + Qdy + Rdz = 0$.

15.3 SPECIAL FIRST ORDER DIFFERENTIAL EQUATIONS

i. Direct Separation

Equation	Solution
$y' = f(x)$	$x = \int f(x)\, dx + C$
$y' = g(y)$	$x = \int \frac{dy}{g(y)} + C$
$y' = f(x)\, g(y)$	$\int \frac{dy}{g(y)} = \int f(x)\, dx + C$
$y' = \frac{f(x)}{g(y)}$	$\int g(y)\, dy = \int f(x)\, dx + C$
$y' = \frac{g(y)}{f(x)}$	$\int \frac{dy}{g(y)} = \int \frac{dx}{f(x)} + C$

ii. Separation by Substitution

Equation, substitution	Solution
$yf(x, y)\, dx + xg(x, y)\, dy = 0$ $y = rx$	$\int \frac{dx}{x} = \int \frac{g(r)\, dr}{f(r) + rg(r)} + C$
$yf(x, y)\, dx + xg(x, y)\, dy = 0$ $y = \frac{r}{x}$	$\int \frac{dx}{x} = \int \frac{f(r)\, dr}{r[g(r) - f(r)} + C$

15.4 LINEAR DIFFERENTIAL EQUATIONS OF FIRST ORDER

i. Variable Coefficients $\quad y' + P(x)y = Q(x)$

Condition	Solution
$Q(x) = 0$	$y = C \exp\left[-\int P(x)\, dx\right]$
$Q(x) \neq 0$	$y = \exp\left[-\int P(x)\, dx\right] \times \left\{\int Q(x) \exp\left[\int P(x)\, dx\right] dx + C\right\}$

ii. Constant Coefficients (A, B, α, β = constants) $\quad y' + \beta y = Q(x)$

Condition	Solution
$Q(x) = 0$	$y = Ce^{-Bx}$
$Q(x) = A$	$y = Ce^{-Bx} + \frac{A}{B}$

$Q(x) = Q(x)$	$y = Ce^{-Bx} - \frac{Q(x)}{B} - \frac{Q'(x)}{B^2} - \frac{Q^n(x)}{B^3} - \dots$
$Q(x) = Ae^{\alpha x}$	$y = Ce^{-Bx} + \frac{A}{\alpha + \beta} e^{\alpha x}$
$Q(x) = A \sin \beta x$	$y = Ce^{-Bx} + \frac{A(\beta \cos \beta x - \beta \sin \beta x)}{\beta^2 + B^2}$
$Q(x) = A \cos \beta x$	$y = Ce^{-Bx} + \frac{A(\beta \cos \beta x + \beta \sin \beta x)}{\beta^2 + B^2}$
$Q(x) = Ae^{\alpha x} \sin \beta x$	$y = Ce^{-Bx} - \frac{A[(\alpha + \beta) \sin \beta x + \beta \cos \beta x]}{\beta^2 + (\alpha + B)^2}$
$Q(x) = Ae^{\alpha x} \cos \beta x$	$y = Ce^{-Bx} + \frac{A[\beta \sin \beta x + (\alpha + \beta) \cos \beta x]}{\beta^2 + (\alpha + \beta)^2}$

iii. Bernoullis's Equation $y' + P(x)y = Q(x)y^n$

Substitution: $y = z^{1/(1-n)}$ $\quad y' = \frac{1}{1-n} z^{n/(1-n)} z'$

Reduced equation: $z' + (1-n) P(x) z = (1-n) Q(x)$

Solution: $y^{1-n} = \exp\left[(n-1)\int P(x)\,dx\right] \times \left\{(1-n)\int Q(x) \exp\left[(1-n)\int P(x)\,dx\right] dx + C\right\}$

15.5 SECOND-ORDER DIFFERENTIAL EQUATIONS, SPECIAL CASES

i. Direct Solution (a = constant)

Equation	Solution
$y'' = a$	$y = \frac{1}{2} ax^2 + C_1 x + C_2$
$y'' = f(x)$	$y = \int\int f(x)\,dx\,dx + C_1 x + C_2$

ii. Substitution ($y' = \psi$)

Equation	Solution
$y'' = f(y)$	$\int \psi\,d\psi = \int f(y)\,dy + C$
$y'' = f(y')$	$\int \frac{d\psi}{f(\psi)} = \int dx + C$
$y'' = f(x, y')$	$\psi' = f(x, \psi)$
$y'' = f(y, y')$	$\psi \frac{d\psi}{dy} = f(y, \psi)$

15.6 nth-ORDER DIFFERENTIAL EQUATION, SPECIAL CASE

Equation	Solution
$\dfrac{d^{(n)} y}{dx^n} = f(x)$	$y = \dfrac{1}{(n-1)!}\int_0^x f(\tau)(x-\tau)^{n-1}\, d\tau + g(x)$ $g(x) = C_0 + C_1 x + C_2 x^2 + \ldots + C_{n-1} x^{n-1}$

15.7 EXACT DIFFERENTIAL EQUATION

If M and N are functions of (x, y) and $\partial M/\partial y = \partial N/\partial x$, then $M\, dx + N\, dy = 0$ is an exact differential equation, the solution of which is

$$\int M\, dx + \int\left(N - \int \frac{\partial M}{\partial y}\, dx\right) dy + C = 0 \quad \text{or} \quad \int N\, dy + \int\left(M - \int \frac{\partial N}{\partial x}\, dy\right) dx + C = 0$$

If the condition $\partial M/\partial y = \partial N/\partial x$ is not satisfied, there exists a function $\psi(x, y) = \psi$ (integrating factor) such that

$$\frac{\partial(\psi M)}{\partial y} = \frac{\partial(\psi N)}{\partial x}$$

15.8 nth-ORDER DIFFERENTIAL EQUATIONS, CONSTANT COEFFICIENTS

i. Standard Form

A linear differential equation of order n with constant coefficients is given as

$$y^{(n)} + a_1 y^{(n-1)} + a_2 y^{(n-2)} + \ldots + a_n y = f(x)$$

where $f(x)$ is an *arbitrary function* of x.

ii. General Solution

For such an equation, the general solution is

$$y = y_C + y_P = (\text{CF} + \text{PS})$$

where y_C = *complementary function* and y_P = *particular solution.*

iii. Complementary Function

By the substitution $y = e^{\lambda x}$ the reduced differential equation transforms into

$$f(\lambda)\, e^{\lambda x} = (\lambda^n + a_1\lambda^{n-1} + a_2\lambda^{n-2} + \ldots + a_n)\, e^{\lambda x} = 0)$$

where $f(\lambda) = 0$ is the *characteristic equation*, the roots of which take one of the forms given below (or their combination) and yield the *coefficients of the complementary functions* given below.

Roots	*Complementary function*
Real, distinct $\lambda_1 \neq \lambda_2 \neq \ldots \neq \lambda_{n-1} \neq \lambda_n$	$y_c = C_1 e^{\lambda 1 x} + C_2 e^{\lambda 2 x} + \ldots + C_n e^{\lambda n x}$
Real, repeated $\lambda_1 = \lambda_2 = \ldots = \lambda_{n-1} = \lambda_n$	$y_c = (C_1 + C_2 x + C_3 x + \ldots + C_n x^{n-1})\, e^{\lambda x}$

Roots		*Complementary function*
Complex, distinct $\lambda_1 = \alpha + \beta i$ .... $\lambda_{n-1} = \gamma + \delta i$	$\lambda_2 = \alpha - \beta i$ .... $\lambda_n = \gamma - \delta i$	$y_c = e^{\alpha x}(C_1 \cos \beta x + C_2 \sin \beta x) + ... + e^{\gamma x}(C_{n-1} \cos \delta x + C_n \sin \delta x)$
Complex, repeated $\lambda_1 = \lambda_3 = ... \lambda_{n-1} = \alpha + \beta i$ $\lambda_2 = \lambda_4 = ... \lambda_n = \alpha - \beta i$		$y_c = e^{\alpha x}(C_1 + C_3 x + ... + C_{n-1} x^{n/2-1}) \cos \beta x + e^{\alpha x}(C_2 + C_4 x + ... + C_n x^{n/2}) \sin \beta x$

iv. Particular Solution

$$y_p = e^{\lambda}\, n^x \int e^{(\lambda_{n-1} - \lambda_n)x} \int e^{(\lambda_{n-2} - \lambda_{n-1})x.....} \int e^{(\lambda_1 - \lambda_2)x} \int e^{-\lambda_1 x} f(x)\,(dx)^n$$

15.9 SECOND-ORDER DIFFERENTIAL EQUATION

$y'' + ay = f(x)$ $\qquad \lambda^2 + a = 0$

i. Complementary Solution

$a < 0$	$a > 0$
$\lambda_{1,2} = \pm\sqrt{-a} = \pm\alpha$	$\lambda_{1,2} = \pm\sqrt{-a} = \pm i\alpha$
$yc = A \cosh \alpha x + B \sinh \alpha x$	$yc = A \cos \alpha x + B \sin \alpha x$

ii. Particular Solution $(\omega = \text{constant})$

$a < 0$	$a > 0$
$y_P = \dfrac{1}{\alpha}\int_0^x \sin[\alpha(x-\tau)]\, f(\tau)\, d\tau$	$y_P = \dfrac{1}{\alpha}\int_0^x \sin[\alpha(x-\tau)]\, f(\tau)\, d\tau$

$$y_P = \frac{1}{a}\left[f(x) - \frac{f''(x)}{a} + \frac{f^{\omega}(x)}{a^2} - ...\right]$$

$f(x)$	$e^{\omega x}$	$\cos \omega x$	$\sin \omega x$
y_P	$\dfrac{e^{\omega x}}{a + \omega^2}$	$\dfrac{\cos \omega x}{a - \omega^2}$	$\dfrac{\sin \omega x}{a - \omega^2}$

15.10 SECOND-ORDER DIFFERENTIAL EQUATION

$y'' + py' + qy = f(x)$ $\qquad \lambda^2 + p\lambda + q = 0$

i. Complementary Solution $(D = p^2 - 4q)$

$D > 0$	$D = 0$	$D < 0$
$\lambda_{1,2} = \dfrac{-p \pm \sqrt{D}}{2} = \alpha, \beta$	$\lambda_{1,2} = -\dfrac{p}{2} = \lambda$	$\lambda_{1,2} = \dfrac{-p \pm \sqrt{D}}{2} = \alpha \pm i\beta$
$yc = Ae^{\alpha x} + Be^{\beta x}$	$yc = (A + Bx)\, e^{\lambda x}$	$yc = (A \cos \beta x + B \sin \beta x)\, e^{\alpha x}$

ii. Particular Solution (ω = constant)

$$y_p = \begin{cases} \dfrac{1}{\sqrt{D}}\left[e^{\alpha x}\int f(x)\,e^{-\alpha x}\,dx - e^{\beta x}\int f(x)\,e^{-\beta x}\,dx\right] & D>0 \\ e^{\lambda x}\left[x\int f(x)\,e^{-\lambda x}\,dx - \int xf(x)\,e^{-\lambda x}\,dx\right] & D=0 \\ \dfrac{e^{\alpha x}}{\beta}\left[\sin\beta x\int f(x)\,e^{-\alpha x}\cos\beta x\,dx - \cos\beta x\int f(x)\,e^{-\alpha x}\sin\beta x\,dx\right] & D<0 \end{cases}$$

$$y_p = \frac{1}{q}\left[f(x) - \frac{p}{q}f'(x) + \frac{p^2-q}{q^2}f''(x) - \frac{p^3-2pq}{q^3}f'''(x) + \ldots\right]$$

$f(x)$	$e^{\omega x}$	$\sin \omega x$	$\cos \omega x$
y_p	$\dfrac{e^{\omega x}}{\omega^2+\omega p+q}$	$\dfrac{(q-\omega^2)\sin\omega x - p\omega\cos\omega x}{(q-\omega^2)^2+(\omega p)^2}$	$\dfrac{(q-\omega^2)\cos\omega x + p\omega\sin\omega x}{(q-\omega^2)^2+(\omega p)^2}$

15.11 FOURTH-ORDER DIFFERENTIAL EQUATION

$$y^{iv} + ay = f(x) \qquad \lambda^4 + a = 0$$

i. Complementary Solution $(a > 0)$

$$a = 4\alpha^4 \qquad \lambda_{1,2} = (1 \pm i)\,\alpha \qquad \lambda_{3,4} = -(1 \pm i)\,\alpha$$

$$y_C = A\Phi_1(\alpha x) + B\Phi_2(\alpha x) + C\Phi_3(\alpha x) + D\Phi_4(\alpha x)$$

$$\Phi_1(\alpha x) = \cosh\alpha x\cos\alpha x \qquad \Phi_2(\alpha x) = \frac{1}{2}(\cosh\alpha x\sin\alpha x + \sinh\alpha x\cos\alpha x)$$

$$\Phi_3(\alpha x) = \frac{1}{2}\sinh\alpha x\sin\alpha x \qquad \Phi_4(\alpha x) = \frac{1}{4}(\cosh\alpha x\sin\alpha x - \sinh\alpha x\cos\alpha x)$$

ii. Particular Solution

$$y_P = \Phi_0(\alpha x) = \frac{1}{\alpha^3}\int_0^x \Phi_4(\alpha x - \alpha\tau)\,f(\tau)\,d\tau$$

iii. Properties of $\Phi_n(\alpha x)$

$n = 1, 2, 3, 4$	$\Phi_1(\alpha x)$	$\Phi_2(\alpha x)$	$\Phi_3(\alpha x)$	$\Phi_4(\alpha x)$
$d\Phi_n(\alpha x)/dx$	$-4\alpha\Phi_4(\alpha x)$	$\alpha\Phi_1(\alpha x)$	$\alpha\Phi_2(\alpha x)$	$\alpha\Phi_3(\alpha x)$
$d^2\Phi_n(\alpha x)/dx^2$	$-4\alpha^2\Phi_3(\alpha x)$	$-4\alpha^2\Phi_4(\alpha x)$	$\alpha^2\Phi_1(\alpha x)$	$\alpha^2\Phi_2(\alpha x)$
$d^3\Phi_n(\alpha x)/dx^3$	$-4\alpha^3\Phi_2(\alpha x)$	$-4\alpha^3\Phi_3(\alpha x)$	$-4\alpha^3\Phi_4(\alpha x)$	$\alpha^3\Phi_1(\alpha x)$
$d^4\Phi_n(\alpha x)/dx^4$	$-4\alpha^4\Phi_1(\alpha x)$	$-4\alpha^4\Phi_2(\alpha x)$	$-4\alpha^4\Phi_3(\alpha x)$	$-4\alpha^4\Phi_4(\alpha x)$

15.12 FOURTH-ORDER DIFFERENTIAL EQUATION

$$y^{iv} - ay = f(x) \qquad \lambda^4 - a = 0$$

i. Complementary Solution (a > 0)

$a = \beta^4 \qquad \lambda_{1,2} = \pm\beta \qquad \lambda_{3,4} = \pm i\beta$

$y_C = A\Phi_1(\beta x) + B\Phi_2(\beta x) + C\Phi_3(\beta x) + D\Phi_4(\beta x)$

$$\Phi_1(\beta x) = \frac{1}{2}(\cosh\beta x + \cos\beta x) \qquad \Phi_2(\beta x) = \frac{1}{2}(\sinh\beta x + \sin\beta x)$$

$$\Phi_3(\beta x) = \frac{1}{2}(\cosh\beta x - \cos\beta x) \qquad \Phi_4(\beta x) = \frac{1}{2}(\sinh\beta x - \sin\beta x)$$

ii. Particular Solution

$$y_P = \Phi_0(\beta x) = \frac{1}{\beta^3}\int_0^x \Phi_4(\beta x - \beta\tau) f(\tau)\, d\tau$$

iii. Properties of $\Phi_n(\beta x)$

$n = 1, 2, 3, 4$	$\Phi_1(\beta x)$	$\Phi_2(\beta x)$	$\Phi_3(\beta x)$	$\Phi_4(\beta x)$
$d\Phi_n(\beta x)/dx$	$\beta\Phi_4(\beta x)$	$\beta\Phi_1(\beta x)$	$\beta\Phi_2(\beta x)$	$\beta\Phi_3(\beta x)$
$d^2\Phi_n(\beta x)/dx^2$	$\beta^2\Phi_3(\beta x)$	$\beta^2\Phi_4(\beta x)$	$\beta^2\Phi_1(\beta x)$	$\beta^2\Phi_2(\beta x)$
$d^3\Phi_n(\beta x)/dx^3$	$\beta^3\Phi_2(\beta x)$	$\beta^3\Phi_3(\beta x)$	$\beta^3\Phi_4(\beta x)$	$\beta^3\Phi_1(\beta x)$
$d^4\Phi_n(\beta x)/dx^4$	$\beta^4\Phi_1(\beta x)$	$\beta^4\Phi_2(\beta x)$	$\beta^4\Phi_3(\beta x)$	$\beta^4\Phi_4(\beta x)$

15.13 EULER'S DIFFERENTIAL EQUATION OF ORDER *N*

i. Standard Form

Euler's differential equation of order n is given as

$$a_0x^ny^{(n)} + a_1x^{n-1}y^{(n-1)} + a_2x^{n-2}y^{(n-2)} + \ldots + a_ny = f(x)$$

where $f(x)$ is an arbitrary function of n.

ii. Complementary Solution

By the substitution $y = x^\lambda$ the reduced differential equation transforms into

$$f(\lambda)\,x^\lambda = \left[a_0\frac{\lambda!}{(\lambda-n)!} + a_1\frac{\lambda!}{(\lambda-n+1)!} + a_2\frac{\lambda!}{(\lambda-n+2)!} + \ldots + a_n\right]x^\lambda = 0$$

where $f(\lambda) = 0$ is the *characteristic equation*, the roots of which take one of the forms (or their combinations).

iii. Particular Solution

In general, there is no *general method of finding a particular solution* of this differential equation. A method which sometimes yields a solution is to assume a series,

$$y_p = \frac{1}{a_n}[f(x) + A_1 f'(x) + A_2 f''(x) + \ldots]$$

where $A_1, A_2, \ldots$ are functions given by the conditions that follows.

$$A_1 = -\ x\bar{a}_{n-1}$$

$$A_2 = -\ x^2\bar{a}_{n-2} - x\bar{a}_{n-1}\,A_1$$

$$A_3 = -\ x^3\bar{a}_{n-3} - x^2\bar{a}_{n-2}\,A_1 - xa_{n-1}\,A_2$$

....

$$A_n = -\ x^n\bar{a}_0 - x^{n-1}\bar{a}_1A_1 - x^{n-2}\bar{a}_2A_2 - x^{n-3}\bar{a}_3A_3 - \ldots$$

$$A_{n+1} = 0 - x^n\bar{a}_0A_1 - x^{n-1}\bar{a}_1A_2 - x^{n-2}\bar{a}_2A_3 - \ldots$$

.... $\bar{a}_j = \dfrac{a_j}{a_n}$

15.14 SECOND ORDER EULER'S DIFFERENTIAL EQUATION

$x^2y'' + bxy' + cy = f(x)$ $\lambda^2 + p\lambda + q = 0$

i. Complementary Solution $(D = p^2 - 4q)$

$D > 0$	$D = 0$	$D < 0$
$\lambda_{1,2} = \dfrac{-p \pm \sqrt{D}}{2} = \alpha, \beta$ $y_C = Ax^\alpha + Bx^\beta$	$\lambda_{1,2} = -\dfrac{p}{2} = \lambda$ $y_C = x^\lambda\,(A \ln \lvert x \rvert + B)$	$\lambda_{1,2} = \dfrac{-p \pm \sqrt{D}}{2} = \alpha \pm i\beta$ $y_C = x^\alpha\,[A \cos(\beta \ln \lvert x \rvert) + B \sin(\beta \ln \lvert x \rvert)]$

ii. Particular Solution

The method of function series frequently yields a particular solution.

15.15 SOLUTION BY POWER SERIES

i. Concept

Some differential equations, particularly homogeneous linear equations with variable coefficients, can be solved by assuming a solution in the form of an *infinte power series* such as

$$y = b_0 + b_1x + b_2x^2 + \ldots + b_rx^r + \ldots = \sum_{r=0}^{\infty} b_rx^r$$

the first and higher derivatives of which are

$$y' = \sum_{r=0}^{\infty} rb_rx^{r-1} \qquad y^n = \sum_{r=0}^{\infty} r\,(r-1)\,b_rx^{r-2}$$

After these expressions have been substituted in the given differential equation, the terms in like power of x are combined, and the coefficients of each power of x are set equal to zero. The system of equations thus obtained yields the values of the constants of the series of which n (corresponding to the order) must remain unknown (arbitrary) as the constants of integration.

ii. Transformation

Frequently it serves to an advantages to transform the given differential equation

$y'' + a(x)\, y' + b(x)\, y = 0$

by the substitution of

$$y = y(x) = u(x) \exp\left[-\frac{1}{2}\int a(x)\,dx\right] \quad \text{to} \quad u'' + \left[b(x) - \frac{a(x)'}{2} - \left(\frac{a(x)}{2}\right)^2\right] u = 0$$

iii. Orthogonal Polynomials

A *set of polynomials* $p_n(x)$ $(n = 0, 1, 2, ...)$ of degree n in x is orthogonal in the interval (a, b) with respect to the weight function $w(x)$ if

$$\int_a^b w(x)\, p_m(x)\, p_n(x)\, dx = \begin{cases} 0 & \text{for } m \neq n \\ a_n & \text{for } m = n \end{cases} \qquad m, n = 0, 1, 2, ...$$

Under certain conditions, this relationship admits the representation of a function in the form

$$f(x) = \sum_{n=0}^{\infty} C_n\, p_n(x) \qquad \text{with} \qquad C_n = \frac{1}{a_n}\int_a^b f(x)\, w(x)\, p_n(x)\, dx$$

iv. Classical Orthogonal Polynomials

Classical orthogonal polynomials of particular interest are designated by the names of their discoverers. The *Legendre, Chebyshev, Laguerre,* and *Hermite polynomials have two typical properties* :

a. $p_n(x)$ *satisfy the differential equation*

$a(x)\, y'' + b(x)\, y' + cy = 0$

where $a(x)$, $b(x)$ are independent of n, c and n, c are independent of x.

b. The polynomial *can be represented by the generalized Rodrigues formula* as

$$p_n(x) = \frac{1}{k_n\, w(x)} \frac{d^n}{dx^n} \{w(x)\, [\phi(x)]^n\}$$

where $\phi(x)$ is a polynomial of the first or second degree.

15.16 HYPERGEOMETRIC DIFFERENTIAL EQUATION

i. Gauss's Differential Equation (α, β, γ = constants)

$$x(1 - x)\, y'' - [(\alpha + \beta + 1)\, x - \gamma]\, y' - \alpha\beta\gamma = 0$$

ii. Solution ($y = A_1 y_1 + A_2 y_2$)

$y_1 = F(\alpha, \beta, \gamma, x)$

$$= 1 + \frac{\alpha\beta}{\gamma}\frac{x}{1!} + \frac{\alpha(\alpha+1)\beta(\beta+1)}{\gamma(\gamma+1)}\frac{x^2}{2!} + \frac{\alpha(\alpha+1)(\alpha+2)\beta(\beta+1)(\beta+2)}{\gamma(\gamma+1)(\gamma+2)}\frac{x^3}{3!} + ...$$

$$y_2 = x^{1-\gamma} F(\alpha - \gamma + 1, \beta - \gamma + 1, 2 - \gamma, x)$$

$$= x^{1-\gamma}\left[1 + \frac{(\alpha-\gamma+1)(\beta-\gamma+1)}{(2-\gamma)}\frac{x}{1!} + \frac{(\alpha-\gamma+1)(\alpha-\gamma+2)(\beta-\gamma+1)(\beta-\gamma+2)}{(2-\gamma)(3-\gamma)}\frac{x^2}{2!} + ...\right]$$

$$|x| < 1, \gamma \neq 0, 1, 2, ...$$

$$y_1 = F(\alpha, \beta, \alpha + \beta - \gamma + 1, 1 - x)$$

$$y_2 = (1-x)^{\gamma-\alpha-\beta} F(\gamma - \beta, \gamma - \alpha, \gamma - \alpha - \beta + 1, 1 - x) \qquad |x - 1| < 1, \alpha + \beta - \gamma \neq 0, 1, 2, ...$$

$$y_1 = x^{-\alpha} F(\alpha, \alpha - \gamma + 1, \alpha - \beta + 1, x^{-1})$$

$$y = x^{-\beta} F(\beta, \beta - \gamma + 1, \beta - \alpha + 1, x^{-1}) \qquad |x| > 1, \alpha - \beta \neq 0, 1, 2, ...$$

15.17 COFLUENT HYPERGEOMETRIC DIFFERENTIAL EQUATION

i. Kummer's Differential Equation (β, γ = constants)

$$xy'' + (\gamma - x)y' - \beta y = 0$$

ii. Solution ($y = A_1 y_1 + A_2 y_2$)

$$y_1 = F(\beta, \gamma, x)$$

$$= 1 + \frac{\beta}{\gamma}\frac{x}{1!} + \frac{\beta(\beta+1)}{\gamma(\gamma+1)}\frac{x^2}{2!} + ...$$

$$y_2 = x^{1-\gamma} F(\beta - \gamma + 1, 2 - \gamma, x)$$

$$= x^{1-\gamma}\left[1 + \frac{\beta-\gamma+1}{2-\gamma}\frac{x}{1!} + \frac{(\beta-\gamma+1)(\beta-\gamma+2)}{(2-\gamma)(3-\gamma)}\frac{x^2}{2!} + ...\right] \qquad \gamma \neq 0, 1, 2, ...$$

15.18 LEGENDRE POLYNOMIALS $P_n(x)$

i. Differential Equation (n = constant)

$$(1 - x^2)y'' - 2xy' + n(n+1)y = 0$$

Interval: $[-1, +1]$
Weight: $w(x) = 1$

ii. Solution ($y = A_1 y_1 + A_2 y_2$)

$$y_1 = P_n(x) = \frac{(1)(3)...(2n-1)}{n!}\left[x^n - \frac{n(n-1)}{2(2n-1)}x^{n-2} + \frac{n(n-1)(n-2)(n-3)}{(2)(4)(2n-1)(2n-3)}x^{n-4} + ...\right]$$

$$y_2 = Q_n(x) = \frac{1}{2}P_n(x)\ln(x)\frac{x+1}{x-1} - \sum_{k=0}^{\infty}\frac{1}{k}P_{k-1}(x)P_{n-k}(x) \qquad n = 0, 1, 2, ...,$$

iii. Relations

$$P_n(x) = \frac{1}{2^n\, n!}\frac{d^n}{dx^n}(x^2 - 1)^n \qquad P_{n+1}(x) = \frac{2n+1}{n+1}xP_n(x) - \frac{n}{n+1}P_{n-1}(x)$$

$$\int_{-1}^{+1} P_m(x)\,P_n(x)\,dx = \begin{cases} 0 & \text{for } m \neq n \\ \dfrac{2}{2n+1} & \text{for } m = n \end{cases}$$

iv. First Five Terms

$P_0(x) = +1;$ $\quad P_1(x) = x$ $\quad P_2(x) = \frac{1}{2}(3x^2 - 1)$ $\quad P_3(x) = \frac{1}{2}(5x^3 - 3x)$

$P_4(x) = \frac{1}{8}(35x^4 - 30x^2 + 3)$

15.19 CHEBYSHEV POLYNOMIALS $T_n(x)$

Interval: [– 1, + 1]

i. Differential Equation $(n = \text{constant})$

$(1 - x^2)\,y'' - xy' + n^2 y = 0$ Weight: $w(x) = \dfrac{1}{\sqrt{1 - x^2}}$

ii. Solution $(y = A_1 y_1 + A_2 y_2)$

$$y_1 = T_{2n}(x) = \left[1 - \frac{n^2}{2!}x^2 + \frac{n^2(n^2 - 4)}{4!}x^4 - \frac{n^2(n^2 - 4)(n^2 - 16)}{6!}x^6 + \ldots\right]$$

$$y_2 = T_{2n+1}(x) = \left[x - \frac{(n^2 - 1)}{3!}x^3 + \frac{(n^2 - 1)(n^2 - 9)}{5!}x^5 - \frac{(n^2 - 1)(n^2 - 9)(n^2 - 25)}{7!}x^7 + \ldots\right] \text{ all } n$$

For $n = 0$,

$$y = A_1 + A_2 \sin^{-1} x$$

iii. Relations

$$T_n(x) = \frac{(-2)^n\, n!}{(2n)!}\sqrt{1 - x^2}\,\frac{d^n}{dx^2}(1 - x^2)^{n - 1/2} \qquad T_{n+1}(x) = 2xT_n(x) - T_{n-1}(x)$$

$$\int_{-1}^{+1} \frac{T_m(x)\,T_n(x)\,dx}{\sqrt{1 - x^2}} = \begin{cases} 0 & \text{for } m \neq n \\ \dfrac{\pi}{2} & \text{for } m = n \neq 0 \\ \pi & \text{for } m = n = 0 \end{cases}$$

iv. First Five Terms

$T_0(x) = 0$ $\quad T_1(x) = x$ $\quad T_2(x) = 2x^2 - 1$ $\quad T_3(x) = 4x^3 - 3x$ $\quad T_4(x) = 8x^4 - 8x^3 + 1$

15.20 LAGUERRE POLYNOMIALS $L_n(x)$

i. Differential Equation (n = constant)

$$xy'' + (1-x)y' + ny = 0$$

Interval: $[0, \infty]$

Weight: $w(x) = e^{-x}$

ii. Solution ($y = A_1y_1 + A_2y_2$)

$$y_1 = L_n(x) = n!\left[1 - nx + \frac{n(n-1)}{(2!)^2}x^2 - \frac{n(n-1)(n-2)}{(3!)^2}x^3 + \ldots + \frac{x^n}{n!}\right]$$

$$y_2 = L_n(x)\ln x + \left[(1+2n)x + \frac{(1+n-3n^2)}{4}x^2 + \frac{(2-4n-24n^2-11n^3)}{108}x^3 + \ldots\right] \quad n = 0, 1, 2, \ldots,$$

iii. Relations

$$L_n(x) = e^x \frac{d^n}{dx^n}(x^n e^{-x}), \qquad L_{n+1}(x) = (2n+1-x)L_n(x) - n^2 L_{n-1}(x)$$

$$\int_0^\infty e^{-x} L_m(x) L_n(x) dx = \begin{cases} 0 & \text{for } m \neq n \\ (n!)^2 & \text{for } m = n \end{cases}$$

iv. First Five Terms

$L_0(x) = 1$ $\qquad$ $L_1(x) = -x + 1$ $\qquad$ $L_2(x) = x^2 - 4x + 2$

$L_3(x) = -x^3 + 9x^2 - 18x + 6$ $\qquad$ $L_4(x) = x^4 - 16x^3 + 72x^2 - 96x + 24$

15.21 HERMITE POLYNOMIALS $H_n(x)$

i. Differential Equation (n = constant)

$$y'' - 2xy + 2ny = 0$$

Interval: $[-\infty, +\infty]$

Weight: $w(x) = e^{x^2}$

ii. Solution $y = (A_1y_1 + A_2y_2)$

$$y_1 = H_{2n}(x) = 1 - \frac{2n}{2!}x^2 + \frac{2^2 n(n-2)}{4!}x^4 - \frac{2^3 n(n-2)(n-4)}{6!}x^6 + \ldots$$

$$y_2 = H_{2n+1}(x) = x - \frac{2(n-1)}{3!}x^3 + \frac{2^2(n-1)(n-3)}{5!}x^5 - \frac{2^3(n-1)(n-3)(n-5)}{7!}x^7 + \ldots \text{ all } n$$

iii. Relations

$$H_n(x) = (-1)^n e^{x^2} \frac{d^n}{dx^n}\left(e^{-x^2}\right) \qquad H_{n+1}(x) = 2xH_n(x) - 2nH_{n-1}(x)$$

$$\int_{-\infty}^{+\infty} e^{-x^2} H_m(x) H_n(x)\, dx = \begin{cases} 0 & \text{for } m \neq n \\ 2^n n! \sqrt{\pi} & \text{for } m = n \end{cases}$$

iv. First Five Terms

$H_0(x) = 1$ $\quad H_1(x) = 2x$ $\quad H_2(x) = 4x^2 - 2$

$H_3(x) = 8x^3 - 12x$ $\quad H_4(x) = 16x^4 - 48x^2 + 12$

15.22 LEGENDRE POLYNOMIALS

$P_n(x)$ $\quad x = 0.00 - 0.49$

x	$P_1(x)$	$P_2(x)$	$P_3(x)$	$P_4(x)$	$P_5(x)$	$P_6(x)$	$P_7(x)$	x
0.00	+ 0.0000	– 0.5000	+ 0.0000	+ 0.3750	+ 0.0000	– 0.3125	+ 0.0000	0.00
0.01	+ 0.0100	– 0.4998	– 0.0150	+ 0.3746	+ 0.0187	– 0.3118	– 0.0219	0.01
0.02	+ 0.0200	– 0.4994	– 0.0300	+ 0.3735	+ 0.0374	– 0.3099	– 0.0436	0.02
0.03	+ 0.0300	– 0.4986	– 0.0449	+ 0.3716	+ 0.560	– 0.3066	– 0.0651	0.03
0.04	+ 0.0400	– 0.4976	– 0.0598	+ 0.3690	+ 0.744	– 0.3021	– 0.0862	0.04
0.05	+ 0.0500	– 0.4962	– 0.0747	+ 0.3657	+ 0.0927	– 0.2962	– 0.1069	0.05
0.06	+ 0.0600	– 0.4946	– 0.0895	+ 0.3616	+ 0.1106	– 0.2891	– 0.1270	0.06
0.07	+ 0.0700	– 0.4926	– 0.1041	+ 0.2567	+ 0.1283	– 0.2808	– 0.1464	0.07
0.08	+ 0.0800	– 0.4904	– 0.1187	+ 0.3512	+ 0.1455	– 0.2713	– 0.1651	0.08
0.09	+ 0.0900	– 0.4848	– 0.1332	+ 0.3449	+ 0.1624	– 0.2606	– 0.1828	0.09
0.10	+ 0.1000	– 0.4850	– 0.1475	+ 0.3379	+ 0.1788	– 0.2488	– 0.1995	0.10
0.11	+ 0.1100	– 0.4818	– 0.1617	+ 0.3303	+ 0.1947	– 0.2360	– 0.2151	0.11
0.12	+ 0.1200	– 0.4784	– 0.1757	+ 0.3219	+ 0.2101	– 0.2220	– 0.2296	0.12
0.13	+ 0.1300	– 0.4746	– 0.1895	+ 0.3129	+ 0.2248	– 0.2071	– 0.2427	0.13
0.14	+ 0.1400	– 0.4706	– 0.2031	+ 0.3032	+ 0.2389	– 0.1913	– 0.2345	0.14
0.15	+ 0.1500	– 0.4663	– 0.2166	+ 0.2928	+ 0.2523	– 0.1746	– 0.2649	0.15
0.16	+ 0.1600	– 0.4616	– 0.2298	+ 0.2819	+ 0.2650	– 0.1572	– 0.2738	0.16
0.17	+ 0.1700	– 0.4567	– 0.2427	+ 0.2703	+ 0.2769	– 0.1389	– 0.2812	0.17
0.18	+ 0.1800	– 0.4514	– 0.2554	+ 0.2581	+ 0.2880	– 0.1201	– 0.2870	0.18
0.19	+ 0.1900	– 0.4458	– 0.2679	+ 0.2453	+ 0.2982	– 0.1006	– 0.2911	0.19
0.20	+ 0.2000	– 0.4400	– 0.2800	+ 0.2320	+ 0.3075	– 0.0806	– 0.2935	0.20
0.21	+ 0.2100	– 0.4348	– 0.2918	+ 0.2181	+ 0.3159	– 0.0601	– 0.2943	0.21
0.22	+ 0.2200	– 0.4274	– 0.3034	+ 0.2037	+ 0.3234	– 0.0394	– 0.2933	0.22
0.23	+ 0.2300	– 0.4206	– 0.3146	+ 0.1889	+ 0.3299	– 0.0183	– 0.2906	0.23
0.24	+ 0.2400	– 0.4136	– 0.3254	+ 0.1735	+ 0.3355	+ 0.0029	– 0.2861	0.24
0.25	+ 0.2500	– 0.4062	– 0.3359	+ 0.1577	+ 0.3397	+ 0.0243	– 0.2799	0.25
0.26	+ 0.2600	– 0.3986	– 0.3461	+ 0.1415	+ 0.3431	+ 0.0456	– 0.2720	0.26
0.27	+ 0.2700	– 0.3906	– 0.3558	+ 0.1249	+ 0.3453	+ 0.0669	– 0.2625	0.27
0.28	+ 0.2800	– 0.3824	– 0.3651	+ 0.1079	+ 0.3465	+ 0.0879	– 0.2512	0.28
0.29	+ 0.2900	– 0.3788	– 0.3740	+ 0.0906	+ 0.3465	+ 0.1087	– 0.2384	0.29
0.30	+ 0.3000	– 0.3650	– 0.3825	+ 0.0729	+ 0.3454	+ 0.1292	– 0.2241	0.30
0.31	+ 0.3100	– 0.3558	– 0.3905	+ 0.0550	+ 0.3431	+ 0.1429	– 0.2082	0.31
0.32	+ 0.3200	– 0.3464	– 0.3981	+ 0.0369	+ 0.3397	+ 0.1686	– 0.1910	0.32

x	$P_1(x)$	$P_2(x)$	$P_3(x)$	$P_4(x)$	$P_5(x)$	$P_6(x)$	$P_7(x)$	x
0.33	+ 0.3300	– 0.3366	– 0.4052	+ 0.0185	+ 0.3351	+ 0.1873	– 0.1724	0.33
0.34	+ 0.3400	– 0.3266	– 0.4117	– 0.0000	+ 0.3294	+ 0.2053	– 0.1527	0.34
0.35	+ 0.3500	– 0.3162	– 0.4178	– 0.0187	+ 0.3225	+ 0.2255	– 0.1318	0.35
0.36	+ 0.3600	– 0.3056	– 0.4243	– 0.0375	+ 0.3144	+ 0.2388	– 0.1098	0.36
0.37	+ 0.3700	– 0.2946	– 0.4284	– 0.0564	+ 0.3051	+ 0.2540	– 0.0870	0.37
0.38	+ 0.3800	– 0.2834	– 0.4328	– 0.0753	+ 0.2948	+ 0.2681	– 0.0635	0.38
0.39	+ 0.3900	– 0.2718	– 0.4367	– 0.0942	+ 0.2833	+ 0.2810	– 0.0393	0.39
0.40	+ 0.4000	– 0.2600	– 0.4400	– 0.1130	+ 0.2706	+ 0.2926	– 0.0146	0.40
0.41	+ 0.4100	– 0.2478	– 0.4427	– 0.1317	+ 0.2569	+ 0.3029	+ 0.0104	0.41
0.42	+ 0.4200	– 0.2354	– 0.4448	– 0.1504	+ 0.2421	+ 0.3118	+ 0.0356	0.42
0.43	+ 0.4300	– 0.2226	– 0.4462	– 0.1668	+ 0.2263	+ 0.3119	+ 0.0859	043
0.44	+ 0.4400	– 0.2096	– 0.4470	– 0.1870	+ 0.2095	+ 0.3249	+ 0.0859	0.44
0.45	+ 0.4500	– 0.1963	– 0.4472	– 0.2050	+ 0.1917	+ 0.3290	+ 0.1106	0.45
0.46	+ 0.4600	– 0.1826	– 0.4467	– 0.2226	+ 0.1730	+ 0.3314	+ 0.1348	0.46
0.47	+ 0.4700	– 0.1687	– 0.4454	– 0.2399	+ 0.1534	+ 0.3321	+ 0.1584	0.47
0.48	+ 0.4800	– 0.1544	– 0.4435	– 0.2568	+ 0.1330	+ 0.3310	+ 0.1811	0.48
0.49	+ 0.4900	– 0.1398	– 0.4409	– 0.2732	+ 0.1118	+ 0.3280	+ 0.2027	0.49
x	$P_1(x)$	$P_2(x)$	$P_3(x)$	$P_4(x)$	$P_5(x)$	$P_6(x)$	$P_7(x)$	x

15.23 LEGENDRE POLYNOMIALS

$P_n(x)$ $x = 0.50 – 1.00$

x	$P_1(x)$	$P_2(x)$	$P_3(x)$	$P_4(x)$	$P_5(x)$	$P_6(x)$	$P_7(x)$	x
0.50	+ 0.5000	– 0.1250	– 0.4375	– 0.2891	+ 0.0898	+ 0.3232	+ 0.2231	0.50
0.51	+ 0.5100	– 0.1098	– 0.4334	– 0.3044	+ 0.0673	+ 0.3166	+ 0.2442	0.51
0.52	+ 0.5200	– 0.0944	– 0.4285	– 0.3191	+ 0.0441	+ 0.3080	+ 0.2596	0.52
0.53	+ 0.5300	– 0.0786	– 0.4228	– 0.3332	+ 0.0204	+ 0.2975	+ 0.2753	0.53
0.54	+ 0.5400	– 0.0626	– 0.4163	– 0.3465	– 0.0037	+ 0.2851	+ 0.2891	0.54
0.55	+ 0.5500	– 0.0462	– 0.4091	– 0.3590	– 0.0282	+ 0.2708	+ 0.3007	0.55
0.56	+ 0.5600	– 0.0296	– 0.4010	– 0.3707	– 0.0529	+ 0.2546	+ 0.3102	0.56
0.57	+ 0.5700	– 0.0126	– 0.3920	– 0.3815	– 0.0779	+ 0.2366	+ 0.3172	0.57
0.58	+ 0.5800	– 0.0046	– 0.3822	– 0.3914	– 0.1028	+ 0.2168	+ 0.3217	0.58
0.59	+ 0.5900	– 0.0222	– 0.3716	– 0.4002	– 0.1278	+ 0.1953	+ 0.3235	0.59
0.60	+ 0.6000	+ 0.0400	– 0.3600	– 0.4080	– 0.1526	+ 0.1721	+ 0.3226	0.60
0.61	+ 0.6100	+ 0.0582	– 0.3476	– 0.4146	– 0.1772	+ 0.1473	+ 0.3188	0.61
0.62	+ 0.6200	+ 0.0766	– 0.3342	– 0.4200	– 0.2014	+ 0.1211	+ 0.3121	0.62
0.63	+ 0.6300	+ 0.0954	– 0.3199	– 0.4242	– 0.2251	+ 0.0935	+ 0.3023	0.63
0.64	+ 0.6400	+ 0.1144	– 0.3046	– 0.4270	– 0.2482	+ 0.0646	+ 0.2895	0.64
0.65	+ 0.6500	+ 0.1338	– 0.2884	– 0.4284	– 0.2705	+ 0.0347	+ 0.2737	0.65

x	$P_1(x)$	$P_2(x)$	$P_3(x)$	$P_4(x)$	$P_5(x)$	$P_6(x)$	$P_7(x)$	x
0.66	+ 0.6600	+ 0.1534	– 0.2713	– 0.4284	– 0.2919	+ 0.0038	+ 0.2548	0.66
0.67	+ 0.6700	+ 0.1734	– 0.2531	– 0.4268	– 0.3122	– 0.0278	+ 0.2329	0.67
0.68	+ 0.6800	+ 0.1936	– 0.2339	– 0.4236	– 0.3313	– 0.0601	+ 0.2081	0.68
0.69	+ 0.6900	+ 0.2142	– 0.2137	– 0.4187	– 0.3490	– 0.0926	+ 0.1805	0.69
0.70	+ 0.7000	+ 0.2350	– 0.1925	– 0.4121	– 0.3652	– 0.1253	– 0.1502	0.70
0.71	+ 0.7100	+ 0.2562	– 0.1702	– 0.4036	– 0.3796	– 0.1578	– 0.1173	0.71
0.72	+ 0.7200	+ 0.2776	– 0.1469	– 0.3933	– 0.3922	– 0.1899	– 0.0822	0.72
0.73	+ 0.7300	+ 0.2994	– 0.1225	– 0.3810	– 0.4026	– 0.2214	– 0.0450	0.73
0.74	+ 0.7400	+ 0.3214	– 0.0969	– 0.3666	– 0.4107	– 0.2581	– 0.0061	0.74
0.75	+ 0.7500	+ 0.3438	– 0.0703	– 0.3501	– 0.4164	– 0.2808	– 0.0342	0.75
0.76	+ 0.7600	+ 0.3664	– 0.0426	– 0.3314	– 0.4193	– 0.3081	– 0.0754	0.76
0.77	+ 0.7700	+ 0.3894	– 0.0137	– 0.3104	– 0.4193	– 0.3333	– 0.1171	0.77
0.78	+ 0.7800	+ 0.4126	+ 0.0164	– 0.2871	– 0.4162	– 0.3559	– 0.1588	0.78
0.79	+ 0.7900	+ 0.4362	+ 0.0476	– 0.2613	– 0.4097	– 0.3756	– 0.1999	0.79
0.80	+ 0.8000	+ 0.4600	+ 0.0800	– 0.2330	– 0.3995	– 0.3918	– 0.2397	0.80
0.81	+ 0.8100	+ 0.4842	+ 0.1136	– 0.2021	– 0.3855	– 0.4091	– 0.2774	0.81
0.82	+ 0.8200	+ 0.5086	+ 0.1484	– 0.1685	– 0.3674	– 0.4119	– 0.3134	0.82
0.83	+ 0.8300	+ 0.5334	+ 0.1845	– 0.1312	– 0.3449	– 0.4147	– 0.3437	0.83
0.84	+ 0.8400	+ 0.5584	+ 0.2218	– 0.0928	– 0.3177	– 0.3177	– 0.3703	0.84
0.85	+ 0.8500	+ 0.5838	+ 0.2603	– 0.0506	– 0.2857	– 0.2857	– 0.3913	0.85
0.86	+ 0.8600	+ 0.6094	+ 0.3001	– 0.0053	– 0.2484	– 0.2484	– 0.4055	0.86
0.87	+ 0.8700	+ 0.6354	+ 0.3413	+ 0.0431	– 0.2056	– 0.2356	– 0.4116	0.87
0.88	+ 0.8800	+ 0.6616	+ 0.3837	+ 0.0947	– 0.1570	– 0.1570	– 0.4083	0.88
0.89	+ 0.8900	+ 0.6882	+ 0.4274	+ 0.1496	– 0.1023	– 0.1023	– 0.3942	0.89
0.90	+ 0.9000	+ 0.7150	+ 0.4725	+ 0.2079	– 0.0411	– 0.0411	– 0.3678	0.90
0.91	+ 0.9100	+ 0.7422	+ 0.5189	+ 0.2698	+ 0.0268	+ 0.0268	– 0.3274	0.91
0.92	+ 0.9200	+ 0.7696	+ 0.5657	+ 0.3352	+ 0.1017	+ 0.1017	– 0.2713	0.92
0.93	+ 0.9300	+ 0.7974	+ 0.6159	+ 0.4044	+ 0.1842	+ 0.1842	– 0.1975	0.93
0.94	+ 0.9400	+ 0.8254	+ 0.6665	+ 0.4773	+ 0.2744	+ 0.2744	– 0.1040	0.94
0.95	+ 0.9500	+ 0.8535	+ 0.7184	+ 0.5541	+ 0.3727	+ 0.3727	+ 0.0112	0.95
0.96	+ 0.9600	+ 0.8824	+ 0.7718	+ 0.6349	+ 0.4796	+ 0.4796	+ 0.1506	0.96
0.97	+ 0.9700	+ 0.9114	+ 0.8267	+ 0.7198	+ 0.5954	+ 0.5954	+ 0.3165	0.97
0.98	+ 0.9800	+ 0.9406	+ 0.8830	+ 0.8089	+ 0.7204	+ 0.7204	+ 0.5115	0.98
0.99	+ 0.9900	+ 0.9702	+ 0.9407	+ 0.9022	+ 0.8552	+ 0.8552	+ 0.7384	0.99
1.00	+ 1.0000	+ 1.0000	+ 1.0000	+ 1.0000	+ 1.0000	+ 1.0000	+ 1.0000	1.00
x	$P_1(x)$	$P_2(x)$	$P_3(x)$	$P_4(x)$	$P_5(x)$	$P_6(x)$	$P_7(x)$	x

15.24 BESSEL'S DIFFERENTIAL EQUATION

i. Differential Equation

$$x^2y'' + xy' + (x^2 - n^2)\, y = 0$$

ii. Solution

$y = A_1 J_n(x) + A_2 J_{-n}(x) \qquad n \neq 0, 1, 2, \ldots \qquad y = A_1 J_n(x) + A_2 Y_n(x)$ for all n

iii. Bessel Functions of the First Kind of Order n

$$J_n(x) = \sum_{k=0}^{\infty} \frac{(-1)^k (x/2)^{2k+n}}{k!\,\Gamma(k+1+n)}$$

$$= \frac{x^n}{2^n\,\Gamma(1+n)}\left[1 - \frac{x^2}{2(2+2n)} + \frac{x^4}{(2)(4)(2+2n)(4+2n)} - \ldots\right]$$

$$J_{-n}(x) = \sum_{k=0}^{\infty} \frac{(-1)^k (x/2)^{2k-n}}{k!\,\Gamma(k+1-n)}$$

$$= \frac{x^{-n}}{2^{-n}\,\Gamma(1-n)}\left[1 - \frac{x^2}{2(2-2n)} + \frac{x^4}{(2)(4)(2-2n)(4-2n)} - \ldots\right]$$

$$= (-1)^n J_n(x) \qquad n = 0, 1, 2, \ldots,$$

$$J_0(x) = 1 - \frac{(x/2)^2}{(1!)^2} + \frac{(x/2)^4}{(2!)^2} - \frac{(x/2)^6}{(3!)^2} + \ldots$$

$$J_1(x) = \frac{x}{2}\left[1 - \frac{(x/2)^2}{2(1!)^2} + \frac{(x/2)^4}{3(2!)^2} - \frac{(x/2)^6}{4(3!)^2} + \ldots\right] = -\frac{d}{dx}[J_0(x)]$$

iv. Bessel Functions of the Second Kind of Order n

$$Y_n(x) = \begin{cases} \dfrac{J_n(x)\cos n\pi - J_{-n}(x)}{\sin n\pi} & n \neq 0, 1, 2, \ldots. \\ \lim\limits_{p \to n} \dfrac{J_p(x)\cos p\pi - J_{-p}(x)}{\sin p\pi} & n = 0, 1, 2, \ldots. \end{cases}$$

$$Y_{-n}(x) = (-1)^n Y_n(x) \qquad n = 0, 1, 2, \ldots$$

$$Y_0(x) = \frac{2}{\pi}\left[\left(I_n \frac{x}{2} + C\right) J_0(x) + \frac{2}{1} J_2(x) - \frac{2}{2} J_4(x) + \frac{2}{3} J_6(x) - \ldots.\right]$$

$$Y_1(x) = \frac{2}{\pi}\left[\left(I_n \frac{x}{2} + C\right) J_1(x) - \frac{1}{x} - \frac{1}{2} J_1(x) + \frac{9}{4} J_3(x) - \ldots.\right] = -\frac{d}{dx}[Y_0(x)]$$

where $\qquad C = 0.57721\ldots.$

15.25 PROPERTIES OF BESSEL FUNCTION

i. Recurrence Relations

$$J_{n+1}(x) = \frac{2n}{x} J_n(x) - J_{n-1}(x) \qquad Y_{n-1}(x) = \frac{2n}{x} Y_n(x) - Y_{n-1}(x)$$

$$= \frac{n}{x} J_n(x) - \frac{d}{dx}\left[J_n(x)\right] \qquad = \frac{n}{x} Y_n(x) - \frac{d}{dx}\left[Y_n(x)\right]$$

$$= -x^n \frac{d}{dx}\left[x^{-n} J_n(x)\right] \qquad = -x^n \frac{d}{dx}\left[x^{-n} Y_n(x)\right]$$

$$J_{n-1}(x) = \frac{2n}{x} J_n(x) - J_{n+1}(x) \qquad Y_{n-1}(x) = \frac{2n}{x} Y_n(x) - Y_{n+1}(x)$$

$$= \frac{n}{x} J_n(x) + \frac{d}{dx}\left[J_n(x)\right] \qquad = \frac{n}{x} Y_n(x) + \frac{d}{dx}\left[Y_n(x)\right]$$

$$= x^{-n} \frac{d}{dx}\left[x^n J_n(x)\right] \qquad = x^{-n} \frac{d}{dx}\left[x^n Y_n(x)\right]$$

$$\frac{d}{dx}\left[J_n(x)\right] = \frac{1}{2}\left[J_{n-1}(x) - J_{n+1}(x)\right] \qquad \frac{d}{dx}\left[Y_n(x)\right] = \frac{1}{2}\left[Y_{n-1}(x) - Y_{n+1}(x)\right]$$

$$\int x^{n+1} J_n(x)\,dx = x^{n+1} J_{n+1}(x) \qquad \int x^{n+1} Y_n(x)\,dx = x^{n+1} Y_{n+1}(x)$$

ii Half-odd Integers

$$J_{1/2}(x) = \sqrt{\frac{2}{\pi x}} \sin x \qquad Y_{1/2}(x) = -\sqrt{\frac{2}{\pi x}} \cos x$$

$$J_{3/2}(x) = \sqrt{\frac{2}{\pi x}}\left(\frac{\sin x}{x} - \cos x\right) \qquad Y_{3/2}(x) = -\sqrt{\frac{2}{\pi x}}\left(\frac{\cos x}{x} + \sin x\right)$$

$$J_{5/2}(x) = \sqrt{\frac{2}{\pi x}}\left[\left(\frac{3}{x^2} - 1\right)\sin x - \frac{3}{2}\cos x\right] \qquad Y_{5/2}(x) = -\sqrt{\frac{2}{\pi x}}\left[\left(\frac{3}{x^2} - 1\right)\cos x + \frac{3}{x}\sin x\right]$$

$$J_{-1/2}(x) = \sqrt{\frac{2}{\pi x}} \cos x \qquad Y_{-1/2}(x) = \sqrt{\frac{2}{\pi x}} \sin x$$

$$J_{-3/2}(x) = -\sqrt{\frac{2}{\pi x}}\left(\frac{\cos x}{x} + \sin x\right) \qquad J_{-3/2}(x) = -\sqrt{\frac{2}{\pi x}}\left(\frac{\sin x}{x} + \cos x\right)$$

$$J_{-5/2}(x) = \sqrt{\frac{2}{\pi x}}\left[\left(\frac{3}{x^2} - 1\right)\cos x + \frac{3}{x}\sin x\right] \qquad Y_{-5/2}(x) = \sqrt{\frac{2}{\pi x}}\left[\left(\frac{3}{x^2} - 1\right)\sin x - \frac{3}{x}\cos x\right]$$

iii. Hankel Function of Order n

$$H^{(1)}{}_n(x) = J_n(x) + iY_n(x) \qquad H^{(2)} n\,(x) = J_n(x) - iY_n(x)$$

$$J_n(x) = \frac{H_n^{(1)}(x) + H_n^{(2)}(x)}{2} \qquad Y_n(x) = \frac{H_n^{(1)}(x) - H_n^{(2)}(x)}{2i}$$

15.26 REPRESENTATION OF $J_n(x)$

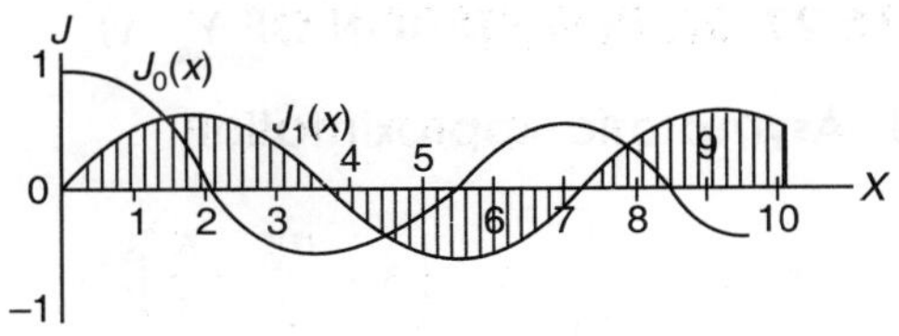

Fig. 15.1

i. Asymptotic Approximation

$$J_n(x) \approx \sqrt{\frac{2}{\pi x}} \cos\left(x - \frac{n\pi}{2} - \frac{\pi}{4}\right), x > 25$$

ii. Numerical Values

$J_0(x) \qquad x = 0 - 10$

x	0	1	2	3	4	5	6	7	8	9	x
0.	1.0000	0.9975	0.9900	0.9776	0.9604	0.9385	0.9120	0.8812	0.8463	0.8075	0.
1.	0.7652	0.7196	0.6711	0.6201	0.5669	0.5118	0.4554	0.3980	0.3400	0.2818	1.
2.	0.2239	0.1667	0.1104	0.0555	0.0025	– 0.0484	– 0.0968	– 0.1424	– 0.1850	– 0.2243	2.
3.	– 0.2601	– 0.2921	– 0.3202	– 0.3443	– 0.3643	– 0.3801	– 0.3918	– 0.3992	– 0.4026	– 0.4018	3.
4.	– 0.3971	– 0.3887	– 0.3766	– 0.3610	– 0.3423	– 0.3205	– 0.2961	– 0.2693	– 0.2404	– 0.2097	4.
5.	– 0.1776	– 0.1443	– 0.1103	– 0.0758	– 0.0412	– 0.0068	0.0270	0.0599	0.0971	0.1220	5.
6.	– 0.1506	0.1773	0.2017	0.2238	0.2433	0.2601	0.2740	0.2851	0.2931	0.2981	6.
7.	– 0.3001	0.2991	0.2951	0.2882	0.2786	0.2663	0.2516	0.2346	0.2154	0.1944	7.
8.	– 0.1717	0.1475	0.1222	0.0960	0.0692	0.0419	0.0146	– 0.0125	– 0.0392	– 0.0653	8.
9.	– 0.0903	– 0.1142	– 0.1367	– 0.1577	– 0.1768	– 0.1939	– 0.2090	– 0.2218	– 0.2323	– 0.2403	9.
10.	– 0.2459										

iii. Numerical Values

$J_1(x) \qquad x = 0 - 10$

x	0	1	2	3	4	5	6	7	8	9	x
0.	0.0000	0.0499	0.0995	0.1483	0.1960	0.2423	0.2867	0.3290	0.3688	0.4059	0.
1.	0.4401	0.4709	0.4983	0.5220	0.5419	0.5579	0.5699	0.5778	0.5815	0.5812	1.
2.	0.5767	0.5683	0.5560	0.5399	0.5202	0.4971	0.4708	0.4416	0.4097	0.3754	2.
3.	0.3391	0.3009	0.2613	0.2207	0.1792	0.1374	0.0955	0.0538	0.0128	– 0.0272	3.
4.	– 0.0660	– 0.1033	– 0.1386	– 0.1719	– 0.2028	– 0.2311	– 0.2566	– 0.2791	– 0.2985	– 0.3147	4.
5.	– 0.3276	– 0.3371	– 0.3432	– 0.3460	– 0.3453	– 0.3414	0.3343	– 0.3241	– 0.3110	– 0.2951	5.
6.	– 0.2767	– 0.2559	– 0.2329	– 0.2081	– 1816	– 0.1538	– 0.1250	– 0.0953	– 0.0652	– 0.0349	6.
7.	– 0.0047	0.0252	0.0543	0.0826	0.1096	0.1352	0.1592	0.1813	0.2014	0.2192	7.
8.	0.2346	0.2476	0.2580	0.2657	0.2708	0.2731	0.2728	0.2697	0.2614	0.2559	8.
9.	0.2453	0.2324	0.2174	0.2004	0.1816	0.1631	0.1395	0.1166	0.0928	0.0684	9.
10.	0.0435										

iv. Asymptotic Series for Large x

$$J_n(x) \approx \sqrt{\frac{2}{\pi x}}\left[\cos\psi\left(1 - \frac{\alpha_1\,\alpha_3}{2!} + \frac{\alpha_1\,\alpha_3\,\alpha_5\,\alpha_7}{4!} - ....\right) - \sin\psi\left(\frac{\alpha_1}{1!} - \frac{\alpha_1\,\alpha_3\,\alpha_5}{3!}\right)\right]$$

$$\psi = x - \frac{n\pi}{2} - \frac{\pi}{4} \qquad \alpha_k = \frac{4n^2 - k^2}{8x} \qquad k = 1, 2..... \qquad x > 15$$

15.27 REPRESENTATION OF $Y_n(x)$

i. Asymptotic Approximation

$$Y_n(x) \approx \sqrt{\frac{2}{\pi x}} \sin\left(x - \frac{n\pi}{2} - \frac{\pi}{4}\right), \qquad x > 25$$

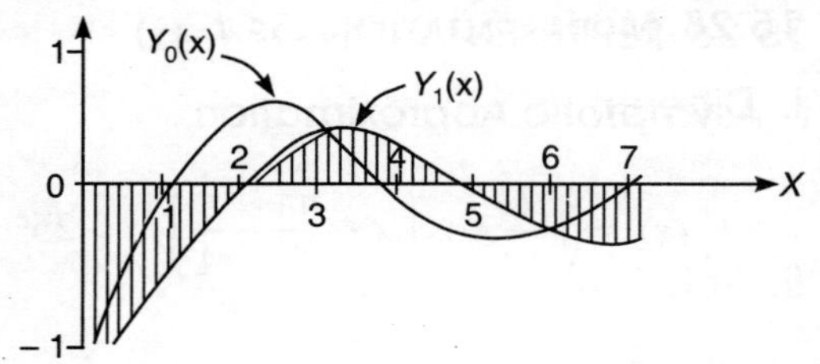

Fig. 15.2

ii. Numerical Values

$Y_0(x) \qquad x = 0 - 10$

x	0	1	2	3	4	5	6	7	8	9	x
0.	– ∞	– 1.5342	– 1.0811	– 0.8273	– 0.6060	– 0.4445	– 0.3085	– 0.1907	– 0.0688	0.0056	0.
1.	0.0883	0.1622	0.2281	0.2865	0.3379	0.3824	0.4204	0.4520	0.4774	0.4968	1.
2.	0.5104	0.5183	0.5208	0.5181	0.5104	0.4981	0.4813	0.4605	0.4359	0.4079	2.
3.	0.3769	0.3431	0.3071	0.2691	0.2296	0.1890	0.1477	0.1061	0.0645	0.0234	3.
4.	– 0.0169	– 0.0561	– 0.1296	– 0.1296	– 0.1633	– 0.1947	– 0.2235	– 0.2494	– 0.2723	– 0.2921	4.
5.	– 0.3085	– 0.3216	– 0.3313	– 0.3374	– 0.3402	– 0.3395	– 0.3354	– 0.3282	– 0.3177	– 0.3044	5.
6.	– 0.2882	– 0.2694	– 0.2483	– 0.2251	– 0.1999	– 0.1732	– 0.1452	– 0.1162	– 0.0864	– 0.0563	6.
7.	– 0.0259	0.0042	0.0339	0.0628	0.0907	0.1173	0.1424	0.1658	0.1872	0.2065	7.
8.	0.2235	0.2381	0.2501	0.2595	0.2662	0.2702	0.2715	0.2700	0.2659	0.0804	8.
9.	0.2499	0.2383	0.2245	0.2086	0.1907	0.1712	0.1502	0.1279	0.1045	0.0804	9.
10.	0.0557										

iii. Numerical Values

$Y_1(x) \qquad x = 0 - 10$

x	0	1	2	3	4	5	6	7	8	9	x
0.	– ∞	– 0.4590	– 3.3328	– 2.2931	– 1.7809	– 1.4715	– 1.2604	– 1.1032	– 0.9781	– 0.8731	0.
1.	– 0.7812	– 0.6981	– 0.6211	– 0.5486	– 0.4791	– 0.4123	– 0.3476	– 0.2847	– 0.2237	– 0.1644	1.
2.	– 0.1070	– 0.0517	0.0015	0.0523	0.1005	0.1459	0.1884	0.2276	0.2635	0.2959	2.
3.	0.3247	0.3496	0.3707	0.3879	0.4010	0.4102	0.4154	0.4167	0.4141	0.4078	3.
4.	0.3979	0.3846	0.3680	0.3484	0.3260	0.3010	0.2737	0.2445	0.2136	0.1812	4.
5.	0.1479	0.1137	0.0792	0.0445	0.0101	– 0.0238	– 0.0568	– 0.0887	– 0.1192	– 0.1481	5.
6.	– 0.1750	– 0.1998	– 0.2223	– 0.2422	– 0.2596	– 0.2741	– 0.2857	– 0.2945	– 0.3002	– 0.3029	6.
7.	– 0.3027	– 0.2995	– 0.2934	– 0.2846	– 0.2731	– 0.2591	– 0.2428	– 0.2243	– 0.2039	– 0.1817	7.
8.	– 0.1581	– 0.1331	– 0.1072	– 0.0806	– 0.0535	– 0.0262	0.0011	– 0.0280	0.0544	0.0799	8.
9.	0.1043	0.1275	0.1491	0.1691	0.1871	0.2032	0.2171	0.2287	0.2379	0.2447	9.
10.	0.2490										

iv. Asymptotic Series for Large x

$$Y_n(x) \approx \sqrt{\frac{2}{\pi x}} \left[\sin\psi\left(1 - \frac{\alpha_1\alpha_3}{2!} + \frac{\alpha_1\alpha_3\alpha_5\alpha_7}{4!} - \ldots\right) + \cos\psi\left(\frac{\alpha_1}{1} - \frac{\alpha_1\,\alpha_3\,\alpha_5}{3!} + \ldots\right)\right]$$

$$\psi = x - \frac{n\pi}{2} - \frac{\pi}{4} \qquad \alpha_k = \frac{4n^2 - k^2}{8x} \qquad k = 1, 2, \ldots \qquad x > 15$$

15.28 MODIFIED BESSEL'S DIFFERENTIAL EQUATION

i. Differential Equation

$$x^2y'' + xy' - (x^2 + n^2)\,y = 0$$

ii. Solution

$y = A_1 I_n(x) + A_2 I_{-n}(x) \qquad n \neq 0, 1, 2, \ldots \qquad\qquad y = A_1 I_n(x) + A_2 K_n(x) \ldots$ all n

iii. Modified Bessel Functions of the First Kind of Order n

$$I_n(x) = i^{-n} J_n(ix) = \sum_{k=0}^{\infty} \frac{(x/2)^{2k+n}}{k!\,\Gamma(k+1+n)}$$

$$= \frac{x^n}{2^n\Gamma(1+n)}\left[1 + \frac{x^2}{2(2+2n)} + \frac{x^4}{(2)(4)(2+2n)(4+2n)} + \ldots\right]$$

$$I_{-n}(x) = i^n J_{-n}(ix) = \sum_{k=0}^{\infty} \frac{(x/2)^{2k-n}}{k!\,\Gamma(k+1-n)}$$

$$= \frac{x^{-n}}{2^{-n}\Gamma(1-n)}\left[1 + \frac{x^2}{2(2-2n)} + \frac{x^4}{(2)(4)(2-2n)(4-2n)} + \ldots\right]$$

$$I_{-n}(x) = I_n(x) \qquad n = 0, 1, 2, \ldots$$

$$I_0(x) = 1 + \frac{(x/2)^2}{(1!)^2} + \frac{(x/2)^4}{(2!)^2} + \frac{(x/2)^6}{(3!)^2} + \ldots.$$

$$I_1(x) = \frac{x}{2}\left[1 + \frac{(x/2)^2}{2(1!)^2} + \frac{(x/2)^4}{3(2!)^2} + \frac{(x/2)^6}{4(3!)^2} + \ldots\right] = \frac{d}{dx}[I_0(x)]$$

iv. Modified Bessel Functions of the Second Kind of Order n

$$K_n(x) = \begin{cases} \dfrac{\pi}{2}\,\dfrac{I_{-n}(x) - I_n(x)}{\sin n\pi} & n \neq 0, 1, 2, \ldots. \\[2ex] \lim\limits_{p\to n} \dfrac{\pi}{2}\,\dfrac{I_{-p}(x) - I_p(x)}{\sin p\pi} & n = 0, 1, 2, \ldots \end{cases}$$

$$K_{-n}(x) = K_n(x) \qquad n = 0, 1, 2, \ldots$$

$$K_0(x) = -\left[\left(I_n\frac{x}{2} + C\right)I_0(x) - \frac{2}{1}I_2(x) - \frac{2}{2}I_4(x) - \frac{2}{3}I_6(x) - \ldots\right]$$

$$-K_1(x) = -\left[\left(I_n\frac{x}{2} + C\right)I_1(x) - \frac{1}{x} - \frac{1}{2}I_1(x) - \frac{9}{4}I_3(x) - \ldots\right] = \frac{d}{dx}[K_0(x)]$$

where $C = 0.57721\ldots.$

15.29 PROPERTIES OF MODIFIED BESSEL FUNCTIONS

i. Recurrence Relations

$$I_{n+1}(x) = -\frac{2n}{x} I_n(x) + I_{n-1}(x)$$

$$= -\frac{n}{x} I_n(x) + \frac{d}{dx}[I_n(x)]$$

$$= x^n \frac{d}{dx}[x^{-n} I_n(x)]$$

$$I_{n-1}(x) = \frac{2n}{x} I_n(x) + I_{n+1}(x)$$

$$= \frac{n}{x} I_n(x) + \frac{d}{dx}[I_n(x)]$$

$$= x^{-n} \frac{d}{dx}[x^n I_n(x)]$$

$$\frac{d}{dx}[I_n(x)] = \frac{1}{2}[I_{n-1}(x) + I_{n-1}(x)]$$

$$\int x^{n+1} I_n(x)\, dx = x^{n+1} I_{n+1}(x)$$

$$K_{n+1}(x) = \frac{2n}{x} K_n(x) + K_{n-1}(x)$$

$$= \frac{n}{x} K_n(x) - \frac{d}{dx}[K_n(x)]$$

$$= -x^n \frac{d}{dx}[x^{-n} K_n(x)]$$

$$K_{n-1} = -\frac{2n}{x} K_n(x) + K_{n+1}(x)$$

$$= -\frac{n}{x} K_n(x) - \frac{d}{dx}[K_n(x)]$$

$$= -x^{-n} \frac{d}{dx}[x^n K_n(x)]$$

$$\frac{d}{dx}[K_n(x)] = -\frac{1}{2}[K_{n+1}(x) + K_{n-1}(x)]$$

$$\int x^{n+1} K_n(x)\, dx = -x^{n+1} K_{n+1}(x)$$

ii. Half-odd Integers

$$I_{1/2}(x) = \sqrt{\frac{2}{\pi x}} \sinh x$$

$$I_{3/2}(x) = -\sqrt{\frac{2}{\pi x}}\left(\frac{\sinh x}{x} - \cosh x\right)$$

$$I_{5/2}(x) = \sqrt{\frac{2}{\pi x}}\left[\left(\frac{3}{x^2} + 1\right)\sinh x - \frac{3}{x}\cosh x\right]$$

$$I_{-1/2}(x) = \sqrt{\frac{2}{\pi x}} \cosh x$$

$$I_{-3/2}(x) = -\sqrt{\frac{2}{\pi x}}\left(\frac{\cosh x}{x} - \sinh x\right)$$

$$I_{-5/2}(x) = \sqrt{\frac{2}{\pi x}}\left[\left(\frac{3}{x^2} + 1\right)\cosh x - \frac{3}{x}\sinh x\right]$$

$$K_{1/2}(x) = e^{-x}\sqrt{\frac{\pi}{2x}}$$

$$K_{3/2}(x) = e^{-x}\sqrt{\frac{\pi}{2x}}\left(\frac{1}{x} + 1\right)$$

$$K_{5/2}(x) = e^{-x}\sqrt{\frac{\pi}{2x}}\left(\frac{2}{x^2} + \frac{2}{x} + 1\right)$$

$$K_{-1/2}(x) = e^{-x}\sqrt{\frac{\pi}{2x}}$$

$$K_{-3/2}(x) = e^{-x}\sqrt{\frac{\pi}{2x}}\left(\frac{1}{x} + 1\right)$$

$$K_{-5/2}(x) = e^{-x}\sqrt{\frac{\pi}{2x}}\left(\frac{2}{x^2} + \frac{2}{x} + 1\right)$$

iii. Transformations ($n = p \pm 1/2, p = 0, 1, 2,....$)

$$J_{p+1/2}(x) = (-1)^p Y_{-p-1/2}(x) = (-1)^p \sqrt{\frac{2}{\pi x}}\, x^{p+1} \left(\frac{1}{x}\frac{d}{dx}\right)^p \frac{\sin x}{x}$$

$$Y_{p+1/2}(x) = -(-1)^p J_{-p-1/2}(x) = -(-1)^p \sqrt{\frac{2}{\pi x}}\, x^{p+1} \left(\frac{1}{x}\frac{d}{dx}\right)^p \frac{\cos x}{x}$$

$$I_{p+1/2}(x) = \sqrt{\frac{2}{\pi x}}\, x^{p+1} \left(\frac{1}{x}\frac{d}{dx}\right)^p \frac{\sinh x}{x}$$

$$K_{p+1/2}(x) = K_{-p-1/2}(x) = (-1)^p \sqrt{\frac{2}{\pi x}}\, x^{p+1} \left(\frac{1}{x}\frac{d}{dx}\right)^p \frac{e^{-x}}{x}$$

15.30 REPRESENTATION OF $I_n(x)$

i. Asymptotic Approximation

$$I_n(x) = \sqrt{\frac{1}{2\pi x}}\, e^x \qquad x > 25$$

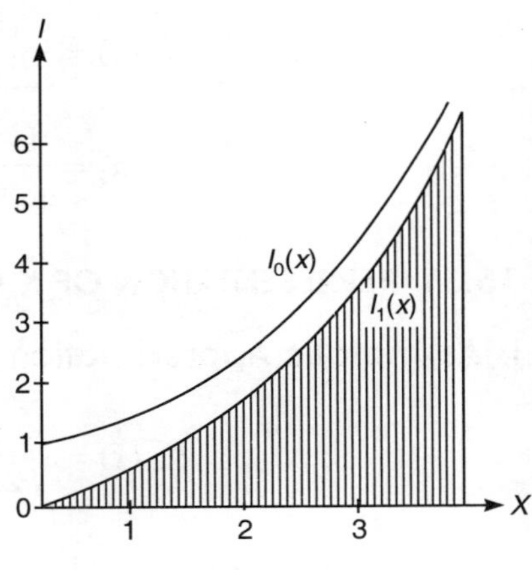

Fig. 15.3

ii. Numerical Values $I_0(x)$ $x = 0 - 10$

x	0	1	2	3	4	5	6	7	8	9	x
0.	1.000	1.003	1.010	1.023	1.040	1.063	1.092	1.126	1.167	1.213	0.
1.	1.266	1.326	1.394	1.469	1.553	1.647	1.750	1.864	1.990	2.128	1.
2.	2.280	2.446	2.629	2.830	3.049	3.290	3.553	3.842	4.157	4.503	2.
3.	4.881	5.294	5.747	6.243	6.785	7.378	8.028	8.739	9.517	10.37	3.
4.	11.30	12.32	13.44	14.67	16.01	17.48	19.09	20.86	22.79	24.91	4.
5.	27.24	29.79	32.58	35.65	39.01	42.69	46.74	51.17	56.04	61.38	5.
6.	67.23	73.66	80.72	88.46	96.96	106.3	116.5	127.8	140.1	153.7	6.
7.	168.6	185.0	202.9	222.7	244.3	268.3	294.3	323.1	354.7	389.4	7.
8.	427.6	469.5	515.6	566.3	621.9	683.2	750.2	824.4	905.8	995.2	8.
9.	1,094	1,202	1,321	1,451	1,595	1,753	1,927	2,119	2,329	2,561	9.
10	2,816										

iii. Numerical Values $I_1(x)$ $x = 0 - 10$

x	0	1	2	3	4	5	6	7	8	9	x
0.	0.000	1.0501	0.1005	0.1517	0.2040	0.2579	0.3137	0.3719	0.4329	0.4971	0.
1.	0.5656	0.6375	0.7147	0.7973	0.8861	1.085	1.085	1.196	1.317	1.448	1.
2.	1.591	1.745	1.914	2.098	2.298	2.755	2.755	3.016	3.301	3.613	2.
3.	3.953	4.326	4.734	5.181	5.670	6.206	6.793	7.436	8.140	8.913	3.
4.	9.759	10.69	11.71	12.82	14.05	15.39	16.86	18.48	20.25	22.20	4.
5.	24.34	25.68	29.25	32.08	35.18	38.59	42.33	46.44	50.95	55.90	5.
6.	61.34	67.32	73.89	81.10	89.03	97.74	107.3	117.8	129.4	142.1	6.
7.	156.0	171.4	188.3	206.8	227.2	249.6	274.2	301.3	381.1	363.9	7.
8.	399.9	439.5	483.0	531.0	583.7	641.6	705.4	775.5	852.7	937.5	8.
9.	1,031	1,134	1,247	1,371	1,508	1,658	1,824	2,006	2,207	2,428	9.
10	2,671										

iv. Asymptotic Series for Large x

$$I_n(x) \approx \frac{e^x}{\sqrt{2\pi x}}\left(1 - \frac{\alpha_1}{1!} + \frac{\alpha_1\alpha_3}{2!} - ....\right)$$

$$\alpha_k = \frac{4n^2 - k^2}{8x} \qquad k = 1, 2,.... \qquad x > 15$$

15.31 REPRESENTATION OF $K_n(x)$

i. Asymptotic Approximation

$$K_n(x) = \sqrt{\frac{\pi}{2x}}\, e^{-x} \qquad x > 25$$

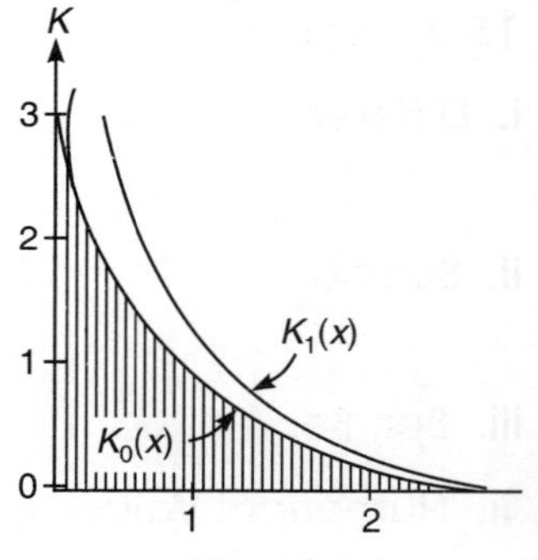

Fig. 15.4

ii. Numerical Values $K_0(x)$ $x = 0 - 10$

x	0	1	2	3	4	5	6	7	8	9	x
0.	∞	2.4271	1.7527	1.3725	1.1145	0.9244	0.7775	0.6605	0.5653	04867	0.
1.	0.4210	0.3656	0.3185	0.2782	0.2437	0.2138	0.1880	0.1655	0.1459	0.1288	1.
2.	0.1139	0.1008	0.08927	0.07914	0.07022	0.06235	0.05540	0.04926	0.04382	0.03901	2.
3.	0.03474	0.03095	0.02759	0.02461	0.02196	0.01960	0.01750	0.01563	0.01397	0.01248	3.
4.	0.01116	$0.0^{2}9980$	$0.0^{2}8927$	$0.0^{2}7988$	$0.0^{2}7149$	$0.0^{2}6400$	$0.0^{2}5730$	$0.0^{2}5132$	$0.0^{2}4597$	$0.0^{2}4119$	4.
5.	$0.0^{2}3691$	$0.0^{2}3308$	$0.0^{2}2966$	$0.0^{2}2659$	$0.0^{2}2385$	$0.0^{2}2139$	$0.0^{2}1918$	$0.0^{2}1721$	$0.0^{2}1544$	$0.0^{2}1386$	5.
6.	$0.0^{2}1244$	$0.0^{2}1117$	$0.0^{2}1003$	$0.0^{3}9001$	$0.0^{3}8083$	$0.0^{3}7259$	$0.0^{3}6520$	$0.0^{3}5857$	$0.0^{3}5262$	$0.0^{3}4728$	6.
7.	$0.0^{3}4248$	$0.0^{3}3817$	$0.0^{3}3431$	$0.0^{3}3084$	$0.0^{3}2772$	$0.0^{3}2492$	$0.0^{3}2240$	$0.0^{3}2014$	$0.0^{3}1811$	$0.0^{3}1629$	7.
8.	$0.0^{3}1465$	$0.0^{3}1317$	$0.0^{3}1185$	$0.0^{3}1066$	$0.0^{4}9588$	$0.0^{4}8626$	$0.0^{4}7761$	$0.0^{4}6983$	$0.0^{4}6983$	$0.0^{4}5654$	8.
9.	$0.0^{4}5088$	$0.0^{4}4579$	$0.0^{4}4121$	$0.0^{4}3710$	$0.0^{4}3339$	$0.0^{4}3006$	$0.0^{4}2706$	$0.0^{4}2436$	$0.0^{4}2193$	$0.0^{4}1975$	9.
10	$0.0^{4}1778$										

iii. Numerical Values $K_1(x)$ $x = 0 - 10$

x	0	1	2	3	4	5	6	7	8	9	x
0.	∞	9.8538	4.7760	3.0560	2.1844	1.6564	1.3028	1.0503	0.8619	0.7165	0.
1.	0.6019	0.5095	0.4346	0.3725	0.3208	0.2774	0.2406	0.2094	0.1826	0.1597	1.
2.	0.1399	0.1227	0.1079	0.09498	0.08372	0.07389	0.06528	0.05774	0.05111	0.044529	2.
3.	0.04016	0.03563	0.029938	0.02812	0.02500	0.02224	0.01979	0.01763	0.01571	0.01400	3.
4.	0.01248	0.01114	$0.0^{2}9938$	$0.0^{2}8872$	$0.0^{2}7983$	0.0^{7078}	$0.0^{2}6325$	$0.0^{4}5654$	$0.0^{4}5055$	$0.0^{2}4521$	4.
5.	$0.0^{2}4045$	$0.0^{2}3619$	$0.0^{2}3239$	$0.0^{2}2900$	$0.0^{2}2597$	$0.0^{2}2326$	$0.0^{2}2083$	$0.0^{2}1866$	$0.0^{2}1673$	$0.0^{2}1499$	5.
6.	$0.0^{2}1344$	$0.0^{2}1205$	$0.0^{2}1081$	$0.0^{3}9691$	$0.0^{3}8693$	$0.0^{3}7799$	$0.0^{2}2083$	$0.0^{3}6280$	$0.0^{3}5636$	$0.0^{3}5059$	6.
7.	$0.0^{3}4542$	$0.0^{3}4078$	$0.0^{3}3662$	$0.0^{3}3288$	$0.0^{3}2953$	$0.0^{3}2653$	$0.0^{3}6998$	$0.0^{3}2141$	$0.0^{3}1924$	$0.0^{3}1729$	7.
8.	$0.0^{3}1554$	$0.0^{3}1396$	$0.0^{3}1255$	$0.0^{3}1128$	$0.0^{3}1014$	$0.0^{4}9120$	$0.0^{3}2383$	$0.0^{4}7374$	$0.0^{4}6631$	$0.0^{4}5964$	8.
9.	$0.0^{4}5364$	$0.0^{4}4825$	$0.0^{4}4340$	$0.0^{4}3904$	$0.0^{4}3512$	$0.0^{4}3160$	$0.0^{4}2843$	$0.0^{4}2559$	$0.0^{4}2302$	$0.0^{4}2072$	9.
10	$0.0^{4}1865$										

iv. Asymptotic Series for Large x

$$K_n(x) \approx \sqrt{\frac{\pi}{2x}}\, e^{-x}\left(1 + \frac{\alpha_1}{1!} + \frac{\alpha_1\alpha_3}{2!} + ...\right)$$

$$\alpha_k = \frac{4n^2 - k^2}{8x} \qquad k = 1, 2,.... \qquad x > 15$$

15.32 Ber, Bei, Ker, Kei DIFFERENTIAL EQUATION

i. Differential Equation

$$x^2y'' + xy' \pm ia^2x^2y = 0$$

ii. Solution

$$y = A_1\,[\text{Ber}\,(ax) \mp i\,\text{Bei}\,(ax)] + A_2\,[\text{Ker}\,(ax) \mp i\,\text{Kei}\,(ax)]$$

iii. Ber, Bei Functions

$$\text{Ber}\,(ax) = 1 - \frac{(ax/2)^4}{(2!)^2} + \frac{(ax/2)^8}{(4!)^2} - ....$$

$$\text{Bei}\,(ax) = \frac{(ax/2)^2}{(1!)^2} - \frac{(ax/2)^6}{(3!)^2} + \frac{(ax/2)^{10}}{(5!)^2} - ...$$

iv. Ker, Kei Function

$$\text{Ker}\,(ax) = -\left(I_n\,\frac{ax}{2} + C\right)\text{Ber}\,(ax) + \frac{\pi}{4}\,\text{Bei}\,(ax) - \lambda_2 + \lambda_4 - ....$$

$$\text{Kei}\,(ax) = -\left(I_n\,\frac{ax}{2} + C\right)\text{Bei}\,(ax) - \frac{\pi}{4}\,\text{Ber}\,(ax) + \lambda_1 + \lambda_3 - ....$$

where $C = 0.57721....$ $\qquad \lambda_k = \frac{(ax/2)^{2k}}{(k!)^2}\left(1 + \frac{1}{2} + \frac{1}{3} + ... + \frac{1}{k}\right)$

v. Ber′, Bei′ Functions

$$\text{Ber}'(ax) = -\frac{2a(ax/2)^3}{(2!)^2} + \frac{4a(ax/2)^7}{(4!)^2} - \ldots\ldots = \frac{d}{dx}[\text{Ber}(ax)]$$

$$\text{Bei}'(ax) = \frac{a(ax/2)}{(1!)^2} - \frac{3a(ax/2)^5}{(3!)^2} + \ldots = \frac{d}{dx}[\text{Bei}(ax)]$$

vi. Ker′, Kei′ Functions

$$\text{Ker}'(ax) = -\left(I_n\frac{ax}{2} + C\right)\text{Ber}'(ax) - \frac{1}{x}\text{Ber}(ax) + \frac{\pi}{4}\text{Bei}'(ax) - \lambda'_2 + \lambda'_4 - \ldots\ldots = \frac{d}{dx}[\text{Ker}(ax)]$$

$$\text{Kei}'(ax) = -\left(I_n\frac{ax}{2} + C\right)\text{Bei}'(ax) - \frac{1}{x}\text{Bei}(ax) - \frac{\pi}{4}\text{Ber}'(ax) + \lambda'_1 - \lambda'_3 + \ldots\ldots = \frac{d}{dx}[\text{Kei}(ax)]$$

where $\lambda'_k = \dfrac{d\lambda_k}{dx}$

vii. Relations

$\text{Ber}(x) + i\,\text{Bei}(x) = J_0\left(xi\sqrt{i}\right) = I_0\left(x\sqrt{i}\right)$

$\text{Ker}(x) + i\,\text{Kei}(x) = K_0\left(x\sqrt{i}\right)$

$$\int x\,\text{Ber}(x)\,dx = x\,\text{Bei}'(x) \qquad \int x\,\text{Bei}(x)\,dx = -x\,\text{Ber}'(x)$$

$$\int x\,\text{Ker}(x)\,dx = x\,\text{Kei}'(x) \qquad \int x\,\text{Kei}(x)\,dx = -x\,\text{Ker}'(x)$$

15.33 Ber$_n$, Bei$_n$, Ker$_n$, Kei$_n$ DIFFERENTIAL EQUATION

i. Differential Equation

$x^2y'' + xy' \pm (ix^2 + n^2)y = 0$

ii. Solution

$y = A_1[\text{Ber}_n(x) \mp i\,\text{Bei}_n(x)] + A_2[\text{Ker}_n(x) \mp i\,\text{Kei}_n(x)]$

iii. Ber$_n$, Bei$_n$ Functions

$$\text{Ber}_n(x) = \sum_{k=0}^{\infty}\frac{(-1)^{n+k}(x/2)^{n+2k}}{k!(n+k)!}\cos\frac{(n+2k)\pi}{4}$$

$$\text{Bei}_n(x) = \sum_{k=0}^{\infty}\frac{(-1)^{n+k+1}(x/2)^{n+2k}}{k!(n+k)!}\sin\frac{(n+2k)\pi}{4} \qquad n = 0, 1, 2, \ldots. \qquad k = 0, 1, 2, \ldots.$$

iv. Ker_n, Kei_n Functions

$$\text{Ker}_n(x) = -\left(I_n \frac{x}{2} + C\right)\text{Ber}_n(x) + \frac{\pi}{4}\text{Bei}_n(x) + \frac{1}{2}\sum_{k=0}^{n-1}\frac{(n-k-1)!(x/2)^{2k-n}}{k!}\cos\frac{(3n+2k)\pi}{4}$$

$$+\frac{1}{2}\sum_{k=0}^{\infty}\frac{(x/2)^{2k+n}}{k!(n+k)!}(\tau_k + \tau_{k+n})\cos\frac{(3n+2k)\pi}{4}$$

$$\text{Kei}_n(x) = -\left(I_n \frac{x}{2} + C\right)\text{Bei}_n(x) - \frac{\pi}{4}\text{Ber}_n(x) - \frac{1}{2}\sum_{k=0}^{n-1}\frac{(n-k-1)!(x/2)^{2k-n}}{k!}\sin\frac{(3n+2k)\pi}{4}$$

$$+\frac{1}{2}\sum_{k=0}^{\infty}\frac{(x/2)^{2k+n}(\tau_k + \tau_{k+n})}{k!(n+k)!}\sin\frac{(3n+2k)\pi}{4}$$

where $\tau_k = 1 + \frac{1}{2} + \frac{1}{3} + \ldots\ldots + \frac{1}{k}$ $\qquad t_{k+n} = 1 + \frac{1}{2} + \frac{1}{3} + \ldots\ldots + \frac{1}{k+n}$

$C = 0.57721 \ldots.$ $\qquad n = 0, 1, 2, \ldots.$ $\qquad k = 0, 1, 2, \ldots$

v. Relations

$$\text{Ber}_n(x) + i\,\text{Bei}_n(x) = J_n(xi\sqrt{i}) = i^n I_n(x\sqrt{i})$$

$$\text{Ber}_n(x) + \text{Bei}_n(x) = -\frac{x}{n\sqrt{2}}[\text{Bei}_{n-1}(x) + \text{Bei}_{n+1}(x)]$$

$$\text{Ber}_n(x) - \text{Bei}_n(x) = -\frac{x}{n\sqrt{2}}[\text{Ber}_{n-1}(x) + \text{Ber}_{n+1}(x)]$$

$$\text{Ker}_x(x) + i\,\text{Kei}_n(x) = i^{-n}\,\text{K}_n(x\sqrt{i})$$

$$\text{Ker}_n(x) + \text{Kei}_n(x) = -\frac{x}{n\sqrt{2}}[\text{Kei}_{n-1}(x) + \text{Kei}_{n+1}(x)]$$

$$\text{Ker}_n(x) - \text{Kei}_n(x) = -\frac{x}{n\sqrt{2}}[\text{Ker}_{n-1}(x) + \text{Kei}_{n+1}(x)]$$

15.34 REPRESENTATION OF Ber (x) AND Bei (x)

i. Asymptotic Approximation

$$\text{Ber}(x) \approx \frac{e^{x/\sqrt{2}}}{\sqrt{2\pi x}}\cos\left(\frac{x}{\sqrt{2}} - \frac{\pi}{8}\right)$$

$$\text{Bei}(x) \approx \frac{e^{x/\sqrt{2}}}{\sqrt{2\pi x}}\sin\left(\frac{x}{\sqrt{2}} - \frac{\pi}{8}\right), \qquad x > 25$$

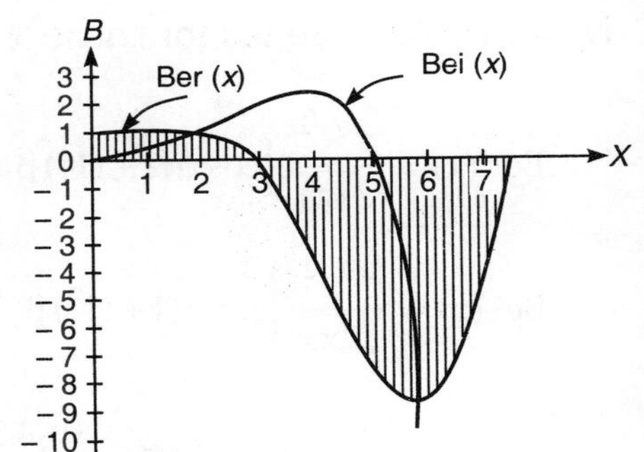

Fig. 15.5

ii. Numerical Values Ber (x) $x = 0 - 10$

x	0	1	2	3	4	5	6	7	8	9	x
0.	1.0000	1.0000	1.0000	0.9999	0.9996	0.9990	0.9980	0.9962	0.9936	0.9890	0.
1.	0.9844	0.9771	0.9676	0.9554	0.9401	0.9211	0.8979	0.8700	0.8367	0.7975	1.
2.	0.7517	0.6987	0.6377	0.5680	0.4890	0.4000	0.3001	0.1887	0.0651	– 0.0714	2.
3.	– 0.2214	– 0.3855	– 0.5644	– 0.7584	– 0.9680	– 1.1936	– 1.4353	– 1.6933	– 1.9674	– 2.2576	3.
4.	– 2.5364	– 2.8843	– 3.3295	– 3.5679	– 3.9283	– 4.2991	– 4.6784	– 5.0639	– 5.4531	– 5.8429	4.
5.	– 6.2301	– 6.6107	– 6.9803	– 7.3344	– 7.6674	– 7.9736	– 8.2466	– 8.4794	– 8.6644	– 8.7937	5.
6.	– 8.8583	– 8.8491	– 8.7561	– 8.5688	– 8.2762	– 7.8669	– 7.3287	– 6.6492	– 5.8155	– 4.8146	6.
7.	– 3.6329	– 2.2571	– 0.6737	1.13.8	3.1695	5.4550	7.9994	10.814	13.909	17.293	7.
8.	20.974	24.957	29.245	33.840	38.738	43.936	49.423	55.187	61.210	67.469	8.
9.	73.936	80.576	87.350	94.208	101.10	107.95	114.70	121.26	127.54	133.43	9.
10.	138.84										

iii. Numerical Values Bei (x) $x = 0 - 10$

x	0	1	2	3	4	5	6	7	8	9	x
0.	0.0000	0.022500	0.01000	0.022500	0.04000	0.06249	0.08998	0.1224	0.1599	0.2023	0.
1.	0.2469	0.3017	0.3587	0.3017	0.4867	0.5576	0.6327	0.7120	0.7953	0.8821	1.
2.	0.9723	1.0654	1.1610	1.0654	1.3575	1.4572	1.5569	1.6557	1.7529	1.8472	2.
3.	1.9376	2.0228	2.1016	2.0228	2.2324	2.2832	2.3199	2.3413	2.3454	2.3300	3.
4.	2.2927	2.2309	2.1422	2.2039	1.8726	1.6860	1.4610	1.1946	0.8837	0.5251	4.
5.	0.1160	– 0.3467	– 0.8658	– 0.3467	– 2.0845	– 2.7890	– 3.5597	– 4.3986	– 5.3068	– 6.2854	5.
6.	– 7.3347	– 8.4545	– 9.6437	– 8.4545	– 12.223	– 13.607	– 15.047	– 16.538	– 18.074	– 19.644	6.
7.	– 21.239	– 22.848	– 24.456	– 22.848	– 27.609	– 29.116	– 30.548	– 30.882	– 33.092	– 34.147	7.
8.	– 35.017	– 35.667	– 36.061	– 35.667	– 35.920	– 35.298	– 34.246	– 32.714	– 30.651	– 28.003	8.
9.	– 24.713	– 20.724	– 15.976	– 10.412	– 3.9693	3.4106	11.787	21.218	31.758	43.459	9.
10.	56.371										

iv. Asymptotic Series for Large x

$$\text{Ber}\,(x) \approx \frac{e^{x/\sqrt{2}}}{\sqrt{2\pi x}}\left\{\cos\phi\,[1+(1^2)\,\beta_1+(1^2)\,(3^2)\beta_2+....]+\sin\phi\,[(1^2)\gamma_1+(1^2)\,(3^2)\gamma_2+....]\right\}$$

$$\text{Bei}\,(x) \approx \frac{e^{x/\sqrt{2}}}{\sqrt{2\pi x}}\left\{\sin\phi\,[1+(1^2)\,\beta_1+(1^2)\,(3^2)\beta_2+....]-\cos\phi\,[(1^2)\gamma_1+(1^2)\,(3^2)\gamma_2+....]\right\}$$

$$\phi = \frac{x}{\sqrt{2}} - \frac{\pi}{8} \qquad \beta_k = \frac{\cos(k\pi/4)}{k!\,(8x)^k} \qquad \gamma_k = \frac{\sin(k\pi/4)}{k!\,(8x)^k} \qquad k = 1, 2, ... \; x > 15$$

15.35 REPRESENTATION OF Ker (x) AND Kei (x)

i. Asymptotic Approximation

$$\text{Ker}(x) \approx \sqrt{\frac{\pi}{2x}}\, e^{-x/\sqrt{2}} \cos\left(\frac{x}{\sqrt{2}} + \frac{\pi}{8}\right)$$

$$\ldots x > 25$$

$$\text{Kei}(x) \approx -\sqrt{\frac{\pi}{2x}}\, e^{-x/\sqrt{2}} \sin\left(\frac{x}{\sqrt{2}} + \frac{\pi}{8}\right)$$

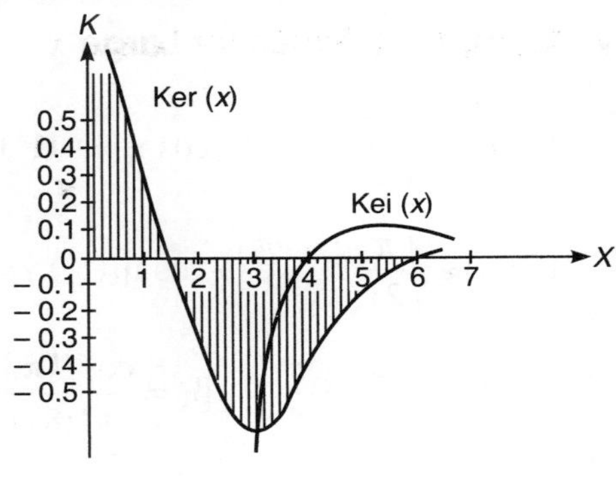

Fig. 15.6

ii. Numerical Values Ker (x) $x = 0 - 10$

x	0	1	2	3	4	5	6	7	8	9	x
0.	∞	2.4205	1.7331	1.3372	1.0626	0.8559	0.6931	0.5614	0.4529	0.3625	0.
1.	0.2867	0.2228	0.1689	0.1235	0.08513	0.05293	0.02603	0.023691	– 0.01470	– 0.02966	1.
2.	– 0.04166	– 0.05111	– 0.05834	– 0.06367	– 0.06737	– 0.06969	– 0.07083	– 0.07097	– 0.07030	– 0.06894	2.
3.	– 0.06703	– 0.06468	– 0.06198	– 0.05903	– 0.05590	– 0.05264	– 0.04932	– 0.04597	– 0.04265	– 0.03937	3.
4.	– 0.03618	– 0.03308	– 0.03011	– 0.02726	– 0.02456	– 0.02200	– 0.01960	– 0.01734	– 0.01525	– 0.01330	4.
5.	– 0.01151	$-0.0^{2}9865$	$-0.0^{2}8359$	$-0.0^{2}6989$	$-0.0^{2}5749$	$-0.0^{2}4632$	$-0.0^{2}3632$	$-0.0^{2}2740$	$-0.0^{2}1952$	$-0.0^{2}1258$	5.
6.	$-0.0^{2}6530$	$-0.0^{3}1295$	$-0.0^{3}3191$	$-0.0^{3}6991$	$-0.0^{2}1017$	$-0.0^{2}1278$	$-0.0^{2}1488$	$-0.0^{2}1653$	$-0.0^{2}1777$	$-0.0^{2}1866$	6.
7.	$0.0^{2}1922$	$0.0^{2}1951$	$0.0^{2}1956$	$0.0^{2}1940$	$0.0^{2}1907$	$0.0^{2}1860$	$0.0^{2}1800$	$0.0^{2}1731$	$0.0^{2}1655$	$0.0^{2}1572$	7.
8.	$0.0^{2}1486$	$0.0^{2}1397$	$0.0^{2}1306$	$0.0^{2}1216$	$0.0^{2}1126$	$0.0^{2}1037$	$0.0^{3}9511$	$0.0^{3}8675$	$0.0^{3}7871$	$0.0^{3}7102$	8.
9.	$0.0^{3}6372$	$0.0^{3}5681$	0.035030	$0.0^{3}4422$	$0.0^{3}3855$	$0.0^{3}3330$	$0.0^{3}9511$	$0.0^{3}8675$	$0.0^{3}7871$	$0.0^{3}1628$	9.
10.	$0.0^{3}1295$										

iii. Numerical Values Kei (x) $x = 0 - 10$

x	0	1	2	3	4	5	6	7	8	9	x
0.	– 0.7854	– 0.7769	– 0.7581	– 0.7331	– 0.7038	– 0.6716	– 0.6374	– 0.6022	– 0.5664	– 0.5305	0.
1.	– 0.4950	– 0.4601	– 0.4262	– 0.3933	– 0.3617	– 0.3314	– 0.3026	– 0.2752	– 0.2494	– 0.2551	1.
2.	– 0.2024	– 0.1812	– 0.1614	– 0.1431	– 0.1262	– 0.1107	– 0.09644	– 0.08342	– 0.07157	– 0.06083	2.
3.	– 0.05112	– 0.04240	– 0.03458	– 0.02762	– 0.02145	– 0.01600	– 0.01123	$0.0^{2}7077$	$-0.0^{2}3487$	$0.0^{3}4108$	3.
4.	$0.0^{2}2198$	$0.0^{2}4386$	$0.0^{2}6194$	$0.0^{2}7661$	$0.0^{2}8826$	$0.0^{2}9721$	0.01038	0.01083	0.01110	0.01121	4.
5.	0.01119	0.01105	0.01082	0.01051	0.01014	$0.0^{2}9716$	$0.0^{2}9255$	$0.0^{2}8766$	$0.0^{2}8258$	$0.0^{2}7739$	5.
6.	$0.0^{2}7216$	$0.0^{2}6696$	$0.0^{2}6183$	$0.0^{2}5681$	$0.0^{2}5194$	$0.0^{2}4724$	$0.0^{2}4274$	$0.0^{2}3846$	$0.0^{2}3440$	$0.0^{2}3058$	6.
7.	$0.0^{2}2700$	$0.0^{2}2366$	$0.0^{2}2057$	$0.0^{2}1770$	$0.0^{2}1507$	$0.0^{2}1267$	$0.0^{2}1048$	$0.0^{3}8498$	$0.0^{3}6714$	$0.0^{3}5117$	7.
8.	$0.0^{3}3696$	$0.0^{3}2440$	$0.0^{3}1339$	$0.0^{4}3809$	$-0.0^{4}4449$	$-0.0^{3}1149$	$-0.0^{3}7742$	$-0.0^{3}2233$	$-0.0^{3}2632$	$-0.0^{3}2949$	8.
9.	$-0.0^{3}3192$	$-0.0^{3}3368$	$-0.0^{3}3486$	$-0.0^{3}3552$	$-0.0^{3}3574$	$-0.0^{3}3557$	$-0.0^{3}3508$	$-0.0^{3}3430$	$-0.0^{3}3329$	$-0.0^{3}3210$	9.
10.	$-0.0^{3}3075$										

iv. Asymptotic Series for Large x

$$\text{Ker}(x) \approx \sqrt{\frac{\pi}{2x}}\, e^{-x/\sqrt{2}} \left\{\cos\eta\,[1-(1^2)\beta_1+(1^2)(3^2)\beta_2-....]+\sin\eta\,[(1^2)\gamma_1-(1^2)(3^2)\gamma_2+....]\right\}$$

$$\text{Kei}(x) \approx \sqrt{\frac{\pi}{2x}}\, e^{-x/\sqrt{2}} \left\{-\cos\eta\,[(1^2)\gamma_1-(1^2)(3^2)\gamma_2+....]-\sin\eta\,[1-(1^2)\beta_1+(1^2)(3^2)\beta_2-....]\right\}$$

$$\eta = \frac{x}{\sqrt{2}}+\frac{\pi}{8}, \quad \beta_k = \frac{\cos(k\pi/4)}{k!\,(8x)^k}, \quad \gamma_k = \frac{\sin(k\pi/4)}{k!\,(8x)^k}, \qquad k=1,2,... \quad x>15$$

15.36 INFINITE SERIES INVOLVING BESSEL FUNCTIONS

$$J_n(x+y) = \sum_{k=-\infty}^{\infty} j_k(x)\, J_{n-k}(y) \qquad n = 0, \pm 1, \pm 2, ...$$

$$\begin{aligned}
1 &= J_0(x) + 2J_2(x) + ... + 2J_{2n}(x) + ...\\
x &= 2\,[J_1(x) + 3J_3(x) + ... + (2n+1)\,J_{2n+1}(x) + ...]\\
x^2 &= 8\,[J_2(x) + 4J_4(x) + ... + n^2 J_{2n}(x) + ...]\\
\sin x &= 2[J_1(x) - J_3(x) + J_5(x) - ....]\\
\cos x &= J_0(x) - 2[J_2(x) - J_4(x) + ...]\\
\sinh x &= 2[I_1(x) + I_3(x) + I_5(x) + ...]\\
\cosh x &= I_0(x) + 2\,[I_2(x) + I_4(x) + ...]\\
\sin(x\sin\omega) &= 2[J_1(x)\sin\omega + J_3(x)\sin 3\omega + ...]\\
\sin(x\cos\omega) &= 2[J_1(x)\cos\omega - J_3(x)\cos 3\omega - ...]\\
\cos(x\sin\omega) &= J_0(x) + 2\,[J_2(x)\cos 2\omega + J_4(x)\cos 4\omega + ...]\\
\cos(x\cos\omega) &= J_0(x) - 2\,[J_2(x)\cos 2\omega - J_4(x)\cos 4\omega + ...]
\end{aligned}$$

15.37 DEFINITE INTEGRALS INVOLVING BESSEL FUNCTIONS

$$\int_0^\infty J_n(\alpha x)\,dx = \frac{1}{\alpha} \qquad n > -1 \qquad\qquad \int_0^\infty J_n(\alpha x)\frac{dx}{x} = \frac{1}{n} \qquad n = 1, 2, ...$$

$$\int_0^\infty e^{-ax} J_0(\beta x)\,dx = \frac{1}{\sqrt{\alpha^2+\beta^2}} \qquad\qquad \int_0^\infty e^{-x} J_1(\beta x)\,dx = \frac{1}{\beta}\left(1-\frac{\alpha}{\sqrt{\alpha^2+\beta^2}}\right)$$

$$\int_0^\infty J_n(\alpha x)\sin\beta x\,dx = \left\{\begin{array}{ll} \dfrac{\sin(n\sin^{-1}(\beta/\alpha)}{\sqrt{\alpha^2-\beta^2}} & 0<\beta<\alpha \\[2ex] \dfrac{\alpha^n\cos(n\pi/2)}{\sqrt{b^2-\alpha^2}\,(\beta+\sqrt{\beta^2-\alpha^2})^n} & 0<\alpha<\beta \end{array}\right\} \int_0^\infty \qquad n > -2$$

$$\int_0^\infty J_n(\alpha x)\cos\beta x\,dx = \begin{cases} \dfrac{\cos[n\cos^{-1}(\beta/\alpha)]}{\sqrt{\alpha^2-\beta^2}} & 0<\beta<\alpha \\ \dfrac{-\alpha^n \sin(n\pi/2)}{\sqrt{\beta^2-\alpha^2}\,(\beta+\sqrt{\beta^2-\alpha^2})^n} & 0<\alpha<\beta \end{cases} \quad n>-1$$

$$\int_0^\infty \frac{J_m(x)\,J_n(x)}{x}\,dx = \begin{cases} \dfrac{2}{(m^2-n^2)\pi}\sin\dfrac{(m-n)\pi}{2} & m\neq n \\ \dfrac{1}{2m} & m=n \end{cases} \quad m+n>0$$

Power Series Method

The power series method is a general method for solving linear differential equation $y'' + p(x)\,y' + q(x)\,y = r(x)$ with variables $p(x)$, $q(x)$ and $r(x)$; it applies to higher order equation. It gives solution in the form of power series.

Operation of Power Series

In the power series method, the operation of differentiation, addition and multiplication are permissible.

Termwise Differentiation

A power series may be differentiated term by term. More precisely, if

$$y(x) = \sum_{m=0}^{\infty} a_m (x-x_0)^m$$

converges for $|x-x_0| < R$ where $R > 0$, then the series obtained by differentiating term by term also converges for these x and represents the derivative y' of y for those x, i.e.

$$y'(x) = \sum_{m=1}^{\infty} m\,a_m (x-x_0)^{m-1} \ldots \quad (|x-x_0|<R)$$

Similarly

$$y''(x) = \sum_{m-2}^{\infty} m(m-1)\,a_m (x-x_0)^{m-2} \ldots \quad (|x-x_0|\,R)$$

Termwise Addition

The power series may be added term by term. More precisely if the series

$$\sum_{m0}^{\infty} a_m (x-x_0)^m \quad \text{and} \quad \sum_{m=0}^{\infty} b_m (x-x_0)^m$$

have positive radii of convergence and their sums are $f(x)$ and $g(x)$, then the series

$$\sum_{m=0}^{\infty}(a_m+b_m)\,(x+x_0)^m,$$

converges and represents $f(x)+g(x)$ for each x that lies in the interior of convergence interval of each of the given series.

Termwise Multiplication

Two power series may be multiplied term by term. Let the series have positive radii of convergence and let $f(x)$ and $g(x)$ be their sum. Then the series obtained by a multiplying each term of the first series by each term of the second series and collecting the powers of $(x-x_0)$, that is

$$\sum_{m=0}^{\infty}(a_0\,b_m+a_1\,b_{m-1}+\ldots a_m\,b_0)\,(x-x_0)^m$$

$$= a_0\,b_0+(a_0\,b_1+a_1\,b_0)\,(x-x_0)+(a_0\,b_2+a_1\,b_1+a_2\,b_0)\,(x-x_0)^2+\ldots$$

converges and represents $f(x)\,g(x)$ for each x in the interior of the convergence interval of each of the given series.

Legendre's Equation

The Legendre's differential equation:

$$(1-x^2)\,y''-2x\,y'+n\,(n+1)\,y=0$$

The parameter n in the equation is a given real number. Solution of this equation is called Legendre function.

Legendre's Polynomials

The solution of Legendre's differential equation is called Legendre's polynomial of degree n and is denoted by $P_n(x)$ as:

$$P_n(x)=\sum_{m=0}^{m}(-1)^m\,\frac{(2n-2m)\qquad x^{n-2m}}{2^n\,m!(n-m)!\,(n-2m)!}$$

$$\frac{(2n)!}{2^n\,(n!)^2}-\frac{(2n-2)!\qquad x^{n-2}}{2^n!\cdot(n-1)!\,(n-2)!}$$

Frobenius Method

Any differential equation of the form

$$y''+\frac{b(x)}{x}y'+\frac{c(x)}{x^2}y=0$$

Where the function $b(x)$ and $c(x)$ are analytic at $x=0$. has at least one solution that can be reprinted in the form that follows.

$$y(x) + x^r \sum_{m=0}^{\infty} a_m x^m = x^r \left(a_0 + a_1 x + a_2 x^2 + \ldots\right),$$

where the exponent r may be any (real or complex) number (and r is chosen so that $a \neq 0$)

Bessel's Equation

Bessel's differential equation

$x^2 y'' + xy' + (x^2 - y^2)\, y = 0$

It appears in connection with electrical fields, vibration, heat conduction, etc.

Bessel's Function

$J_n(x)$: Bessel's equation can be solved by Frobenius method. Solution of Bessel's equation is given by J_n:

$$J_n(x) + x^n \sum_{m=0}^{\infty} \frac{(-1)^m\, x^{2m}}{2^{2m+n}\, m!(n+m)}$$

It is called Bessel's function of the first kind of order n.

CHAPTER 16

Partial Differential Equations

16.1 GENERAL CONCEPT

i. Definition

A partial differential equation of order n is a *functional equation* involving at least one n^{th} partial derivative of the unknown function $\Phi(x_1, x_2, ..., x_m)$ of two or more variables $x_1, x_2, ..., x_m$.

ii. Solution

A partial differential equation of order n given in general form as

$$F\left(x_1, x_2, ..., x_m, \Phi; \frac{\partial \Phi}{\partial x_1}, \frac{\partial \Phi}{\partial x_2}, ..., \frac{\partial \Phi}{\partial x_m}; \frac{\partial^2 \Phi}{\partial x_1^2}, ..., \frac{\partial^2 \Phi}{\partial x_m^2}, ...\right) = 0$$

It has a *general solution* (general integral), which involves *arbitrary functions. A particular solution* (particular integral) is a special case of the general solution with *specific functions* substituted for the arbitrary functions. These specific functions are generated by the given conditions (boundary conditions, initial conditions). Many partial differential equations admit *singular solutions* (singular integrals) unrelated to the general integral.

iii. Separation of Variables

Although the separation of variables method is not universally applicable, it offers a convenient and simple solution of many partial differential equations in engineering and the physical sciences. The underlying idea is to *separate* the given partial differential equation into *a set of ordinary differential equations*, the solution of which is known.

The *assumed solution* is

$$\Phi(x_1, x_2, ..., x_m) = \Phi_1(x_1)\, \Phi_2(x_2) \ldots \Phi_m(x_m)$$

The linear combination of which is the general solution. This procedure is applicable when the substitution of the assumed solution in the given partial differential equation transforms this equation into

$$\psi_1\left(x_1;\frac{d\Phi_1}{dx_1};\ldots\right)+\psi_2\left(x_2,x_3,\ldots,xm;\frac{d\Phi_2}{dx_2};\ldots\right)=0$$

A setting, $\Psi_1 = C$ and $\Psi_2 = -C$ breaks the equation $\Psi_1 + \Psi_2 = 0$ into an ordinary differential equation and a partial differential equation, the second of which (if necessary) can be the subject of a repeated separation process.

iv. Some useful Formulae

a. The Lacobain of the function $u(x, y)$ and $v(x, y)$ with respect to x and y is given by

$$\frac{\partial (u,v)}{\partial (x,y)}=\begin{vmatrix}\partial u/\partial x & \partial u/\partial y\\ \partial v/\partial x & \partial v/\partial y\end{vmatrix}$$

b. *Lagrange's method* and Lagrangis Subsidairy equations. The Lagrange's subsidiary equations of the quasi linear partial differential equation

$Pp + Qq = R$ are

$$\frac{dx}{P}=\frac{dy}{Q}+\frac{dz}{R}$$

Its general solution is $F(u, v) = 0$ where $u(x, y, z) = c_1$ and $v(x, y, z) = c_2$ are two linearly independent solutions of $\frac{dx}{P}+\frac{dy}{Q}=\frac{dz}{R}$

c. The Charpit's auxiliary equations of $f(x, y, z, p, q) = 0$ are

$$\frac{dx}{fp}=\frac{dy}{fq}=\frac{dz}{pfp+qfq}=\frac{dp}{-(fx+pfz)}=\frac{dq}{-(fy+qfz)}$$

d. The Jacobi's auxiliary equation of $f(x_1, x_2, x_3, p_1, p_2, p_3) = 0$ and

$$\frac{dx_1}{fp_1}=\frac{dx_2}{fp_2}=\frac{dx_3}{fp_3}=\frac{dp_1}{-fx_1}=\frac{dp_2}{-fx_2}=\frac{dp_3}{-fx_3}$$

e. $(\alpha D+\beta D^1+\gamma)z=0 \Rightarrow z=e^{-\frac{\gamma x}{\alpha}}\phi(\beta x-\alpha y)$ if $\alpha \neq 0$; and $z = e^{\frac{-\gamma x}{\beta}}\phi(\beta x-\alpha y)$ if $\beta \neq 0$.

In particular,

(a) $(D - mD^1)z = 0 \Rightarrow z = \phi(y + mx)$ and $D(z) = 0 \Rightarrow z = \phi(y)$

(b) $(D^1 - mD)z = 0 \Rightarrow z = \phi(x + my)$ and $D^1(z) = 0 \Rightarrow z = \phi(x)$

(c) $(D - mD^1 - a)z = 0 \Rightarrow z = e^{ax}\phi(y + mx)$

(d) $(D^1 - mD - a)z = 0 \Rightarrow z = e^{ay}\phi(x + my)$

f. $(\alpha D + \beta D^1 + \gamma)^2 z = 0 \Rightarrow z = e^{\frac{-\gamma x}{\alpha}}\phi_1(\beta x-\alpha y)+x\phi_2(\beta x-\alpha y)$, $\alpha \neq 0$

In particular,

(a) $(D - mD^1)^2 z = 0 \Rightarrow z = \phi(y + mx) + x\,\phi_2(y + mx)$

(b) $(D - mD^1 - a)^2 z = 0 \Rightarrow z = e^{ax}\phi_1(y + mx) + x\,\phi_2(y + mx)$

g. $(\alpha D + \beta D^1 + \gamma)^3 y = 0$

$$\Rightarrow z=e^{\frac{-\gamma/x}{\alpha}}\left\{\phi_1(\beta x-\alpha y)+x\,\phi_2(\beta x-\alpha y)+x^2\phi_3(\beta x-\alpha y)\right\}$$

h. $\dfrac{1}{(\alpha D+\beta D^1+\gamma)}f(x,y)=\dfrac{e^{-\gamma x/\alpha}}{\alpha}\int e^{\gamma x/\alpha}f\left(x,\dfrac{\alpha+\beta x}{\alpha}\right)dx$, $\alpha \neq 0$ and $a = \alpha y - \beta x$

i. $\dfrac{1}{F(D,D^1)}e^{ax+by} = \dfrac{1}{F(a,b)}e^{ax+by}$, if $F(a, b) \neq 0$

j. $\dfrac{1}{F(D,D^1)}e^{ax+by}, v(x,y) = e^{ax+by}\dfrac{1}{F(D+a,D^1+b)}v(x,y)$

k. $\dfrac{1}{F(D^2,DD^1,D'^2)}\sin(ax+by) = \dfrac{1}{F(-a^2,-ab,-b^2)}\sin(ax+by)$ if $F(-a^2,-ab,-b^2) \neq 0$

l. $\dfrac{1}{F(D^2,DD^1,D'^2)}\cos(ax+by) = \dfrac{1}{F(-a^2,-ab,-b^2)}\cos(ax+by)$ if $F(-a^2,-ab,-b^2) \neq 0$

m. $\dfrac{1}{(bD-aD^1)^r}\phi(ax+by)=\dfrac{x^r}{b^r r!}\phi(ax+by)$

v. Classification

Partial differential equations are classified according to the type of conditions imposed by the problem. Linear differential equations of the second order in the two variables given as

$$Au_{xx} + Bu_{xy} + Cu_{yy} + Du_x + Eu_y + F_u + G = 0$$

where u_{xx}, u_{xy}, ... are the partial derivatives with respect to x and/or y and A, B, ..., are functions of x and y, are classified as *elliptic, hyperbolic, or parabolic* according to whether

$$\Delta = \begin{vmatrix} A & B \\ B & C \end{vmatrix}$$

is *positive, negative, or zero*. The concept of these definitions can be extended for differential equations with more than two variables.

16.2 LAPLACE'S DIFFERENTIAL EQUATION IN TWO DIMENSIONS

a. *Rectangular coordinates* $\dfrac{\partial^2\Phi}{\partial x^2}+\dfrac{\partial^2\Phi}{\partial y^2}=0$

Solution:

$$\Phi_\alpha = (A_1e^{i\alpha x} + A_2 e^{-i\alpha x})(B_1 e^{\alpha y} + B_2 e^{-\alpha y}) \qquad \alpha \neq 0$$
$$\Phi_0 = (A_1 + A_2x)(B_1 + B_2y) \qquad \alpha = 0$$

b. *Polar coordinates* $\dfrac{\partial^2\Phi}{\partial r^2}+\dfrac{l}{r}\dfrac{\partial\Phi}{\partial r}+\dfrac{1}{r^2}\dfrac{\partial^2\Phi}{\partial\Phi^2}=0$

Solution:

$$\Phi_\alpha = (A_1r^\alpha + A_2r^{-\alpha})(B_1 e^{\alpha i\phi} + B_2 e^{-\alpha i\phi}) \qquad \alpha \neq 0$$
$$\Phi_0 = (A_1 + A_2I_n r)(B_1 + B_2\phi) \qquad \alpha = 0$$

16.3 LAPLACE'S DIFFERENTIAL EQUATION IN THREE DIMENSIONS

a. *Rectangular coordinates* $\dfrac{\partial^2 \Phi}{\partial x^2}+\dfrac{\partial^2 \Phi}{\partial y^2}+\dfrac{\partial^2 \Phi}{\partial z^2}=0$

Solution: $(\alpha^2+\beta^2+\gamma^2=0)$

$$\phi_{\alpha\beta\gamma}=(A_1 e^{\alpha x}+A_2 e^{-\alpha x})(B_1 e^{\beta y}+B_2 e^{-\beta y})(C_1 e^{\gamma z}+C_2 e^{-\gamma z}) \quad \alpha\neq 0, \beta\neq 0, \gamma\neq 0$$

$$\phi_{000}=(A_1+A_2 x)(B_1+B_2 y)(C_1+C_2 z) \quad \alpha=0, \beta=0, \gamma=0$$

b. *Cylindrical coordinates** $\dfrac{\partial^2 \Phi}{\partial r^2}+\dfrac{l}{r}\dfrac{\partial \Phi}{\partial r}+\dfrac{1}{r^2}\dfrac{\partial^2 \Phi}{\partial \phi^2}+\dfrac{\partial^2 \Phi}{\partial z^2}=0$

Solution:

$$\phi_{\alpha\beta}=[A_1 J_{\alpha}+(i\beta r)+A_2 Y\alpha(i\beta r)](B_1 e^{i\alpha\phi}+B_2 e^{-i\alpha\phi})(C_1 e^{i\beta z}+C_2 e^{-i\beta z}) \quad \alpha\neq 0, \beta\neq 0$$

$$\phi_{00}=(A_1+A_2 I_n r)(B_1+B_2\phi)(C_1+C_2 z) \quad \alpha=0, \beta=0$$

$$\phi_{\alpha 0}=(A_1+A_2 I_n r)(B_1 e^{i\alpha\phi}+B_2 e^{-i\alpha\phi})(C_1+C_{2z}) \quad \alpha\neq 0, \beta=0$$

$$\phi_{0\beta}=J_0(i\beta r)(B_1+B_2\phi)(C_1 e^{i\beta z}+C_2 e^{-i\beta z}) \quad \alpha=0, \beta\neq 0$$

16.4 THE WAVE EQUATION

Wave equation in two dimension is of the form

$$\frac{\partial^2\phi}{\partial x^2}+\frac{\partial^2\phi}{\partial y^2}=\frac{1}{c^2}\frac{\partial^2\phi}{\partial t^2}$$

Wave equation in three dimension is of the form

$$\frac{\partial^2\phi}{\partial x^2}+\frac{\partial^2\phi}{\partial y^2}+\frac{\partial^2\phi}{\partial z^2}=\frac{-1}{c^2}\frac{\partial^2\phi}{\partial t^2}$$

If we assume a solution of the wave equation of the form $\phi=\phi(x, y, z)\,e^{\pm ikt}$ then the function ϕ must satisfy the equation $(\nabla^2+k^2)\phi=0$, which is called the space form of wave equation or Helmholtz's equation.

16.5 HELMHOLTZ'S DIFFERENTIAL EQUATION IN TWO DIMENSIONS

a. *Rectangular coordinates* $\dfrac{\partial^2 \Phi}{\partial x^2}+\dfrac{\partial^2 \Phi}{\partial y^2}+k^2\Phi=0$

Solution: $(\alpha^2+\beta^2=k^2)$

$$\Phi_{\alpha\beta}=(A_1 e^{i\alpha x}+A_2 e^{-i\alpha x})(B_1 e^{i\beta y}+B_2 e^{-i\beta y}) \quad \alpha\neq 0$$

$$\Phi_{0k}=(A_1+A_2 x)(B_1 e^{iky}+B_2 e^{-iky}) \quad \alpha=0$$

b. *Polar coordinates* $\dfrac{\partial^2 \Phi}{\partial r^2}+\dfrac{1}{r}\dfrac{\partial\Phi}{\partial r}+\dfrac{1}{r^2}\dfrac{\partial^2\Phi}{\partial\phi}+k^2\Phi=0$

Solution:

$$\Phi_{\alpha k}=[A_1 J_{\alpha}(ikr)+A_2 Y_{\alpha}(ikr)](B_1 e^{i\alpha\phi}+B_2 e^{-i\alpha\phi}) \quad \alpha\neq 0$$

$$\Phi_{0k}=J_0(ikr)(B_1+B_2\phi) \quad \alpha=0$$

* $J_{\alpha}()$, $Y_{\alpha}()$, $J_0()$ are Bessel functions.

16.6 HELMHOLTZ'S DIFFERENTIAL EQUATION IN THREE DIMENSIONS

a. *Rectangular coordinates* $\frac{\partial^2\Phi}{\partial x^2}+\frac{\partial^2\Phi}{\partial y^2}+\frac{\partial^2\Phi}{\partial z^2}+k^2\,\Phi=0$

Solution: $(\alpha^2+\beta^2+\gamma^2=k^2)$

$$\Phi_{\alpha\beta\gamma}=(A_1e^{i\alpha x}+A_2e^{-i\alpha x})\,(B_1e^{i\beta y}+B_2e^{-i\beta y})\,(C_1e^{i\gamma z}+C_2e^{-i\gamma z})\qquad \alpha\neq 0,\beta\neq 0,\gamma\neq 0$$

$$\Phi_{00k}=(A_1+A_2x)\,(B_1+B_2y)\,(C_1e^{ikz}+C_2e^{-ikz})\qquad \alpha=0,\beta=0$$

b. *Cylindrical coordinates* $\frac{\partial^2\Phi}{\partial r^2}+\frac{1}{r}\frac{\partial\Phi}{\partial r}+\frac{1}{r^2}\frac{\partial^2\Phi}{\partial\phi^2}+\frac{\partial^2\Phi}{\partial z^2}+k^2\,\Phi=0$

Solution:

$$\Phi_{\alpha\beta k}=[A_1J_\alpha(i\gamma r)+A_2Y_\alpha(i\gamma r)]\,(B_1e^{i\alpha\phi}+B_2e^{-i\alpha\phi})\,(C_1e^{i\beta z}+C_2e^{-i\beta z})\qquad \alpha\neq 0,\beta\neq 0$$

$$\Phi_{00k}=J_0(ikr)\,(B_1+B_2\phi)\,(C_1+C_2z)\qquad \alpha=0,\beta=0$$

$$\gamma=\sqrt{k^2-\beta^2}$$

[1]$J_\alpha(\,),Y_\alpha(\,),J_0(\,)$ = Bessel functions.

16.7 DIFFUSION EQUATION OF ONE DIMENSION

It is a typical parabolic equation

$$k\frac{\partial^2Q}{\partial x^2}=\frac{\partial Q}{\partial t}$$

where Q is called the heat function of the diffusion equation. Its solution is given by

$$Q\,(x,t)=\sum_{n=0}^{\infty}c_n\cos\,(nx+\zeta_n)e^{-n^2kt}$$

16.8 DIFFUSION EQUATION OF TWO DIMENSION

$$\frac{\partial^2Q}{\partial x^2}+\frac{\partial^2Q}{\partial y^2}=\frac{1}{k}\frac{\partial Q}{\partial t}$$

The solution is

$$Q(x,y,t)=\sum_\lambda\sum_\mu C_\lambda\cos\,(\lambda x+\zeta\varepsilon_\lambda)\cos\,(\mu y+\zeta'_\mu)\,e^{-(\lambda^2+\mu^2)\,kt}$$

16.9 DIFFUSION EQUATION OF THREE DIMENSION

$$\frac{\partial^2Q}{\partial x^2}+\frac{\partial^2Q}{\partial y^2}+\frac{\partial^2Q}{\partial z^2}=\frac{1}{k}\frac{\partial Q}{\partial t}$$

CHAPTER 17

Laplace Transforms

17.1 LAPLACE TRANSFORMS—PROPERTIES

i. Definitions

Laplace Transforms: The Laplace transform of the function $f(t)$, denoted by $F(s)$ or $\mathcal{L}\{f(t)\}$ is defined by $F(s) = \mathcal{L} f(t) = \int_0^\infty f(t)\, e^{-st} dt$ provided the integration may be validily performed.

The Laplace transform of $g(t)$ is denoted by $\mathcal{L}\{g(t)\}$ or $G(s)$.

where s = real or complex and $e = 2.71828...$

The inverse Laplace transform of $F(s)$ is

$$\mathcal{L}^{-1}\{F(s)\} = f(t)$$

Not every function $F(s)$ has an inverse Laplace transform.

ii. Basic Relationship (*a*, *b*, *n* = constants)

$$\mathcal{L}\{af_1(t) + bf_2(t)\} = \mathcal{L}\{af_1(t)\} + \mathcal{L}\{bf_2(t)\} = aF_1(s) + bF_2(s)$$

$\mathcal{L}\{e^{at} f(t)\} = F(s-a)$	$\mathcal{L}\{e^{-at} f(t)\} = F(s+a)$
$\mathcal{L}\{f(at)\} = \dfrac{1}{a} F\left(\dfrac{s}{a}\right)$	$\mathcal{L}\left\{f\left(\dfrac{t}{a}\right)\right\} = aF(as)$

iii. Laplace Transform of Derivatives of Differential Equations

$$\mathcal{L}\{f'(t)\} = s\,F(s) - f(+0)$$
$$\mathcal{L}\{f''(t)\} = s^2 F(s) - sf(+0) - f'(+0)$$
$$\mathcal{L}\{f'''(t)\} = s^3 F(s) - s^2 f(0) - sf'(0) - f''(0)$$

Laplace transform of the derivative of any order n

$$\mathcal{L}\{f^{(n)}(t)\} = s^n F(s) - s^{n-1} f(+0) - s^{n-2} f'(+0) - \dots - f^{(n-1)}(+0)$$

iv. Transform of Integrals of a Function

$$\mathcal{L}\left\{\int_0^1 f(u)\,du\right\} = \frac{F(s)}{s}$$

Integrals [$u = u(t)$, $v = v(t)$]

$$\mathcal{L}\left\{\int_0^t \int_0^u f(v)\,dv\,du\right\} = \frac{F(s)}{s^2}$$

$$\mathcal{L}\left\{\int_0^t f(t-\tau)\,g(\tau)\,d\tau\right\} = \mathcal{L}\left\{\int_0^t f(t)\,g(t-\tau)\,d\tau\right\} = F(s)\,.\,G(s)$$

First Shifting Theorem: If $f(t)$ has the transform $F(s)$ then $e^{at} f(t)$ has the function $F(s-a)$ $\mathcal{L}\,\{e^{at} f(t)\} = F(s-a)$

Shifting Theorem: If $f(t)$ has the transform $F(s)$ if $t < a$ then shifted function

$$f(t) = f(t-a)\,\mu(t-a) = \begin{cases} 0 & \\ f(t-a) & \text{if } t > a \end{cases}$$

has the transform $e^{-as} F(s)$. That is

$$\mathcal{L}\,\{f(t-a)\,\mu(t-a)\} = e^{-as} F(s).$$

Dirac Delta Function: Laplace transform of dirac delta function $\delta(t-a)$ is :

$$\mathcal{L}\,\{\delta(t-a)\} = e^{-as}.$$

v. Periodic Functions [$G(T + t) = G(t)$, $H(T + t) = -H(t)$]

$$\mathcal{L}\,\{g(T+t)\} = \frac{\int_0^T e^{-st}\,g(t)\,dt}{1-e^{-sT}} \qquad \mathcal{L}\,\{h(T+t)\} = \frac{\int_0^T e^{-st}\,h(t)\,dt}{1+e^{-sT}}$$

17.2 OPERATION OF LAPLACE TRANSFORM

PROPERTIES OF LAPLACE TRANSFORMS

S. No.	$f(t)$	$F(s) = \int_0^\infty f(t)\,e^{-st}\,dt$
1.	$af(t) + bg(t)$	$aF(s) + bG(s)$
2.	$f'(t)$	$sF(s) - f(0)$
3.	$f''(t)$	$s^2 F(s) - sf(0) - f'(0)$
4.	$f^{(n)}(t)$	$s^n F(s) - s^{n-1} f(0) - s^{n-2} f'(0) \ldots f^{n-1}(0)$
5.	$t\,f(t)$	$-F'(s)$
6.	$t^n f(t)$	$(-1)^n F^{(n)}(s)$
7.	$e^{at} f(t)$	$F(s-a)$
8.	$\int_0^t f(t-\beta \cdot g(\beta)\,d\beta)$	$F(s), G(s)$
9.	$f(t-a)$	$e^{-as} F(s)$
10.	$f(t/a)$	$a\,F(as)$
11.	$\int_0^t g(\beta)\,d\beta$	$\frac{1}{s} G(s)$

S. No.	$f(t)$	$F(s) = \int_0^{\infty} f(t)\, e^{-st}\, dt$
12.	$f(t-c)\,\delta(t-c)$ where $\delta(t-c) = 0$ *if* $0 \le t < c = 1$ *if* $t \ge c$	$e^{-cs}\, F(s)\; c > 0$
13.	$f(t) = f(t+w)$ *(Periodic)*	$\int_0^{w} e - sT\, f(\tau) d\tau$ $(1 - e^{-s\omega})$

LAPLACE TRANSFORMS OF FUNCTIONS

S. No.	$\mathcal{L}\, f(t)$	$F(s)$
1.	1	$\frac{1}{s}$
2.	t	$\frac{1}{s^2}$
3.	$\frac{t^{n-1}}{(n-1)!}, \quad 0! = 1$	$\frac{1}{s^n} \quad n = 1, 2, 3, \ldots$
4.	$\frac{t^{n-1}}{\Gamma(n)}$	$\frac{1}{s^n} \quad n > 0$
5.	e^{at}	$\frac{1}{s-a}$
6.	$\frac{t^{n-1}\, e^{at}}{(n-1)!}, \quad 0! = 1$	$\frac{1}{(s-a)^n} \quad n = 1, 2, 3, \ldots$
7.	$\frac{t^{n-1}\, e^{at}}{\Gamma(n)}$	$\frac{1}{(s-a)^n} \quad n > 0$
8.	$\frac{\sin at}{a}$	$\frac{1}{s^2 + a^2}$
9.	$\cos at$	$\frac{s}{s^2 + a^2}$
10.	$\frac{e^{bt} \sin at}{a}$	$\frac{1}{(s-b)^2 + a^2}$
11.	$e^{bt} \cos at$	$\frac{s-b}{(s-b)^2 + a^2}$
12.	$\frac{\sinh at}{a}$	$\frac{1}{s^2 - a^2}$

S. No.	$\mathcal{L} f(t)$	$F(s)$	
13.	$\cosh at$	$\frac{s}{s^2 - a^2}$	
14.	$\frac{e^{bt} \sinh at}{a}$	$\frac{1}{(s-b)^2 - a^2}$	
15.	$e^{bt} \cosh at$	$\frac{s-b}{(s-b)^2 - a^2}$	
16.	$\frac{e^{bt} - e^{at}}{b - a}$	$\frac{1}{(s-a)(s-b)}$	$a \neq b$
17.	$\frac{be^{bt} - ae^{at}}{b-a}$	$\frac{s}{(s-a)(s-\text{b})}$	$a \neq b$
18.	$\frac{\sin at - at \cos at}{2a^3}$	$\frac{1}{(s^2 + a^2)^2}$	
19.	$\frac{t \sin at}{2a}$	$\frac{s}{(s^2 + a^2)^2}$	
20.	$\frac{\sin at + at \cos at}{2a}$	$\frac{s^2}{(s^2 + a^2)^2}$	
21.	$\cos at - \frac{1}{2} at \sin at$	$\frac{s^3}{(s^2 + a^2)^2}$	
22.	$t \cos at$	$\frac{s^2 - a^2}{(s^2 + a^2)^2}$	
23.	$\frac{at \cosh at - \sinh at}{2a^3}$	$\frac{1}{(s^2 - a^2)^2}$	
24.	$\frac{t \sinh at}{2a}$	$\frac{s}{(s^2 - a^2)^2}$	
25.	$\frac{\sinh at + at \cosh at}{2a}$	$\frac{s^2}{(s^2 - a^2)^2}$	
26.	$\cosh at + \frac{1}{2} at \sinh at$	$\frac{s^3}{(s^2 - a^2)^2}$	
27.	$t \cosh at$	$\frac{s^2 + a^2}{(s^2 - a^2)^2}$	
28.	$\frac{(3 - a^2t^2) \sin at - 3at \cos at}{8a^5}$	$\frac{1}{(s^2 + a^2)^3}$	

S. No.	$\mathcal{L} f(t)$	$F(s)$
29.	$\dfrac{t \sin at - at^2 \cos at}{8a^3}$	$\dfrac{s}{(s^2 + a^2)^3}$
30.	$\dfrac{(1 + a^2t^2) \sin at - at \cos at}{8a^3}$	$\dfrac{s^2}{(s^2 + a^2)^3}$
31.	$\dfrac{3t \sin at + at^2 \cos at}{8a}$	$\dfrac{s^3}{(s^2 + a^2)^3}$
32.	$\dfrac{(3 - a^2t^2) \sin at + 5at \cos at}{8a}$	$\dfrac{s^4}{(s^2 + a^2)^3}$
33.	$\dfrac{(8 - a^2t^2) \cos at - 7at \sin at}{8}$	$\dfrac{s^5}{(s^2 + a^2)^3}$
34.	$\dfrac{t^2 \sin at}{2a}$	$\dfrac{3s^2 - a^2}{(s^2 + a^2)^3}$
35.	$\dfrac{1}{2}\ t^2 \cos at$	$\dfrac{s^3 - 3a^2s}{(s^2 + a^2)^3}$
36.	$\dfrac{1}{6} t^3 \cos at$	$\dfrac{s^4 - 6a^2s^2 + a^4}{(s^2 + a^2)^4}$
37.	$\dfrac{t^3 \sin at}{24a}$	$\dfrac{s^2 - a^2s}{(s^2 + a^2)^4}$
38.	$\dfrac{(3 + a^2t^2) \sinh at - 3at \cosh at}{8a^5}$	$\dfrac{1}{(s^2 - a^2)^3}$
39.	$\dfrac{at^2 \cosh at - t \sinh at}{8a^3}$	$\dfrac{s}{(s^2 - a^2)^3}$
40.	$\dfrac{at \cosh at + (a^2t^2 - 1) \sinh at}{8a^3}$	$\dfrac{s^2}{(s^2 - a^2)^3}$
41.	$\dfrac{3t \sinh at + at^2 \cosh at}{8a}$	$\dfrac{s^3}{(s^2 - a^2)^3}$
42.	$\dfrac{(3 + a^2t^2) \sinh at + 5at \cosh at}{8a}$	$\dfrac{s^4}{(s^2 - a^2)^3}$
43.	$\dfrac{(8 + a^2t^2) \cosh at + 7at \sinh at}{8}$	$\dfrac{s^5}{(s^2 - a^2)^3}$
44.	$\dfrac{t^2 \sinh at}{2a}$	$\dfrac{3a^2 + a^2}{(s^2 - a^2)^3}$

S. No.	$\mathcal{L} f(t)$	$F(s)$
45.	$\frac{1}{2}\ t^2 \cosh at$	$\frac{s^3+3a^2s}{(s^2-a^2)^3}$
46.	$\frac{1}{6}\ t^3 \cosh at$	$\frac{s^4+6a^2s^2+a^4}{(s^2-a^2)^4}$
47.	$\frac{t^3 \sinh at}{24a}$	$\frac{s^3+a^2s}{(s^2-a^2)^4}$
48.	$\frac{e^{at/2}}{3a^2}\left\{\sqrt{3}\sin\frac{\sqrt{3}\,at}{2}-\cos\frac{\sqrt{3}\,at}{2}+e^{-3at/2}\right\}$	$\frac{1}{s^3+a^3}$
49.	$\frac{e^{at/2}}{3a}\left\{\cos\frac{\sqrt{3}\,at}{2}+\sqrt{3}\sin\frac{\sqrt{3}\,at}{2}-e^{-3at/2}\right\}$	$\frac{s}{s^3+a^3}$
50.	$\frac{1}{3}\left(e^{-at}+2e^{at/2}\cos\frac{\sqrt{3}\,at}{2}\right)$	$\frac{s^2}{s^3+a^3}$
51.	$\frac{e^{-at/2}}{3a^2}\left\{e^{3at/2}-\cos\frac{\sqrt{3}\,at}{2}-\sqrt{3}\sin\frac{\sqrt{3}\,at}{2}\right\}$	$\frac{1}{s^3-a^3}$
52.	$\frac{e^{-at/2}}{3a}\left\{\sqrt{3}\sin\frac{\sqrt{3}\,at}{2}-\cos\frac{\sqrt{3}\,at}{2}+e^{3at/2}\right\}$	$\frac{s}{s^3-a^3}$
53.	$\frac{1}{3}\left(e^{at}+2e^{-at/2}\cos\frac{\sqrt{3}\,at}{2}\right)$	$\frac{s^2}{s^3-a^3}$
54.	$\frac{1}{4a^3}(\sin at \cosh at-\cos at \sinh at)$	$\frac{1}{s^4+4a^4}$
55.	$\frac{\sin at \sinh at}{2a^2}$	$\frac{s}{s^4+4a^4}$
56.	$\frac{1}{2a}(\sin at \cosh at+\cos at \sinh at)$	$\frac{s^2}{s^4+4a^4}$
57.	$\cos at \cosh at$	$\frac{s^3}{s^4+4a^4}$
58.	$\frac{1}{2a^3}(\sinh at-\sin at)$	$\frac{1}{s^4-a^4}$
59.	$\frac{1}{2a^2}(\cosh at-\cosh at)$	$\frac{s}{s^4-a^4}$

S. No.	$\mathcal{L} f(t)$	$F(s)$
60.	$\frac{1}{2a}(\sinh at + \sin at)$	$\frac{s^2}{a^4 - a^4}$
61.	$\frac{1}{2}(\cosh at + \cos at)$	$\frac{s^3}{s^4 - a^4}$
62.	$\frac{e^{-bt} - e^{-at}}{2(b-a)\sqrt{\pi t^3}}$	$\frac{1}{\sqrt{s+a} + \sqrt{s+b}}$
63.	$\frac{\text{erf}\sqrt{at}}{\sqrt{a}}$	$\frac{1}{s\sqrt{s+a}}$
64.	$\frac{e^{at}\,\text{erf}\sqrt{at}}{\sqrt{a}}$	$\frac{1}{\sqrt{s}(s-a)}$
65.	$e^{at}\left\{\frac{1}{\sqrt{\pi t}} - b e^{b^2 t}\,\text{erfc}\left(b\sqrt{t}\right)\right\}$	$\frac{1}{\sqrt{s-a} + b}$
66.	$J_0(at)$	$\frac{1}{\sqrt{s^2 + a^2}}$
67.	$I_0(at)$	$\frac{1}{\sqrt{s^2 - a^2}}$
68.	$a^n J_n(at)$	$\frac{\left(\sqrt{s^2 + a^2} - s\right)^n}{\sqrt{s^2 + a^2}}$ $n > -1$
69.	$a^n I_n(at)$	$\frac{\left(s - \sqrt{s^2 - a^2}\right)^n}{\sqrt{s^2 - a^2}}$ $n > -1$
70.	$J_0\left(a\sqrt{t(t+2b)}\right)$	$\frac{e^{b\left(s - \sqrt{s^2 - a^2}\right)}}{\sqrt{s^2 + a^2}}$
71.	$\begin{cases} J_0\left(a\sqrt{t^2 - b^2}\right) & t > b \\ 0 & t < b \end{cases}$	$\frac{e^{-b\sqrt{s^2 + a^2}}}{\sqrt{s^2 + a^2}}$
72.	$\frac{t J_1(at)}{a}$	$\frac{1}{(s^2 + a^2)^{3/2}}$
73.	$t J_0(at)$	$\frac{s}{(s^2 + a^2)^{3/2}}$
74.	$J_0(at) - at\, J_1(at)$	$\frac{s^2}{(s^2 + a^2)^{3/2}}$

S. No.	$\mathcal{L} f(t)$	$F(s)$
75.	$\dfrac{t\,I_1(at)}{a}$	$\dfrac{1}{(s^2-a^2)^{3/2}}$
76.	$t\,I_0(at)$	$\dfrac{s}{(s^2-a^2)^{3/2}}$
77.	$I_0\,(at) + at\,I_1\,(at)$	$\dfrac{s^2}{(s^2-a^2)^{3/2}}$
78.	$f(t) = n \le t < n+1, \quad n = 0, 1, 2, \ldots$	$\dfrac{1}{s(e^s-1)} = \dfrac{e^{-s}}{s(1-e^{-s})}$
79.	$f(t) = \sum_{k=1}^{[t]} r^k$ where (t) = greatest integer $\le t$	$\dfrac{1}{s(e^s-r)} = \dfrac{e^{-s}}{s(1-re^{-s})}$
80.	$f(t) = r^n, n \le t < n+1, \quad n = 0, 1, 2,$	$\dfrac{e^s-1}{s(e^s-r)} = \dfrac{1-e^{-s}}{s(1-re^{-s})}$
81.	$\dfrac{\cos 2\sqrt{at}}{\sqrt{\pi t}}$	$\dfrac{e^{-a/s}}{\sqrt{s}}$
82.	$\dfrac{\sin 2\sqrt{at}}{\sqrt{\pi a}}$	$\dfrac{e^{-a/s}}{s^{3/2}}$
83.	$\left(\dfrac{t}{a}\right)^{n/2} J_n(2\sqrt{at})$	$\dfrac{e^{-a/s}}{s^{n+1}}$ $n > -1$
84.	$\dfrac{e^{-a^2/4t}}{\sqrt{\pi t}}$	$\dfrac{e^{-a\sqrt{s}}}{\sqrt{s}}$
85.	$\dfrac{a}{2\sqrt{\pi t^3}}\, e^{-a^2/4t}$	$e^{-a\sqrt{s}}$
86.	$\operatorname{erf}\,(a/2\sqrt{t})$	$\dfrac{1-e^{-a\sqrt{s}}}{s}$
87.	$\operatorname{erfc}\,(a/2\sqrt{t})$	$\dfrac{e^{-a\sqrt{s}}}{s}$
88.	$e^{b(bt+a)}\operatorname{erfc}\left(b\sqrt{t} + \dfrac{a}{2\sqrt{t}}\right)$	$\dfrac{e^{-a\sqrt{s}}}{\sqrt{s}(\sqrt{s}+b)}$
89.	$\dfrac{1}{\sqrt{\pi t}a^{2n+1}} \int_0^\infty u^n\, e^{-u^2/4a^2t} J_{2n}(2\sqrt{u})\, du$	$\dfrac{e^{-a/\sqrt{s}}}{s^{n+1}}$ $n > -1$

S. No.	$\mathcal{L} f(t)$	$F(s)$
90.	$\dfrac{e^{-bt} - e^{-at}}{t}$	$\ln\left(\dfrac{s+a}{s+b}\right)$
91.	$Ci\,(at)$	$\dfrac{\ln[(s^2+a^2)/a^2]}{2s}$
92.	$Ei\,(at)$	$\dfrac{\ln[(s+a)/a]}{s}$
93.	$\ln t$	$-\dfrac{(\gamma + \ln s)}{s}$ γ = Euler's constant = 0.5772156...
94.	$\dfrac{2(\cos at - \cos bt)}{t}$	$\ln\left(\dfrac{s^2+a^2}{s^2+b^2}\right)$
95.	$\ln^2 t$	$\dfrac{\pi^2}{6s} + \dfrac{(\gamma + \ln s)^2}{s}$ γ = Euler's constant = 0.5772156...
96.	$-(\ln t + \gamma)$ γ is Euler's constant = 0.5772156...	$\dfrac{\ln s}{s}$
97.	$(\ln t + \gamma)^2 - \dfrac{1}{6}\pi^2$ γ is Euler's constant = 0.5772156...	$\dfrac{\ln^2 s}{s}$
98.	$t^n \ln t$	$\dfrac{\Gamma''(n+1) - \Gamma''(n+1)\ln a}{s^{n+1}}$ $\quad n > -1$
99.	$\dfrac{\sin at}{t}$	$\tan^{-1}(a/s)$
100.	$Si\,(at)$	$\dfrac{\tan^{-1}(a/s)}{s}$
101.	$\dfrac{e^{-2\sqrt{at}}}{\sqrt{\pi t}}$	$\dfrac{e^{a/s}}{\sqrt{s}}\,\text{erfc}\,(\sqrt{a/s})$
102.	$\dfrac{2a}{\sqrt{\pi}}\,e^{-a^2t^2}$	$e^{s^2/4a^2}\,\text{erfc}\,(s/2a)$
103.	$\text{erf}\,(at)$	$\dfrac{e^{s^2/4a^2}\,\text{erfc}\,(s/2a)}{s}$
104.	$\dfrac{1}{\sqrt{\pi(t+a)}}$	$\dfrac{e^{as}\,\text{erfc}\,\sqrt{as}}{\sqrt{s}}$

S. No.	$\mathcal{L} f(t)$	$F(s)$
105.	$\frac{1}{t+a}$	$e^{as}\, Ei\,(as)$
106.	$\frac{1}{t^2+a^2}$	$\frac{1}{a}\left[\cos as\left\{\frac{\pi}{2}-Si\,(as)\right\}-\sin as\, Ci\,(as)\right]$
107.	$\frac{t}{t^2+a^2}$	$\sin as\left\{\frac{\pi}{2}-Si\,(as)\right\}+\cos as\, Ci\,(as)$
108.	$\tan^{-1}(t/a)$	$\frac{\cos as\left\{\frac{\pi}{2}-Si\,(as)\right\}-\sin as\, Ci\,(as)}{s}$
109.	$\frac{1}{2} I_n\left(\frac{t^2+a^2}{a^2}\right)$	$\frac{\sin as\left\{\frac{\pi}{2}-Si\,(as)\right\}+\cos as\, Ci\,(as)}{s}$
110.	$\frac{1}{t} I_n\left(\frac{t^2+a^2}{a^2}\right)$	$\left[\frac{\pi}{2}-Si\,(as)\right]^2+Ci^2(as)$
111.	$\mathcal{N}(t)$	0
112.	$\delta(t)$	1
113.	$\delta(t-a)$	e^{-as}
114.	$\mathcal{U}(t-a)$	$\frac{e^{-as}}{s}$
115.	$\frac{x}{a}+\frac{2}{\pi}\sum_{n=1}^{\infty}\frac{(-1)^n}{n}\sin\frac{n\pi x}{a}\cos\frac{n\pi t}{a}$	$\frac{\sinh sx}{s\sinh sa}$
116.	$\frac{4}{\pi}\sum_{n=1}^{\infty}\frac{(-1)^n}{2n-1}\sin\frac{(2n-1)}{2a}\pi x\sin\frac{(2n-1)\pi t}{2a}$	$\frac{\sinh sx}{s\cosh sa}$
117.	$\frac{t}{a}+\frac{2}{\pi}\sum_{n=1}^{\infty}\frac{(-1)^n}{n}\cos\frac{n\pi x}{a}\sin\frac{n\pi t}{a}$	$\frac{\cosh sx}{s\sinh sa}$
118.	$1+\frac{4}{\pi}\sum_{n=1}^{\infty}\frac{(-1)^n}{2n-1}\cos\frac{(2n-1)\pi x}{2a}\cos\frac{(2n-1)\pi t}{2a}$	$\frac{\cosh sx}{s\cosh sa}$
119.	$\frac{xt}{a}+\frac{2a}{\pi^2}\sum_{n=1}^{\infty}\frac{(-1)^n}{n^2}\sin\frac{n\pi x}{a}\sin\frac{n\pi t}{a}$	$\frac{\sinh sx}{s^2\sinh sa}$
120.	$x+\frac{8a}{\pi^2}\sum_{n=1}^{\infty}\frac{(-1)^n}{(2n-1)^2}\sin\frac{(2n-1)}{2a}\cos\frac{(2n-1)\pi t}{2a}$	$\frac{\sinh sx}{s^2\cosh sa}$

S. No.	$\mathcal{L} f(t)$	$F(s)$
121.	$\frac{t^2}{2a}+\frac{2a}{\pi^2}\sum_{n=1}^{\infty}\frac{(-1)^n}{n^2}\cos\frac{n\pi x}{a}\left(1-\cos\frac{n\pi t}{a}\right)$	$\frac{\cosh sx}{s^2\sinh sa}$
122.	$t+\frac{8a}{\pi^2}\sum_{n=1}^{\infty}\frac{(-1)^n}{(2n-1)^2}\cos\frac{(2n-1)\pi x}{2a}\sin\frac{(2n-1)\pi t}{2a}$	$\frac{\cosh sx}{s^2\cosh sa}$
123.	$\frac{1}{2}(t^2+x^2-a^2)-\frac{16a^2}{\pi^2}\sum_{n=1}^{\infty}\frac{(-1)^n}{(2n-1)^3}\cos\frac{(2n-1)\pi x}{2a}\cos\frac{(2n-1)\pi t}{2a}$	$\frac{\cosh sx}{s^3\cosh sa}$
124.	$\frac{2\pi}{a^2}\sum_{n=1}^{\infty}(-1)^n n e^{-n^2\pi^2 t/a^2}\sin\frac{n\pi x}{a}$	$\frac{\sinh x\sqrt{s}}{\sinh a\sqrt{s}}$
125.	$\frac{\pi}{a^2}\sum_{n=1}^{\infty}(-1)^{n-1}(2n-1)e^{-(2n-1)^2 t/4a^2}\cos\frac{(2n-1)\pi x}{2a}$	$\frac{\cosh x\sqrt{s}}{\cosh a\sqrt{s}}$
126.	$\frac{2}{a}\sum_{n=1}^{\infty}(-1)^{a-1}e^{-(2n-1)^2\pi^2 t/4a^4}\sin\frac{(2n-1)\pi x}{2a}$	$\frac{\sinh x\sqrt{s}}{\sqrt{s}\cosh a\sqrt{s}}$
127.	$\frac{1}{a}+\frac{2}{a}\sum_{n=1}^{\infty}(-1)^n e^{-n^2\pi^2 t/a^2}\cosh\frac{n\pi x}{a}$	$\frac{\cosh x\sqrt{s}}{\sqrt{s}\sinh a\sqrt{s}}$
128.	$\frac{x}{a}+\frac{2}{\pi}\sum_{n=1}^{\infty}\frac{(-1)^n}{n}e^{-n^2\pi^2 t/a^2}\sin\frac{n\pi x}{a}$	$\frac{\sinh x\sqrt{s}}{s\sinh a\sqrt{s}}$
129.	$1+\frac{4}{\pi}\sum_{n=1}^{\infty}\frac{(-1)^n}{2n-1}e^{-(2a-1)^2\pi^2 t/4a^2}\cosh\frac{(2n-1)\pi x}{2a}$	$\frac{\cosh x\sqrt{s}}{s\cosh a\sqrt{s}}$
130.	$\frac{xt}{a}+\frac{2a^2}{\pi^3}\sum_{n=1}^{\infty}\frac{(-1)^n}{n^3}(1-e^{-n^2\pi^2 t/a^2})\sin\frac{n\pi x}{a}$	$\frac{\sinh x\sqrt{s}}{s^2\sinh a\sqrt{s}}$
131.	$\frac{1}{2}(x^2-a^2)+t-\frac{16a^2}{\pi^3}\sum_{n=1}^{\infty}\frac{(-1)^n}{(2n-1)^3}e^{-(2n-1)^2\pi^2 t/4a^2}\cos\frac{(2n-1)\pi x}{2a}$	$\frac{\cosh x\sqrt{s}}{s^2\cosh a\sqrt{s}}$
132.	$1-2\sum_{n=1}^{\infty}\frac{e^{-\lambda_n^2 t/a^2}J_0(\lambda_n x/a}{\lambda_n J_1(\lambda_n)}$ where $\lambda_1, \lambda_2, \ldots$ are the positive roots of $J_0(\lambda)=0$	$\frac{J_0(ix\sqrt{s})}{s\,J_0(ia\sqrt{s})}$
133.	$\frac{1}{4}(x^2-a^2)+t+2a^2\sum_{n=1}^{\infty}\frac{e^{-\lambda_n^2/a^2}J_0(\lambda_n x/a)}{\lambda_n^3 J_1(\lambda_n)}$ where $\lambda_1, \lambda_2, \ldots$ are the positive roots of $J_0(\lambda)=0$	$\frac{J_2(ix\sqrt{s})}{s^2 J_0(ia\sqrt{s})}$

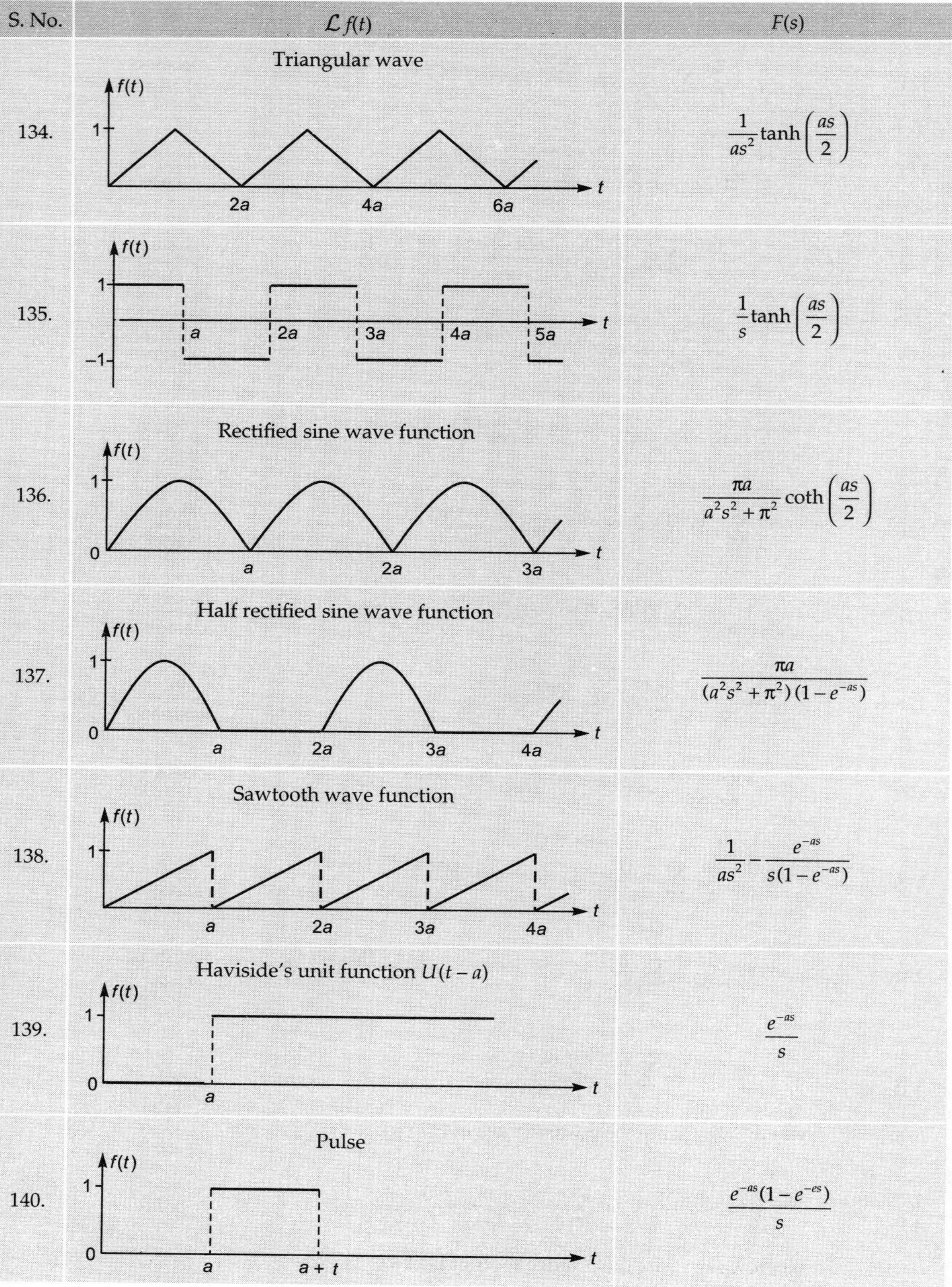

S. No.	$\mathcal{L} f(t)$	$F(s)$
134.	Triangular wave	$\frac{1}{as^2}\tanh\left(\frac{as}{2}\right)$
135.		$\frac{1}{s}\tanh\left(\frac{as}{2}\right)$
136.	Rectified sine wave function	$\frac{\pi a}{a^2s^2+\pi^2}\coth\left(\frac{as}{2}\right)$
137.	Half rectified sine wave function	$\frac{\pi a}{(a^2s^2+\pi^2)(1-e^{-as})}$
138.	Sawtooth wave function	$\frac{1}{as^2}-\frac{e^{-as}}{s(1-e^{-as})}$
139.	Haviside's unit function $U(t-a)$	$\frac{e^{-as}}{s}$
140.	Pulse	$\frac{e^{-as}(1-e^{-es})}{s}$

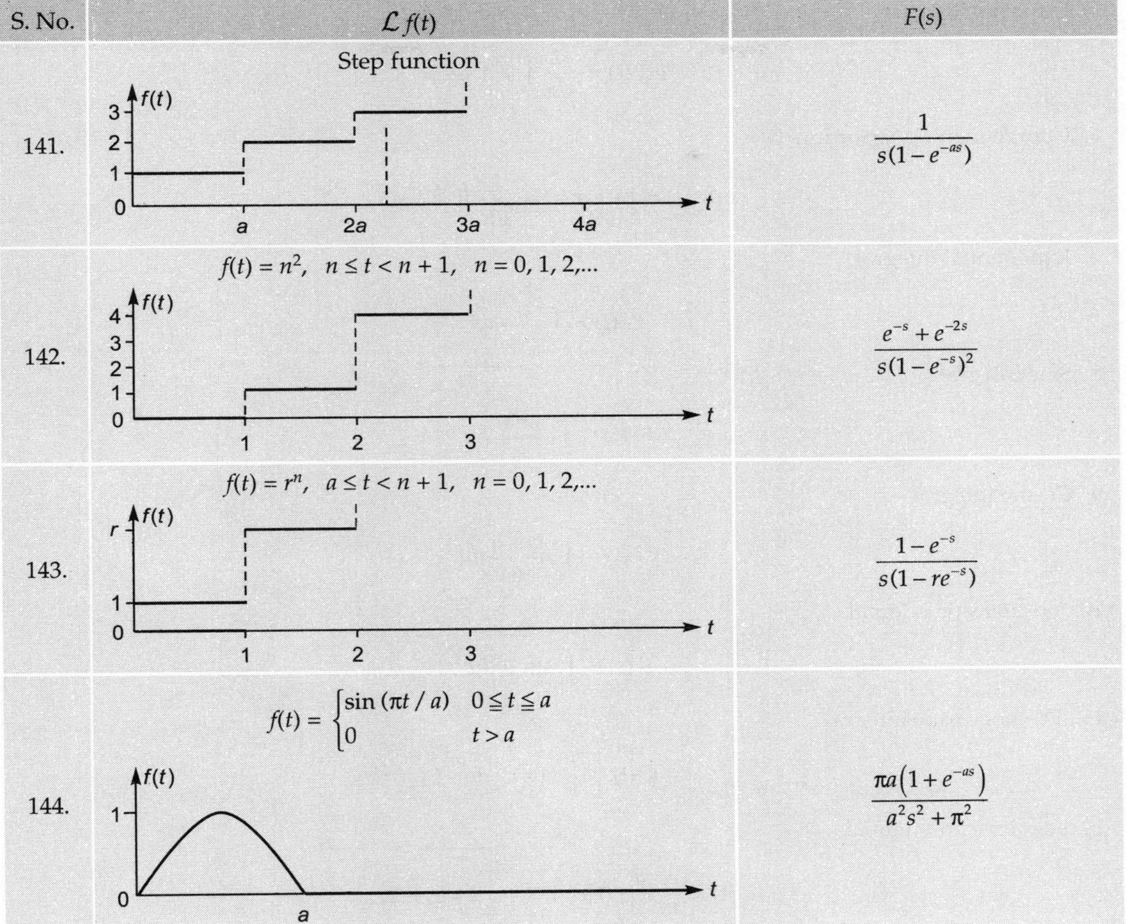

S. No.	$\mathcal{L} f(t)$	$F(s)$
141.	Step function	$\dfrac{1}{s(1-e^{-as})}$
142.	$f(t) = n^2, \quad n \le t < n+1, \quad n = 0, 1, 2,...$	$\dfrac{e^{-s}+e^{-2s}}{s(1-e^{-s})^2}$
143.	$f(t) = r^n, \quad a \le t < n+1, \quad n = 0, 1, 2,...$	$\dfrac{1-e^{-s}}{s(1-re^{-s})}$
144.	$f(t) = \begin{cases} \sin(\pi t/a) & 0 \leqq t \leqq a \\ 0 & t > a \end{cases}$	$\dfrac{\pi a\left(1+e^{-as}\right)}{a^2s^2+\pi^2}$

TABLE OF SPECIAL FUNCTIONS

1. Gamma function

$$\Gamma(n) = \int_0^\infty u^{n-1} e^{-u}\, du, \qquad n > 0$$

2. Beta function

$$B(m, n) = \int_0^1 u^{m-1}(1-u)^{n-1}\, du = \frac{\Gamma(m)\,\Gamma(n)}{\Gamma(m+n)}, \qquad m, n > 0$$

3. Bessel function

$$J_n(x) = \frac{x^n}{2^n\,\Gamma(n+1)}\left\{1 - \frac{x^2}{2(2n+2)} + \frac{x^4}{2\cdot 4(2n+2)(2n+4)} - \dots\right\}$$

4. Modified Bassel function

$$I_n(x) = i^{-n} J_n(ix) = \frac{x^n}{2^n\,\Gamma(n+1)}\left\{1 + \frac{x^2}{2(2n+2)} + \frac{x^4}{2\cdot 4(2n+2)(2n+4)} + \dots\right\}$$

5. Error function $$\text{erf}(t) = \frac{2}{\sqrt{\pi}} \int_0^1 e^{-u^2}\, du$$
6. Complementary error function $$\text{erfc}(t) = 1 - \text{erf}(t) = \frac{2}{\sqrt{\pi}} \int_1^\infty e^{-u^2}\, du$$
7. Exponential integral $$Ei(t) = \int_1^\infty \frac{e^{-u}}{u}\, du$$
8. Sine integral $$Si(t) = \int_0^t \frac{\sin u}{u}\, du$$
9. Cosine integral $$Ci(t) = \int_t^\infty \frac{\cos u}{u}\, du$$
10. Fresnel sine integral $$S(t) = \int_0^t \sin u^2 du$$
11. Fresnel cosine integral $$C(t) = \int_0^t \cos u^2\, du$$
12. Laguerre polynomials $$L_n(t) = \frac{e^t}{n!}\frac{d^n}{dt^n}(t^n e^{-t}), \qquad n = 0, 1, 2, \ldots$$

17.3 *z*-TRANSFORM

For the real-valued sequence $\{F(k)\}$ and complex variable z, the z-transform, $f(z) = Z\{f(k)\}$ is defined by

$$Z\{F(k)\} = f(z) = \sum_{k=0}^{\infty} f(k) z^{-k}$$

i. *z*-Transform and the Laplace Transform

For function $U(t)$ the output of the ideal sampler $U^*(t)$ is a set of values $U(kT)$, $k = 0, 1, 2, \ldots$, that is,

$$U^*(t) = \sum_{k=0}^{\infty} U(t)\, \delta(t - kT)$$

The Laplace transform of the output is

$$\mathcal{L}\,(U^*(t)\} = \int_0^\infty e^{-st} U^*(t)\, dt = \int_0^\infty e^{-st} \sum_{k=0}^{\infty} U(t)\, \delta(t - kT)\, dt$$

$$= \sum_{k=0}^{\infty} e^{-skT} U(kT)$$

Defining $z = e^{sT}$ gives

$$\mathcal{L}\{U^*(t)\} = \sum_{k=0}^{\infty} U(kT) z^{-k}$$

which is the z-transform of the sampled signal $U(kT)$.

ii. Properties

Linearity: $Z\{af_1(k) + bf_2(k)\} = aZ\{f_1(k)\} + bZ\{f_2(k)\} = aF_1(z) + bF_2(z)$

Right-shifting property: $Z\{f(k-n)\} = z^{-n} F(z)$

Left-shifting property: $Z\{f(k+n)\} = z_n F(z) - \sum_{k=0}^{n-1} f(k) z^{n-k}$

Time scaling: $Z\{a^k f(k)\} = F(z)/a$

Multiplication by k: $Z\{kf(k)\} = -zdF(z)/dz$

Initial value: $f(0) = \lim_{z \to \infty} (1 - z^{-1}) F(z) = f(\infty)$

Final value: $\lim_{k \to \infty} F(k) = \lim_{z \to 1} (1 - z^{-1}) F(z)$

Convolution: $Z\{f_1(k)^* f_2(k)\} = F_1(z) F_2(z)$

iii. *z*-Transforms of Sampled Function

$f(k)$	$Z\{F(kT)\} = f(z)$
1 at k; else 0	z^{-k}
1	$\frac{z}{z-1}$
kT	$\frac{Tz}{(z-1)^2}$
$(kT)^2$	$\frac{T^2 z(z+1)}{(z-1)^3}$
$\sin \omega kT$	$\frac{z \sin \omega T}{z^2 - 2z \cos \omega T + 1}$
$\cos \omega T$	$\frac{z(z - \cos \omega T)}{z^2 - 2z \cos \omega T + 1}$
e^{-akT}	$\frac{z}{z - e^{-aT}}$

$kT\,e^{-akT}$	$\dfrac{zT\,e^{-aT}}{(z-e^{-aT})^2}$
$(kT)^2\,e^{-akT}$	$\dfrac{T^2e^{-aT}\,z(z+e^{-aT})}{(z-e^{-aT})^3}$
$e^{-akT}\sin\omega kT$	$\dfrac{ze^{-aT}\sin\omega T}{z^2-2ze^{-aT}\cos\omega T+e^{-2aT}}$
$e^{-akT}\cos\omega kT$	$\dfrac{z(z-e^{-aT}\cos\omega T)}{z^2-2ze^{-aT}\cos\omega T+e^{-2aT}}$
$a^k\sin\omega kT$	$\dfrac{az\sin\omega T}{z^2-2az\cos\omega T+a^2}$
$a^k\cos\omega kT$	$\dfrac{z(z-a\cos\omega T)}{z^2-2az\cos\omega T+a^2}$

17.4 FOURIER SERIES

The periodic function $f(t)$, with period 2π may be represented by the trignometric series

$$a_0+\sum_1^{\infty}(a_n\cos nt+b_n\sin nt)$$

where the coefficients are determined from

$$a_0=\frac{1}{2\pi}\int_{-\pi}^{\pi}f(t)\,dt$$

$$a_n=\frac{1}{\pi}\int_{-\pi}^{\pi}f(t)\cos nt\,dt$$

$$b_n=\frac{1}{\pi}\int_{-\pi}^{\pi}f(t)\sin nt\,dt \qquad (n=1,2,3\ldots)$$

Such a trigonometric series is called the Fourier series corresponding to $f(t)$ and the coefficients are termed Fourier coefficients of $f(t)$. If the function is piecewise continuous in the interval $-\pi\le t\le\pi$, and has left and right-hand derivatives at each point in that interval, then the series is convergent with sum $f(t)$ except at points t_i at which $f(t)$ is discontinuous. At such points of discontinuity, the sum of the series is the arithmetic mean of the right- and left-hand limits of $f(t)$ at t_i. The integrals in the formulas for the Fourier coefficients can have limits of integration that span a length of 2π.

i. Functions with Period Other Than 2π

If $f(t)$ has period P the Fourier series is

$$f(t)\sim a_0+\sum_1^{\infty}\left(a_n\cos\frac{2\pi n}{P}t+b_n\sin\frac{2\pi n}{P}t\right)$$

where

$$a_0 = \frac{1}{P}\int_{-P/2}^{P/2} f(t)\,dt$$

$$a_n = \frac{2}{P}\int_{-P/2}^{P/2} f(t)\cos\frac{2\pi n}{P}t\,dt$$

$$b_n = \frac{2}{P}\int_{-P/2}^{P/2} f(t)\sin\frac{2\pi n}{P}t\,dt$$

Again, the interval of integration in these formulas may be replaced by an interval of length P, for example, 0 to P.

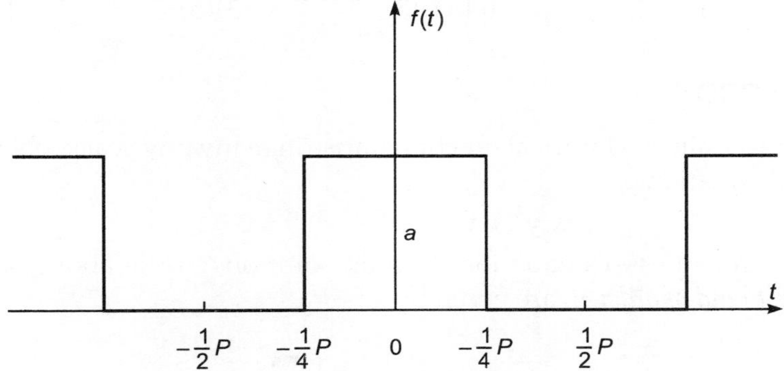

Fig. 17.1 Square wave

$$f(t) \sim \frac{a}{2} + \frac{2a}{\pi}\left(\cos\frac{2\pi t}{P} - \frac{1}{3}\cos\frac{6\pi t}{P} + \frac{1}{5}\cos\frac{10\pi t}{P} + \ldots\right)$$

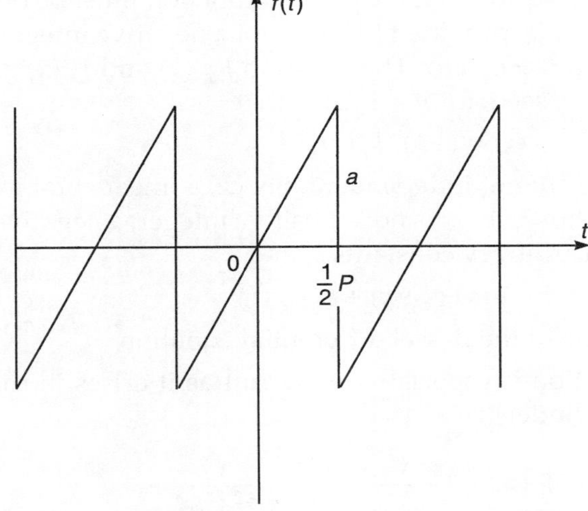

Fig. 17.2 Sawtooth wave

$$f(t) \sim \frac{2a}{\pi}\left(\sin\frac{2\pi t}{P} - \frac{1}{2}\sin\frac{4\pi t}{P} + \frac{1}{3}\sin\frac{6\pi t}{P} - \ldots\right)$$

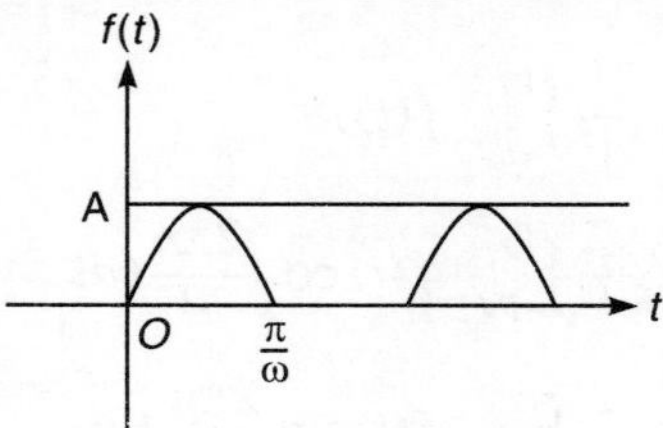

Fig. 17.3 Halfwave rectifier

$$f(t) \sim \frac{A}{\pi} + \frac{A}{2}\sin \omega t - \frac{2A}{\pi}\left(\frac{1}{(1)(3)}\cos 2\omega t + \frac{1}{(3)(5)}\cos 4\omega t + \ldots\right)$$

17.5 BESSEL FUNCTION

Bessel functions, also called cylindrical functions, arise in many physical problems as solutions of the differential equation

$$x^2 y'' + xy' + (x^2 - n^2)\, y = 0$$

which is known as Bessel's equation. Certain solutions of the above, known as *Bessel functions of the first kind of order n*, are given by

$$J_n(x) = \sum_{k=0}^{\infty} \frac{(-1)^k}{k!\Gamma(n+k+1)}\left(\frac{x}{2}\right)^{n+2k}$$

$$J_{-n}(x) = \sum_{k=0}^{\infty} \frac{(-1)^k}{k!\Gamma(-n+k+1)}\left(\frac{x}{2}\right)^{-n+2k}$$

In the above it is noteworthy that the gamma function must be defined for the negative argument q: $\Gamma(q) = \Gamma(q+1)/q$, provided that q is not a negative integer. When q is a negative integer, $1/\Gamma(q)$ is defined to be zero. The function $J_{-n}(x)$ and $J_n(x)$ are solution of Bessel's equation for all real n. It is seen, for $n = 1,2,3,\ldots$ that

$$J_{-n}(x) = (-1)^n J_n(x)$$

and, therefore, these are not independent; hence, a linear combination of these is not a general solution. When, however, n is not a positive integer, a negative integer, nor zero, the linear combination with arbitrary constants c_1 and c_2.

$$y = c_1 J_n(x) + c_2 J_{-n}(x)$$

is the general solution of the Bessel differential equation.

The zero order function is especially important as it arises in the solution of the heat equation (for a "long" cylinder):

$$J_0(x) = 1 - \frac{x^2}{2^2} + \frac{x^4}{2^2 4^2} - \frac{x^6}{2^2 4^2 6^2} + \ldots$$

while the following relations show a connection to the trigonometric functions:

$$J_{\frac{1}{2}}(x) = \left[\frac{2}{\pi x}\right]^{1/2} \sin x$$

$$J_{-\frac{1}{2}}(x) = \left[\frac{2}{\pi x}\right]^{1/2} \cos x$$

The following recursion formula gives $J_{n+1}(x)$ for any order in terms of lower order functions:

$$\frac{2n}{x} J_n(x) = J_{n-1}(x) + J_{n+1}(x)$$

17.6 LEGENDRE POLYNOMIALS

If Laplace's equation, $\nabla^2 V = 0$, is expressed in spherical coordinates, it is

$$r^2 \sin\theta \frac{\delta^2 V}{\delta r^2} + 2r\sin\theta\frac{\delta V}{\delta r} + \sin\theta\frac{\delta^2 V}{\delta\theta^2} + \cos\theta\frac{\delta V}{\delta\theta} + \frac{1}{\sin\theta}\frac{\delta^2 V}{\delta\phi^2} = 0$$

and any of its solutions, $V(r, \theta, \phi)$, are known as *spherical harmonics*. The solution as a product

$$V(r, \theta, \phi) = R(r)\,\Theta(\theta)$$

which is independent of ϕ, leads to

$$\sin^2\theta\Theta'' + \sin\theta\cos\theta\Theta' + [n(n+1)\sin^2\theta]\,\Theta = 0$$

Rearrangement and substitution of $x = \cos\theta$ leads to

$$(1 \Theta x^2)\frac{d^2\Theta}{dx^2} - 2x\frac{d\Theta}{dx} + n(n+1)\Theta = 0$$

known as *Legendre's equation*. Important special cases are those in which n is zero or a positive integer, and for such cases, Legendre's equation is satisfied by polynomials called Legendre polynomials, $P_n(x)$. A short list of Legendre polynomials, expressed in terms of x and $\cos\phi$, is given below. These are given by the following general formula:

$$P_n(x) = \sum_{j=0}^{L} \frac{(-1)^j (2n-2j)!}{2^n j!(n-j)!(n-2j)!} x^{n-2j}$$

where $L = n/2$ if n is even and $L = (n-1)/2$ if n is odd. Some are given below:

$$P_0(x) = 1$$

$$P_j(x) = x$$

$$P_2(x) = \frac{1}{2}(3x^2 - 1)$$

$$P_3(x) = \frac{1}{2}(5x^3 - 3x)$$

$$P_4(x) = \frac{1}{8}(35x^4 - 30x^2 + 3)$$

$$P_5(x) = \frac{1}{8}(63x^5 - 70x^3 + 15x)$$

$$P_0(\cos\theta) = 1$$

$$P_1(\cos\theta) = \cos\theta$$

$$P_2(\cos\theta) = \frac{1}{4}(3\cos 2\theta + 1)$$

$$P_3(\cos\theta) = \frac{1}{8}(5\cos 3\theta + 3\cos\theta)$$

$$P_4(\cos\theta) = \frac{1}{64}(35\cos 4\theta + 20\cos 2\theta + 9)$$

Additional Legendre polynomials may be determined from the *recursion formula*

$$(n+1)P_{n+1}(x) - (2n+1)xP_n(x) + nP_{n-1}(x) = 0 \quad (n = 1, 2, ...)$$

or the *Rodregues formula*

$$P_n(x) = \frac{1}{2^n n!}\frac{d^n}{dx^n}(x^2-1)^n$$

17.7 LAGUERRE POLYNOMIALS

Laguerre polynomials, denoted $L_n(x)$, are solutions of the differential equation

$$xy'' + (1-x)y' + ny = 0$$

and are given by

$$L_n(x) = \sum_{j=0}^{n}\frac{(-1)^j}{j!}C_{(n,j)}x^j \quad (n = 0, 1, 2, ...)$$

Thus,

$$L_0(x) = 1$$
$$Lj(x) = 1 - x$$
$$L_2(x) = 1 - 2x + \frac{1}{2}x^2$$
$$L_3(x) = 1 - 3x + \frac{3}{2}x^2 - \frac{1}{6}x^3$$

Additional Laguerre polynomials may be obtained from the recursion formula

$$(n+1)L_{n+1}(x) - (2n+1-x)L_n(x) + nL_{n-1}(x) = 0$$

17.8 HERMITE POLYNOMIALS

The Hermite polynomials, denoted $H_n(x)$, are given by

$$H_0 = 1, \quad H_n(x) = (-1)^n e^{x^2}\frac{d^n e^{-x^2}}{dx^n} \quad (n = 1, 2, ...)$$

and are solutions of the differential equation

$$y'' - 2xy' + 2ny = 0 \quad (n = 0, 1, 2, ...)$$

The first few Hermite polynomials are

$$H_0 = 1 \qquad H_1(x) = 2x$$

$$H_2(x) = 4x^2 - 2 \qquad H_3(x) = 8x^3 - 12x$$

Additional Hermite polynomials may be obtained from the relation

$$H_{n+1}(x) = 2xH_n(x) - H_n'(x),$$

where prime denotes differentiation with respect to x.

17.9 ORTHOGONALITY

A set of functions $(f_n(x)]\}$ $(n = 1, 2, \ldots)$ is orthogonal in an interval (a, b) with respect to a given weight function $w(x)$ if

$$\int_a^b w(x)\, f_m(x)\, f_n(x)\, dx = 0 \quad \text{when } m \neq n$$

The following polynomials are orthogonal on the given interval for the given $w(x)$:

Legendre polynomials: $P_n(x)$ $w(x) = 1$
$a = -1, b = 1$

Laguerre polynomials: $L_n(x)$ $w(x) = \exp(-x)$
$a = 0, b = \infty$

Hermite polynomials: $H_n(x)$ $w(x) = \exp(-x^2)$
$a = -\infty, b = \infty$

The Bessel function *of order* n, $J_n(\lambda_1 x), J_n(\lambda_2, x), \ldots$, are orthogonal with respect to $w(x) = x$ over the interval $(0, c)$ provided that the λ_i are the positive roots of $J_n(\lambda c) = 0$:

$$\int_0^c xJ_n(\lambda_1 x)\, J_n(\lambda_k x)\, dx = 0 \qquad (j \neq k)$$

where n is fixed and $n \geq 0$.

17.10 DIRAC DELTA FUNCTION

i. Definition

The Dirac Delta function $\delta(x - t)$ is not a function in the true sense but a definition of a distribution.

$$\delta(x-t) = \lim_{\Delta \to 0} \begin{cases} 0 & \text{for } x < \left(t - \dfrac{\Delta}{2}\right) \\ \dfrac{1}{\Delta} & \text{for } \left(t - \dfrac{\Delta}{2}\right) < x < \left(t + \dfrac{\Delta}{2}\right) \\ 0 & \text{for } x > \left(t + \dfrac{\Delta}{2}\right) \end{cases}$$

ii. Operations

$$\delta(x) = \delta(-x) \qquad \delta(x \neq 0) = 0$$

$$a\,\delta(x) = \delta\left(\frac{x}{a}\right) \qquad x\,\delta(x) = 0$$

iii. Properties

$$\int_{-\infty}^{+\infty} \delta(x-t)\,dt = 1 \qquad \int_{0}^{+\infty} \delta(t)\,dt = 1$$

$$\int_{-\infty}^{+\infty} f(t)\,\delta(x-t)\,dt = f(x) \qquad \int_{0}^{+\infty} f(t)\,\delta(t)\,dt = f(0)$$

$$\int_{-\infty}^{+\infty} \delta(x-t)s\,\delta(y-t)\,dt = \delta(x-y)$$

iv. Derviatives

$$\frac{d\delta(x)}{dx} = -\frac{\delta(x)}{x}$$

$$\frac{d^n\,\delta(x)}{dx^n} = (-1)^n \frac{n!\,\delta(x)}{x^n}$$

v. Integrals

$$\int_{-\infty}^{+\infty} f(t)\frac{d\delta(x-1)}{dx}\,dt = \frac{df(x)}{dx}$$

$$\int_{-\infty}^{+\infty} f(t)\frac{d^n\,\delta(x-t)}{dx^n} = (-1)^n \frac{d^n\,f(x)}{dx^n}$$

vi Laplace Transforms

$\mathcal{L}\,\{[\delta\,(t)]\} = 1 \qquad \mathcal{L}\,\{[\mathrm{D}\,(t-a)]\} = e^{-as}$

CHAPTER

18

Numerical Methods

18.1 BASIC CONCEPTS

i. Methods

Whenever the solution of an engineering problem leads to an expression, equation, or system of equations which cannot be evaluated or solved in closed form, *numerical methods* must be employed. The best-known numerical techniques, *approximate evaluation of functions, numerical solution of equations, finite differences*, and *numerical integration*, are given in this chapter.

ii. Errors

The *absolute error* ε in the result is the difference between the result a (assumed to be known) and the approximate result $\bar{a}$. The *relative error* $\bar{\varepsilon}$ is the absolute error ε divided by $\bar{a}$. Aside from possible outright mistakes there are there basic types of errors in numerical calculations: *inherent errors* (due to initial data error), *truncation errors* (due to finite approximation of limiting processes), and *round-off errors* (due to use of finite number of digits).

18.2 APPROXIMATIONS (BASED ON SERIES EXPANSION)

i. Algebraic Functions

	$\varepsilon_T \le$ *0.1 percent*	$\varepsilon_T \le$ *1 percent*
$\dfrac{1}{1+x} = 1 - x$	$-0.03 \le x \le +0.03$	$-0.10 \le x \le +0.10$
$= 1 - x + x^2$	$-0.10 \le x \le +0.10$	$-0.21 \le x \le +0.21$
$\sqrt{1+x} = 1 + \dfrac{x}{2}$	$-0.08 \le x \le +0.10$	$-0.24 \le x \le +0.32$
$= 1 + \dfrac{x}{2} - \dfrac{x^2}{8}$	$-0.22 \le x \le +0.27$	$-0.44 \le x \le +0.66$

$\frac{1}{\sqrt{1+x}} = 1 - \frac{x}{2}$	$-0.04 \le x \le +0.06$	$-0.15 \le x \le +0.17$
$= 1 - \frac{x}{2} + \frac{3x^2}{8}$	$-0.14 \le x \le +0.15$	$-0.30 \le x \le +0.32$

ii. Transcendent Functions

	$\varepsilon_T \le$ *0.1 percent*	$\varepsilon_T \le$ *1 percent*
$\varepsilon^x = 1 + x$	$-0.04 \le x \le +0.04$	$-0.13 \le x \le +0.14$
$= 1 + x + \frac{x^2}{2}$	$-0.17 \le x \le +0.19$	$-0.35 \le x \le +0.43$
$\sin x = x$	$\left.\begin{matrix}-0.077\\-4.4°\end{matrix}\right\} \le x \le \left\{\begin{matrix}+0.077\\+4.4°\end{matrix}\right.$	$\left.\begin{matrix}-0.244\\-14.0°\end{matrix}\right\} \le x \le \left\{\begin{matrix}+0.244\\+14.0°\end{matrix}\right.$
$= x - \frac{x^2}{6}$	$\left.\begin{matrix}-0.578\\-33.1°\end{matrix}\right\} \le x \le \left\{\begin{matrix}+0.578\\+33.1°\end{matrix}\right.$	$\left.\begin{matrix}-1.032\\-59.0°\end{matrix}\right\} \le x \le \left\{\begin{matrix}+1.032\\+59.0°\end{matrix}\right.$
$\cos x = 1$	$\left.\begin{matrix}-0.045\\-2.6°\end{matrix}\right\} \le x \le \left\{\begin{matrix}+0.045\\+2.6°\end{matrix}\right.$	$\left.\begin{matrix}-0.141\\-8.1°\end{matrix}\right\} \le x \le \left\{\begin{matrix}+0.141\\+8.1°\end{matrix}\right.$
$= 1 - \frac{x^2}{2}$	$\left.\begin{matrix}-0.384\\-22.0°\end{matrix}\right\} \le x \le \left\{\begin{matrix}+0.384\\+22.0°\end{matrix}\right.$	$\left.\begin{matrix}-0.650\\-37.2°\end{matrix}\right\} \le x \le \left\{\begin{matrix}+0.650\\+37.2°\end{matrix}\right.$
$\tan x = x$	$\left.\begin{matrix}-0.054\\-3.1°\end{matrix}\right\} \le x \le \left\{\begin{matrix}+0.054\\+3.1°\end{matrix}\right.$	$\left.\begin{matrix}-0.183\\-10.5°\end{matrix}\right\} \le x \le \left\{\begin{matrix}+0.183\\+10.5°\end{matrix}\right.$
$= x + \frac{x^3}{3}$	$\left.\begin{matrix}-0.385\\-22.0°\end{matrix}\right\} \le x \le \left\{\begin{matrix}+0.385\\+22.0°\end{matrix}\right.$	$\left.\begin{matrix}-0.533\\-30.5°\end{matrix}\right\} \le x \le \left\{\begin{matrix}+0.533\\+30.5°\end{matrix}\right.$

18.3 NUMERICAL SOLUTION OF ALGEBRAIC EQUATIONS

i. General Properties

Every algebraic equation of the nth degree (with n real and complex roots),

$$f(x) = a_0 x^n + a_1 x^{n-1} + \dots + a^{n-1} x + a_n = 0 \qquad a_0 \neq 0$$

can be representd as a product of n linear factors. If $x_1, x_2, \dots, x_n$ are the roots of this equation, then

$$f(x) = (x - x_1)(x - x_2) \dots (x - x_n) = 0$$

ii. Relations between the Roots and Coefficients

$$x_1 + x_2 + \dots + x_n = \sum_{i=1}^{n} x_i - \frac{a_1}{a_0}$$

$$x_1 x_2 + x_1 x_3 + \dots + x_{n-1} x_n = \sum_{i,j=1}^{n} x_i x_j = \frac{a_2}{a_0}$$

$$x_1x_2x_3 + x_1x_2x_4 + \dots + x_{n-2}x_{n-1}x_n = \sum_{i,j,k=1}^{n} x_ix_jx_k = -\frac{a_3}{a_0}$$

....

$$(-1)^n x_1x_2x_3 \dots x_n = \frac{a_n}{a_0}$$

iii. Methods of Solution

If $n > 4$, there is no formula which gives the roots of this general equation. The following methods are useful;

a. *Roots by trial*

Find a number x_1 that satisfies $f(x_1) = 0$. Then divide $f(x)$ by $x - x_1$, thus obtaining an equation of degree one less than that of the original equation. Repeat the same procedure with the reduced equation.

b. *Roots by regular falsi approximation*

If coefficients $a_0a_1, \dots$ are real, introduce $x = a$ and $x = b$ so that $f(a)$ and $f(b)$ have opposite signs. Then

$$c_1 = a - f(a)\frac{b-a}{f(b)-(a)}$$

is the first approximation of x_1. By repeating the application of this idea, real roots may be obtained to any degree of accuracy.

c. *Roots by Newton's approximation*

Assume c_1 as an approxmate root, calculate a better approximation by means of

$$c_2 = c_1 - \frac{f(c_1)}{f'(c_1)}$$

and repeat this process to a desired accuracy.

d. *Roots by Steinman's approximation*

First locate the roots approximately, and write the equation in the form

$$a_0x^n = a_1x^{n-1} + a_2x^{n-2} + \dots + a_n$$

Then, with $x = d_1$,

$$d_2 = \frac{a_1 + 2a_2 / d_1 + 3a_3 / d_1^2 + 4a_4 / d_1^3 + \dots}{a_0 + a_2 / d_1^2 + 2a_3 / d_1^3 + 3a_4 / d_1^4 + \dots}$$

and repeat this process to a desired accuracy.

18.4 NUMERICAL SOLUTION OF SYSTEMS OF LINEAR EQUATIONS

i. General

The solution of a system of linear equations by the method outlined earlier becomes impractical if $n > 4$. In such a case, the employment of the numerical methods outlined in the following discussion offers a workable solution.

ii. Gauss's Elimination Method

It involves replacing the given system by combination and modification of the initial equations leading to the following *triangular system*.

$$x_1 + \alpha_{12} x_2 + \ldots + \alpha_{1,\, n-1} x_{n-1} + \alpha_{1,\, n} x_n = \beta_1$$

$$x_2 + \ldots + \alpha_{2,\, n-1} x_{n-1} + \alpha_{2,\, n} x_n = \beta_2$$

....

$$x_{n-1} + a_{n-1,\, n} x_n = \beta_{n-1}$$

$$x_n = \beta_n$$

The last equation (n) yields the value of x_n, which is then substituted in the preceding equation (n – 1) from which x_{n-1} is determined, etc.

iii. Gauss-Seidel Iteration Method

This method involves rearrangement and / or modification of the given system to obtain the *largest diagonal coefficients* possible. Then dividing each equation by the respective diagonal coefficient leads to the carryover form

$$x_j = r_{j1} x_1 + r_{j2} x_2 + \ldots + m_j + \ldots + r_{j,\, n-1} x_{n-1} + r_{jn} xn$$

where $r_{j1} = \bar{a}_{j1} / a_{jj}, r_{j2} = -a_{j2}/a_{jj}, \ldots$ are the *carryover factors* and $m_j = b_j / a_{jj}$ is the starting value.

Starting with the *trial soltuion*

$$x_1^{(1)} = m_1, x_2^{(1)} = m_2, \ldots$$

The successive approximation become

$$x_1^{(2)} = x_1^{(1)} + \sum_{j=1}^{n} r_{1j}\, x_j^{(1)}, x_2^{(2)} = x_2^{(1)} + \sum_{j=1}^{n} r_{2j} x_j^{(1)}, \ldots$$

$$x_1^{(3)} = x_1^{(1)} + x_1^{(2)} + \sum_{j=1}^{n} r_1 j\, x_j^{(2)}, x_2^{(3)} = x_2^{(1)} + x_2^{(2)} + \sum_{j=1}^{n} r_{2j} x_j^{(2)}, \ldots$$

....

Under certain conditions this procedure is rapidly convergent, but it may also converge slowly or not at all.

iv. Matrix Iterative Method

The given system is written in a *matrix carryover form* as

$$X = rX + m$$

where $X = \{x_1, x_2, \ldots, x_j, \ldots, x_{n-1}, x_n\}$

$$r = \begin{bmatrix} 0 & r_{12} & \ldots r_{1,n-1} & r_{1n} \\ r_{21} & 0 & \ldots r_{2,n-1} & r_{2n} \\ & \ldots\ldots\ldots\ldots & & \end{bmatrix}$$

$$m = \{m_1, m_2, \ldots, m_j, \ldots, m_{n-1}, m_n\}$$

The successive approximations become

$$X^{(1)} = m,\ X^{(2)} = m + rm,\ X^{(3)} = m + rm + r^2m, \ldots$$

and the final solution takes the form of the series,

$$X = [1 + r + r^2 + \ldots]\ m$$

where the sum of the power series in the brackets equals the inverse of A.

18.5 FINITE DIFFERENCES, FORMULAS

i. Forward Differences

For a given discrete function $y = f(x)$, a set of *equally spaced arguments* $x_n = x_0 + n\,\Delta x$ ($n = 0, \pm 1, \pm 2, \ldots; \Delta x = h > 0$) and a corresponding set of values $y_n = y\,(x_0 + n\,\Delta\,x)$ define the *forward differences* as follows:

$\Delta y_n = y_{n+1} - y_n$ First-order difference

$\Delta^2 y_n = \Delta\,y_{n+1} \ldots \Delta\,y_n = y_{n+2} - 2y_{n+1} + y_n$ Second-order differences

....

$\Delta^k y_n = \Delta^{k-1} y_{n+1} - \Delta^{k-1} y_n = \sum_{j=0}^{k} (-1)^j \binom{k}{j} y_{n+k-j}$ kth-order difference $k = 2, 3, \ldots$

ii. Backward Differences

The same set of values defines the *backward differences* as follows;

$\nabla\,y_n = y_n - y_{n-1}$ First-order difference

$\nabla^2 y_n = \nabla y_n - \nabla y_{n-1} = y_n - 2\,y_{n-1} + y_{n-2}$ Second-order difference

....

$\nabla^k y_n = \nabla^{k-1} y_n - \nabla^{k-1} y_{n-1} = \sum_{j=0}^{k} (-1)^j \binom{k}{j} y_{n+j}$ kth-order difference $k = 2, 3, \ldots$

iii. Central Differences

For the same discrete function, a similar set of values y_n corresponding to $x_n = x_0 + n\,\Delta x$ defines the *central differences* as follows:

$\delta\,y_n = y_{n+1/2} - y_{n-1/2}$ First-order difference

$\delta^2 y_n = \delta y_{n+1} - \delta y_{n-1} = y_{n+1} - 2\,y_n + y_{n-1}$ Second-order difference

....

$\delta^k y_n = \delta^{k-1} y_{n+1/2} - \delta^{k-1} y_{n-1/2} = \sum_{j=0}^{k} (-1)^j \binom{k}{j} y_{n-j+k/2}$ kth-order difference

$$\left(n = \pm\frac{1}{2}, \frac{3}{2}, \ldots \text{ if } k \text{ is odd, and } n = 0, \pm 1, \pm 2, \ldots \text{ if } k \text{ is even}\right)$$

18.6 FINITE DIFFERENCES, TABLES

i. Forward Differences

Table

x_{-2}	y_{-2}	Δ_{y-2}			
x_{-1}	y_{-1}	Δ_{y-1}	Δ^2_{y-2}	Δ^3_{y-2}	
x_0	y_0	Δ_{y0}	Δ^2_{y-1}	Δ_{y-2}	Δ^4_{y-2}
x_1	y_1	Δ_{y1}	Δ^2_{y0}		
x_2	y_2				

Table

– 4	200	– 180	162		
– 2	20	– 18		654	
0	2	798	816	– 414	– 1.068
2	800	1,200	402		
4	2,000				

ii. Backward Differences

Table

x_{-2}	y_{-2}	∇_{y-1}			
x_{-1}	y_{-1}	∇_{y0}	∇^2_{y0}		
x_0	y_0	∇_{y1}	∇^2_{y1}	∇^3_{y1}	∇^4_{y2}
x_1	y_1		Δ^2_{y0}	Δ^3_{y2}	
x_2	y_2	Δ_{y1}			

Example

– 4	200	– 180	162		
– 2	20	– 18		654	
0	2	798	816	– 414	– 1.068
2	800		402		
4	2,000	1,200			

iii. Central Differences

Table

x_{-2}	y_{-2}	$\delta_{y-3/2}$	δ^2_{y-1}		
x_{-1}	y_{-1}	$\delta_{y-1/2}$		δ^3_{y-1}	
x_0	y_0	$\delta_{y1/2}$	δ^2_{y0}	$\Delta^3_{y1/2}$	δ^4_{y0}
x_1	y_1	$\delta_{y3/2}$	δ^2_{y1}		
x_2	y_2				

Example

– 4	200	– 180			
– 2	20	162	162	654	
0	2	798	816	– 414	– 1.068
2	800	1,200	402		
4	2,000				

18.7 INTERPOLATION, GENERAL SPACING

i. General

The *process of interpolation* is used for finding in-between values of a tabulated function or for the development of a substitute function $\overline{y}(x)$ closely approximating a more complicated function $y(x)$ in a given interval.

ii. Lagrange's Interpolation Formula

From the tabulated $n + 1$ values given as

$$\frac{|x_0, x_1, x_2, \ldots, x_n}{|y_0, y_1, y_2, \ldots, y_n}$$

and not necessarily equally spaced, the *substitute (approximate) function* is

$$\bar{y}(x) = \frac{(x-x_0)(x-x_1)\ldots(x-x_n)}{(x_0-x_1)(x_0-x_2)\ldots(x_0-x_n)}y_0 + \frac{(x-x_0)(x-x_2)\ldots(x-x_n)}{(x_1-x_0)(x_1-x_2)\ldots(x_1-x_n)}y_1 + \ldots + \frac{(x-x_0)(x-x_1)\ldots(x-x_{n-1})}{(x_n-x_0)(x_n-x_1)\ldots(x_n-x_{n-1})}y_n$$

iii. Newton's Interpolation Formula

From the same values the alternate form of the *substitute function* is

$\bar{y}(x) = y_0 + (x-x_0)\,\Delta_1(x_1) + (x-x_0)(x-x_1)\,\Delta_2(x_2) + \ldots + [(x-x_0)(x-x_1)\ldots(x-x_{n-1})]\,\Delta_n(x_n)$

where the *divided differences* are

$$\Delta_1(x_1) = \frac{y_1-y_0}{x_1-x_0}$$ First-order divided difference

$$\Delta_2(x_2) = \frac{\Delta_1(x_2)-\Delta_1(x_1)}{x_2-x_0}$$ Second-order divided differences

....

$$\Delta_n(x_n) = \frac{\Delta_{n-1}(x_n)-\Delta_{n-1}(x_{n-1})}{x_n-x_0}$$ nth-order divided difference

Divided differences can be again computed in *tabular form* as shown below.

Table					Example				
x_0	y_0				6	1,000			
x_1	y_1	$\Delta_1(x_1)$			8	2,000	500		
x_2	y_2	$\Delta_1(x_2)$	$\Delta_2(x_2)$		16	6,000	500	0	
x_3	y_3	$\Delta_1(x_3)$	$\Delta_2(x_3)$	$\Delta_3(x_3)$	26	7,400	320	10	1
.	.	.	.	.	.	.	.	.	.
.	.	.	.	.	.	.	.	.	.

18.8 INTERPOLATION, EQUAL SPACING

i. General

If the in increment $\Delta x = h$ is a fixed value, the interpolation functions may be taken in *polynomial form*, the coefficients of which can be expressed in *differences of ascending order.*

ii. Newton's Formula, Forward Interpolation

$$\bar{y}(x) = y_0 + \frac{u}{1!}\Delta y_0 + \frac{u(u-1)}{2!}\Delta^2 y_0 + \frac{u(u-1)(u-2)}{3!}\Delta^3 y_0 + \ldots + \frac{u(u-1)(u-2)\ldots(u-n+1)}{n!}\Delta^n y_0$$

$$u = \frac{u-x_0}{h}$$

iii. Newton's Formula, Backward Interpolation

$$\bar{y}(x) = y_n + \frac{u}{1!}\nabla y_n + \frac{u(u+1)}{2!}\nabla^2 y_n + \frac{u(u+1)(u+2)}{3!}\nabla^3 y_n + \dots + \frac{u(u+1)(u+2)\dots(u+n-1)}{n!}\nabla^n y_n \qquad u = \frac{x - x_n}{h}$$

iv. Stirling's Interpolation Formula*

$$\bar{y}(x) = y_0 + u\frac{\Delta y_0 + \nabla y_0}{2} + \frac{u^2}{2!}\Delta\nabla y_0 + \frac{u(u^2-1)}{3!}\frac{\Delta^2\nabla y_0 + \Delta\nabla^2 y_0}{2} + \frac{u^2(u^2-1)}{4!}\Delta^2\nabla^2 y_0 + \dots \qquad u = \frac{x - x_0}{h}$$

v. Bessel's Interpolation Formula* $(v \neq 0)$

$$\bar{y}(x) = \frac{y_0 + y_1}{2} + v\,\Delta y_0 + \frac{v^2 - \frac{1}{4}}{2}\,\frac{\Delta\nabla y_0 + \Delta\nabla y_1}{2} + \frac{v\left(v^2 - \frac{1}{4}\right)}{3!}\Delta^2\nabla y_0 + \frac{\left(v^2 - \frac{1}{4}\right)\left(v^2 - \frac{9}{4}\right)}{4!}\,\frac{\Delta^2\nabla^2 y_0 + \Delta^2\nabla^2 y_1}{2} + \dots \qquad v = u - \frac{1}{2} = \frac{x - x_0}{h} - \frac{1}{2}$$

vi. Bessel's Interpolation Formula* $(v = 0)$

$$\bar{y}(x) = \frac{1}{2}\left[(y_0 + y_1) - \frac{1}{8}(\Delta\nabla y_0 + \Delta\nabla y_1) + \frac{3}{128}\left(\Delta^2\nabla^2 y_0 + \Delta^2\nabla^2 y_1\right) - \frac{5}{1{,}024}\left(\Delta^3\nabla^3 y_0 + \Delta^3\nabla^3 y_1\right) + \dots\right]$$

$$\nabla y_i = \Delta y_{i-1},\ \Delta\nabla y_j = \Delta^2 y_{j-2},\ \Delta^p\nabla^q y_j = \Delta^{p+q} y_{j-q}$$

18.9 NUMERICAL INTEGRATION, DIFFERENCE POLYNOMIALS

i. Concept

Whenever the closed-form integration becomes too involved or is not feasible, the numerical value of a definite integral can be found (to any degree of accuracy) by means of any of several *quadrature formulas* which express the given integral as a linear combination of a selected set of integrands. The basis of these formulas are *difference polynomials or orthogonal polynomials*.

ii. Trapezoidal Rule [n even or odd; $h = (b - a)/n$]

$$\int_a^b y(x)\,dx = \int_a^b \bar{y}(x)\,dx = \frac{h}{2}(y_0 + 2y_1 + 2y_2 + \dots + 2y_{n-2} + 2y_{n-1} + y_n) + \varepsilon_T$$

Truncation error: $\varepsilon_T \approx -\dfrac{n(h)^3 f''(\xi)}{12}$ $\qquad a \le \xi \le b$

* $\nabla y_j = \Delta y_{j-1},\ \Delta\nabla y_j = \Delta^2 y_{j-2},\ \Delta^p\nabla^q y_j = \Delta^{p+q} y_{j-q}$.

iii. Simpson's Rule [n even ; $h = (b - a) / n$]

$$\int_a^b y(x)\,dx = \int_a^b \overline{y}(x)\,dx = \frac{h}{3}\,(y_0 + 4y_1 + 2y_2 + 4y_3 + \ldots + 4y_{n-3} + 2y_{n-2} + 4y_{n-1} + y_n) + \varepsilon_T$$

Truncation error: $\varepsilon_T \approx -\dfrac{n(h)^5 f^{iv}(\xi)}{180}$ $\quad a \le \xi \le b$

iv. Weddle's Rule [n must be a multiple of 6; $h = (b - a) / n$]

$$\int_a^b y(x)\,dx = \int_a^b \overline{y}(x)\,dx = \frac{3h}{10}\,[(y_0 + 5y_1 + y_2 + 6y_3 + y_4 + 5y_5 + y_6) + (y_6 + 5y_7 + y_8 + 6y_9 + y_{10} + 5y_{11} + y_{12}) + \ldots + (y_{n-6} + 5y_{n-5} + y_{n-4} + 6y_{n-4} + 6y_{n-3} + y_{n-2} + 5y_{n-1} + y_n)] + \varepsilon_T$$

Truncation error: $\varepsilon_T = -\dfrac{n(h)^7 f^{vi}(\xi)}{140}$ $\quad a \le \xi \le b$

v. Euler's Quadrtature Formula [n even or odd; $h = (b - a) / n$]

$$\int_a^b y(x)\,dx = \int_a^b \overline{y}(x)\,dx = \frac{h}{2}\,(y_0 + 2y_1 + 2y_2 + \ldots + 2y_{n-2} + 2y_{n-1} + y_n)$$

$$- B_2\,\frac{h^2}{2!}\big[y'(b) - y'(a)\big] - B_4\,\frac{h^4}{4!}\big[y'''(b) - y'''(a)\big] - \ldots + \varepsilon_T$$

Truncation error: $\varepsilon_T = -\,n\,B_{2m}\,\dfrac{h^{2m+1}}{(2m)!}\,f^{(2m)}\,(\xi)$ $\quad a \le \xi \le b$

18.10 NUMERICAL INTEGRATION, ORTHOGONAL POLYNOMIALS

i. Gauss–Chebyshev Quadrature Formula

$$\int_a^b y(x)dx = \frac{b-a}{n}\,[y(X_1) + y(X_2) + \ldots + y(X_n)] + \varepsilon_T$$

$X_j = \dfrac{a+b}{2} + \dfrac{b-a}{2}x_j$ $\quad n = 2, 3, 4, 5, 6, 7, 9$

Truncation error for $n = 3$: $\varepsilon_T \approx \dfrac{1}{360}\left(\dfrac{b-a}{2}\right)^5 y^{iv}(X)$ $\quad a < X < b$

ii. Gauss–Legendre Quadrature Formula

$$\int_a^b y(x)\,dx = (b - a)\,[C_1 y(X_1) + C_2 y(X_2) + \ldots + C_n y(X_n)] + \varepsilon_T$$

$X_j = (a + b - a)\,x_j$ $\quad n = 1, 2, 3, \ldots$

Truncation error for $n = n$: $\varepsilon_T \approx \dfrac{(n!)^4 (b-a)^{2n+1}}{(2n+1)\big[(2n)!\big]^3}\,y^{(2n)}(X)$ $\quad a < X < b$

iii. Table of x_j for the Gauss–Chebyshew Formula

n	x_1	x_2	x_3	x_4	x_5
2	0.577350	– 0.577350			
3	0.707107	0	– 0.707107		
4	0.794654	0.187592	– 0.187592	– 0.794654	
5	0.832497	0.374541	0	0.374541	– 0.832497

iv. Table of x_j for the Gauss–Legendre Formula

n	x_1	x_2	x_3	x_4	x_5
1	0.500000				
2	0.211325	0.788675			
3	0.112702	0.500000	0.887298		
4	0.069432	0.330009	0.669991	0.930568	
5	0.046910	0.230765	0.500000	0.769235	0.953090

$B_2, B_4,...B_{2m}$ = Bernoulilli's numbers

v. Table of C_j for the Gauss–Legendre Formula

n	C_1	C_2	C_3	C_4	C_5
1	1.000000				
2	0.500000	0.500000			
3	0.277778	0.444444	0.277778		
4	0.173927	0.326073	0.326073	0.173927	
5	0.118463	0.299314	0.284444	0.239314	0.118463

18.11 NUMERICAL COEFFICIENTS, ORTHOGONAL POLYNOMIALS

i. Chebyshew Polynomials T_n (x)

n \ x	0.2	0.4	0.6	0.8
0	+ 1.00000 00000	+ 1.00000 00000	+ 1.00000 00000	+ 1.00000 00000
1	+0.20000 00000	+ 0.40000 00000	+ 0.60000 00000	+ 0.80000 00000
2	– 0.92000 00000	– 0.68000 00000	– 0.28000 00000	+ 0.28000 00000
3	– 0.56800 00000	– 0.94400 00000	– 0.93600 00000	– 0.35200 00000
4	+ 0.69280 00000	– 0.07520 00000	– 0.84320 00000	– 0.84320 00000
5	+ 0.84512 00000	+ 0.88384 00000	– 0.07584 00000	– 0.99712 00000
6	– 0.35475 20000	+ 0.78227 20000	+ 0.75219 20000	– 0.75219 20000
7	– 0.98702 08000	– 0.25802 24000	+ 0.97847 04000	– 0.20638 72000
8	– 0.04005 63200	– 0.98868 99200	+ 0.42197 24800	+ 0.42197 24800
9	+ 0.97099 82720	– 0.53292 95360	– 0.47210 34240	+ 0.88154 31680
10	+ 0.42845 56288	+ 0.56234 62912	– 0.98849 65888	+ 0.98849 65888
11	– 0.79961 60205	+ 0.98280 65690	– 0.71409 24826	+ 0.70005 12741

ii. Laguerre Polynomials $L_n(x)$

n \ x	0.5	1.0	3.0	5.0
0	+ 1.00000 00000	+ 1.00000 00000	+ 1.00000 00000	+ 1.00000 00000
1	+ 0.50000 00000	+ 0.00000 00000	+ 0.20000 00000	– 4.00000 00000
2	+ 0.12500 00000	– 0.50000 00000	– 0.50000 00000	+ 3.50000 00000
3	– 0.14583 33333	– 0.66666 66667	+ 1.00000 00000	+ 2.66666 66667
4	– 0.33072 91667	– 0.62500 00000	+ 1.37500 00000	– 1.29166 66667
5	– 0.44557 29167	– 0.46666 66667	+ 0.85000 00000	– 3.16666 66667
6	– 0.50414 49653	– 0.25694 44444	– 0.01250 00000	– 2.09027 77778
7	– 0.51833 92237	– 0.04047 61905	– 0.74642 85714	+ 0.32539 68254
8	– 0.49836 29984	+ 0.15399 30556	– 1.10870 53571	+ 2.23573 90873
9	– 0.45291 95204	+ 0.30974 42681	– 1.06116 07143	+ 2.69174 38272
10	– 0.38937 44141	+ 0.41894 59325	– 0.70002 23214	+ 1.75627 61795
11	– 0.31390 72988	+ 0.48013 41791	– 0.18079 95130	+ 0.10754 36909
12	– 0.23164 96389	+ 0.49621 22235	+ 0.34035 46063	– 1.44860 42948

iii. Hermite Polynomials $H_n(x)$

n \ x	0.5	1.0	3.0	5.0
0	+ 1.00000	+ 1.00000	+ 1.00000 00	+ 1.00000 00000
1	+ 1.00000	+ 2. 00000	+ 6. 00000 00	(1) 1. 00000 00000
2	– 1.00000	+ 2. 00000	(1) + 3.40000 00	(1) 9.80000 00000
3	– 5. 00000	– 4. 00000	(2) + 1.80000 00	(2) 9.40000 00000
4	+ 1. 00000	(1) – 2. 00000	(2) + 8.76000 00	(3) 8.81200 00000
5	(1) + 4.10000	(0) – 8. 00000	(3) + 3.81600 00	(4) 8.06000 00000
6	(1) + 3.10000	(2) + 1.84000	(4) + 1.41360 00	(5) 7.17880 00000
7	(2) – 4.61000	(2) + 4.64000	(4) + 3.90240 00	(6) 6.21160 00000
8	(2) – 8.95000	(3) – 1.64800	(4) + 3.62400 00	(7) 5.20656 80000
9	(3) + 6.48100	(4) – 1.07200	(5) – 4.06944 00	(8) 4.21271 20000
10	(4) + 2.25910	(3) + 8.22400	(6) – 3.09398 40	(9) 3.27552 97600
11	(5) – 1.07029	(5) + 2.30848	(7) – 1.04250 24	(10) 2.43298 73600

18.12 NUMERICAL METHODS IN LINEAR LEGEBRA

Numerical methods are important for solving linear systems of equation for fitting straight lines or parabolas and for matrix eigen value problems. Many engineering or other problems for instance, in statistics, lead to mathematical methods whose solution requires methods for numerical linear algebra.

Linear Systems: A linear system of n equation in n unknowns $x_1, x_2 \ldots x_n$ is a set of equations of the form

$$E_1: \quad \begin{aligned} a_{11} x_1 + \ldots\ldots\ldots a_{1n} x_n &= b_1 \\ a_{21} x_1 + \ldots\ldots\ldots a_{2n} x_n &= b_2 \\ \ldots\ldots\ldots\ldots\ldots\ldots\ldots\ldots\ldots \\ a_{n1x1} + \ldots\ldots\ldots a_{nm} x_n &= b_n \end{aligned}$$

Where the coefficients a_{jk} and the b_j are given numbers. The system is called homogenous if all the b_j are zero, otherwise it is nonhomogeneous using matrix multiplication, a single vector equation can be represented by $Ax = b$

Where the coefficients matrix $A = [a_{jk}]$ is the $n \times n$ matrix

$$A = \begin{bmatrix} a_{11} & a_{12} & \cdots & a_{1n} \\ a_{21} & a_{22} & \cdots & a_{2n} \\ \cdots & \cdots & \cdots & \cdots \\ a_{n1} & a_{n2} & \cdots & a_{nn} \end{bmatrix} \text{ and } x = \begin{bmatrix} x_1 \\ \cdots \\ \cdots \\ x_n \end{bmatrix} \text{ and } b = \begin{bmatrix} b_1 \\ \cdots \\ \cdots \\ b_n \end{bmatrix}$$

are column vectors. The augmented matrix A of the system is

$$A = [A\ b] = \begin{bmatrix} a_{11} & \cdots & a_{1n} & b_1 \\ a_{21} & \cdots & a_{2n} & b_2 \\ \cdots & \cdots & \cdots & \cdots \\ a_{n1} & \cdots & a_{nn} & b_n \end{bmatrix}$$

A solution of above eqn is a set of numbers $x_1 \ldots x_n$ that satisfy all the n equation and a solution vector of the eqn is a vector whose components constitute a solution of eqn. Method for solution of a Linear system called Gauss Elimination.

18.13 CHOLESKY'S METHOD

The popular method for solving

$$Ax = b$$

based on this factorisation $A = LL^T$ is called Cholesky's method:

Theorem: Stability of the Cholesky factorisation.

The Cholesky LLT factorisation is numerically stable.

CHAPTER 19

Probability

19.1 PROBABILITY

Given a sample space S, with each event A of S (subset of S) there is an associated number $P(A)$, called the probability of A, such that the following axioms of probability are satisfied.

1. For every A in S,

$$0 \leq P(A) \leq A.$$

If $P(A) = 1$ then A is called a certain event and A is called an impossible event, if

$$P(A) = 0$$

2. The entire sample space S has the probability $P(S) = 1$
3. For mutually exclusive events A and B $(A \cap B = \phi)$

$\mathbf{P(A \cup B) = P(A) + P(B)}$

i. Classification of Events

An *observation* (experiment, trial) may have (theoretically) a class (set) S of possible results (states, events) $A_1, A_2,..., A_n$ permitting the following classification:

a. The union $A_1 \cup A_2 \cup ... \cup A_n$ of S is the single event of realizing at least one of the events in S.

b. The intersection $A_i \cap A_j$ of S is the joint event realizing two events A_i and A_j.

c. The complement $\overline{A}$ of S is the event not included in S.

ii. Probability

If A can occur (happen) in m ways (m times) out of a total of n mutually exclusive and equally likely ways, then we have the following conclusions.

a. The probability of occurrence (*called success*) of A is

$$P(A) = \frac{m}{n} = p$$

$$0 \leq P(A) \leq 1$$

b. The probability of nonoccurrence (*called failure*) of A is

$$(\overline{A}) = 1 - \frac{m}{n} = q$$

$$0 \leq P(\overline{A}) \leq 1$$

iii. Complementation Rule

For an event A and its complement A^c in a sample space S,

$$P(A^c) = 1 - P(A)$$

iv. Addition Rule for Mutually Exclusive Events

For mutually exclusive events $A_1 \ldots A_m$ in a sample space S

$$P(A_1 \cup A_2 \cup A_m) = P(A_1) + P(A_2) + \ldots P(A_m).$$

v. Addition Rule for Arbitary Events

For events A and B in a sample space

$$P(A \cup B) = P(A) + P(B) - P(A \cap B).$$

vi. Conditional Probability, Independent Events

Often it is required to find the probability of an even B under the condition that an event A occurs. This probability is called the conditional probability of B given A and is denoted by $P(B/A)$. In this case A serves as a new (reduced) sample space, and that probability is the fraction of $P(A)$ which corresponds to $A \cap B$. Thus

$$P(B/A) = \frac{P(A \cap B)}{P(A)} \qquad \ldots [P(A) \neq 0]$$

Similarly, the conditional probability of A given B is

$$P(A/B) = \frac{P(A \cap B)}{P(B)} \qquad \ldots [P(B) \neq 0].$$

vii. Multiplication Rules

If A and B are events in a sample space and $P(A) \neq 0$, $P(B) \neq 0$, then

$$P(A \cap B) = P(A)\, P(B/A) = P(B)\, P(A/B).$$

viii. Independent Event

If events A and B are such that

$$P(A \cap B) = P(A)\, P(B)$$

They are called independent events.

19.2 PROBABILITY DISTRIBUTION

i. Frequency Distribution

A set of numbers $x_1, x_2, ..., x_n$ corresponding to the events $E_1, E_2, ..., E_n$ which occur with frequencies $f_1, f_2, ..., f_n$, respectively, so that $f_i \geq 0$ and $x_1 < x_2 < ... < x_n$ is called the frequency distribution

ii. Discrete Random Variable

If a variable X can assume a discrete set of values $x_1, x_2, ..., x_n$ with respect to probabilities $p_1, p_2 + ... + p_n = 1$, then these values define the discrete *probability distribution*. Because x takes specific values with given probabilities, it is designated as the discrete random variable (change variable, stochastic variable). The probability that X takes the value x_j is

$$P(X = x_j) = f(x_j) \qquad j = 1, 2, ..., n$$

where $f(x)$ is the *probability function of the random variable* X, with $f(x_j) \geq 0$, $\sum_1^n f(x_j) = 1$.

The *joint probability* that X and Y take the values x_j and y_k, respectively, is

$$P(X = x_j, Y = y_j) = f(x_j, y_k) \qquad j = 1, 2, ..., n \quad k = 1, 2, ..., r$$

where $f(x, y)$ is the *probability function of a two-dimensional random variable*

$$\left[f(x_j, y_j) \geq 0, \ \sum_1^n f(x_j) = 1, \ \sum_1^r f(y_k) = 1\right].$$

iii. Some Discrete Random Variables

i. *Binomial Random Variable*

Suppose that n independent trials, each of which results in a success with probability p and in a failure with probability $1 - p$, are to be performed. If X represents the number of successes that occur in n trials, then X is said to be a binomial random variable.

$$P(x = r) = {}^nC_r \, p^r (1 - p)^{n-r}; \qquad r = 0, 1, ... n$$

ii. *Bernoulli Random Variable*

Let X be a random variable that equals 1 if the experiment is a success and 0 if it is a failure. Then, X is a Bernoulli random variable such that

$$p(X = 0) = 1 - p$$
$$p(X = 1) = p$$

iii. *Geometric Random Variable*

Let X be a random variable that denotes the number of trials required until the first success occurs, then X is a geometric random variable

$$p(X = n) = (1 - p)^{n-1} p$$

iv. *Poisson Random Variable*

When n is large and p is small in binomial distribution then $\lambda = np$ is a parameter that is used to describe the distribution.

i.e. $$p(X = r) = e^{-\lambda}\frac{\lambda^r}{r!},\ r = 0, 1, 2, \ldots.$$

iv. Continuous Random Variable

The random variable X is denoted as a continuous random variable if $f(x)$ is continuously differentiable.

$$\frac{df(x)}{dx} = \lim_{\Delta x \to 0} \frac{P(X < x \le X + \Delta x)}{\Delta x} = \phi(x)$$

where $\phi(x)$ is the *probability density*, and the *cumulative probability distributions* are

$$P(x \le X) = \int_{-\infty}^{X} \phi(x)\,dx$$

$$P(a < x \le b) = \int_a^b \phi(x)\,dx = f(b) - f(a)$$

$$P(-\infty < x \le \infty) = \int_{-\infty}^{\infty} \phi(x)\,dx = 1$$

A *two-dimensional random variable* is a *continuous random variable* if $f(x, y)$ is continuous for all x_j and y_k, and the *joint probability density*

$$\phi(x, y) = \frac{\partial^2 f(x,y)}{\partial x \partial y}$$

exists and is piecewise continuous.

v. Some Continuous Random Variables

i. *Uniform Random Variable*

A random variable is said to be uniformly distributed over (0, 1) if its probability density function is as follows

$$f(x) = \begin{cases} 1 & 0 < x < 1 \\ 0 & \text{otherwise} \end{cases}$$

ii. *Exponential Random Variable*

A continuous random variable whose probability density function is given, for some $\lambda > 0$, by

$$f(x) = \begin{cases} \lambda e^{-\lambda x} & x \ge 0 \\ 0 & x < 0 \end{cases}$$

iii. *Gamma Random Variable*

A continuous random variable whose density is

$$f(x) = \begin{cases} \dfrac{\lambda^{e-\lambda x}(\lambda x)^{\infty - 1}}{\Gamma(\infty)} & x \ge 0 \\ 0 & x < 0 \end{cases}$$

for some $\lambda > 0,\quad \infty > 0$

$$\Gamma(\infty) = \int_0^{\infty} e^{-x} \ x^{\infty-1} \ dx$$

$$\Gamma(n) = (n-1)!$$

iv. *Normal Random Variable*

X is a normal random variable with parameters μ and σ^2 if the density is given by

$$f(x) = \frac{1}{\sqrt{2\pi}\,\sigma} \quad e^{-\frac{(x-\mu)^2}{2\sigma^2}} \ ; \quad -\infty < x < \infty$$

Moment generating functions, mean and variance of different distributions

Table I

Discrete probability distribution	*Probability mass function, p(x)*	*Moment generating function, $\phi(t)$*	*Mean*	*Variance*
Binomial with parameters n, p, $0 \le p \le 1$	$\binom{n}{x} p^x (1-p)^{n-x}$ $x = 0, 1, ..., n$	$(pe^t + (1-p))^n$	np	$np(1-p)$
Poisson with parameter $\lambda > 0$	$e^{-\lambda} \frac{\lambda^x}{x!}$, $x = 1, 2, ...$	$\exp[\lambda(e^t - 1)]$	λ	λ
Geometric with parameter $0 \le p \le 1$	$p(1-p)^{x-1}$, $x = 1, 2, ...$	$\frac{pe^t}{1-(1-p)e^t}$	$\frac{1}{p}$	$\frac{1-p}{p^2}$

Table II

Continuous probability distribution	*Probability density function, f(x)*	*Moment generating function, $\phi(t)$*	*Mean*	*Variance*
Uniform over (a, b)	$f(x) = \begin{cases} \frac{1}{b-a}, & a<x<b \\ 0 & \text{otherwise} \end{cases}$	$\frac{e^{tb} - e^{ta}}{t(b-a)}$	$\frac{a+b}{2}$	$\frac{(b-a)^2}{12}$
Exponential with parameter $\lambda > 0$	$f(x) = \begin{cases} \lambda e^{-\lambda x}, & x>0 \\ 0, & x<0 \end{cases}$	$\frac{\lambda}{\lambda - t}$	$\frac{1}{\lambda}$	$\frac{1}{\lambda^2}$
Gamma with parameters (n, λ) $\lambda > 0$	$f(x) = \begin{cases} \frac{\lambda e^{-\lambda x} (\lambda x)^{n-1}}{(n-1)!}, & x>0 \\ 0, & x<0 \end{cases}$	$\left(\frac{\lambda}{\lambda-1}\right)^n$	$\frac{n}{\lambda}$	$\frac{n}{\lambda^2}$
Normal with parameters (μ, σ^2)	$f(x) = \frac{1}{\sqrt{2\pi\sigma}}$ $x \exp. \{-(x-\mu^2)/2\sigma^2\}$ $-\infty < x < \infty$	$\exp\left\{\mu t + \frac{\sigma^2 + t^2}{2}\right\}$	μ	σ^2

19.3 LIMIT THEOREMS

i. Markov's Inequality

If X is a random variable that takes only nonnegative values then for any value $a > 0$

$$P\{x \geq a\} \leq \frac{E[X]}{a}$$

ii. Chebyshev's Inequality

If X is a random variable with mean μ and variance σ^2, then for any value $k > 0$,

$$P\{|X - \mu| \geq k\} \leq \frac{\sigma^2}{k^2}$$

iii. Strong Law of Large Numbers

Let X_1, X_2, ... be a sequence of independent random variables having a common distribution, and let $E[X_i] = \mu$. Then, with probability 1,

$$\frac{X_1 + X_2 + \ldots + X_n}{n} \to \mu \quad \text{as } n \to \infty.$$

iv. Central Limit Theorem

Let X_1, X_2, ... be a sequence of independent, identically distributed random variables, each with mean μ and variance σ^2. Then, the distribution of

$$\frac{X_1 + X_2 + \ldots + X_n - n\mu}{\sigma\sqrt{n}}$$ tends to the standard normal as $n \to \infty$. That is,

$$P\left\{\frac{X_1 + X_2 + \ldots + X_n - n\mu}{\sigma\sqrt{n}} \leq a\right\} \to \frac{1}{\sqrt{2\pi}} \int_{-\infty}^{a} e^{-x^2/2}\, dx.$$

CHAPTER

20

Statistics

20.1 MEAN AND VARIANCE OF DISTRIBUTION

The mean μ and variance σ^2 of a random variable X and of its distribution are the theoretical counterparts of the mean $\bar{x}$ and variance s^2 of a frequency distribution. Mean characterizes the central location and the variance the spread (the variability) of the distribution.

i. Mean μ is defined by

i. $\mu = \sum_j x_j \, f(x_j)$ **(Discrete distribution)**

ii. $\mu = \int_{-\infty}^{\infty} x \, f(x) \, dx$ **(Continuous distribution)**

and

ii. Variance σ^2

i. $\sigma^2 = \sum_j (x_j - \mu)^2 \, f(x_j)$ **(Discrete distribution)**

ii. $\sigma^2 = \int_{-\infty}^{\infty} \left(x - \mu^2\right)^2 f(x) \, dx$ **(Continuous distribution)**

σ (positive square root of σ^2) is called the standard deviation of X and its distribution f is the probability function or density.

The mean μ is also denoted by $E(x)$ and is called the *expectation of* X. Quantities such as μ and σ^2 the measure certain properties of a distribution are called parameters.

20.2 MEASURES OF CENTRAL TENDENCY: STATISTICAL CONSTANTS

The commonly used measures of central tendency are:

i. Arithmetic mean (AM)
ii. Geometric mean (GM)
iii. Harmonic mean (HM)
iv. Median
v. Mode

i. Measures of Tendency and Frequency

a. An average is a value which is typical for a set of numbers X ($x_1, x_2, \ldots x_n$). Since it tends to lie centrally within the set arranged according to magnitude, it is also a measure of central tendency.

b. A median of the same set is the average of the two middle values if n is even.

c. A mode is the value of x_m having a maximum frequency f_m.

ii. Mean

If $x_1, x_2, x_3, \ldots x_n$ are n values of a variable X, then

a. *The arithmetic mean of a set of n numbers is*

$$\bar{X} = \frac{x_1 + x_2 + \ldots + x_n}{n} = \frac{\Sigma x}{n}$$

If the numbers occur $f_1, f_2 \ldots, f_n$ times, respectively, then

$$\bar{X} = \frac{f_1 x_1 + f_2 x_2 + \ldots + f_n x_n}{f_1 + f_2 + \ldots + f_n} = \frac{\Sigma f x}{\Sigma f}$$

If the numbers are associated with a *weighting factor* $\omega_j \geq 0$, then

$$\bar{X} = \frac{\omega_1 x_1 + \omega_2 x_2 + \ldots + \omega_n x_n}{\omega_1 + \omega_2 + \ldots + \omega_n} = \frac{\Sigma \omega x}{\Sigma \omega}$$

b. *The geometric mean (GM) of a set of n numbers is*

$$\bar{G} = \sqrt[n]{x_1 x_2 \ldots x_n}$$

If the numbers occur $f_1, f_2 \ldots f_n$ times, respectively, then

$$\bar{G} = \sqrt[n]{x_1^{f_1} x_2^{f_2} \ldots x_n^{f_n}}$$

where $n = f_1 + f_2 + \ldots + f_n$.

c. *The harmonic mean (HM) of a set of n numbers is*

$$\bar{H} = \frac{1}{(1/n)(1/x_1 + 1/x_2 + \ldots + 1/x_n)} = \frac{n}{\Sigma(1/x)}$$

If the numbers occur $f_1, f_2, \ldots f_n$ times, respectively, then

$$\bar{H} = \frac{1}{(1/n)(f_1/x_1 + f_2/x_2 + \ldots + f_n/x_n)} = \frac{n}{\Sigma(f/x)}$$

where $n = f_1 + f_2 + \ldots + f_n$

d. *The Quadratic Mean of a set of n numbers is*

$$\bar{Q} = \sqrt{\frac{x_1^2 + x_2^2 + \ldots + x_n^2}{n}} = \sqrt{\frac{\Sigma x^2}{n}}$$

e. *Relation between Arithmetic Mean ($\bar{X}$) Geometric Mean ($\bar{G}$) and Harmonic Mean ($\bar{H}$) is*

$$\bar{H} \leq \bar{G} \leq \bar{X}$$

f. *Variance* (σ^2)

The mean of the sum of squares of deviations from the mean (μ) is the population variance, denoted σ^2:

$$\sigma^2 = \Sigma (X_i - \mu)^2 / N$$

The sample variance, s^2 for sample size n is

$$s^2 = \Sigma (X_i - \bar{X})^2 / (n-1)$$

A simpler computational form is

$$s^2 = \frac{\Sigma X_i^2 - \dfrac{(\Sigma X_i)^2}{n}}{n-1}$$

g. Standard Deviation

The positive square root of population variance is the standard deviation. For a population

$$\sigma = \left[\frac{\Sigma X_i - (\Sigma X_i)^2}{N}\right]^{1/2}$$

for a sample

$$s = \left[\frac{\Sigma X_i^2 - \dfrac{(\Sigma X_i)^2}{N}}{N}\right]^{1/2}$$

h. *Coefficient of Variation*

$$V = s / \bar{X}.$$

20.3 PROBABILITY

For the sample space U, with subsets A of U (called "event"), we consider the probability measure of an event A to be a real-valued function p defined over all subsets of U such that:

$$0 \leq p(A) \leq 1$$

$$p(U) = 1 \text{ and } p(\Phi) = 0$$

If A_1 and A_2 are subsets of U

$$p(A_1 \cup A_2) = p(A_1) + p(A_2) - p(A_1 \cap A_2)$$

Two events A_1 and A_2 are called mutually exclusive if and only if $A_1 \cap A_2 = \phi$ (null set). These events are said to be independent if and only if $p(A_1 \cap A_2) = p(A_1)\, p(A_2)$.

Conditional Probability and Bayes' Rule

The probability of an event A, given that an event B has occurred, is called the conditional probability and is denoted $p(A/B)$. Further

$$p(A/B) = \frac{p(A \cap B)}{p(B)}$$

Bayes' rule permits a calculation of a *posteriori* probability from given a *priori* probabilities and is stated below:

If $A_1, A_2, \ldots, A_n$ are n mutually exclusive events, and $p(A_1) + p(A_2) + \ldots + p(A_n) = 1$, and B is any event such that $p(B)$ is not 0, then the conditional probability $p(A_i/B)$ for any one of the events A_i, given that B has occurred is

$$p(A_i/B) = \frac{p(A_i)\,p(B/A_i)}{p(A_i)\,p(B/A_i) + p(A_2)\,p(B/A_2) + \ldots + p(A_n)\,p(B/A_n)}$$

20.4 BINOMIAL DISTRIBUTION

In an experiment consisting of n independent trials in which an event has probability p in a single trial, the probability P_X of obtaining X successes is given by

$$P_X = C_{(n,\,X)}\, p^X q^{(n-X)}$$

where

$$q = (1 - p) \quad \text{and} \quad C_{(n,\,X)} = \frac{n!}{X!(n-X)!}$$

The probability of between a and b successes (both) a and b included is $P_a + P_{a+1} + \ldots + P_b$, so if $a = 0$ and $b = n$, this sum is

$$\sum_{X=0}^{n} C_{(n,\,x)}\, p^X q^{(n-X)} = q^n + C_{(n,\,1)}\, q^{n-1} p$$

$$+ C_{(n,\,2)}\, q^{n-2} p^2 + \ldots + p^n = (q + p)^n = 1.$$

20.5 MEAN OF BINOMIALLY DISTRIBUTED VARIABLE

The mean number of successes in n independent trials is $m = np$ with standard deviation

$$\sigma = \sqrt{npq}.$$

20.6 NORMAL DISTRIBUTION

In the binomial distribution, as n increases the histogram of heights is approximated by the bell-shaped curve (normal curve)

$$Y = \frac{1}{\sigma\sqrt{2\pi}} e^{-(x-m)^2/2\sigma^2}$$

where m = the mean of the binomial distribution = np_1 and $\sigma = \sqrt{npq}$ is the standard deviation. For any normally distributed random variable X with mean m and standard deviation σ the probability function (density) is given by the above. The standard normal probability curve is given by

$$y = \frac{1}{\sqrt{2\pi}} e^{-z^2/2}$$

and has mean = 0 and standard deviation = 1. The total area under the standard normal curve is 1. Any normal variable X can be put into standard form by defining $Z = (X - m)/\sigma$; thus the probability of X between a given X_1 and X_2 is the area under the standard normal curve between the corresponding Z_1 and Z_1. The standard normal curve is often used instead of the binomial distribution in experiments with discrete outcomes.

20.7 POISSON DISTRIBUTION

$$P = \frac{e^{-m} \cdot m^r}{r!}$$

is an approximation to the binomial probability for r successes in n trials when $m = np$ is small (< 5) and the normal curve is not recommended to approximate binomial probabilities. The variance σ^2 in the Poisson distribution is np, the same value as the mean.

20.8 LEAST SQUARES REGRESSION

A set of n values (X_i, Y_i) that display a linear trend is described by the linear equation $\hat{Y}_i = a + \beta X_i$. Variables a and β are constants (population parameters) and are the intercept and slope, respectively. The rule for determining the line is one minimizing the sum of the squared deviations

$$\sum_{i=1}^{n} (Y_i - \hat{Y}_i)^2$$

and with this criterion the parameters a and β are best estimated from a and b calculated as

$$b = \frac{\dfrac{\sum X_i Y_i - (\sum X_i)(\sum Y_i)}{n}}{\dfrac{\sum X_i^2 - (\sum X_i)^2}{n}}$$

and

$$a = \overline{Y} - b\overline{X}$$

where $\overline{X}$ and $\overline{Y}$ are mean values, assuming that for any value of X the distribution of Y values is normal with variances that are equal for all X and the latter (X) are obtained with negligible error. The null hypothesis, H_0: $\beta = 0$, is tested with analysis of variance:

Source	**SS**	**DF**	**MS**
Total $(Y_i - \overline{Y})$	$\Sigma(Y_i - \overline{Y})^2$	$n - 1$	
Regression $(\hat{Y}_i - \overline{Y})$	$\Sigma(\hat{Y}_i - \overline{Y})^2$	1	
Residual $(Y_i - \hat{Y}_i)$	$\Sigma(Y_i - \hat{Y}_i)^2$	$n - 2$	$\dfrac{SS_{resid.}}{(n-2)} = S_{Y.X}^2$

Computing forms for SS terms are

$$SS_{total} = \Sigma(Y_i - \overline{Y})^2 = \Sigma Y_i^2 - (\Sigma Y_i^2)/n$$

$$SS_{\text{regr.}} = \Sigma(\hat{Y}_i - \bar{Y})^2 = \frac{[\Sigma X_i Y_i - (\Sigma X_i)(\Sigma Y_i)/n]^2}{\Sigma X_i^2 - (\Sigma X_i)^2/n}$$

$F = MS_{regr}/MS_{resid}$ is calculated and compared with the critical value of F for the desired confidence level for degrees of freedom 1 and $n - 2$. The coefficient of determination, denoted by r^2, is

$$r^2 = SS_{regr}/SS_{total}$$

r is the *correlation coefficient*. The *standard error of estimate* is $\sqrt{S^2_{Y \cdot X}}$ and is used to calculate confidence intervals for α and β. For the confidence limits of β and α

$$b \pm tS_{Y.X}\sqrt{\frac{1}{\Sigma(X_i - \bar{X})^2}}, \quad a \pm tS_{Y.X}\sqrt{\frac{1}{n} + \frac{\bar{X}^2}{\Sigma(X_i - \bar{X})^2}}$$

where t has $n - 2$ degrees of freedom.

The null hypothesis H_0: $\beta = 0$, can also be tested with the t statistics:

$$t = \frac{b}{S_b}$$

where s_b is the standard error of b

$$S_b = \frac{S_{Y.X}}{[\Sigma(X_i - \bar{X}^2]^{1/2}}$$

- ***Standard Error of $\hat{Y}_0$***

An estimate of the mean value of Y for a given value of X, say X_0, is given by the regression equation

$$\hat{Y}_0 = a + bX_0.$$

The standard error of this predicted value is given by

$$S_{\hat{Y}_0} = S_{Y,X}\left[\frac{1}{n} + \frac{(X_0 - \bar{X})^2}{\Sigma(X_i - \bar{X})^2}\right]^{\frac{1}{2}}$$

and is *a* minimum when $X_0 = \bar{X}$ and increases as X_0 moves away from $\bar{X}$ in either direction.

20.9 SUMMARY OF PROBABILITY DISTRIBUTIONS

i. Continuous Distributions

a. *Normal Distribution*

$$y = \frac{1}{\sigma\sqrt{2\pi}} \exp\,[-(x - m)^2/2\sigma^2]$$

$$\text{Mean} = m$$

$$\text{Variance} = \sigma^2$$

b. *Standard normal*

$$y = \frac{1}{\sqrt{2\pi}} \exp(-z^2/2)$$

Mean = 0

Variance = 1

c. *F-distribution*

$$y = A \frac{F^{\frac{f_1 - 2}{2}}}{(f_2 + f_1 F)^{\frac{f_1 + f_2}{2}}}$$

where
$$A = \frac{\Gamma\left(\frac{(f_1 + f_2)}{2}\right)}{\Gamma\left(\frac{f_1}{2}\right)\Gamma\left(\frac{f_2}{2}\right)} f_1^{f^{1/2}} f_2^{f^{2/2}}$$

$$Mean = \frac{f_2}{f_2 - 2}$$

$$Variance = \frac{2 f_2^2 (f_1 + f_2 - 2)}{f_1 (f_2 - 2)^2 (f_2 - 4)}$$

d. *Chi-square*

$$y = \frac{1}{2^{f/2} \Gamma(f/2)} \exp\left(-\frac{1}{2} x^2\right) (x^2)^{\frac{f-2}{2}}$$

$$Mean = f$$

$$Variance = 2f$$

e. *Students t*

$$y = A(1 + t^2/f)^{-(f+1)/2}; \quad \text{where } A = \frac{\Gamma(f/2 + 1/2)}{\sqrt{f\pi}\,\Gamma(f/2)}$$

$$Mean = 0$$

$$Variance = \frac{f}{f - 2} \quad (\text{for } f > 2)$$

ii. Discrete Distributions

a. *Binomial distribution*

$$y = C_{(n,x)}\, p^x (1 - p)^{n - x}$$

Mean = np

Variance = $np\,(1 - p)$

b. Poisson distribution

$$y = \frac{e^{-m}\, m^x}{x!}$$

$$\text{Mean} = m$$

$$\text{Variance} = m$$

20.10 MEASURES OF DISPERSION, SKEWNESS, AND KURTOSIS

i. Dispersion

a. The degree to which a frequency distribution of a set of numbers *tends to spread* about a point of central tendency is called the dispersion (spread, variance).

b. The *range* of a set ofs numbers is the *difference* between the *largest* and *smallest numbers* in the set.

ii. Deviation

a. The deviation from the arithmetic mean $\bar{X}$ of each number x_j in a set of numbers $x_1, x_2, \ldots x_n$ is

$$D_j = x_j - \bar{X} \qquad j = 1, 2 \ldots n$$

b. The mean deviation of the same set is

$$\bar{D} = \frac{|x_1 + x_2 + \ldots + x_n - n\bar{X}|}{n} = \frac{\Sigma|x - \bar{X}|}{n}$$

where $|x_j - X|$ is the absolute value of D_j.

If the numbers occur $f_1, f_2, \ldots, f_n$ times, then $n = f_1 + f_2 + \ldots + f_n$ and

$$\bar{D} = \frac{f_1|x_1 - \bar{X}| + f_2|x_2 - \bar{X}| + \ldots + f_n|x_n - \bar{X}|}{n} = \frac{\Sigma f|x - \bar{X}|}{n}$$

c. The standard deviation of the same set is

$$\sigma = \sqrt{\frac{(x_1 - \bar{X})^2 + (x_2 - \bar{X})^2 + \ldots + (x_n - \bar{X})^2}{n}} = \sqrt{\frac{\Sigma(x - \bar{X})^2}{n}}$$

If the numbers occur $f_1, f_2, \ldots, f_n$ times, then $n = f_1 + f_2 + \ldots + f_n$ and

$$\sigma = \sqrt{\frac{f_1(x_1 - \bar{X})^2 + f_2(x_2 - \bar{X})^2 + \ldots + f_n(x_n - \bar{X})^2}{n}} = \sqrt{\frac{\Sigma f(x - \bar{X})^2}{n}}$$

d. The variance of the same set is defined as

$$\sigma^2 = \frac{\Sigma(x - \bar{X})^2}{n} \quad \text{or} \quad \sigma^2 = \frac{\Sigma f(x - \bar{X})^2}{n}$$

e. The covariance of two sets $X\,(x_1, x_2, \ldots, x_n)$ and $Y\,(y_1, y_2, \ldots y_n)$, the arithmetic means of which are $\bar{X}$ and $\bar{Y}$, respectively is

$$\sigma_{xy} = \frac{\Sigma(x - \bar{X})(y - \bar{Y})}{n}$$

iii. Skewness and Kurtosis

a. The skewness of the smoothed frequency polygon is the departure of the curve from symmetry.

b. The kurtosis of the curve is the degree of peakedness of the same curve.

c. Moments are defined by

$$\mu_k = \frac{\Sigma f(x-\overline{X})^k}{n}$$

where k is the constant indicating the *degree of the moment*. The first moment is $\mu_1 = 0$, the *second* is $\mu_2 = \sigma^2$, the *coefficient of skewness* is $\gamma_1 = \mu_3/\sigma^3$, the *coefficient of excess* is $\gamma_2 = \mu_4/\sigma^4 - 3$, and the kurtosis $\beta_2 = \mu_4/\sigma^4 = \gamma_2 + 3$.

20.11 DISCRETE PROBABILITY DISTRIBUTION

i. Binomial Distribution

a. Probability function

$$P(X = x) = f(x) = \frac{n!}{x!(n-x)!}\, p^x (1-p)^{n-x} \qquad x = 1, 2, \ldots, n$$

$$0 \le p \le 1$$

b. Properties

$$\text{Mean } \overline{X} = np$$

$$\text{Variance } \sigma^2 = np(1-p)$$

$$\text{Standard deviation, } \sigma = \sqrt{np(1-p)}$$

$$\text{Coefficient of skewness, } \gamma_1 = \frac{1-2p}{\sqrt{np(1-p)}}$$

$$\text{Coefficient of excess, } \gamma_2 = \frac{1-6p(1-p)}{np(1-p)}$$

ii. Poission Distribution

a. The discrete distribution with many possible values and probability function

$$P(X = x) = \frac{e^{-\lambda}\lambda^x}{x!} \qquad x = 0, 1, 2, \ldots$$

$$\lambda > 0$$

$$e = 2.71828$$

b. Properties

$$\text{Mean } \overline{X}, = \lambda$$

$$\text{Variance, } \sigma^2 = \lambda$$

$$\text{Standard deviation, } \sigma = \sqrt{\lambda}$$

Coefficient of skewness, $\gamma_1 = \frac{1}{\sqrt{\lambda}}$

Coefficient of excess, $\gamma_2 = \frac{1}{\lambda}$

iii. Multinomial Distribution

A multinomial distribution is defined by

$$P(X_1 = x_1, X_2 = x_2, \ldots, X_n = x_n) = f(x_1, x_2, \ldots, x_n) = \frac{n!}{x_1!\, x_2! \ldots x_n!} p_1^{x_1} p_2^{x_2} \ldots p_n^{x_n}$$

where $\sum_{j=1}^{n} pj = 1 \qquad \sum_{j=1}^{n} x_j = n \qquad p_j = 0$

20.12 CONTINUOUS PROBABILITY DISTRIBUTIONS

i. Normal Distribution

a. Gauss distribution is defined as the distribution with Density function

$$\phi_N(x) = \frac{1}{\sigma\sqrt{2\pi}} e^{-(x-\mu)^2/2\sigma^2} \qquad \phi = 3.14159 \ldots, e = 2.71828$$

where μ is mean, σ, standard deviation of the random variable X, and $\phi_N(x)$ is the *normal distribution density function* (gaussian function).

b. Properties

μ is mean $\qquad \gamma_1 = 0$ = coefficient of skewness

σ^2 is variance $\qquad \gamma_2 = 0$ = coefficient of excess

σ is standard deviation

Moments about $x = \mu$:

$\mu_1 = 0$

$\mu_2 = \sigma^2$

$\mu_3 = 0$

$\mu_4 = 3\sigma^4$

Moments about $x = 0$:

$v_1 = \mu$

$v_2 = \mu^2 + \sigma^2$

$v_3 = \mu(\mu^2 + 3\sigma^2)$

$v_4 = \mu^4 + 6\mu^2\sigma^2 + 3\sigma^4$

c. Probability function

$$P(X = x) = \int_{-\infty}^{x} \phi_N(x)\, dx = F_N(x)$$

is the *cumulative normal distrubution function.*

ii. Standard Normal Distribution

a. Density function

$$\phi_N(t) = \frac{1}{\sqrt{2\pi}} e^{-t^2/2} \qquad t = \frac{x-\mu}{\sigma}$$

is the *standard normal distribution density function.*

b. Properties

$$\text{mean } \mu = 0$$
$$\text{variance } \sigma^2 = 1$$

c. Probability function

$$P(T = t) = \frac{1}{\sqrt{2\pi}} \int_{-\infty}^{t} \phi_N(t)dt = F_N(t)$$

is the *cumulative standard normal distribution function.*

Theorem 1:

The distribution function $F(x)$ of the normal distribution with any μ and σ is related to the standardized distribution function φ(t) by the formula

$$F(x) = \phi\left(\frac{x-\mu}{\sigma}\right)$$

Theorem 2 (Normal probabilities):

The probability that a normal random variable X with mean μ and standard deviation σ assume any value in an interval $a < x \leq b$ is

$$P(a < x \leq b = F(b) - F(a) = \phi\left(\frac{b-\mu}{\sigma}\right) - \phi\left(\frac{a-\mu}{\sigma}\right)$$

A two dimentional random variable (x,y) occurs if two quantities are observed simultaneously. Its distribution function is

$F(x,y) = P(x \leq x, Y \leq y)$, X, Y have the distribution function

$F_1(x) = P(X \leq x, Y \text{ arbitary})$ and $F_2(y) = P(x \text{ arbitary}, Y \leq y)$ respectively, their distributions are called marginal distribution. If both X and Y are discrete then (X, Y) has a probability function $f(x, y) = P(X = x, Y = y)$.

If both X and Y are continues then (X, Y) has a density $f(x, y)$.

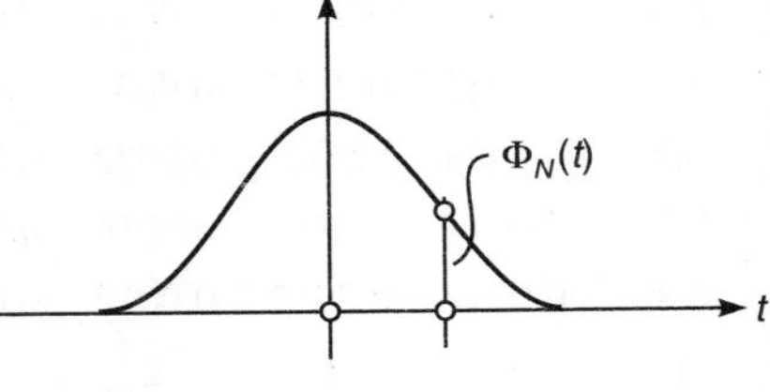

Fig. 20.1

20.13 ORDINATES Φ_N (T) OF THE STANDARD NORMAL CURVE

$$\phi_N(t) = \frac{1}{\sqrt{2\pi}} e^{-t^2/2}$$

t	0	1	2	3	4	5	6	7	8	9	t
0.0	0.3989	0.3989	0.3989	0.3988	0.3986	0.3984	0.3982	0.3980	0.3977	0.3973	0.0
0.1	0.3970	0.3965	0.3961	0.3956	0.3951	0.3945	0.3939	0.3932	0.3925	0.3918	0.1
0.2	0.3910	0.3902	0.3894	0.3885	0.3876	0.3867	0.3857	0.3847	0.3836	0.3825	0.2
0.3	0.3814	0.3802	0.3790	0.3778	0.3765	0.3752	0.3739	0.3726	0.3712	0.3697	0.3
0.4	0.3683	0.3668	0.3653	0.3637	0.3621	0.3605	0.3589	0.3572	0.3555	0.3538	0.4
0.5	0.3521	0.3503	0.3485	0.3467	0.3448	0.3429	0.3410	0.3391	0.3372	0.3352	0.5

t	0	1	2	3	4	5	6	7	8	9	t
0.6	0.3332	0.3321	0.3292	0.3271	0.3251	0.3230	0.3209	0.3187	0.3166	0.3144	0.6
0.7	0.3123	0.3101	0.3079	0.3056	0.3034	0.3011	0.2989	0.2966	0.2943	0.2920	0.7
0.8	0.2897	0.2874	0.2850	0.2827	0.2803	0.2780	0.2756	0.2732	0.2709	0.2685	0.8
0.9	0.2661	0.2637	0.2613	0.2589	0.2565	0.2541	0.2516	0.2492	0.2468	0.2444	0.9
1.0	0.2420	0.2396	0.2371	0.2347	0.2323	0.2299	0.2275	0.2251	0.2227	0.2203	1.0
1.1	0.2179	0.2155	0.2131	0.2107	0.2083	0.2059	0.2036	0.2012	0.1989	0.1965	1.1
1.2	0.1942	0.1919	0.1895	0.1872	0.1849	0.1826	0.1804	0.1781	0.1758	0.1736	1.2
1.3	0.1714	0.1691	0.1669	0.1647	0.1626	0.1604	0.1582	0.1561	0.1539	0.1518	1.3
1.4	0.1497	0.1476	0.1456	0.1435	0.1415	0.1394	0.1374	0.1354	0.1334	0.1315	1.4
1.5	0.1295	0.1276	0.1257	0.1238	0.1219	0.1200	0.1182	0.1163	0.1145	0.1127	1.5
1.6	0.1109	0.1092	0.1074	0.1057	0.1040	0.1023	0.1006	0.0989	0.0973	0.0957	1.6
1.7	0.0940	0.0925	0.0909	0.0893	0.0878	0.0863	0.0848	0.0833	0.0818	0.0804	1.7
1.8	0.0790	0 0775	0.0761	0.0748	0.0734	0.0721	0.0707	0.0694	0.0681	0.0669	1.8
1.9	0.0656	0.0644	0.0632	0.0620	0.0608	0.0596	0.0584	0.0573	0.0562	0.0551	1.9
2.0	0.0540	0.0529	0.0519	0.0508	0.0498	0.0488	0.0478	0.0468	0.0459	0.0449	2.0
2.1	0.0440	0.0431	0.0422	0.0413	0.0404	0.0396	0.0387	0.0379	0.0371	0.0361	2.1
2.2	0.0355	0.0347	0.0339	0.0332	0.0325	0.0317	0.0310	0.0303	0.0297	0.0290	2.2
2.3	0.0283	0.0277	0.0270	0.0264	0.0258	0.0252	0.0246	0.0241	0.0235	0.0229	2.3
2.4	0.0224	0.0219	0.0213	0.0208	0.0203	0.0198	0.0194	0.0189	0.0184	0.0180	2.4
2.5	0.0175	0.0171	0.0167	0.0163	0.0158	0.0155	0.0151	0.0147	0.0143	0.0139	2.5
2.6	0.0136	0.0132	0.0129	0.0126	0.0122	0.0119	0.0116	0.0113	0.0110	0.0107	2.6
2.7	0.0104	0.0101	0.0099	0.0096	0.0093	0.0091	0.0088	0.0086	0.0084	0.0081	2.7
2.8	0.0079	0.0077	0.0075	0.0073	0.0071	0.0069	0.0067	0.0065	0.0063	0.0061	2.8
2.9	0.0060	0.0058	0.0056	0.0055	0.0053	0.0051	0.0050	0.0048	0.0047	0.0046	2.9
3.0	0.0044	0.0043	0.0042	0.0040	0.0039	0.0038	0.0037	0.0036	0.0035	0.0034	3.0
3.1	0.0033	0.0032	0.0031	0.0030	0.0029	0.0028	0.0027	0.0026	0.0025	0.0025	3.1
3.2	0.0024	0.0023	0.0022	0.0022	0.0021	0.0020	0.0020	0.0019	0.0018	0.0018	3.2
3.3	0.0017	0.0017	0.0016	0.0016	0.0015	0.0015	0.0014	0.0014	0.0013	0.0013	3.3
3.4	0.0012	0.0012	0.0012	0.0011	0.0011	0.0010	0.0010	0.0010	0.0009	0.0009	3.4
t	0	1	2	3	4	5	6	7	8	9	t

20.14 AREAS $F_N(t)$ UNDER THE STANDARD NORMAL CURVE

$$F_N(t) = \frac{2}{\sqrt{2\pi}} \int_{-\infty}^{t} e^{-t^2/2}\, dt$$

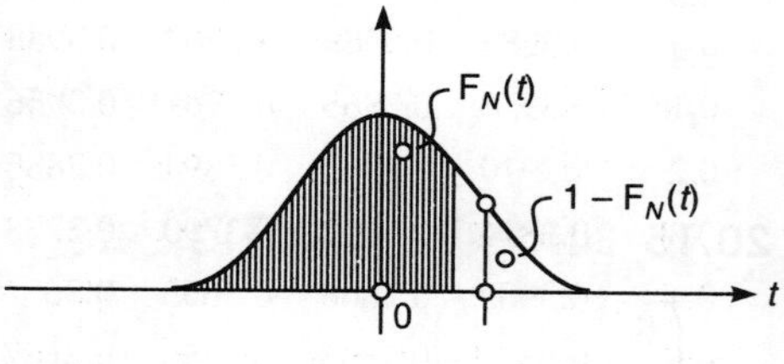

Fig. 20.2

t	0	1	2	3	4	5	6	7	8	9	t
0.0	0.5000	0.5040	0.5080	0.5120	0.5160	0.5199	0.5239	0.5279	0.5319	0.5359	0.0
0.1	0.5398	0.5438	0.5478	0.5517	0.5557	0.5596	0.5636	0.5675	0.5714	0.5754	0.1
0.2	0.5793	0.5832	0.5871	0.5910	0.5948	0.5987	0.6026	0.6064	0.6103	0.6141	0.2
0.3	0.6179	0.6217	0.6255	0.6293	0.6331	0.6368	0.6406	0.6443	0.6480	0.6517	0.3
0.4	0.6554	0.6591	0.6628	0.6664	0.6700	0.6736	0.6772	0.6808	0.6844	0.6879	0.4
0.5	0.6915	0.6950	0.6985	0.7019	0.7054	0.7088	0.7123	0.7157	0.7190	0.7224	0.5
0.6	0.7258	0.7291	0.7324	0.7357	0.7389	0.7422	0.7454	0.7486	0.7517	0.7549	0.6
0.7	0.7580	0.7612	0.7642	0.7673	0.7704	0.7734	0.7764	0.7794	0.7823	0.7852	0.7
0.8	0.7881	0.7910	0.7939	0.7967	0.7996	0.8023	0.8051	0.8078	0.8106	0.8133	0.8
0.9	0.8159	0.8186	0.8212	0.8238	0.8264	0.8289	0.8315	0.8340	0.8365	0.8389	0.9
1.0	0.8413	0.8438	0.8461	0.8485	0.8508	0.8531	0.8554	0.8577	0.8599	0.8621	1.0
1.1	0.8643	0.8665	0.8686	0.8708	0.8729	0.8749	0.8770	0.8790	0.8810	0.8830	1.1
1.2	0.8849	0.8869	0.8888	0.8907	0.8925	0.8944	0.8962	0.8980	0.8997	0.9015	1.2
1.3	0.9032	0.9049	0.9066	0.9082	0.9099	0.9115	0.9131	0.9147	0.9162	0.9177	1.3
1.4	0.9192	0.9207	0.9222	0.9236	0.9251	0.9265	0.9279	0.9292	0.9306	0.9319	1.4
1.5	0.9332	0.9345	0.9357	0.9370	0.9382	0.9394	0.9406	0.9418	0.9429	0.9441	1.5
1.6	0.9452	0.9463	0.9474	0.9484	0.9495	0.9505	0.9515	0.9525	0.9535	0.9545	1.6
1.7	0.9554	0.9564	0.9573	0.9582	0.9591	0.9599	0.9608	0.9616	0.9625	0.9633	1.7
1.8	0.9641	0.9649	0.9656	0.9664	0.9671	0.9678	0.9686	0.9693	0.9699	0.9706	1.8
1.9	0.9713	0.9719	0.9726	0.9732	0.9738	0.9744	0.9750	0.9756	0.9761	0.9767	1.9
2.0	0.9773	0.9778	0.9783	0.9788	0.9793	0.9798	0.9803	0.9808	0.9812	0.9817	2.0
2.1	0.9821	0.9826	0.9830	0.9834	0.9838	0.9842	0.9846	0.9850	0.9854	0.9857	2.1
2.2	0.9861	0.9864	0.9868	0.9871	0.9875	0.9878	0.9881	0.9884	0.9887	0.9890	2.2
2.3	0.9893	0.9896	0.9898	0.9901	0.9904	0.9906	0.9909	0.9911	0.9913	0.9916	2.3
2.4	0.9918	0.9920	0.9922	0.9925	0.9927	0.9929	0.9931	0.9932	0.9934	0.9936	2.4
2.5	0.9938	0.9940	0.9941	0.9943	0.9945	0.9946	0.9948	0.9949	0.9951	0.9952	2.5
2.6	0.9953	0.9955	0.9956	0.9957	0.9959	0.9960	0.9961	0.9962	0.9963	0.9964	2.6
2.7	0.9965	0.9966	0.9967	0.9968	0.9969	0.9970	0.9971	0.9972	0.9973	0.9974	2.7
2.8	0.9974	0.9975	0.9976	0.9977	0.9977	0.9978	0.9979	0.9979	0.9980	0.9981	2.8
2.9	0.9981	0.9982	0.9982	0.9983	0.9984	0.9984	0.9985	0.9985	0.9986	0.9986	2.9
3.0	0.9987	0.9987	0.9987	0.9988	0.9988	0.9989	0.9989	0.9989	0.9990	0.9990	3.0
3.1	0.9990	0.9991	0.9991	0.9991	0.9992	0.9992	0.9992	0.9992	0.9993	0.9993	3.1
3.2	0.9993	0.9993	0.9994	0.9994	0.9994	0.9994	0.9994	0.9995	0.9995	0.9995	3.2
3.3	0.9995	0.9995	0.9996	0.9996	0.9996	0.9996	0.9996	0.9996	0.9996	0.9997	3.3
3.4	0.9997	0.9997	0.9997	0.9997	0.9997	0.9997	0.9997	0.9997	0.9997	0.9998	3.4
t	0	1	2	3	4	5	6	7	8	9	t

20.15 BINOMIAL COEFFICIENTS

$$\binom{n}{0}=1,\ \binom{n}{1}=n,\ \binom{n}{n}=1 \qquad \binom{n}{k}=\frac{n!}{k!(n-k)!}=\frac{n(n-1)\dots(n-k+1)}{k!}=\binom{n}{n-k}$$

n	$\binom{n}{0}$	$\binom{n}{1}$	$\binom{n}{2}$	$\binom{n}{3}$	$\binom{n}{4}$	$\binom{n}{5}$	$\binom{n}{6}$	$\binom{n}{7}$	$\binom{n}{8}$	$\binom{n}{9}$	$\binom{n}{10}$	n
0	1											0
1	1	1										1
2	1	2	1									2
3	1	3	3	1								3
4	1	4	6	4	1							4
5	1	5	10	10	5	1						5
6	1	6	15	20	15	6	1					6
7	1	7	21	35	35	21	7	1				7
8	1	8	28	56	70	56	28	8	1			8
9	1	9	36	84	126	126	84	36	9	1		9
10	1	10	45	120	210	252	210	120	45	10	1	10
11	1	11	55	165	330	462	462	330	165	55	11	11
12	1	12	66	220	495	792	924	792	495	220	66	12
13	1	13	78	286	715	1,287	1,716	1,716	1,287	715	286	13
14	1	14	91	364	1,001	2,002	3,003	3,432	3,003	2,002	1,001	14
15	1	15	105	455	1,365	3,003	5,005	6,435	6,435	5,005	3,003	15
16	1	16	120	560	1,820	4,368	8,008	11,440	12,870	11,440	8,808	16
17	1	17	136	680	2,380	6,188	12,376	19,448	24,310	24,310	19,448	17
18	1	18	153	816	3,060	8,568	18,564	31,824	43,758	43,620	43,758	18
19	1	19	171	969	3,876	11,628	27,132	50,388	75,582	92,378	92.378	19
20	1	20	190	1,140	4,845	15,504	38,760	77,520	125,970	167,960	184,756	20
n	$\binom{n}{0}$	$\binom{n}{1}$	$\binom{n}{2}$	$\binom{n}{3}$	$\binom{n}{4}$	$\binom{n}{5}$	$\binom{n}{6}$	$\binom{n}{7}$	$\binom{n}{8}$	$\binom{n}{9}$	$\binom{n}{10}$	n

Note: For coefficients not given above use $\binom{n}{k} = \binom{n}{n-k}$

20.16 CORRELATION AND REGRESSION

i. The coefficient of correlation between x and y is given by

$$r_{xy} = \frac{\sum xy - \frac{1}{N}\left(\sum x\right)\left(\sum y\right)}{\sqrt{\left(\sum x^2\right) - \frac{1}{N}\left(\sum x\right)^2}\sqrt{\left(\sum y^2\right) - \left(\frac{1}{N}\right)\left(\sum y\right)^2}}$$

ii. Two random variates u and v are independent if
$\text{cov}\,(u, v) = E\,\{u - \bar{u}\,(v - \bar{v})\} = 0$

iii. The regression coefficients are

$$b_{yx} = \frac{r\sigma_y}{\sigma_x}, \qquad b_{xy} = \frac{r\sigma_x}{\sigma_y}$$

iv. r, b_{yx} and b_{xy} have the same sign and $r = \sqrt{b_{yx} \times b_{xy}}$.

v. a. The line of regression of y on x is

$$y - \bar{y} = \frac{r\sigma_y}{\sigma_x}(x - \bar{x}).$$

b. The line of regression of x on y is

$$x - \bar{x} = \frac{r\sigma_x}{\sigma_y}(y - \bar{y})$$

20.17 SAMPLING AND TEST OF SIGNIFICANCE

i. *Test of significance for single proportion*

The test statistic is $Z = \dfrac{X - nP}{\sqrt{nPQ}}$

ii. *Test of significance for the difference of proportions*

Under the null hypothesis: There is no significant difference between the sample proportions and the test statistic

or $$Z = \frac{P_1 - P_2}{\sqrt{PQ\left(\dfrac{1}{n_1} + \dfrac{1}{n_2}\right)}}, \quad P = \frac{x_1 + x_2}{n_1 + n_2}. \quad Q = 1 - P.$$

Under the null hypothesis: The difference in population preparations is likely to be hidden in sampling, the test statistic is

$$Z = \frac{P_1 - P_2}{\sqrt{(P_1Q_1 / n_1) + (P_2Q_2 / n_2)}}$$

iii. *Test of significance for single mean*

Under the null hypothesis: The sample has been drawn from a population with mean μ and $s.d.$ σ, the test statistic is

$$Z = \frac{\bar{x} - \mu}{\sigma / \sqrt{n}}.$$

iv. *Test of significance for difference of means*

Under the null hypothesis: There is no significant difference between the sample means, the test statistic is

$$Z = \frac{\bar{x}_1 - \bar{x}_2}{\sqrt{\sigma_1^2 / n_1 + \sigma_2^2 / n2}}.$$

If the sample are drawn from the populations with common $s.d.\sigma$, then the above test statistic becomes

$$Z = \frac{\bar{x}_1 - \bar{x}_2}{\sigma\sqrt{1 / n_1 + 1 / n_2}}.$$

v. *Test of significance for difference of standard deviations*

Under the null hypothesis: There is no significant difference between the sample standard deviations, the test statistic is

$$Z = \frac{s_1 - s_2}{\sqrt{\left(\sigma_1^2 / 2n_1\right) + \left(\sigma_2^2 / 2n_2\right)}}.$$

How to Accept or Reject the Null Hypothesis?

If $|Z| > 3$, the difference is highly significant and we reject the null hypothesis

If $|Z| < 1.96$, the null hypothesis may be accepted at 5% level of significance.

t-Test

i. Under the null hypothesis: There is no significant difference between the sample mean and the population mean, the test statistic is

$$t = \frac{\bar{x} - \mu}{S / \sqrt{n}}, \quad S^2 = \frac{1}{n-1} \sum_{i=1}^{n} (x_i - \bar{x})^2,$$

and t has $(n - 1)$ *d.f.*

ii. Under the null hypothesis: There is no significant difference between the sample means, the t-statistic is

$$t = \frac{\bar{x}_1 - \bar{x}_2}{S\sqrt{1/n_1 + 1/n_2}}, \ \bar{x}_1 = \frac{1}{n_1} \sum x_1, \bar{x}_2 = \frac{1}{n_2} \sum x_2,$$

$$S^2 = \frac{1}{(n_1 + n_2 - 2)} \left[\sum (x_1 - \bar{x}_1)^2 + \sum (x_2 - \bar{x}_2)^2\right]$$

and t has $(n_1 + n_2 - 2)$ d.f.

How to Accept or Reject the Null Hypothesis?

Let t_{cal} denote the calculated value of t by means of the above formulae and $t_{0.05}$ the tabulated value of t for $(n - 1)$ or $(n_1 + n_2 - 2)$ d.f. (as the case may be).

If $t_{cal} < t_{0.05}$, the null hypothesis is accepted at 5% level of significance.

If $t_{cal} > t_{0.05}$, the null hypothesis is rejected at 5% level of signicance.

F-Test

Under the null hypothes: The population variances are equal, the F-statistics is

$$F = \frac{S_x^{\ 2}}{S_y^{\ 2}} (S_x^{\ 2} > S_y^{\ 2}),$$

where $S_x^{\ 2} = \frac{1}{n_1 - 1} \sum_{i=1}^{n_1} (x_i - \bar{x})^2, \ S_Y^{\ 2} = \frac{1}{n_2 - 1} \sum_{j=1}^{n_2} (y_i - \bar{y})^2$

and F has $(n - 1, n_2 - 1)$ d.f.

Let $F_{0.05}$ denote the tabulated value of F for $(n_1 - 1, n_2 - 1)$ d.f.

If $F_{cal} < F_{0.05}$, the null hypothesis is accepted at 5% level of significance.

If $F_{cal} > F_{0.05}$, the null hypothesis is rejected at 5% level of significance.

χ^2-Test

If O_i $(i = 1, 2, ..., n)$ are observed frequencies and E_i $(i = 1, 2, ..., n)$ are the corresponding expected frequencies, then the statistic χ^2 is defined as

$$\chi^2 = \sum_{i=1}^{n} \frac{(O_i - E_i)^2}{E_i}$$

and has $(n - 1)$ d.f.

Under the null hypothesis: The two attributes are independent, calculate χ^2 by the above formula and find its tabulated value $\chi^2_{0.05}$ for $(n - 1)$ d.f.

If $\chi^2_{cal} < \chi^2_{0.05}$ accept the null hypothesis at 5% level of significance.

If $\chi^2_{cal} > \chi^2_{0.05}$ reject the null hypothesis at 5% level of significance.

CHAPTER

21

Tables of Indefinite Integrals

21.1 BASIC CONCEPTS

i. Notation

The more frequently encountered indefinite integrals and their solutions are tabulated in this chapter. Particular symbols used in the following are

a, b, c, d, e, f are constants — α, β, γ are constant equivalents

m, n, p, q, r are integers — x is independent variable

$A = a + bx$	$M = a^2 + b^2x$
$B = a + bx + cx^2$	$N = a^2 - b^2x$
$C = a^2 + x^2$	$E = a^2 - x^2$
$F = x^2 - a^2$	$G = a^3 \pm x^3$
$H = a^4 + x^4$	$L = a^4 - x^4$
$P = a + bx^q$	$R = ax^q + bx^{q+r}$

In these tables, the constant of integration is omitted but implied, logarithmic expressions are for the absolute value of the respective argument, all angles are in radians, and all inverse functions represent principal values (angles).

ii. Table of Basic Forms of Indefinite Integrals

1.	$\int dx = x + c$
2.	$\int \frac{dx}{x} = \log x + c$
3.	$\int x^n\, dx = \frac{x^{n+1}}{n+1} + c \ldots (n \neq -1)$

4.	$\int e^x \, dx = e^x + c$
5.	$\int a^x \, dx = \dfrac{a^x}{\log a} + c$
6.	$\int \sin x dx = -\cos x + c$
7.	$\int \cos x dx = \sin x + c$
8.	$\int \tan x dx = -\log \cos x + c$
9.	$\int \sec^2 x dx = \tan x + c$
10.	$\int \cos^2 x dx = -ctn\ x + c$
11.	$\int \sec x \ \tan x \, dx = \sec x + c$
12.	$\sin^2 x \, dx = \dfrac{1}{2} x - \dfrac{1}{2} \sin x \cos x + c$
13.	$\cos^2 x \, dx = \dfrac{1}{2} x + \dfrac{1}{2} \sin x \cos x + c$
14.	$\int \log x \, dx = x \log x - x + c$
	Form = $ax + b$ **Note:** In the following list, a constant of integration C should be added to the result of each integration.
15.	$\int (ax+b)^m \, dx = \dfrac{(ax+b)^{m+1}}{a(m+1)}, \qquad m \neq -1$
16.	$\int x(ax+b)^m \, dx = \dfrac{(ax+b)^{m+2}}{a^2(m+2)} - \dfrac{b(ax+b)^{m+1}}{a^2(m+1)}$
17.	$\int \dfrac{dx}{ax+b} = \dfrac{1}{a} \log(ax+b)$
18.	$\int \dfrac{dx}{(ax+b)^2} = -\dfrac{1}{a(ax+b)}$

	Trignometric Forms
19.	$\int \sin x\, dx = -\frac{1}{a}\cos ax$
20.	$\sin^2 ax\, dx = -\frac{1}{2}\cos ax \sin ax + \frac{1}{2}x = \frac{1}{2}x - \frac{1}{4a}\sin 2\, ax$
21.	$\int \sin(a+bx)\, dx = -\frac{1}{b}\cos(a+bx)$
22.	$\int \frac{dx}{1 \pm \sin ax} = \mp\frac{1}{a}\tan\left(\pi/4 \mp \frac{ax}{2}\right)$
23.	$\int \frac{\sin ax}{1 \pm \sin ax}\, dx = \pm x + \frac{1}{a}\tan\left(\frac{\pi}{4} \mp \frac{ax}{2}\right)$
24.	$\int (\cos ax)\, dx = \frac{1}{a}\sin ax$
25.	$\int \cos^2 ax\, dx = \frac{1}{2a}\sin ax \cos ax + \frac{1}{2}x = \frac{1}{2}x + \frac{1}{4a}\sin 2ax$
26.	$\int (\sin mx \sin nx)dx = \frac{\sin(m-n)x}{2(m-n)} - \frac{\sin(m+n)x}{2(m+n)} \qquad m^2 \neq n^2$
27.	$\int (\cos mx)(\cos nx)\, dx = \frac{\sin(m-n)x}{2(m-n)} + \frac{\sin(m+n)x}{2(m+n)} \qquad m^2 \neq n^2$
28.	$\int (\sin ax)(\cos ax)\, dx = \frac{1}{2a}\sin^2 ax$
29.	$\int (\sin mx)(\cos nx)\, dx = -\frac{\cos(m-n)x}{2(m-n)} - \frac{\cos(m+n)x}{2(m+n)} \qquad m^2 \neq n^2$
	Logarithmic Forms
30.	$\int (\log x)\, dx = x \log x - x$
31.	$\int (x \log x)\, dx = \frac{x^2}{2}\log x - \frac{x^2}{4}$
32.	$\int x^2 (\log x)\, dx = \frac{x^3}{3}\log x - \frac{x^3}{9}$

	Exponential Forms	
33.	$\int e^x\,dx = e^x$	
34.	$\int e^{-x}\,dx = -e^{-x}$	
35.	$\int e^{ax}\,dx = \frac{e^{ax}}{a}$	
36.	$\int x^{ax}\,dx = \frac{e^{ax}}{a^2}(ax-1)$	
37.	$\int \frac{dx}{1+e^x} = x - \log(1+e^x) = \log\left(\frac{e^x}{1+e^x}\right)$	
	Definite Integrals	
38.	$\int_1^\infty \frac{dx}{x^m} = \frac{1}{m-1}$	... $(m > 1)$
39.	$\int_0^\infty \frac{dx}{(1+x)\sqrt{x}} = \pi$	
40.	$\int_0^\infty \frac{a\,dx}{a^2+x^2} = \pi/2$	if $a > 0$
	$= 0$	if $a = 0$
	$= -\pi/2$	if $a < 0$
41.	$\int_0^\infty e^{-ax}dx = \frac{1}{a}$	$(a > 0)$
42.	$\int_0^\infty \frac{e^{-ax}-e^{-bx}}{x}\,dx = \log\frac{b}{a}$	$(a, b > 0)$

iii. Indefinite Integrals Involving $f(x^m, A^n)$ $A = a + bx$ $a \neq 0$

$\int x^m\,dx = \frac{x^{m+1}}{m+1}$ $m \neq -1$

$\int A^n\,dx = \frac{A^{n+1}}{b(n+1)}$ $n \neq -1$

$$\int xA^n\,dx = \frac{[(n+1)A-(n+2)a]A^{n+1}}{(n+1)(n+2)b^2} \qquad n \neq -1, -2$$

$$\int x^m A^n \, dx = \frac{1}{b(m+n+1)}\left(x^m A^{n+1} - ma\int x^{m-1} A^n \, dx\right)$$

$$= \frac{1}{m+n+1}\left(x^{m+1} A^n + na\int x^m A^{n-1} \, dx\right) \qquad m > 0,\ m+n+1 \neq 0$$

---

iv. Indefinite Integrals Involving $f(x) = \dfrac{x^m}{A^n}$ $\quad A = a + bx \quad a \neq 0$

$$\int \frac{dx}{A} = \frac{\mathrm{In}\, A}{b}$$

$$\int \frac{dx}{A^2} = -\frac{1}{bA}$$

$$\int \frac{dx}{A^n} = -\frac{1}{(n-1)\, bA^{n-1}} \qquad n \neq 0, 1$$

$$\int \frac{x\, dx}{A} \frac{1}{b^2}(A - a\ln A)$$

$$\int \frac{x\, dx}{A^2} \frac{1}{b^2}\left(\frac{a}{A} - \ln A\right)$$

$$\int \frac{x\, dx}{A^n} = -\frac{1}{(n-2)\, b^2 A^{n-1}}\left(A - \frac{n-2}{n-1}a\right) \qquad n \neq 0, 1, 2$$

$$\int \frac{x^2\, dx}{A} = \frac{1}{b^3}\left(\frac{A^2}{2} - 2aA + a^2 \ln A\right)$$

$$\int \frac{x^2\, dx}{A^2} = \frac{1}{b^3}\left(A - \frac{a^2}{A} - 2a\ln A\right)$$

$$\int \frac{x^2\, dx}{A^3} = \frac{1}{b^3}\left(\frac{2a}{A} - \frac{a^2}{2A^2} + \ln A\right)$$

$$\int \frac{x^2\, dx}{A^n} = -\frac{1}{(n-3)\, b^3 A^{n-1}}\left[A^2 - \frac{2(n-3)a}{n-2}A + \frac{n-3}{n-1}a^2\right] \qquad n \neq 0, 1, 2, 3$$

$$\int \frac{x^m}{A^n} dx = \frac{1}{b^{m+1}} \int \frac{(A-a)^m}{A^n} dA$$

$$= \frac{1}{b^{m+1}} \sum_{k=0}^{m} \binom{m}{k} \frac{A^{m-n-k+1}(-a)^k}{m-n-k+1}$$

with terms $m-n-k+1=0$ replaced by $\binom{m}{n-1}(-a)^{m-n+1}\ln A$

v. Indefinite Integrals Involving $f(x)=\dfrac{1}{x^m A^n}$ $\quad A=a+bx \quad a\neq 0$

$$\int\frac{dx}{xA}=-\frac{1}{a}\ln\frac{A}{x}$$

$$\int\frac{dx}{xA^2}=-\frac{1}{a^2}\ln\frac{A}{x}+\frac{1}{aA}$$

$$\int\frac{dx}{xA^n}=\frac{-1}{a^n}\ln\frac{A}{x}+\frac{1}{a^n}\sum_{k=1}^{n-1}\binom{n-1}{k}\frac{(-bx)^k}{kA^k} \qquad n\neq 0$$

$$\int\frac{dx}{x^2A^n}=\frac{b}{a^2}\ln\frac{A}{x}-\frac{1}{ax}$$

$$\int\frac{dx}{x^2A^2}=\frac{2b}{a^3}\ln\frac{A}{x}-\frac{1}{a^2x}-\frac{b}{a^2A}$$

$$\int\frac{dx}{x^2A^n}=\frac{nb}{a^{n+1}}\ln\frac{A}{x}-\frac{A}{a^{n+1}x}+\frac{A}{a^{n+1}x}\sum_{k=2}^{n}\binom{n}{k}\frac{(-bx)^k}{(k-1)A^k} \qquad n\neq 0,1$$

$$\int\frac{dx}{x^3A}=-\frac{b^2}{a^3}\ln\frac{A}{x}+\frac{2bA}{a^3x}-\frac{A^2}{2a^3x^2}$$

$$\int\frac{dx}{x^3A^2}=-\frac{3b^2}{a^4}\ln\frac{A}{x}+\frac{3bA}{a^4x}-\frac{A^2}{2a^4x^2}-\frac{b^3x}{a^4A}$$

$$\int\frac{dx}{x^3A^n}=-\frac{n(n+1)b^2}{2a^{n+2}}\ln\frac{A}{x}+\frac{(n+1)bA}{a^{n+2}x}-\frac{b^2A^2}{2a^{n+2}x^2}+\frac{A^2}{a^{n+2}x^2}\sum_{k=3}^{n+1}\binom{n+1}{k}\frac{(-bx)^k}{(k-2)A^k} \qquad n\neq 0,1,2$$

$$\int\frac{dx}{x^mA^n}=\frac{-1}{a^{m+n-1}}\sum_{k=0}^{m+n-2}\binom{m+n-2}{k}\frac{A^{m-n-1}(-b)^k}{(m-n-1)^{m-n-1}}$$

The term $m-k-1=0$ is replaced by $\binom{m+n-2}{m-1}(-b)^{m-1}\ln\frac{A}{x}$

vi. Indefinite Integrals Involving $f(x)=f(A,D)$ $\quad A=a+x \quad a\neq 0$ $\quad \alpha=a-b\neq 0 \quad D=b+x \quad b\neq 0$

$$\int A^m D^n\,dx=\frac{A^{m+1}D^n}{m+n+1}+\frac{n\alpha}{m+n+1}\int A^mD^{n-1}\,dx \qquad m+n\neq -1$$

$$\int \frac{dx}{A^m D^n} = \frac{-1}{(n-1)\,\alpha A^{m-1} D^{n-1}} - \frac{m+n-2}{(n-1)\,\alpha} \int \frac{dx}{A^m D^{n-1}} \qquad n \neq 1$$

$$\int \frac{dx}{AD} = \frac{1}{\alpha} \ln \frac{D}{A}$$

$$\int \frac{x\,dx}{AD} = \frac{1}{\alpha}(a \ln A - b \ln D)$$

$$\int \frac{x^2\,dx}{AD} = x - \frac{a+b}{2} \ln AD + \frac{a^2+b^2}{2} \ln \frac{D}{A}$$

$$\int \frac{dx}{AD^2} = \frac{-1}{\alpha D} + \frac{1}{\alpha^2} \ln \frac{A}{D}$$

$$\int \frac{x\,dx}{AD^2} = \frac{b}{\alpha D} - \frac{a}{\alpha^2} \ln \frac{A}{D}$$

$$\int \frac{x^2\,dx}{AD^2} = \frac{b^2}{\alpha D} + \frac{a^2}{\alpha^2} \ln \frac{A}{D} + \ln D$$

$$\int \frac{dx}{A^2 D^2} = \frac{-1}{\alpha^2}\left(\frac{1}{A} + \frac{1}{D}\right) + \frac{2}{\alpha^2} \ln \frac{A}{D}$$

$$\int \frac{x dx}{A^2 D^2} = \frac{1}{\alpha^2}\left(\frac{a}{A} + \frac{b}{D}\right) + \frac{a+b}{\alpha^3} \ln \frac{A}{D}$$

$$\int \frac{x^2\,dx}{A^2 D^2} = -\frac{1}{\alpha^2}\left(\frac{a^2}{A} + \frac{b^2}{D}\right) + \frac{2ab}{\alpha^3} \ln \frac{A}{D}$$

$$\int \frac{A^m}{D^n}\,dx = \frac{-A^{m+1}}{(n-1)\,aD^{n-1}} - \frac{n-m-2}{(n-1)\alpha} \int \frac{A^m}{D^{n-1}}\,dx \qquad n \neq 1$$

$$= \frac{-A^m}{(n-m-1)\,D^{n-1}} - \frac{m\alpha}{n-m-1} \int \frac{A^{m-1}}{D^n}\,dx \qquad n-m \neq 1$$

vii. Indefinite Integrals Involving $f(x) = \dfrac{x^m}{B^n}$ $\quad f(x) = \dfrac{1}{x^m B^n}$ $\quad B = a + bx + cx^2 \quad a \neq 0$

$\gamma = 4ac - b^2$

$$\int \frac{dx}{B} = \begin{cases} \dfrac{2}{\sqrt{\gamma}} \tan^{-1} \dfrac{2cx+b}{\sqrt{\gamma}} & \gamma > 0 \\ -\dfrac{2}{2cx+b} & \gamma = 0 \\ \dfrac{1}{\sqrt{-\gamma}} \ln \dfrac{2cx+b-\sqrt{-\gamma}}{2cx+b+\sqrt{-\gamma}} & \gamma < 0 \end{cases}$$

$$\int \frac{dx}{B^n} = \frac{2cx+b}{(n-1)\gamma B^{n-1}} + \frac{2(2n-3)c}{(n-1)\gamma} \int \frac{dx}{B^{n-1}} \qquad n \neq 1$$

$$\int \frac{x\,dx}{B} = \frac{1}{2c} \ln B - \frac{b}{2c} \int \frac{dx}{B}$$

$$\int \frac{x^m\,dx}{B^n} = -\frac{x^{m-1}}{(2n-m-1)cB^{n-1}} + \frac{(m-1)a}{(2n-m-1)c} \int \frac{x^{m-2}\,dx}{B^n}$$

$$- \frac{(n-m)b}{(2n-m-1)c} \int \frac{x^{m-1}\,dx}{B^n} \qquad 2n-m-1 \neq 0$$

$$\int \frac{dx}{xB} = \frac{1}{2a} \ln \frac{x^2}{B} - \frac{b}{2a} \int \frac{dx}{B}$$

$$\int \frac{dx}{x^2B} = \frac{b}{2a^2} \ln \frac{B}{x^2} - \frac{1}{ax} + \left(\frac{b^2}{2a^2} - \frac{c}{a} \right) \int \frac{dx}{B}$$

$$\int \frac{dx}{x^mB^n} = \frac{-1}{(m-1)ax^{m-1}B^{n-1}} - \frac{(2n+m-3)c}{(m-1)a} \int \frac{dx}{x^{m-2}B^n} - \frac{(n+m-2)b}{(m-1)a} \int \frac{dx}{x^{m-1}B^n} \qquad m \neq 1$$

$$\int \frac{e+fx}{B}\,dx = \frac{f}{2c} \ln B + \frac{2ce-bf}{2c} \int \frac{dx}{B}$$

$$\int \frac{e+fx}{B^n}\,dx = \frac{f}{2(n-1)aB^{n-1}} + \frac{(2ce-bf)}{2c} \int \frac{dx}{B^n} \qquad n \neq 1$$

---

viii. Indefinite Integrals Involving $f(x) = \frac{x^m}{C^n}$ $\quad f(x) = \frac{1}{CC_1}$ $\quad C = a^2 + x^2 \quad a^2 \neq 0$

$C_1 = b^2 + x^2 \quad b^2 \neq 0$

$$\int \frac{dx}{C} = \frac{1}{a} \tan^{-1} \frac{x}{a}$$

$$\int \frac{dx}{CC_1} = \frac{1}{b^2-a^2} \left(\frac{1}{a} \tan^{-1} \frac{x}{a} - \frac{1}{b} \tan^{-1} \frac{x}{b} \right) \qquad a \neq b$$

$$\int \frac{dx}{C^n} = \frac{x}{2(n-1)a^2C^{n-1}} + \frac{2n-3}{2(n-1)a^2} \int \frac{dx}{C^{n-1}} \qquad n \neq 1$$

$$\int \frac{x\,dx}{C} = \frac{1}{2} \ln C$$

$$\int \frac{x\,dx}{C^2} \frac{-1}{2C}$$

$$\int \frac{x\,dx}{C^n} \frac{-1}{2\,(n-1)\,C^{n-1}} \qquad n \neq 1$$

$$\int \frac{x^2\,dx}{C} = x - a\tan^{-1}\frac{x}{a}$$

$$\int \frac{x^2\,dx}{C^2} = -\frac{x}{2C} + \frac{1}{2a}\tan^{-1}\frac{x}{a}$$

$$\int \frac{x^2\,dx}{C^n} = \frac{-x}{2(n-1)\,C^{n-1}} + \frac{1}{2\,(n-1)}\int \frac{dx}{C^{n-1}} \qquad n \neq 1$$

$$\int \frac{x^3\,dx}{C} = \frac{x^2}{2} - \frac{a^2}{2}\ln C$$

$$\int \frac{x^3\,dx}{C^2} = \frac{a^2}{2C} + \frac{1}{2}\ln C$$

$$\int \frac{x^3\,dx}{C^n} = \frac{1}{2(n-2)\,C^{n-2}} + \frac{a^2}{2(n-1)\,C^{n-1}} \qquad n \neq 1, 2$$

$$\int \frac{x^m\,dx}{C^n} = -\frac{x^{m-1}}{2(n-1)\,C^{n-1}} + \frac{m-1}{2\,(n-1)}\int \frac{x^{m-2}}{C^{n-1}} \qquad n \neq 1$$

ix. Indefinite Integrals Involving $f(x) = \dfrac{1}{x^m C^n}$ $\quad C = a^2 + x^2 \quad a^2 \neq 0$

$$\int \frac{dx}{xC} = \frac{1}{2a^2}\ln\frac{x^2}{C}$$

$$\int \frac{dx}{xC^2} = \frac{1}{2a^2C} + \frac{1}{2a^4}\ln\frac{x^2}{C}$$

$$\int \frac{dx}{xC^3} = \frac{1}{4a^2C^2} + \frac{1}{2a^4C} + \frac{1}{2a^6}\ln\frac{x^2}{C}$$

$$\int \frac{dx}{xC^n} = \frac{1}{2\,(n-1)\,a^2C^{n-1}} + \frac{1}{a^2}\int \frac{dx}{xC^{n-1}} \qquad n \neq 1$$

$$\int \frac{dx}{x^2C} = \frac{-1}{a^2x} - \frac{1}{a^3}\tan^{-1}\frac{x}{a}$$

$$\int \frac{dx}{x^2C^2} = \frac{-1}{a^4\,x} - \frac{x}{2a^4C} - \frac{3}{2a^5}\tan^{-1}\frac{x}{a}$$

$$\int \frac{x^3\,dx}{\sqrt{F}} = \left(\frac{F}{3} + a^2\right)\sqrt{F}$$

$$\int \frac{x^3\,dx}{\sqrt{F^3}} = \frac{F - a^2}{\sqrt{F}}$$

$$\int \frac{x^m\,dx}{\sqrt[p]{F^n}} = -\frac{px^{m+1}}{2a^2(n-p)\sqrt[p]{F^{n-p}}} - \frac{2n - p(m+3)}{2a^2(n-p)}\int \frac{x^m\,dx}{\sqrt[n]{F^{n-p}}} \qquad n \neq p$$

---

xxxiv. Indefinite Integrals Involving $f(x) = \dfrac{1}{x^m \sqrt[p]{F^n}}$ $\quad F = x^2 - a^2 \quad a^2 \neq 0$

$$\int \frac{dx}{x\sqrt{F}} = \frac{1}{a}\cos^{-1}\frac{a}{x}$$

$$\int \frac{dx}{x\sqrt{F^3}} = \frac{-1}{a^2\sqrt{F}} - \frac{1}{a^3}\cos^{-1}\frac{a}{x}$$

$$\int \frac{dx}{x\sqrt{F^n}} = -\frac{1}{a^2(n-2)x\sqrt{F^{n-2}}} - \frac{1}{a^2}\int \frac{dx}{x\sqrt{F^{n-2}}} \qquad n \neq 2$$

$$\int \frac{dx}{x^2\sqrt{F}} = \frac{\sqrt{F}}{a^2 x}$$

$$\int \frac{dx}{x^2\sqrt{F^3}} = -\frac{\sqrt{F}}{a^4 x}\left(1 + \frac{x^2}{F}\right)$$

$$\int \frac{dx}{x^2\sqrt{F^n}} = -\frac{1}{a^2(n-2)x\sqrt{F^{n-2}}} - \frac{n-1}{a^2(n-2)}\int \frac{dx}{x^2\sqrt{F^{n-2}}} \qquad n \neq 2$$

$$\int \frac{dx}{x^3\sqrt{F}} = \frac{\sqrt{F}}{2a^2x^2} + \frac{1}{2a^3}\cos^{-1}\frac{a}{x}$$

$$\int \frac{dx}{x^3\sqrt{F^3}} = \frac{1}{2a^2x^2\sqrt{F}} - \frac{3}{2a^4\sqrt{F}} - \frac{3}{2a^5}\cos^{-1}\frac{a}{x}$$

$$\int \frac{dx}{x^3\sqrt{F^n}} = -\frac{1}{a^2(n-2)x^2\sqrt{F^{n-2}}} - \frac{n}{a^2(n-2)}\int \frac{dx}{x^3\sqrt{F^{n-2}}} \qquad n \neq 2$$

$$\int \frac{dx}{x^4\sqrt{F}} = \frac{\sqrt{F}}{a^4 x}\left(1 - \frac{F^2}{3x^2}\right)$$

$$\int \frac{dx}{x^4\sqrt{F^3}} = -\frac{x}{a^6\sqrt{F}}\left(1+\frac{2F}{x^2}-\frac{F^2}{3x^4}\right)$$

$$\int \frac{dx}{x^m \sqrt[p]{F^n}} = \frac{p}{2a^2(n-p)x^{m-1}\sqrt[p]{F^{n-p}}} - \frac{2n-p(m-3)}{2a^2(n-p)}\int \frac{dx}{x^m \sqrt[p]{F^{n-p}}} \qquad n \neq p$$

xxxv. Indefinite Integrals Involving $f(x) = \frac{\sqrt[p]{F^n}}{x^m}$ $\quad F = x^2 - a^2 \quad a^2 \neq 0$

$$\int \frac{\sqrt{F}}{x}dx = \sqrt{F} - a\cos^{-1}\frac{a}{x}$$

$$\int \frac{\sqrt{F^3}}{x}dx = \frac{1}{3}\sqrt{F^3} - a^2\sqrt{F} + a^3\cos^{-1}\frac{a}{x}$$

$$\int \frac{\sqrt{F^n}}{x}dx = \frac{1}{n}\sqrt{F^n} - a^2\int \frac{\sqrt{F^{n-2}}}{x}dx$$

$$\int \frac{\sqrt{F}}{x^2}dx = -\frac{1}{x}\sqrt{F} + \ln(x+\sqrt{F})$$

$$\int \frac{\sqrt{F^3}}{x^2}dx = -\frac{1}{x}\sqrt{F^3} + \frac{3x}{2}\sqrt{F} - \frac{3a^2}{2}\ln(x+\sqrt{F})$$

$$\int \frac{\sqrt{F^n}}{x^2}dx = \frac{\sqrt{F^n}}{(n-1)x} - \frac{a^2 n}{n-1}\int \frac{\sqrt{F^{n-2}}}{x^2}dx \qquad n \neq 1$$

$$\int \frac{\sqrt{F}}{x^3}dx = \frac{-1}{2x^2}\sqrt{F} + \frac{1}{2a}\cos^{-1}\frac{a}{x}$$

$$\int \frac{\sqrt{F^3}}{x^3}dx = -\frac{-1}{2x^2}\sqrt{F^3} + \frac{3}{2}\sqrt{F} - \frac{3a}{2}\cos^{-1}\frac{a}{x}$$

$$\int \frac{\sqrt{F^n}}{x^3}dx = \frac{\sqrt{F^n}}{(n-2)x^2} - \frac{a^2 n}{n-2}\int \frac{\sqrt{F^{n-2}}}{x^3}dx \qquad n \neq 2$$

$$\int \frac{\sqrt{F}}{x^4}dx = \frac{1}{3a^2x^3}\sqrt{F^3}$$

$$\int \frac{\sqrt{F^3}}{x^4}dx = -\frac{1}{3x^3}\sqrt{F^3} - \frac{1}{x}\sqrt{F} + \ln(x+\sqrt{F})$$

$$\int \frac{\sqrt[p]{F^n}}{x^m}dx = \frac{p\sqrt[p]{F^n}}{(2n-mp+p)x^{m-1}} - \frac{2a^2 n}{2n-mp+p}\int \frac{\sqrt[p]{F^{n-p}}}{x^m}dx \qquad 2n \neq -p(1-m)$$

$$\int \frac{\sqrt{E}}{x^3}\,dx = -\frac{1}{2x^2}\sqrt{E} + \frac{1}{2a}\ln\frac{a+\sqrt{E}}{x}$$

$$\int \frac{\sqrt{E^3}}{x^3}\,dx = -\frac{1}{2x^2}\sqrt{E^3} - \frac{3}{2}\sqrt{E} + \frac{3a}{2}\ln\frac{a+\sqrt{E}}{x}$$

$$\int \frac{\sqrt{E^n}}{x^3}\,dx = \frac{\sqrt{E^n}}{(n-2)x^2} + \frac{a^2 n}{n-2}\int\frac{\sqrt{E^{n-2}}}{x^3} \qquad n \neq 2$$

$$\int \frac{\sqrt{E}}{x^4}\,dx = -\frac{1}{3a^2x^3}\sqrt{E^3}$$

$$\int \frac{\sqrt{E^3}}{x^4}\,dx = -\frac{1}{3x^3}\sqrt{E^3} + \frac{1}{x}\sqrt{E} + \sin^{-1}\frac{x}{a}$$

$$\int \frac{\sqrt[p]{E^n}}{x^m}\,dx = \frac{p\sqrt[p]{E^n}}{(2n-mp+p)x^{m-1}} + \frac{2a^2n}{2n-mp+p}\int\frac{\sqrt[p]{E^{n-p}}}{x^m}\,dx \qquad 2n \neq -p\,(1-m)$$

---

xxxii. Indefinite Integrals Involving $f(x) = x^m \sqrt[p]{F^n}$ $\quad F = x^2 - a^2 \quad a^2 \neq 0$

$$\int \sqrt{F}\,dx = \frac{x}{2}\sqrt{F} - \frac{a^2}{2}\ln(x+\sqrt{F})$$

$$\int \sqrt{F^3}\,dx = \frac{x\sqrt{F^3}}{4} - \frac{3a^2}{4}\int\sqrt{F}\,dx$$

$$\int \sqrt{F^n}\,dx = \frac{x\sqrt{F^n}}{n+1} - \frac{a^2n}{n+1}\int\sqrt{F^{n-2}}\,dx \qquad n \neq -1$$

$$\int x\sqrt{F}\,dx = \frac{1}{3}\sqrt{F^3}$$

$$\int x\sqrt{F^3}\,dx = \frac{1}{5}\sqrt{F^3}$$

$$\int x\sqrt{F^n}\,dx = \frac{x^2\sqrt{F^n}}{n+2} - \frac{a^2n}{n+2}\int x\sqrt{F^{n-2}}\,dx \qquad n \neq -2$$

$$\int x^2\sqrt{F}\,dx = \frac{x}{4}\sqrt{F^3} + \frac{a^2x}{8}\sqrt{F} - \frac{a^4}{8}\ln(x+\sqrt{F})$$

$$\int x^2\sqrt{F^3}\,dx = \frac{x}{6}\sqrt{F^5} + \frac{a^2x}{24}\sqrt{F^3} - \frac{a^4x}{16}\sqrt{F} + \frac{a^6}{16}\ln(x+\sqrt{F})$$

$$\int x^2\sqrt{F^n}\,dx = \frac{x^3\sqrt{F^n}}{n+3} - \frac{a^2 n}{n+3}\int x^2\sqrt{F^{n-2}}\,dx \quad n \neq -3$$

$$\int x^3\sqrt{F}\,dx = \frac{1}{5}\sqrt{F^5} + \frac{a^2}{3}\sqrt{F^3}$$

$$\int x^3\sqrt{F^3}\,dx = \frac{1}{7}\sqrt{F^7} + \frac{a^2}{5}\sqrt{F^5}$$

$$\int x^m\sqrt[p]{F^n}\,dx = \frac{px^{m+1}\sqrt[p]{F^n}}{2n+mp+p} - \frac{2a^2 n}{2n+mp+p}\int x^m\sqrt[p]{F^{n-p}}\,dx \qquad 2n \neq -p\,(m+1)$$

xxxiii. Indefinite Integrals Involving $f(x) = \dfrac{x^n}{\sqrt[p]{F^n}}$ $\quad F = x^2 - a^2 \quad a^2 \neq 0$

$$\int \frac{dx}{\sqrt{F}} = \ln(x+\sqrt{F}) = \cosh^{-1}\frac{x}{a}$$

$$\int \frac{dx}{\sqrt{F^3}} = -\frac{x}{a^2\sqrt{F}}$$

$$\int \frac{dx}{\sqrt{F^n}} = -\frac{x}{a^2(n-2)\sqrt{F^{n-2}}} - \frac{n-3}{a^2(n-2)}\int \frac{dx}{\sqrt{F^{n-2}}} \qquad n \neq 2$$

$$\int \frac{x\,dx}{\sqrt{F}} = \sqrt{F}$$

$$\int \frac{x\,dx}{\sqrt{F^3}} = \frac{-1}{\sqrt{F}}$$

$$\int \frac{x\,dx}{\sqrt{F^n}} = -\frac{x^2}{a^2(n-2)\sqrt{F^{n-2}}} - \frac{n-4}{a^2(n-2)}\int \frac{x\,dx}{\sqrt{F^{n-2}}} \qquad n \neq 2$$

$$\int \frac{x^2\,dx}{\sqrt{F}} = \frac{x}{2}\sqrt{F} + \frac{a^2}{2}\ln(x+\sqrt{F})$$

$$\int \frac{x^2\,dx}{\sqrt{F^3}} = -\frac{x}{\sqrt{F}} + \ln(x+\sqrt{F})$$

$$\int \frac{x^2\,dx}{\sqrt{F^3}} = -\frac{x^3}{a^2(n-2)\sqrt{F^{n-2}}} - \frac{n-5}{a^2(n-2)}\int \frac{x^2\,dx}{\sqrt{F^{n-2}}} \qquad n \neq 2$$

$$\int \frac{x\,dx}{\sqrt{E^3}} = \frac{1}{\sqrt{E}}$$

$$\int \frac{x\,dx}{\sqrt{E^n}} = \frac{x^2}{a^2(n-2)\sqrt{E^{n-2}}} + \frac{n-4}{a^2(n-2)} \int \frac{x\,dx}{\sqrt{E^{n-2}}} \qquad n \neq 2$$

$$\int \frac{x^2\,dx}{\sqrt{E}} = -\frac{x}{2}\sqrt{E} + \frac{a^2}{2}\sin^{-1}\frac{x}{a}$$

$$\int \frac{x^2\,dx}{\sqrt{E^3}} = \frac{x}{\sqrt{E}} - \sin^{-1}\frac{x}{a}$$

$$\int \frac{x^2\,dx}{\sqrt{E^n}} = \frac{x^3}{a^2(n-2)\sqrt{E^{n-2}}} + \frac{n-5}{a^2(n-2)} \int \frac{x^2\,dx}{\sqrt{E^{n-2}}} \qquad n \neq 2$$

$$\int \frac{x^3\,dx}{\sqrt{E}} = \left(\frac{E}{3} - a^2\right)\sqrt{E}$$

$$\int \frac{x^3\,dx}{\sqrt{E^3}} = \frac{E + a^2}{\sqrt{E}}$$

$$\int \frac{x^m\,dx}{\sqrt[p]{E^n}} = \frac{px^{m+1}}{2a^2(n-p)\sqrt[p]{E^{n-p}}} + \frac{2n - p(m+3)}{2a^2(n-p)} \int \frac{x^m\,dx}{\sqrt[p]{E^{n-p}}} \qquad n \neq p$$

XXX. Indefinite Integrals Involving $f(x) = \dfrac{1}{x^m \sqrt[p]{E^n}}$ $\quad E = a^2 - x^2 \quad a^2 \neq 0$

$$\int \frac{dx}{x\sqrt{E}} = \frac{-1}{a}\ln\frac{a - \sqrt{E}}{x}$$

$$\int \frac{dx}{x\sqrt{E^3}} = \frac{1}{a^2\sqrt{E}} - \frac{1}{a^3}\ln\frac{a + \sqrt{E}}{x}$$

$$\int \frac{dx}{x\sqrt{E^n}} = \frac{1}{a^2(n-2)\sqrt{E^{n-2}}} + \frac{n-2}{a^2\,(n-2)} \int \frac{dx}{x\sqrt{E^{n-2}}} \qquad n \neq 2$$

$$\int \frac{dx}{x^2\sqrt{E}} = -\frac{\sqrt{E}}{a^2\,x}$$

$$\int \frac{dx}{x^2\sqrt{E^3}} = -\frac{\sqrt{E}}{a^4 x}\left(1 - \frac{x^2}{E}\right)$$

$$\int \frac{dx}{x^2\sqrt{E^n}} = \frac{1}{a^2(n-2)x\sqrt{E^{n-2}}} + \frac{n-1}{a^2(n-2)}\int \frac{dx}{x^2\sqrt{E^{n-2}}} \qquad n \neq 2$$

$$\int \frac{dx}{x^3\sqrt{E}} = \frac{-\sqrt{E}}{2a^2x^2} - \frac{1}{2a^3}\ln\frac{a+\sqrt{E}}{x}$$

$$\int \frac{dx}{x^3\sqrt{E^3}} = \frac{-1}{2a^2x^2\sqrt{E}} + \frac{3}{2a^4\sqrt{E}} - \frac{3}{2a^5}\ln\frac{a+\sqrt{E}}{x}$$

$$\int \frac{dx}{x^3\sqrt{E^n}} = \frac{1}{a^2(n-2)x^2\sqrt{E^{n-2}}} + \frac{n}{a^2(n-2)}\int \frac{dx}{x^3\sqrt{E^{n-2}}} \qquad n \neq 2$$

$$\int \frac{dx}{x^4\sqrt{E}} = -\frac{\sqrt{E}}{a^4 x}\left(1+\frac{E}{3x^2}\right)$$

$$\int \frac{dx}{x^4\sqrt{E^3}} = \frac{1}{a^6\sqrt{E}}\left(x - \frac{2E}{x} - \frac{E^2}{3x^3}\right)$$

$$\int \frac{dx}{x^m\sqrt[p]{E_n}} = \frac{p}{2a^2(n-p)x^{m-1}\sqrt[p]{E^{n-p}}} + \frac{2n+p(m-3)}{2a^2(n-p)}\int \frac{dx}{x^m\sqrt[p]{E^{n-p}}} \qquad n \neq p$$

xxxi. Indefinite Integrals Involving $f(x) = \frac{\sqrt[p]{E^n}}{x^m}$ $\quad E = a^2 - x^2 \quad a^2 \neq 0$

$$\int \frac{\sqrt{E}}{x}dx = \sqrt{E} - a\ln\frac{a+\sqrt{E}}{x}$$

$$\int \frac{\sqrt{E^3}}{x}dx = \frac{1}{3}\sqrt{E^3} + a^2\sqrt{E} - a^3\ln\frac{a+\sqrt{E}}{x}$$

$$\int \frac{\sqrt{E^n}}{x}dx = \frac{1}{n}\sqrt{E^n} + a^2\int \frac{\sqrt{E^{n-2}}}{x}dx$$

$$\int \frac{\sqrt{E}}{x^2}dx = -\frac{1}{x}\sqrt{E} - \sin^{-1}\frac{x}{a}$$

$$\int \frac{\sqrt{E^3}}{x^2}dx = -\frac{1}{x}\sqrt{E^3} - \frac{3x}{2}\sqrt{E} - \frac{3a^2}{2}\sin^{-1}\frac{x}{a}$$

$$\int \frac{\sqrt{E^n}}{x^2}dx = \frac{\sqrt{E^n}}{(n-1)x} + \frac{a^2 n}{n-1}\int \frac{\sqrt{E^{n-2}}}{x^2}dx \qquad n \neq 1$$

$$\int \frac{\sqrt{C^3}}{x}\,dx = \frac{1}{3}\sqrt{C^3} + a^2\sqrt{C} - a^3 \ln \frac{a+\sqrt{C}}{x}$$

$$\int \frac{\sqrt{C^n}}{x}\,dx = \frac{1}{n}\sqrt{C^n} + a^2 \int \frac{\sqrt{C^{n-2}}}{x}\,dx$$

$$\int \frac{\sqrt{C}}{x^2}\,dx = -\frac{1}{x}\sqrt{C} + \ln(x+\sqrt{C})$$

$$\int \frac{\sqrt{C^3}}{x^2}\,dx = -\frac{1}{x}\sqrt{C^3} + \frac{3x}{2}\sqrt{C} + \frac{3a^2}{2}\ln(x+\sqrt{C})$$

$$\int \frac{\sqrt{C^n}}{x^2}\,dx = \frac{\sqrt{C^n}}{(n-1)\,x} + \frac{a^2 n}{n-1}\int \frac{\sqrt{C^{n-2}}}{x^2}\,dx \qquad n \neq 1$$

$$\int \frac{\sqrt{C}}{x^3}\,dx = -\frac{1}{2x^2}\sqrt{C} - \frac{1}{2a}\ln \frac{a+\sqrt{C}}{x}$$

$$\int \frac{\sqrt{C^3}}{x^3}\,dx = -\frac{1}{2x^2}\sqrt{C^3} + \frac{3}{2}\sqrt{C} - \frac{3a}{2}\ln \frac{a+\sqrt{C}}{x}$$

$$\int \frac{\sqrt{C^n}}{x^3}\,dx = \frac{\sqrt{C^n}}{(n-2)x^2} + \frac{a^2 n}{n-2}\int \frac{\sqrt{C^{n-2}}}{x^3}\,dx \qquad n \neq 2$$

$$\int \frac{\sqrt{C}}{x^4}\,dx = -\frac{\sqrt{C^3}}{3a^2x^3}$$

$$\int \frac{\sqrt{C^3}}{x^4}\,dx = -\frac{\sqrt{C^3}}{3x^3} - \frac{\sqrt{C}}{x} + \ln(x+\sqrt{C})$$

$$\int \frac{\sqrt[p]{C^n}}{x^m}\,dx = \frac{p\sqrt[p]{C^n}}{(2n-mp+p)\,x^{m-1}} + \frac{2a^2 n}{2n-mp+p}\int \frac{\sqrt[p]{C^{n-p}}}{x^m}\,dx \qquad 2n = -p\,(1-m)$$

---

xxviii. Indefinite Integrals Involving $f(x) = x^m \sqrt[p]{E^n}$ $\quad E = a^2 - x^2$ $\quad a^2 \neq 0$

$$\int \sqrt{E}\,dx = \frac{1}{2}\left(x\sqrt{E} + a^2 \sin^{-1}\frac{x}{a}\right)$$

$$\int \sqrt{E^3}\,dx = \frac{1}{8}\left(2x\sqrt{E^3} + 3a^2x\sqrt{E} + 3a^4 \sin^{-1}\frac{x}{a}\right)$$

$$\int \sqrt{E^n}\, dx = \frac{x\sqrt{E^n}}{n+1} + \frac{a^2 n}{n+1}\int \sqrt{E^{n-2}}\, dx \qquad n \neq -1$$

$$\int x\sqrt{E}\, dx = -\frac{1}{3}\sqrt{E^3}$$

$$\int x\sqrt{E^3}\, dx = -\frac{1}{5}\sqrt{E^5}$$

$$\int x\sqrt{E^n}\, dx = \frac{x^2\sqrt{E^n}}{n+2} + \frac{a^2 n}{n+2}\int x\sqrt{E^{n-2}}\, dx \qquad n \neq -2$$

$$\int x^2\sqrt{E}\, dx = -\frac{x}{4}\sqrt{E^3} + \frac{a^2}{8}\left(x\sqrt{E} + a^2 \sin^{-1}\frac{x}{a}\right)$$

$$\int x^2\sqrt{E^3}\, dx = -\frac{x}{6}\sqrt{E^5} + \frac{a^2 x}{24}\sqrt{E^3} + \frac{a^4}{16}\left(x\sqrt{E} + a^2 \sin^{-1}\frac{x}{a}\right)$$

$$\int x^2\sqrt{E^n}\, dx = \frac{x^3\sqrt{E^n}}{n+3} + \frac{a^2 n}{n+3}\int x^2\sqrt{E^{n-2}}\, dx \qquad n \neq -3$$

$$\int x^3\sqrt{E}\, dx = \frac{1}{5}\sqrt{E^5} - \frac{a^2}{3}\sqrt{E^3}$$

$$\int x^3\sqrt{E^3}\, dx = \frac{1}{7}\sqrt{E^7} - \frac{a^2}{5}\sqrt{E^5}$$

$$\int x^m \sqrt[p]{E^n}\, dx = \frac{p x^{m+1}\sqrt[p]{E^n}}{2n + mp + p} + \frac{2a^2 n}{2n + mp + p}\int x^m \sqrt[p]{E^{n-p}}\, dx \qquad 2n \neq -p\,(1+m)$$

---

xxix. Indinite Integrals Involving $f(x) = \frac{x^m}{\sqrt[p]{E^n}}$ $E = a^2 - x^2$ $a^2 \neq 0$

$$\int \frac{dx}{\sqrt{E}} = \sin^{-1}\frac{x}{a}$$

$$\int \frac{dx}{\sqrt{E^3}} = \frac{x}{a^2\sqrt{E}}$$

$$\int \frac{dx}{\sqrt{E^n}} = \frac{x}{a^2(n-2)\sqrt{E^{n-2}}} + \frac{n-3}{a^2(n-2)}\int \frac{dx}{\sqrt{E^{n-2}}} \qquad n \neq 2$$

$$\int \frac{x\, dx}{\sqrt{E}} = -\sqrt{E}$$

$$\int \frac{\sin A}{x^m}\,dx = -\frac{\sin A}{(m-1)\,x^{m-1}} + \frac{b}{m-1}\int \frac{\cos A}{x^{m-1}}\,dx \quad m \neq 1$$

$$\int \frac{\sin^{2n-1}}{x^m}\,dx = \frac{(-1)^{n-1}}{2^{2n-2}}\int\left[\sin(2n-1)A - \binom{2n-1}{1}\sin(2n-3)A + \ldots (-1)^{n-1}\binom{2n-1}{n-1}\sin A\right]\frac{dx}{x^m}$$

$$\int \frac{\sin^{2n} A}{x^m}\,dx = -\frac{\binom{2n}{n}}{(m-1)\,2^{2n}\,x^{m-1}} + \frac{(-1)^n}{2^{2n-1}}\int\left[\cos 2nA - \binom{2n}{1}\cos(2n-2)A + \ldots + (-1)^{n-1}\binom{2n}{n-1}\cos 2A\right]\frac{dx}{x^m}$$

---

xlvi. Indefinite Integrals Involving $f(x) = \dfrac{x^m}{\cos^n A}$ $f(x) = \dfrac{\cos^n A}{x^m}$ $A = bx$

$$\int \frac{x\,dx}{\cos A} = \left[1 + \frac{A^2}{4} + \frac{5A^4}{72} + \frac{61A^6}{2{,}880} + \ldots + \frac{E_k A^{2k}}{(k+1)(2k)!} + \ldots\right]\frac{A^2}{2b^2}$$

$$\int \frac{x^m\,dx}{\cos A} = \left[\frac{1}{(m+1)} + \frac{A^{m+1}}{2(m+3)} + \frac{5A^{m+3}}{24(m+5)} + \ldots + \frac{E_k A^{m+2k-1}}{(2k)!(m+2k+1)} + \ldots\right]\frac{A^m}{b^{m+1}}$$

Note: E_k is Euler's number

$$\int \frac{x\,dx}{\cos^2 A} = \frac{x}{b}\tan A + \frac{1}{b^2}\ln\cos A$$

$$\int \frac{x\,dx}{\cos^n A} = \frac{x\sin A}{(n-1)\,b\cos^{n-1} A} - \frac{1}{(n-1)(n-2)\,b^2\cos^{n-2} A} + \frac{n-2}{n-1}\int \frac{x\,dx}{\cos^{n-2} A} \quad n > 2$$

$$\int \frac{\cos A\,dx}{x} = \ln A - \frac{A^2}{2.2!} + \frac{A^4}{4.4!} - \frac{A^6}{6.6!} + \ldots$$

$$\int \frac{\cos A}{x^m}\,dx = -\frac{\cos A}{(m-1)\,x^{m-1}} - \frac{b}{m-1}\int \frac{\sin A}{x^{m-1}}\,dx \quad m \neq 1$$

$$\int \frac{\cos^{2n-1}}{x^n}\,dx = \frac{1}{2^{2n-2}}\int\left[\cos(2n-1)A + \binom{2n-1}{1}\cos(2n-3)A + \ldots + \binom{2n-1}{n-1}\cos A\right]\frac{dx}{x^m}$$

$$\int \frac{\cos^{2n} A}{x^m}\, dx = -\frac{\binom{2n}{2}}{(m-1)\, 2^{2n}\, x^{m-1}} + \frac{1}{2^{2n-1}} \int \left[\cos 2nA + \binom{2n}{1} \cos (2n-2) A + \ldots + \binom{2n}{n-1} \cos 2A \right] \frac{dx}{x^m}$$

xlvii. Indefinite Integrals Involving $f(x) = f(1 \pm a \sin A) \quad A = bx \quad \alpha, \beta = \text{constants}$

$$\int \frac{dx}{\sin A + \sin \alpha} = \frac{1}{b \cos \alpha} \ln \left[\frac{\sin [(A+\alpha)/2]}{\cos [(A-\alpha)/2]} \right]$$

$$\int \frac{dx}{1 + \sin \alpha \sin A} = \frac{2}{b \cos \alpha} \tan^{-1} \left[\tan \left(\frac{\pi}{4} + \frac{A}{2} \right) \tan \left(\frac{\pi}{4} + \frac{\alpha}{2} \right) \right]$$

$$\int \frac{dx}{1 + \cos \alpha \sin A} = \frac{2}{b \sin \alpha} \tan^{-1} \frac{\cos \alpha + \tan (A/2)}{\tan \alpha}$$

$$\int \frac{dx}{1 + a \sin A} = \begin{cases} \dfrac{2}{b\sqrt{1-a^2}} \tan^{-1} \dfrac{a + \tan (A/2)}{1-a^2} & a^2 < 1 \\ \dfrac{1}{b\sqrt{1-a^2}} \sin^{-1} \dfrac{a + \sin A}{1 + a \sin A} & a^2 < 1 \\ \dfrac{1}{b\sqrt{a^2-1}} \ln \dfrac{\tan (A/2) + a - \sqrt{a^2-1}}{\tan (A/2) + a + \sqrt{a^2-1}} & a^2 > 1 \end{cases}$$

$$\int \frac{dx}{(1 + a \sin A)^2} = \frac{a \cos A}{b\,(1-a^2)\,(1 + a \sin A)} + \frac{1}{1-a^2} \int \frac{dx}{1 + a \sin A}$$

$$\int \frac{dx}{(1 + a \sin A)^n} = \frac{a \cos A}{b\,(n-1)\,(1-a^2)\,(1 + a \sin A)^{n-1}} + \frac{(2n-3)}{(n-1)\,(1-a^2)} \int \frac{dx}{(1 + a \sin A)^{n-1}} - \frac{(n-2)}{(n-1)\,(1-a^2)} \int \frac{dx}{(1 + a \sin A)^{n-2}} \qquad a^2 \neq 1, \quad n \neq 1$$

$$\int \sin \alpha x \sin \beta x \, dx = \frac{\sin (\alpha - \beta) x}{2(\alpha - \beta)} - \frac{\sin (\alpha + \beta) x}{2(\alpha + \beta)} \qquad a^2 \neq \beta^2$$

$$\int \sin \alpha x \cos \beta x \, dx = -\frac{\cos (\alpha - \beta) x}{2(\alpha - \beta)} - \frac{\cos (\alpha + \beta) x}{2(\alpha + \beta)} \qquad a^2 \neq \beta^2$$

$$\int x^2 \sin^2 A\,dx = \frac{1}{2}\int x^2 (1 - \cos 2A)\,dx$$

$$\int x^2 \sin^3 A\,dx = \frac{1}{4}\int x^2 (3\sin A - \sin 3A)\,dx$$

$$\int x^2 \sin^n A\,dx = x^2 D_1 - 2xD_2 + 2D_3$$

$$\int x^3 \sin A\,dx = \frac{-x^3 \cos A}{b} + \frac{3x^2 \sin A}{b^2} + \frac{6x \cos A}{b^3} - \frac{6 \sin A}{b^4}$$

$$\int x^3 \sin^2 A\,dx = \frac{1}{2}\int x^3 (1 - \cos 2A)\,dx$$

$$\int x^3 \sin^3 A\,dx = \frac{1}{4}\int x^3 (3\sin A - \sin 3A)\,dx$$

$$\int x^3 \sin^n A\,dx = x^3 D_1 - 3x^2 D_2 + 6xD_3 - 6D_4$$

$$\int x^m \sin A\,dx = \frac{m!\sin A}{b}\left[\frac{x^{m-1}}{(m-1)!b} - \frac{x^{m-3}}{(m-3)!b^3} + \frac{x^{m-5}}{(m-5)\,b^5} - \ldots\right]$$

$$- \frac{m!\cos A}{b}\left[\frac{x^m}{m!} - \frac{x^{m-2}}{(m-2)!b^2} + \frac{x^{m-4}}{(m-4)!b^4} - \ldots\right]$$

$$\int x^m \sin^n A\,dx = x^m D_1 - mx^{m-1} D_2 + m(m-1)\,x^{m-2} D_3 - \ldots$$

Series terminates with the term *involving* x^{m-m}

---

xliv. Indefinite Integrals Involving $f(x) = x^m \cos^n A \quad A = bx$

$D_1 = \int \cos^n A\,dx \qquad D_2 = \int D_1\,dx \qquad D_k = \int D_{k-1}\,dx$

$$\int x \cos A\,dx = \frac{x \sin A}{b} + \frac{\cos A}{b^2}$$

$$\int x \cos^2 A\,dx = \frac{x^2}{4} + \frac{x \sin 2A}{4b} + \frac{\cos 2A}{8b^2}$$

$$\int x \cos^3 A\,dx = \frac{x \sin 3A}{12b} + \frac{\cos 3A}{36b^2} + \frac{3x \sin A}{4b} + \frac{3\cos A}{4b^2}$$

$$\int x \cos^n A\,dx = xD_1 - D_2$$

$$\int x^2 \cos A\,dx = \frac{x^2 \sin A}{b} + \frac{2x \cos A}{b^2} - \frac{2\sin A}{b^3}$$

$$\int x^2 \cos^2 A\,dx = \frac{1}{2}\int x^2 (1 + \cos 2A)\,dx$$

$$\int x^2 \cos^3 A\,dx = \frac{1}{4}\int x^2 (3\cos A + \cos 3A)\,dx$$

$$\int x^2 \cos^n A\,dx = x^2 D_1 - 2xD_2 + D_3$$

$$\int x^3 \cos A\,dx = \frac{x^3 \sin A}{b} + \frac{3x^2 \cos A}{b^2} - \frac{6x \sin A}{b^3} - \frac{6\cos A}{b^4}$$

$$\int x^3 \cos^2 A\,dx = \frac{1}{2}\int x^3 (1 + \cos 2A)\,dx$$

$$\int x^3 \cos^3 A\,dx = \frac{1}{4}\int x^3 (3\cos A + \cos 3A)\,dx$$

$$\int x^3 \cos^n A\,dx = x^3 D_1 - 3x^2 D_2 + 6xD_3 - 6D_4$$

$$\int x^m \cos A\,dx = \frac{m!\sin A}{b}\left[\frac{x^m}{(m)!} - \frac{x^{m-2}}{(m-2)!b^2} + \frac{x^{m-4}}{(m-4)!b^4} - \ldots\right] - \frac{m!\cos A}{b}\left[\frac{x^{m-1}}{(m-1)!b} - \frac{x^{m-3}}{(m-3)!b^3} + \frac{x^{m-5}}{(m-5)!b^5} - \ldots\right]$$

$$\int x^m \cos^n A\,dx = x^m D_1 - mx^{m-1} D_2 + m(m-1)x^{m-2} D_3 - \ldots$$

Series terminates with the term *involving* x^{m-m}

---

xlv. Indefinite Integrals Involving $f(x) = \dfrac{x^m}{\sin^n A}$ $f(x) = \dfrac{\sin^n A}{x^m}$ $A = bx$

$$\int \frac{x\,dx}{\sin A} = \left[1 + \frac{A^2}{18} + \frac{7A^4}{1,800} + \ldots + \frac{2(2^{2k-1}-1)B_k A^{2k}}{(2k+1)!} + \ldots\right]\frac{A}{b^2}$$

$$\int \frac{x^m\,dx}{\sin A} = \left[\frac{1}{m} + \frac{A^2}{6(m+2)} + \frac{7A^4}{360(m+4)} + \ldots + \frac{2(2^{2k-1}-1)B_k A^{2k}}{(2k)!(m+2k)} + \ldots\right]\frac{A^m}{b^{m+1}}$$

Note: B_k = Bernoulli's number

$$\int \frac{x\,dx}{\sin^2 A} = -\frac{x}{b}\cot A + \frac{1}{b^2}\ln \sin A$$

$$\int \frac{x\,dx}{\sin^n A} = -\frac{x\cos A}{(n-1)b\sin^{n-1} A} - \frac{1}{(n-1)(n-2)b^2 \sin^{n-2} A} + \frac{n-2}{n-1}\int \frac{x\,dx}{\sin^{n-2} A} \quad n > 2$$

$$\int \frac{\sin A}{x}\,dx = A - \frac{A^3}{3.3!} + \frac{A^5}{5.5!} - \frac{A^7}{7.7!} + \ldots$$

xli. Indefinite Integrals Involving $f(x) = \sin^n A$ $f(x) = \dfrac{1}{\sin^n A}$ $A = bx$

$$\int \sin A\,dx = -\frac{\cos A}{b}$$

$$\int \sin^2 A\,dx = -\frac{\sin 2A}{4b} + \frac{x}{2}$$

$$\int \sin^3 A\,dx = -\frac{\cos A}{b} + \frac{\cos^3 A}{3b}$$

$$\int \sin^4 A\,dx = -\frac{\sin 2A}{4b} + \frac{\sin 4A}{32b} + \frac{3x}{8}$$

$$\int \sin^5 A\,dx = -\frac{5\cos A}{8b} + \frac{5\cos 3A}{48b} - \frac{\cos 5A}{80b}$$

$$\int \sin^n A\,dx = -\frac{\cos A}{b}\frac{\sin^{n-1} A}{n} + \frac{n-1}{n}\int \sin^{n-2} A\,dx \qquad n > 0$$

$$\int \frac{dx}{\sin A} = \frac{1}{b}\ln\tan\frac{A}{2}$$

$$\int \frac{dx}{\sin^2 A} = -\frac{\cot A}{b}$$

$$\int \frac{dx}{\sin^3 A} = -\frac{\cos A}{2b\sin^2 A} + \frac{1}{2b}\ln\tan\frac{A}{2}$$

$$\int \frac{dx}{\sin^4 A} = -\frac{\cot A}{b} - \frac{\cot^3 A}{3b}$$

$$\int \frac{dx}{\sin^5 A} = -\frac{\cos A}{4b\sin^4 A} - \frac{3\cos A}{8b\sin^4 A} - \frac{3}{16b}\ln\frac{1+\cos A}{1-\cos A}$$

$$\int \frac{dx}{\sin^n A} = -\frac{\cos A}{b(n-1)\sin^{n-1} A} + \frac{n-2}{n-1}\int \frac{dx}{\sin^{n-2} A} \qquad n > 1$$

xlii. Indefinite Integrals Involving $f(x) = \cos^n A$ $f(x) = \dfrac{1}{\cos^n A}$ $A = bx$

$$\int \cos A\,dx = \frac{\sin A}{b}$$

$$\int \cos^2 A\,dx = \frac{\sin 2A}{4b} + \frac{x}{2}$$

$$\int \cos^3 A\,dx = \frac{\sin A}{b} - \frac{\sin^3 A}{3b}$$

$$\int \cos^4 A\,dx = \frac{\sin 2A}{4b} + \frac{\sin 4A}{32b} + \frac{3x}{8}$$

$$\int \cos^5 A\,dx = \frac{5\sin A}{8b} + \frac{5\sin 3A}{48b} + \frac{\sin 5A}{80b}$$

$$\int \cos^n A\,dx = \frac{\sin A}{b}\,\frac{\cos^{n-1} A}{n} + \frac{n-1}{n}\int \cos n^{n-2} A\,dx \qquad n > 0$$

$$\int \frac{dx}{\cos A} = \frac{1}{b}\ln\tan\left(\frac{\pi}{4} + \frac{A}{2}\right)$$

$$\int \frac{dx}{\cos^2 A} = \frac{1}{b}\tan A$$

$$\int \frac{dx}{\cos^3 A} = \frac{\sin A}{2b\cos^2 A} + \frac{1}{2b}\ln\tan\left(\frac{\pi}{4} + \frac{A}{2}\right)$$

$$\int \frac{dx}{\cos^4 A} = \frac{\tan A}{b} + \frac{\tan^3 A}{3b}$$

$$\int \frac{dx}{\cos^5 A} = \frac{\sin A}{4b\cos^4 A} + \frac{3\sin A}{8\cos^4 A} + \frac{3}{8b}\ln\tan\left(\frac{\pi}{4} + \frac{A}{2}\right)$$

$$\int \frac{dx}{\cos^n A} = \frac{\sin A}{b(n-1)\cos^{n-1} A} + \frac{n-2}{n-1}\int \frac{dx}{\cos^{n-2} A} \qquad n > 1$$

xliii. Indefinite Integrals Involving $f(x) = x^m \sin^n A \quad A = bx$

$D_1 = \int \sin^n A\,dx \qquad D_2 = \int D_1\,dx \qquad D_k = \int D_{k-1}\,dx$

$$\int x\sin A\,dx = -\frac{x\cos A}{b} + \frac{\sin A}{b^2}$$

$$\int x\sin^2 A\,dx = \frac{x^2}{4} - \frac{x\sin 2A}{4b} - \frac{\cos 2A}{8b^2}$$

$$\int x\sin^3 A\,dx = \frac{x\cos 3A}{12b} - \frac{\sin 3A}{36b^2} + \frac{3x\cos A}{4b} + \frac{3\sin A}{4b^2}$$

$$\int x\sin^n A\,dx = xD_1 - D_2$$

$$\int x^2\sin A\,dx = -\frac{x^2\cos A}{b} + \frac{2x\sin A}{b^2} + \frac{2\cos A}{b^3}$$

xxxvi. Indefinite Integrals Involving $f(x) = x^m \sqrt[p]{P^n}$ $\lambda = \dfrac{n}{p}$ $P = a + bx^q$ $a \neq 0$ $b \neq 0$

$$\int P^{\lambda}\,dx = \frac{1}{1+q\lambda}\left(xP^{\lambda} + aq\lambda\int P^{\lambda-1}\,dx\right)$$

$$\int \frac{dx}{P^{\lambda}} = \frac{-1}{aq(1-\lambda)}\left[\frac{x}{P^{\lambda-1}} - (1+q-q\lambda)\int\frac{dx}{P^{\lambda-1}}\right]$$

$$\int x^m\,P^{\lambda}\,dx = \frac{1}{1+m+q\lambda}\left(x^{m+1}\,P^{\lambda} + aq\lambda\int x^m\,P^{\lambda-1}\,dx\right)$$

$$= \frac{-1}{aq(1+\lambda)}\left[x^{m+1}\,P^{\lambda+1} - (1+m+q+q\lambda)\int x^m P^{\lambda+1}\,dx\right]$$

$$= \frac{1}{a(1+m)}\left[x^{m+1}\,P^{\lambda+1} - b(1+m+q+q\lambda)\int x^{m+q}P^{\lambda}\,dx\right]$$

$$\frac{1}{b(1+m+q^{\lambda})}\left[x^{m-q+1}\,P^{\lambda+1} - a(1+m-q)\int x^{m-q}P^{\lambda}\,dx\right]$$

xxxvii. Indefinite Integrals Involving $f(x) = \dfrac{x^{p-1}}{x^{2m+1} \pm a^{2m+1}}$ $a \neq 0$

$$\int\frac{x^{p-1}\,dx}{x^{2m+1}+a^{2m+1}} = \frac{2(-1)^{p-1}}{(2m+1)\,a^{2m-p+1}}\sum_{k=1}^{m}\sin\frac{2kp\pi}{2m+1}\tan^{-1}\left\{\frac{x+a\cos[2k\pi/(2m+1)]}{a\sin[2k\pi/(2m+1)]}\right\}$$

$$-\frac{(-1)^{p-1}}{(2m+1)\,a^{2m-p+1}}\sum_{k=1}^{m}\cos\frac{2kp\pi}{2m+1}\ln\left(x^2+2ax\cos\frac{2k\pi}{2m+1}+a^2\right)$$

$$+\frac{(-1)^{p-1}\ln(x+a)}{(2m+1)\,a^{2m-p+1}} \qquad 0<p\le 2m+1$$

$$\int\frac{x^{p-1}\,dx}{x^{2m+1}-a^{2m+1}} = \frac{-2}{(2m+1)\,a^{2m-p+1}}\sum_{k=1}^{m}\sin\frac{2kp\pi}{2m+1}\tan^{-1}\left\{\frac{x-a\cos[2k\pi/(2m+1)]}{a\sin[2k\pi/(2m+1)]}\right\}$$

$$+\frac{1}{(2m+1)\,a^{2m-p+1}}\sum_{k=1}^{m}\cos\frac{2kp\pi}{2m+1}\ln\left(x^2-2ax\cos\frac{2k\pi}{2m+1}+a^2\right)$$

$$+\frac{\ln(x-a)}{(2m+1)\,a^{2m-p+1}} \qquad 0<p\le 2m+1$$

xxxviii. Indefinite Integrals Involving $f(x) = x^m \sqrt[p]{R^n} \quad \lambda = n/p \quad R = a^{x^q} + b^{x^{q+r}} \quad a \neq 0$

$a \neq 0$

$$\int R^\lambda \, dx = \frac{\lambda}{1+\lambda(q+r)}\left(xR^\lambda + ar\lambda \int x^q R^{\lambda-1}\, dx\right)$$

$$\int \frac{dx}{R^\lambda} = \frac{-1}{ar(1+\lambda)}\left[xR^{\lambda+1} - (1+r+q\lambda+r\lambda)\int \frac{x^q\, dx}{R^{\lambda-1}}\right]$$

$$\int x^m R^\lambda \, dx + \frac{1}{1+m+q\lambda+r\lambda}\left(x^{m+1} R^\lambda + ar\lambda \int x^{m+q} R^{\lambda-1}\, dx\right)$$

$$= \frac{-1}{ar(1+\lambda)}\left[x^{m+1} R^{\lambda+1} - (1+r+q\lambda+r\lambda)\int x^{m-q} R^{\lambda+1}\, dx\right]$$

xxxix. Indefinite Integrals Involving $f(x) = \dfrac{x^m}{(x-a_1)(x-a_2)\ldots(x-a_k)} \quad a_i \neq a_j \neq 0$

$$\int \frac{x^m\, dx}{(x-a_1)(x-a_2)\ldots(x-a_k)} = \frac{a_1{}^m \ln(x-a_1)}{(a_1-a_2)(a_1-a_3)\ldots(a_1-a_k)}$$

$$+ \frac{a_2^m \ln(x-a_2)}{(a_2-a_1)(a_2-a_3)\ldots(a_2-a_k)} + \ldots + \frac{ak^m \ln(x-a_k)}{(a_k-a_1)(a_k-a_2)\ldots(a_k-a_{k-1})}$$

xl. Indefinite Integrals Involving $f(x) = \dfrac{x^{p-1}}{x^{2m} \pm a^{2m}} \quad a \neq 0$

$$\int \frac{x^{p-1}}{x^{2m}+a^{2m}} = \frac{1}{ma^{2m-p}} \sum_{k=1}^{m} \sin\frac{(2k-1)p\pi}{2m} \tan^{-1}\left\{\frac{x + a\cos[(2k-1)\pi/2m]}{a\sin[(2k-1)\pi/2m]}\right\}$$

$$- \frac{1}{2ma^{2m-p}} \sum_{k=1}^{m} \cos\frac{(2k-1)p\pi}{2m} \ln\left[x^2 + 2ax\cos\frac{(2k-1)\pi}{2m} + a^2\right] \qquad 0 < p \leq 2m$$

$$\int \frac{x^{p-1}\, dx}{x^{2m}-a^{2m}} = \frac{1}{2ma^{2m-p}} \sum_{k=1}^{m-1} \cos\frac{kp\pi}{m} \ln\left(x^2 - 2ax\cos\frac{k\pi}{m} + a^2\right)$$

$$- \frac{1}{na^{2m-p}} \sum_{k=1}^{m-1} \sin\frac{kp\pi}{m} \tan^{-1}\frac{x - a\cos(k\pi/m)}{a\sin(k\pi/m)} + \frac{1}{2ma^{2m-p}}[\ln(x-a) + (-1)^p \ln(x+a)]$$

$$0 < p \leq 2m$$

$$\int \frac{\sqrt{C^3}}{x}\,dx = \frac{1}{3}\sqrt{C^3} + a^2\sqrt{C} - a^3 \ln \frac{a+\sqrt{C}}{x}$$

$$\int \frac{\sqrt{C^n}}{x}\,dx = \frac{1}{n}\sqrt{C^n} + a^2 \int \frac{\sqrt{C^{n-2}}}{x}\,dx$$

$$\int \frac{\sqrt{C}}{x^2}\,dx = -\frac{1}{x}\sqrt{C} + \ln(x+\sqrt{C})$$

$$\int \frac{\sqrt{C^3}}{x^2}\,dx = -\frac{1}{x}\sqrt{C^3} + \frac{3x}{2}\sqrt{C} + \frac{3a^2}{2}\ln(x+\sqrt{C})$$

$$\int \frac{\sqrt{C^n}}{x^2}\,dx = \frac{\sqrt{C^n}}{(n-1)\,x} + \frac{a^2 n}{n-1}\int \frac{\sqrt{C^{n-2}}}{x^2}\,dx \qquad n \neq 1$$

$$\int \frac{\sqrt{C}}{x^3}\,dx = -\frac{1}{2x^2}\sqrt{C} - \frac{1}{2a}\ln \frac{a+\sqrt{C}}{x}$$

$$\int \frac{\sqrt{C^3}}{x^3}\,dx = -\frac{1}{2x^2}\sqrt{C^3} + \frac{3}{2}\sqrt{C} - \frac{3a}{2}\ln \frac{a+\sqrt{C}}{x}$$

$$\int \frac{\sqrt{C^n}}{x^3}\,dx = \frac{\sqrt{C^n}}{(n-2)x^2} + \frac{a^2 n}{n-2}\int \frac{\sqrt{C^{n-2}}}{x^3}\,dx \qquad n \neq 2$$

$$\int \frac{\sqrt{C}}{x^4}\,dx = -\frac{\sqrt{C^3}}{3a^2x^3}$$

$$\int \frac{\sqrt{C^3}}{x^4}\,dx = -\frac{\sqrt{C^3}}{3x^3} - \frac{\sqrt{C}}{x} + \ln(x+\sqrt{C})$$

$$\int \frac{\sqrt[p]{C^n}}{x^m}\,dx = \frac{p\sqrt[p]{C^n}}{(2n-mp+p)\,x^{m-1}} + \frac{2a^2 n}{2n-mp+p}\int \frac{\sqrt[p]{C^{n-p}}}{x^m}\,dx \qquad 2n = -p\,(1-m)$$

---

xxviii. Indefinite Integrals Involving $f(x) = x^m \sqrt[p]{E^n}$ $\quad E = a^2 - x^2 \quad a^2 \neq 0$

$$\int \sqrt{E}\,dx = \frac{1}{2}\left(x\sqrt{E} + a^2 \sin^{-1}\frac{x}{a}\right)$$

$$\int \sqrt{E^3}\,dx = \frac{1}{8}\left(2x\sqrt{E^3} + 3a^2x\sqrt{E} + 3a^4 \sin^{-1}\frac{x}{a}\right)$$

$$\int \sqrt{E^n}\,dx = \frac{x\sqrt{E^n}}{n+1} + \frac{a^2 n}{n+1}\int \sqrt{E^{n-2}}\,dx \qquad n \neq -1$$

$$\int x\sqrt{E}\,dx = -\frac{1}{3}\sqrt{E^3}$$

$$\int x\sqrt{E^3}\,dx = -\frac{1}{5}\sqrt{E^5}$$

$$\int x\sqrt{E^n}\,dx = \frac{x^2\sqrt{E^n}}{n+2} + \frac{a^2 n}{n+2}\int x\sqrt{E^{n-2}}\,dx \qquad n \neq -2$$

$$\int x^2\sqrt{E}\,dx = -\frac{x}{4}\sqrt{E^3} + \frac{a^2}{8}\left(x\sqrt{E} + a^2 \sin^{-1}\frac{x}{a}\right)$$

$$\int x^2\sqrt{E^3}\,dx = -\frac{x}{6}\sqrt{E^5} + \frac{a^2 x}{24}\sqrt{E^3} + \frac{a^4}{16}\left(x\sqrt{E} + a^2 \sin^{-1}\frac{x}{a}\right)$$

$$\int x^2\sqrt{E^n}\,dx = \frac{x^3\sqrt{E^n}}{n+3} + \frac{a^2 n}{n+3}\int x^2\sqrt{E^{n-2}}\,dx \qquad n \neq -3$$

$$\int x^3\sqrt{E}\,dx = \frac{1}{5}\sqrt{E^5} - \frac{a^2}{3}\sqrt{E^3}$$

$$\int x^3\sqrt{E^3}\,dx = \frac{1}{7}\sqrt{E^7} - \frac{a^2}{5}\sqrt{E^5}$$

$$\int x^m \sqrt[p]{E^n}\,dx = \frac{p x^{m+1}\sqrt[p]{E^n}}{2n+mp+p} + \frac{2a^2 n}{2n+mp+p}\int x^m \sqrt[p]{E^{n-p}}\,dx \qquad 2n \neq -p\,(1+m)$$

xxix. Indinite Integrals Involving $f(x) = \dfrac{x^m}{\sqrt[p]{E^n}}$ $\quad E = a^2 - x^2 \quad a^2 \neq 0$

$$\int \frac{dx}{\sqrt{E}} = \sin^{-1}\frac{x}{a}$$

$$\int \frac{dx}{\sqrt{E^3}} = \frac{x}{a^2\sqrt{E}}$$

$$\int \frac{dx}{\sqrt{E^n}} = \frac{x}{a^2(n-2)\sqrt{E^{n-2}}} + \frac{n-3}{a^2(n-2)}\int \frac{dx}{\sqrt{E^{n-2}}} \qquad n \neq 2$$

$$\int \frac{x\,dx}{\sqrt{E}} = -\sqrt{E}$$

$$\int \frac{x\,dx}{\sqrt{E^3}} = \frac{1}{\sqrt{E}}$$

$$\int \frac{x\,dx}{\sqrt{E^n}} = \frac{x^2}{a^2(n-2)\sqrt{E^{n-2}}} + \frac{n-4}{a^2(n-2)} \int \frac{x\,dx}{\sqrt{E^{n-2}}} \qquad n \neq 2$$

$$\int \frac{x^2\,dx}{\sqrt{E}} = -\frac{x}{2}\sqrt{E} + \frac{a^2}{2} \sin^{-1}\frac{x}{a}$$

$$\int \frac{x^2\,dx}{\sqrt{E^3}} = \frac{x}{\sqrt{E}} - \sin^{-1}\frac{x}{a}$$

$$\int \frac{x^2\,dx}{\sqrt{E^n}} = \frac{x^3}{a^2(n-2)\sqrt{E^{n-2}}} + \frac{n-5}{a^2(n-2)} \int \frac{x^2\,dx}{\sqrt{E^{n-2}}} \qquad n \neq 2$$

$$\int \frac{x^3\,dx}{\sqrt{E}} = \left(\frac{E}{3} - a^2\right)\sqrt{E}$$

$$\int \frac{x^3\,dx}{\sqrt{E^3}} = \frac{E + a^2}{\sqrt{E}}$$

$$\int \frac{x^m\,dx}{\sqrt[p]{E^n}} = \frac{px^{m+1}}{2a^2(n-p)\sqrt[p]{E^{n-p}}} + \frac{2n - p(m+3)}{2a^2(n-p)} \int \frac{x^m\,dx}{\sqrt[p]{E^{n-p}}} \qquad n \neq p$$

XXX. Indefinite Integrals Involving $f(x) = \dfrac{1}{x^m \sqrt[p]{E^n}}$ $\qquad E = a^2 - x^2 \qquad a^2 \neq 0$

$$\int \frac{dx}{x\sqrt{E}} = \frac{-1}{a} \ln \frac{a - \sqrt{E}}{x}$$

$$\int \frac{dx}{x\sqrt{E^3}} = \frac{1}{a^2\sqrt{E}} - \frac{1}{a^3} \ln \frac{a + \sqrt{E}}{x}$$

$$\int \frac{dx}{x\sqrt{E^n}} = \frac{1}{a^2(n-2)\sqrt{E^{n-2}}} + \frac{n-2}{a^2(n-2)} \int \frac{dx}{x\sqrt{E^{n-2}}} \qquad n \neq 2$$

$$\int \frac{dx}{x^2\sqrt{E}} = -\frac{\sqrt{E}}{a^2 x}$$

$$\int \frac{dx}{x^2\sqrt{E^3}} = -\frac{\sqrt{E}}{a^4 x}\left(1 - \frac{x^2}{E}\right)$$

$$\int \frac{dx}{x^2\sqrt{E^n}} = \frac{1}{a^2(n-2)x\sqrt{E^{n-2}}} + \frac{n-1}{a^2(n-2)}\int \frac{dx}{x^2\sqrt{E^{n-2}}} \qquad n \neq 2$$

$$\int \frac{dx}{x^3\sqrt{E}} = \frac{-\sqrt{E}}{2a^2x^2} - \frac{1}{2a^3}\ln\frac{a+\sqrt{E}}{x}$$

$$\int \frac{dx}{x^3\sqrt{E^3}} = \frac{-1}{2a^2x^2\sqrt{E}} + \frac{3}{2a^4\sqrt{E}} - \frac{3}{2a^5}\ln\frac{a+\sqrt{E}}{x}$$

$$\int \frac{dx}{x^3\sqrt{E^n}} = \frac{1}{a^2(n-2)x^2\sqrt{E^{n-2}}} + \frac{n}{a^2(n-2)}\int \frac{dx}{x^3\sqrt{E^{n-2}}} \qquad n \neq 2$$

$$\int \frac{dx}{x^4\sqrt{E}} = -\frac{\sqrt{E}}{a^4x}\left(1+\frac{E}{3x^2}\right)$$

$$\int \frac{dx}{x^4\sqrt{E^3}} = \frac{1}{a^6\sqrt{E}}\left(x - \frac{2E}{x} - \frac{E^2}{3x^3}\right)$$

$$\int \frac{dx}{x^m\sqrt[p]{E_n}} = \frac{p}{2a^2(n-p)x^{m-1}\sqrt[p]{E^{n-p}}} + \frac{2n+p(m-3)}{2a^2(n-p)}\int \frac{dx}{x^m\sqrt[p]{E^{n-p}}} \qquad n \neq p$$

---

xxxi. Indefinite Integrals Involving $f(x) = \frac{\sqrt[p]{E^n}}{x^m}$ $\quad E = a^2 - x^2 \quad a^2 \neq 0$

$$\int \frac{\sqrt{E}}{x}\,dx = \sqrt{E} - a\ln\frac{a+\sqrt{E}}{x}$$

$$\int \frac{\sqrt{E^3}}{x}\,dx = \frac{1}{3}\sqrt{E^3} + a^2\sqrt{E} - a^3\ln\frac{a+\sqrt{E}}{x}$$

$$\int \frac{\sqrt{E^n}}{x}\,dx = \frac{1}{n}\sqrt{E^n} + a^2\int \frac{\sqrt{E^{n-2}}}{x}\,dx$$

$$\int \frac{\sqrt{E}}{x^2}\,dx = -\frac{1}{x}\sqrt{E} - \sin^{-1}\frac{x}{a}$$

$$\int \frac{\sqrt{E^3}}{x^2}\,dx = -\frac{1}{x}\sqrt{E^3} - \frac{3x}{2}\sqrt{E} - \frac{3a^2}{2}\sin^{-1}\frac{x}{a}$$

$$\int \frac{\sqrt{E^n}}{x^2}\,dx = \frac{\sqrt{E^n}}{(n-1)x} + \frac{a^2n}{n-1}\int \frac{\sqrt{E^{n-2}}}{x^2}\,dx \qquad n \neq 1$$

$$\int \frac{\sqrt{E}}{x^3}\,dx = -\frac{1}{2x^2}\sqrt{E} + \frac{1}{2a}\ln\frac{a+\sqrt{E}}{x}$$

$$\int \frac{\sqrt{E^3}}{x^3}\,dx = -\frac{1}{2x^2}\sqrt{E^3} - \frac{3}{2}\sqrt{E} + \frac{3a}{2}\ln\frac{a+\sqrt{E}}{x}$$

$$\int \frac{\sqrt{E^n}}{x^3}\,dx = \frac{\sqrt{E^n}}{(n-2)x^2} + \frac{a^2 n}{n-2}\int \frac{\sqrt{E^{n-2}}}{x^3} \qquad n \neq 2$$

$$\int \frac{\sqrt{E}}{x^4}\,dx = -\frac{1}{3a^2x^3}\sqrt{E^3}$$

$$\int \frac{\sqrt{E^3}}{x^4}\,dx = -\frac{1}{3x^3}\sqrt{E^3} + \frac{1}{x}\sqrt{E} + \sin^{-1}\frac{x}{a}$$

$$\int \frac{\sqrt[p]{E^n}}{x^m}\,dx = \frac{p\sqrt[p]{E^n}}{(2n-mp+p)x^{m-1}} + \frac{2a^2 n}{2n-mp+p}\int \frac{\sqrt[p]{E^{n-p}}}{x^m}\,dx \qquad 2n \neq -p\,(1-m)$$

---

xxxii. Indefinite Integrals Involving $f(x) = x^m \sqrt[p]{F^n}$ $\quad F = x^2 - a^2 \quad a^2 \neq 0$

$$\int \sqrt{F}\,dx = \frac{x}{2}\sqrt{F} - \frac{a^2}{2}\ln\,(x+\sqrt{F})$$

$$\int \sqrt{F^3}\,dx = \frac{x\sqrt{F^3}}{4} - \frac{3a^2}{4}\int \sqrt{F}\,dx$$

$$\int \sqrt{F^n}\,dx = \frac{x\sqrt{F^n}}{n+1} - \frac{a^2 n}{n+1}\int \sqrt{F^{n-2}}\,dx \qquad n \neq -1$$

$$\int x\sqrt{F}\,dx = \frac{1}{3}\sqrt{F^3}$$

$$\int x\sqrt{F^3}\,dx = \frac{1}{5}\sqrt{F^3}$$

$$\int x\sqrt{F^n}\,dx = \frac{x^2\sqrt{F^n}}{n+2} - \frac{a^2 n}{n+2}\int x\sqrt{F^{n-2}}\,dx \qquad n \neq -2$$

$$\int x^2\sqrt{F}\,dx = \frac{x}{4}\sqrt{F^3} + \frac{a^2 x}{8}\sqrt{F} - \frac{a^4}{8}\ln\,(x+\sqrt{F})$$

$$\int x^2\sqrt{F^3}\,dx = \frac{x}{6}\sqrt{F^5} + \frac{a^2 x}{24}\sqrt{F^3} - \frac{a^4 x}{16}\sqrt{F} + \frac{a^6}{16}\ln\,(x+\sqrt{F})$$

$$\int x^2\sqrt{F^n}\,dx = \frac{x^3\sqrt{F^n}}{n+3} - \frac{a^2 n}{n+3}\int x^2\sqrt{F^{n-2}}\,dx \quad n \neq -3$$

$$\int x^3\sqrt{F}\,dx = \frac{1}{5}\sqrt{F^5} + \frac{a^2}{3}\sqrt{F^3}$$

$$\int x^3\sqrt{F^3}\,dx = \frac{1}{7}\sqrt{F^7} + \frac{a^2}{5}\sqrt{F^5}$$

$$\int x^m\sqrt[p]{F^n}\,dx = \frac{px^{m+1}\sqrt[p]{F^n}}{2n+mp+p} - \frac{2a^2 n}{2n+mp+p}\int x^m\sqrt[p]{F^{n-p}}\,dx \qquad 2n \neq -p\,(m+1)$$

---

xxxiii. Indefinite Integrals Involving $f(x) = \dfrac{x^n}{\sqrt[p]{F^n}}$ $\quad F = x^2 - a^2 \quad a^2 \neq 0$

$$\int \frac{dx}{\sqrt{F}} = \ln(x+\sqrt{F}) = \cosh^{-1}\frac{x}{a}$$

$$\int \frac{dx}{\sqrt{F^3}} = -\frac{x}{a^2\sqrt{F}}$$

$$\int \frac{dx}{\sqrt{F^n}} = -\frac{x}{a^2(n-2)\sqrt{F^{n-2}}} - \frac{n-3}{a^2(n-2)}\int\frac{dx}{\sqrt{F^{n-2}}} \qquad n \neq 2$$

$$\int \frac{x\,dx}{\sqrt{F}} = \sqrt{F}$$

$$\int \frac{x\,dx}{\sqrt{F^3}} = \frac{-1}{\sqrt{F}}$$

$$\int \frac{x\,dx}{\sqrt{F^n}} = -\frac{x^2}{a^2(n-2)\sqrt{F^{n-2}}} - \frac{n-4}{a^2(n-2)}\int\frac{x\,dx}{\sqrt{F^{n-2}}} \qquad n \neq 2$$

$$\int \frac{x^2\,dx}{\sqrt{F}} = \frac{x}{2}\sqrt{F} + \frac{a^2}{2}\ln(x+\sqrt{F})$$

$$\int \frac{x^2\,dx}{\sqrt{F^3}} = -\frac{x}{\sqrt{F}} + \ln(x+\sqrt{F})$$

$$\int \frac{x^2\,dx}{\sqrt{F^3}} = -\frac{x^3}{a^2(n-2)\sqrt{F^{n-2}}} - \frac{n-5}{a^2(n-2)}\int\frac{x^2\,dx}{\sqrt{F^{n-2}}} \qquad n \neq 2$$

$$\int \frac{x^3\,dx}{\sqrt{F}} = \left(\frac{F}{3} + a^2\right)\sqrt{F}$$

$$\int \frac{x^3\,dx}{\sqrt{F^3}} = \frac{F - a^2}{\sqrt{F}}$$

$$\int \frac{x^m\,dx}{\sqrt[p]{F^n}} = -\frac{px^{m+1}}{2a^2(n-p)\sqrt[p]{F^{n-p}}} - \frac{2n - p(m+3)}{2a^2(n-p)} \int \frac{x^m\,dx}{\sqrt[n]{F^{n-p}}} \qquad n \neq p$$

xxxiv. Indefinite Integrals Involving $f(x) = \dfrac{1}{x^m \sqrt[p]{F^n}}$ $\quad F = x^2 - a^2 \quad a^2 \neq 0$

$$\int \frac{dx}{x\sqrt{F}} = \frac{1}{a}\cos^{-1}\frac{a}{x}$$

$$\int \frac{dx}{x\sqrt{F^3}} = \frac{-1}{a^2\sqrt{F}} - \frac{1}{a^3}\cos^{-1}\frac{a}{x}$$

$$\int \frac{dx}{x\sqrt{F^n}} = -\frac{1}{a^2(n-2)x\sqrt{F^{n-2}}} - \frac{1}{a^2}\int \frac{dx}{x\sqrt{F^{n-2}}} \qquad n \neq 2$$

$$\int \frac{dx}{x^2\sqrt{F}} = \frac{\sqrt{F}}{a^2 x}$$

$$\int \frac{dx}{x^2\sqrt{F^3}} = -\frac{\sqrt{F}}{a^4 x}\left(1 + \frac{x^2}{F}\right)$$

$$\int \frac{dx}{x^2\sqrt{F^n}} = -\frac{1}{a^2(n-2)x\sqrt{F^{n-2}}} - \frac{n-1}{a^2(n-2)}\int \frac{dx}{x^2\sqrt{F^{n-2}}} \qquad n \neq 2$$

$$\int \frac{dx}{x^3\sqrt{F}} = \frac{\sqrt{F}}{2a^2x^2} + \frac{1}{2a^3}\cos^{-1}\frac{a}{x}$$

$$\int \frac{dx}{x^3\sqrt{F^3}} = \frac{1}{2a^2x^2\sqrt{F}} - \frac{3}{2a^4\sqrt{F}} - \frac{3}{2a^5}\cos^{-1}\frac{a}{x}$$

$$\int \frac{dx}{x^3\sqrt{F^n}} = -\frac{1}{a^2(n-2)x^2\sqrt{F^{n-2}}} - \frac{n}{a^2(n-2)}\int \frac{dx}{x^3\sqrt{F^{n-2}}} \qquad n \neq 2$$

$$\int \frac{dx}{x^4\sqrt{F}} = \frac{\sqrt{F}}{a^4 x}\left(1 - \frac{F^2}{3x^2}\right)$$

$$\int \frac{dx}{x^4\sqrt{F^3}} = -\frac{x}{a^6\sqrt{F}}\left(1+\frac{2F}{x^2}-\frac{F^2}{3x^4}\right)$$

$$\int \frac{dx}{x^m \sqrt[p]{F^n}} = \frac{p}{2a^2(n-p)x^{m-1}\sqrt[p]{F^{n-p}}} - \frac{2n-p(m-3)}{2a^2(n-p)}\int \frac{dx}{x^m\sqrt[p]{F^{n-p}}} \qquad n \neq p$$

xxxv. Indefinite Integrals Involving $f(x) = \dfrac{\sqrt[p]{F^n}}{x^m}$ $\quad F = x^2 - a^2 \quad a^2 \neq 0$

$$\int \frac{\sqrt{F}}{x}\,dx = \sqrt{F} - a\cos^{-1}\frac{a}{x}$$

$$\int \frac{\sqrt{F^3}}{x}\,dx = \frac{1}{3}\sqrt{F^3} - a^2\sqrt{F} + a^3\cos^{-1}\frac{a}{x}$$

$$\int \frac{\sqrt{F^n}}{x}\,dx = \frac{1}{n}\sqrt{F^n} - a^2\int \frac{\sqrt{F^{n-2}}}{x}\,dx$$

$$\int \frac{\sqrt{F}}{x^2}\,dx = -\frac{1}{x}\sqrt{F} + \ln(x+\sqrt{F})$$

$$\int \frac{\sqrt{F^3}}{x^2}\,dx = -\frac{1}{x}\sqrt{F^3} + \frac{3x}{2}\sqrt{F} - \frac{3a^2}{2}\ln(x+\sqrt{F})$$

$$\int \frac{\sqrt{F^n}}{x^2}\,dx = \frac{\sqrt{F^n}}{(n-1)x} - \frac{a^2 n}{n-1}\int \frac{\sqrt{F^{n-2}}}{x^2}\,dx \qquad n \neq 1$$

$$\int \frac{\sqrt{F}}{x^3}\,dx = \frac{-1}{2x^2}\sqrt{F} + \frac{1}{2a}\cos^{-1}\frac{a}{x}$$

$$\int \frac{\sqrt{F^3}}{x^3}\,dx = -\frac{-1}{2x^2}\sqrt{F^3} + \frac{3}{2}\sqrt{F} - \frac{3a}{2}\cos^{-1}\frac{a}{x}$$

$$\int \frac{\sqrt{F^n}}{x^3}\,dx = \frac{\sqrt{F^n}}{(n-2)x^2} - \frac{a^2 n}{n-2}\int \frac{\sqrt{F^{n-2}}}{x^3}\,dx \qquad n \neq 2$$

$$\int \frac{\sqrt{F}}{x^4}\,dx = \frac{1}{3a^2x^3}\sqrt{F^3}$$

$$\int \frac{\sqrt{F^3}}{x^4}\,dx = -\frac{1}{3x^3}\sqrt{F^3} - \frac{1}{x}\sqrt{F} + \ln(x+\sqrt{F})$$

$$\int \frac{\sqrt[p]{F^n}}{x^m}\,dx = \frac{p\sqrt[p]{F^n}}{(2n-mp+p)x^{m-1}} - \frac{2a^2 n}{2n-mp+p}\int \frac{\sqrt[p]{F^{n-p}}}{x^m}\,dx \qquad 2n \neq -p(1-m)$$

xxxvi. Indefinite Integrals Involving $f(x) = x^m \sqrt[p]{P^n}$ $\lambda = \frac{n}{p}$ $P = a + bx^q$ $a \neq 0$ $b \neq 0$

$$\int P^{\lambda}\, dx = \frac{1}{1+q\lambda}\left(xP^{\lambda} + aq\lambda \int P^{\lambda-1}\, dx\right)$$

$$\int \frac{dx}{P^{\lambda}} = \frac{-1}{aq(1-\lambda)}\left[\frac{x}{P^{\lambda-1}} - (1+q-q\lambda)\int \frac{dx}{P^{\lambda-1}}\right]$$

$$\int x^m P^{\lambda}\, dx = \frac{1}{1+m+q\lambda}\left(x^{m+1}P^{\lambda} + aq\lambda \int x^m P^{\lambda-1}\, dx\right)$$

$$= \frac{-1}{aq(1+\lambda)}\left[x^{m+1}P^{\lambda+1} - (1+m+q+q\lambda)\int x^m P^{\lambda+1}\, dx\right]$$

$$= \frac{1}{a(1+m)}\left[x^{m+1}P^{\lambda+1} - b(1+m+q+q\lambda)\int x^{m+q} P^{\lambda}\, dx\right]$$

$$\frac{1}{b(1+m+q^{\lambda})}\left[x^{m-q+1}P^{\lambda+1} - a(1+m-q)\int x^{m-q} P^{\lambda}\, dx\right]$$

xxxvii. Indefinite Integrals Involving $f(x) = \frac{x^{p-1}}{x^{2m+1} \pm a^{2m+1}}$ $a \neq 0$

$$\int \frac{x^{p-1}\, dx}{x^{2m+1} + a^{2m+1}} = \frac{2(-1)^{p-1}}{(2m+1)\, a^{2m-p+1}} \sum_{k=1}^{m} \sin\frac{2kp\pi}{2m+1} \tan^{-1}\left\{\frac{x + a\cos[2k\pi/(2m+1)]}{a\sin[2k\pi/(2m+1)]}\right\}$$

$$- \frac{(-1)^{p-1}}{(2m+1)\, a^{2m-p+1}} \sum_{k=1}^{m} \cos\frac{2kp\pi}{2m+1} \ln\left(x^2 + 2ax\cos\frac{2k\pi}{2m+1} + a^2\right)$$

$$+ \frac{(-1)^{p-1}\ln(x+a)}{(2m+1)\, a^{2m-p+1}} \qquad 0 < p \leq 2m+1$$

$$\int \frac{x^{p-1}\, dx}{x^{2m+1} - a^{2m+1}} = \frac{-2}{(2m+1)\, a^{2m-p+1}} \sum_{k=1}^{m} \sin\frac{2kp\pi}{2m+1} \tan^{-1}\left\{\frac{x - a\cos[2k\pi/(2m+1)]}{a\sin[2k\pi/(2m+1)]}\right\}$$

$$+ \frac{1}{(2m+1)\, a^{2m-p+1}} \sum_{k=1}^{m} \cos\frac{2kp\pi}{2m+1} \ln\left(x^2 - 2ax\cos\frac{2k\pi}{2m+1} + a^2\right)$$

$$+ \frac{\ln(x-a)}{(2m+1)\, a^{2m-p+1}} \qquad 0 < p \leq 2m+1$$

xxxviii. Indefinite Integrals Involving $f(x) = x^m \sqrt[p]{R^n}$ $\lambda = n/p$ $R = a^{x^q} + b^{x^{q+r}}$ $a \neq 0$

$a \neq 0$

$$\int R^\lambda \, dx = \frac{\lambda}{1+\lambda(q+r)}\left(xR^\lambda + ar\lambda \int x^q R^{\lambda-1}\, dx\right)$$

$$\int \frac{dx}{R^\lambda} = \frac{-1}{ar(1+\lambda)}\left[xR^{\lambda+1} - (1+r+q\lambda+r\lambda)\int \frac{x^q\, dx}{R^{\lambda-1}}\right]$$

$$\int x^m R^\lambda \, dx + \frac{1}{1+m+q\lambda+r\lambda}\left(x^{m+1} R^\lambda + ar\lambda \int x^{m+q} R^{\lambda-1}\, dx\right)$$

$$= \frac{-1}{ar\,(1+\lambda)}\left[x^{m+1} R^{\lambda+1} - (1+r+q\lambda+r\lambda)\int x^{m-q} R^{\lambda+1}\, dx\right]$$

xxxix. Indefinite Integrals Involving $f(x) = \dfrac{x^m}{(x-a_1)(x-a_2)\ldots(x-a_k)}$ $a_i \neq a_j \neq 0$

$$\int \frac{x^m\, dx}{(x-a_1)(x-a_2)\ldots(x-a_k)} = \frac{a_1{}^m \ln(x-a_1)}{(a_1-a_2)(a_1-a_3)\ldots(a_1-a_k)}$$

$$+ \frac{a_2^m \ln(x-a_2)}{(a_2-a_1)(a_2-a_3)\ldots(a_2-a_k)} + \ldots + \frac{ak^m \ln(x-a_k)}{(a_k-a_1)(a_k-a_2)\ldots(a_k-a_{k-1})}$$

xl. Indefinite Integrals Involving $f(x) = \dfrac{x^{p-1}}{x^{2m} \pm a^{2m}}$ $a \neq 0$

$$\int \frac{x^{p-1}}{x^{2m}+a^{2m}} = \frac{1}{ma^{2m-p}} \sum_{k=1}^{m} \sin\frac{(2k-1)p\pi}{2m} \tan^{-1}\left\{\frac{x + a\cos[(2k-1)\pi/2m]}{a\sin[(2k-1)\pi/2m]}\right\}$$

$$- \frac{1}{2ma^{2m-p}} \sum_{k=1}^{m} \cos\frac{(2k-1)p\pi}{2m} \ln\left[x^2 + 2ax\cos\frac{(2k-1)\pi}{2m} + a^2\right] \qquad 0 < p \leq 2m$$

$$\int \frac{x^{p-1}\, dx}{x^{2m}-a^{2m}} = \frac{1}{2ma^{2m-p}} \sum_{k=1}^{m-1} \cos\frac{kp\pi}{m} \ln\left(x^2 - 2ax\cos\frac{k\pi}{m} + a^2\right)$$

$$- \frac{1}{na^{2m-p}} \sum_{k=1}^{m-1} \sin\frac{kp\pi}{m} \tan^{-1}\frac{x - a\cos(k\pi/m)}{a\sin(k\pi/m)} + \frac{1}{2ma^{2m-p}}[\ln(x-a) + (-1)^p \ln(x+a)]$$

$$0 < p \leq 2m$$

xli. Indefinite Integrals Involving $f(x) = \sin^n A$ $f(x) = \dfrac{1}{\sin^n A}$ $A = bx$

$$\int \sin A\,dx = -\frac{\cos A}{b}$$

$$\int \sin^2 A\,dx = -\frac{\sin 2A}{4b} + \frac{x}{2}$$

$$\int \sin^3 A\,dx = -\frac{\cos A}{b} + \frac{\cos^3 A}{3b}$$

$$\int \sin^4 A\,dx = -\frac{\sin 2A}{4b} + \frac{\sin 4A}{32b} + \frac{3x}{8}$$

$$\int \sin^5 A\,dx = -\frac{5\cos A}{8b} + \frac{5\cos 3A}{48b} - \frac{\cos 5A}{80b}$$

$$\int \sin^n A\,dx = -\frac{\cos A}{b}\frac{\sin^{n-1} A}{n} + \frac{n-1}{n}\int \sin^{n-2} A\,dx \qquad n > 0$$

$$\int \frac{dx}{\sin A} = \frac{1}{b}\ln\tan\frac{A}{2}$$

$$\int \frac{dx}{\sin^2 A} = -\frac{\cot A}{b}$$

$$\int \frac{dx}{\sin^3 A} = -\frac{\cos A}{2b\sin^2 A} + \frac{1}{2b}\ln\tan\frac{A}{2}$$

$$\int \frac{dx}{\sin^4 A} = -\frac{\cot A}{b} - \frac{\cot^3 A}{3b}$$

$$\int \frac{dx}{\sin^5 A} = -\frac{\cos A}{4b\sin^4 A} - \frac{3\cos A}{8b\sin^4 A} - \frac{3}{16b}\ln\frac{1+\cos A}{1-\cos A}$$

$$\int \frac{dx}{\sin^n A} = -\frac{\cos A}{b(n-1)\sin^{n-1} A} + \frac{n-2}{n-1}\int \frac{dx}{\sin^{n-2} A} \qquad n > 1$$

xlii. Indefinite Integrals Involving $f(x) = \cos^n A$ $f(x) = \dfrac{1}{\cos^n A}$ $A = bx$

$$\int \cos A\,dx = \frac{\sin A}{b}$$

$$\int \cos^2 A\,dx = \frac{\sin 2A}{4b} + \frac{x}{2}$$

$$\int \cos^3 A\,dx = \frac{\sin A}{b} - \frac{\sin^3 A}{3b}$$

$$\int \cos^4 A\,dx = \frac{\sin 2A}{4b} + \frac{\sin 4A}{32b} + \frac{3x}{8}$$

$$\int \cos^5 A\,dx = \frac{5\sin A}{8b} + \frac{5\sin 3A}{48b} + \frac{\sin 5A}{80b}$$

$$\int \cos^n A\,dx = \frac{\sin A}{b}\,\frac{\cos^{n-1} A}{n} + \frac{n-1}{n}\int \cos n^{n-2} A\,dx \qquad n > 0$$

$$\int \frac{dx}{\cos A} = \frac{1}{b}\ln\tan\left(\frac{\pi}{4} + \frac{A}{2}\right)$$

$$\int \frac{dx}{\cos^2 A} = \frac{1}{b}\tan A$$

$$\int \frac{dx}{\cos^3 A} = \frac{\sin A}{2b\cos^2 A} + \frac{1}{2b}\ln\tan\left(\frac{\pi}{4} + \frac{A}{2}\right)$$

$$\int \frac{dx}{\cos^4 A} = \frac{\tan A}{b} + \frac{\tan^3 A}{3b}$$

$$\int \frac{dx}{\cos^5 A} = \frac{\sin A}{4b\cos^4 A} + \frac{3\sin A}{8\cos^4 A} + \frac{3}{8b}\ln\tan\left(\frac{\pi}{4} + \frac{A}{2}\right)$$

$$\int \frac{dx}{\cos^n A} = \frac{\sin A}{b(n-1)\cos^{n-1} A} + \frac{n-2}{n-1}\int \frac{dx}{\cos^{n-2} A} \qquad n > 1$$

---

xliii. Indefinite Integrals Involving $f(x) = x^m \sin^n A \quad A = bx$

$D_1 = \int \sin^n A\,dx \qquad D_2 = \int D_1\,dx \qquad D_k = \int D_{k-1}\,dx$

$$\int x\sin A\,dx = -\frac{x\cos A}{b} + \frac{\sin A}{b^2}$$

$$\int x\sin^2 A\,dx = \frac{x^2}{4} - \frac{x\sin 2A}{4b} - \frac{\cos 2A}{8b^2}$$

$$\int x\sin^3 A\,dx = \frac{x\cos 3A}{12b} - \frac{\sin 3A}{36b^2} + \frac{3x\cos A}{4b} + \frac{3\sin A}{4b^2}$$

$$\int x\sin^n A\,dx = xD_1 - D_2$$

$$\int x^2\sin A\,dx = -\frac{x^2\cos A}{b} + \frac{2x\sin A}{b^2} + \frac{2\cos A}{b^3}$$

$$\int x^2 \sin^2 A\,dx = \frac{1}{2}\int x^2 (1 - \cos 2A)\,dx$$

$$\int x^2 \sin^3 A\,dx = \frac{1}{4}\int x^2 (3 \sin A - \sin 3A)\,dx$$

$$\int x^2 \sin^n A\,dx = x^2 D_1 - 2xD_2 + 2D_3$$

$$\int x^3 \sin A\,dx = \frac{-x^3 \cos A}{b} + \frac{3x^2 \sin A}{b^2} + \frac{6x \cos A}{b^3} - \frac{6 \sin A}{b^4}$$

$$\int x^3 \sin^2 A\,dx = \frac{1}{2}\int x^3 (1 - \cos 2A)\,dx$$

$$\int x^3 \sin^3 A\,dx = \frac{1}{4}\int x^3 (3 \sin A - \sin 3A)\,dx$$

$$\int x^3 \sin^n A\,dx = x^3 D_1 - 3x^2 D_2 + 6xD_3 - 6D_4$$

$$\int x^m \sin A\,dx = \frac{m! \sin A}{b}\left[\frac{x^{m-1}}{(m-1)!b} - \frac{x^{m-3}}{(m-3)!b^3} + \frac{x^{m-5}}{(m-5)\,b^5} - \ldots\right]$$

$$- \frac{m! \cos A}{b}\left[\frac{x^m}{m!} - \frac{x^{m-2}}{(m-2)!b^2} + \frac{x^{m-4}}{(m-4)!b^4} - \ldots\right]$$

$$\int x^m \sin^n A\,dx = x^m D_1 - mx^{m-1} D_2 + m(m-1)\,x^{m-2} D_3 - \ldots$$

Series terminates with the term *involving* x^{m-m}

---

xliv. Indefinite Integrals Involving $f(x) = x^m \cos^n A \quad A = bx$

$D_1 = \int \cos^n A\,dx \qquad D_2 = \int D_1\,dx \qquad D_k = \int D_{k-1}\,dx$

$$\int x \cos A\,dx = \frac{x \sin A}{b} + \frac{\cos A}{b^2}$$

$$\int x \cos^2 A\,dx = \frac{x^2}{4} + \frac{x \sin 2A}{4b} + \frac{\cos 2A}{8b^2}$$

$$\int x \cos^3 A\,dx = \frac{x \sin 3A}{12b} + \frac{\cos 3A}{36b^2} + \frac{3x \sin A}{4b} + \frac{3 \cos A}{4b^2}$$

$$\int x \cos^n A\,dx = xD_1 - D_2$$

$$\int x^2 \cos A\,dx = \frac{x^2 \sin A}{b} + \frac{2x \cos A}{b^2} - \frac{2 \sin A}{b^3}$$

$$\int x^2 \cos^2 A\,dx = \frac{1}{2}\int x^2 (1 + \cos 2A)\,dx$$

$$\int x^2 \cos^3 A\, dx = \frac{1}{4}\int x^2 (3\cos A + \cos 3A)\, dx$$

$$\int x^2 \cos^n A\, dx = x^2 D_1 - 2xD_2 + D_3$$

$$\int x^3 \cos A\, dx = \frac{x^3 \sin A}{b} + \frac{3x^2 \cos A}{b^2} - \frac{6x \sin A}{b^3} - \frac{6\cos A}{b^4}$$

$$\int x^3 \cos^2 A\, dx = \frac{1}{2}\int x^3 (1 + \cos 2A)\, dx$$

$$\int x^3 \cos^3 A\, dx = \frac{1}{4}\int x^3 (3\cos A + \cos 3A)\, dx$$

$$\int x^3 \cos^n A\, dx = x^3 D_1 - 3x^2 D_2 + 6xD_3 - 6D_4$$

$$\int x^m \cos A\, dx = \frac{m!\sin A}{b}\left[\frac{x^m}{(m)!} - \frac{x^{m-2}}{(m-2)!b^2} + \frac{x^{m-4}}{(m-4)!b^4} - \ldots\right] - \frac{m!\cos A}{b}\left[\frac{x^{m-1}}{(m-1)!b} - \frac{x^{m-3}}{(m-3)!b^3} + \frac{x^{m-5}}{(m-5)!b^5} - \ldots\right]$$

$$\int x^m \cos^n A\, dx = x^m D_1 - mx^{m-1} D_2 + m(m-1)\, x^{m-2} D_3 - \ldots.$$

Series terminates with the term *involving* x^{m-m}

---

xlv. Indefinite Integrals Involving $f(x) = \frac{x^m}{\sin^n A}$ $f(x) = \frac{\sin^n A}{x^m}$ $A = bx$

$$\int \frac{x\, dx}{\sin A} = \left[1 + \frac{A^2}{18} + \frac{7A^4}{1{,}800} + \ldots + \frac{2(2^{2k-1} - 1) B_k A^{2k}}{(2k+1)!} + \ldots\right]\frac{A}{b^2}$$

$$\int \frac{x^m\, dx}{\sin A} = \left[\frac{1}{m} + \frac{A^2}{6(m+2)} + \frac{7A^4}{360(m+4)} + \ldots + \frac{2(2^{2k-1} - 1) B_k A^{2k}}{(2k)!(m+2k)} + \ldots\right]\frac{A^m}{b^{m+1}}$$

Note: B_k = Bernoulli's number

$$\int \frac{x\, dx}{\sin^2 A} = -\frac{x}{b}\cot A + \frac{1}{b^2}\ln \sin A$$

$$\int \frac{x\, dx}{\sin^n A} = -\frac{x\cos A}{(n-1)\, b \sin^{n-1} A} - \frac{1}{(n-1)(n-2)\, b^2 \sin^{n-2} A} + \frac{n-2}{n-1}\int \frac{x\, dx}{\sin^{n-2} A} \quad n > 2$$

$$\int \frac{\sin A}{x}\, dx = A - \frac{A^3}{3.3!} + \frac{A^5}{5.5!} - \frac{A^7}{7.7!} + \ldots$$

$$\int \frac{\sin A}{x^m}\,dx = -\frac{\sin A}{(m-1)\,x^{m-1}} + \frac{b}{m-1}\int \frac{\cos A}{x^{m-1}}\,dx \quad m \neq 1$$

$$\int \frac{\sin^{2n-1}}{x^m}\,dx = \frac{(-1)^{n-1}}{2^{2n-2}}\int\left[\sin(2n-1)A - \binom{2n-1}{1}\sin(2n-3)A + \ldots (-1)^{n-1}\binom{2n-1}{n-1}\sin A\right]\frac{dx}{x^m}$$

$$\int \frac{\sin^{2n} A}{x^m}\,dx = -\frac{\binom{2n}{n}}{(m-1)\,2^{2n}\,x^{m-1}} + \frac{(-1)^n}{2^{2n-1}}\int\left[\cos 2nA - \binom{2n}{1}\cos(2n-2)A + \ldots + (-1)^{n-1}\binom{2n}{n-1}\cos 2A\right]\frac{dx}{x^m}$$

xlvi. Indefinite Integrals Involving $f(x) = \dfrac{x^m}{\cos^n A}$ $\quad f(x) = \dfrac{\cos^n A}{x^m}$ $\quad A = bx$

$$\int \frac{x\,dx}{\cos A} = \left[1 + \frac{A^2}{4} + \frac{5A^4}{72} + \frac{61\,A^6}{2{,}880} + \ldots + \frac{E_k\,A^{2k}}{(k+1)\,(2k)!} + \ldots\right]\frac{A^2}{2b^2}$$

$$\int \frac{x^m\,dx}{\cos A} = \left[\frac{1}{(m+1)} + \frac{A^{m+1}}{2\,(m+3)} + \frac{5A^{m+3}}{24\,(m+5)} + \ldots + \frac{E_k\,A^{m+2k-1}}{(2k)!\,(m+2k+1)} + \ldots\right]\frac{A^m}{b^{m+1}}$$

Note: E_k is Euler's number

$$\int \frac{x\,dx}{\cos^2 A} = \frac{x}{b}\tan A + \frac{1}{b^2}\ln\cos A$$

$$\int \frac{x\,dx}{\cos^n A} = \frac{x\sin A}{(n-1)\,b\cos^{n-1} A} - \frac{1}{(n-1)(n-2)\,b^2\cos^{n-2} A} + \frac{n-2}{n-1}\int \frac{x\,dx}{\cos^{n-2} A} \qquad n > 2$$

$$\int \frac{\cos A\,dx}{x} = \ln A - \frac{A^2}{2.2!} + \frac{A^4}{4.4!} - \frac{A^6}{6.6!} + \ldots$$

$$\int \frac{\cos A}{x^m}\,dx = -\frac{\cos A}{(m-1)\,x^{m-1}} - \frac{b}{m-1}\int \frac{\sin A}{x^{m-1}}\,dx \qquad m \neq 1$$

$$\int \frac{\cos^{2n-1}}{x^n}\,dx = \frac{1}{2^{2n-2}}\int\left[\cos(2n-1)A + \binom{2n-1}{1}\cos(2n-3)A + \ldots + \binom{2n-1}{n-1}\cos A\right]\frac{dx}{x^m}$$

$$\int \frac{\cos^{2n} A}{x^m}\,dx = -\frac{\binom{2n}{2}}{(m-1)\,2^{2n}\,x^{m-1}}$$

$$+\frac{1}{2^{2n-1}}\int\left[\cos 2nA + \binom{2n}{1}\cos(2n-2)\,A + \ldots + \binom{2n}{n-1}\cos 2A\right]\frac{dx}{x^m}$$

xlvii. Indefinite Integrals Involving $f(x) = f(1 \pm a \sin A)$ $A = bx$ α, β = constants

$$\int \frac{dx}{\sin A + \sin\alpha} = \frac{1}{b\cos\alpha}\ln\left[\frac{\sin[(A+\alpha)/2]}{\cos[(A-\alpha)/2]}\right]$$

$$\int \frac{dx}{1+\sin\alpha\sin A} = \frac{2}{b\cos\alpha}\tan^{-1}\left[\tan\left(\frac{\pi}{4}+\frac{A}{2}\right)\tan\left(\frac{\pi}{4}+\frac{\alpha}{2}\right)\right]$$

$$\int \frac{dx}{1+\cos\alpha\sin A} = \frac{2}{b\sin\alpha}\tan^{-1}\frac{\cos\alpha + \tan(A/2)}{\tan\alpha}$$

$$\int \frac{dx}{1+a\sin A} = \begin{cases} \dfrac{2}{b\sqrt{1-a^2}}\tan^{-1}\dfrac{a+\tan(A/2)}{1-a^2} & a^2 < 1 \\ \dfrac{1}{b\sqrt{1-a^2}}\sin^{-1}\dfrac{a+\sin A}{1+a\sin A} & a^2 < 1 \\ \dfrac{1}{b\sqrt{a^2-1}}\ln\dfrac{\tan(A/2)+a-\sqrt{a^2-1}}{\tan(A/2)+a+\sqrt{a^2-1}} & a^2 > 1 \end{cases}$$

$$\int \frac{dx}{(1+a\sin A)^2} = \frac{a\cos A}{b(1-a^2)(1+a\sin A)} + \frac{1}{1-a^2}\int\frac{dx}{1+a\sin A}$$

$$\int \frac{dx}{(1+a\sin A)^n} = \frac{a\cos A}{b(n-1)(1-a^2)(1+a\sin A)^{n-1}} + \frac{(2n-3)}{(n-1)(1-a^2)}\int\frac{dx}{(1+a\sin A)^{n-1}}$$

$$-\frac{(n-2)}{(n-1)(1-a^2)}\int\frac{dx}{(1+a\sin A)^{n-2}} \qquad a^2 \neq 1,\; n \neq 1$$

$$\int \sin\alpha x\sin\beta x\,dx = \frac{\sin(\alpha-\beta)x}{2(\alpha-\beta)} - \frac{\sin(\alpha+\beta)x}{2(\alpha+\beta)} \qquad a^2 \neq \beta^2$$

$$\int \sin\alpha x\cos\beta x\,dx = -\frac{\cos(\alpha-\beta)x}{2(\alpha-\beta)} - \frac{\cos(\alpha+\beta)x}{2(\alpha+\beta)} \qquad a^2 \neq \beta^2$$

xlviii. Indefinite Integrals Involving $f(x) = f(1 \pm a\cos A)$ $A = bx$ $\alpha, \beta =$ constants

$$\int \frac{dx}{\cos A + \cos\alpha} = \frac{1}{b\sin\alpha}\ln\frac{\cos[(A-\alpha)/2]}{\cos[(A+\alpha)/2]}$$

$$\int \frac{dx}{\cos A - \cos\alpha} = \frac{1}{b\sin\alpha}\ln\frac{\sin[(A+\alpha)/2]}{\sin[(A-\alpha)/2]}$$

$$\int \frac{dx}{1+\cos\alpha\cos A} = \frac{1}{b\sin\alpha}\tan^{-1}\frac{\sin\alpha\sin A}{\cos\alpha+\cos A}$$

$$\int \frac{dx}{1+a\cos A} = \begin{cases} \dfrac{2}{b\sqrt{1-a^2}}\tan^{-1}\left(\sqrt{\dfrac{1-a}{1+a}}\tan\dfrac{A}{2}\right) & a^2<1 \\ \dfrac{1}{b\sqrt{1-a^2}}\cos^{-1}\dfrac{a+\cos A}{1+a\cos A} & a^2<1 \\ \dfrac{1}{b\sqrt{a^2-1}}\ln\dfrac{(\sqrt{a+1})+(\sqrt{a-1})\tan(A/2)}{(\sqrt{a+1})-(\sqrt{a-1})\tan(A/2)} & a^2>1 \end{cases}$$

$$\int \frac{dx}{(1+a\cos A)^2} = \frac{-a\sin A}{b(1-a^2)(1+a\cos A)} + \frac{1}{1-a^2}\int\frac{dx}{1+a\cos A}$$

$$\int \frac{dx}{(1+a\cos A)^n} = \frac{-a\sin A}{b(n-1)(1-a^2)(1+a\cos A)^{n-1}} + \frac{(2n-3)}{(n-1)(1-a^2)}\int\frac{dx}{(1+a\cos A)^{n-1}}$$

$$-\frac{(n-2)}{(n-1)(1-a^2)}\int\frac{dx}{(1+a\cos A)^{n-2}} \qquad a^2 \neq 1$$

$$n \neq 1$$

$$\int \cos\alpha x\cos\beta x\,dx = \frac{\sin(\alpha-\beta)x}{2(\alpha-\beta)} + \frac{\sin(\alpha+\beta)x}{2(\alpha+\beta)} \qquad \alpha^2 \neq \beta^2$$

$$\int \cos\alpha x\sin\beta x\,dx = -\frac{\cos(\alpha-\beta)x}{2(\alpha-\beta)} - \frac{\cos(\alpha+\beta)x}{2(\alpha+\beta)} \qquad \alpha^2 \neq \beta^2$$

xlix. Indefinite Integrals Involving $f(x) = f(1 \pm \sin A)$ $A = bx$

$$\int \frac{dx}{1+\sin A} = -\frac{1}{b}\tan\left(\frac{\pi}{4}-\frac{A}{2}\right)$$

$$\int \frac{x\,dx}{1+\sin A} = -\frac{x}{b}\tan\left(\frac{\pi}{4}-\frac{A}{2}\right) + \frac{2}{b^2}\ln\left[\cos\left(\frac{\pi}{4}-\frac{A}{2}\right)\right]$$

$$\int \frac{\sin A\,dx}{1+\sin A} = \frac{1}{b}\tan\left(\frac{\pi}{4}-\frac{A}{2}\right)+x$$

$$\int \frac{dx}{\sin A\,(1+\sin A)} = \frac{1}{b}\tan\left(\frac{\pi}{4}-\frac{A}{2}\right)+\frac{1}{b}\ln\left(\tan\frac{A}{2}\right)$$

$$\int \frac{dx}{1-\sin A} = \frac{1}{b}\tan\left(\frac{\pi}{4}+\frac{A}{2}\right)$$

$$\int \frac{x\,dx}{1-\sin A} = \frac{x}{b}\cot\left(\frac{\pi}{4}+\frac{A}{2}\right)+\frac{2}{b^2}\ln\left[\sin\left(\frac{\pi}{4}-\frac{A}{2}\right)\right]$$

$$\int \frac{\sin A\,dx}{1-\sin A} = \frac{1}{b}\tan\left(\frac{\pi}{4}+\frac{A}{2}\right)-x$$

$$\int \frac{dx}{\sin A\,(1-\sin A)} = \frac{1}{b}\tan\left(\frac{\pi}{4}+\frac{A}{2}\right)+\frac{1}{b}\ln\left(\tan\frac{A}{2}\right)$$

$$\int \frac{dx}{1+\sin^2 A} = \frac{1}{2b\sqrt{2}}\sin^{-1}\frac{3\sin^2 A-1}{\sin^2 A+1} \qquad \int \frac{dx}{1-\sin^2 A} = \frac{\tan A}{b}$$

$$\int \frac{dx}{(1+\sin A)^2} = \frac{-1}{2b}\tan\left(\frac{\pi}{4}-\frac{A}{2}\right)-\frac{1}{6b}\tan^3\left(\frac{\pi}{4}-\frac{A}{2}\right)$$

$$\int \frac{\sin A\,dx}{(1+\sin A)^2} = \frac{-1}{2b}\tan\left(\frac{\pi}{4}-\frac{A}{2}\right)+\frac{1}{6b}\tan^3\left(\frac{\pi}{4}-\frac{A}{2}\right)$$

$$\int \frac{dx}{(1-\sin A)^2} = \frac{1}{2b}\cot\left(\frac{\pi}{4}-\frac{A}{2}\right)+\frac{1}{6b}\cot^3\left(\frac{\pi}{4}-\frac{A}{2}\right)s$$

$$\int \frac{\sin A\,dx}{(1-\sin A)^2} = -\frac{1}{2b}\cot\left(\frac{\pi}{4}-\frac{A}{2}\right)+\frac{1}{6b}\cot^3\left(\frac{\pi}{4}-\frac{A}{2}\right)$$

I. Indefinite Integrals Involving $f(x) = f(1 \pm \cos A)$ $\quad A = bx$

$$\int \frac{dx}{1+\cos A} = \frac{1}{b}\tan\frac{A}{2}$$

$$\int \frac{dx}{1+\cos A} = \frac{x}{b}\tan\frac{A}{2}+\frac{2}{b^2}\ln\left(\cos\frac{A}{2}\right)$$

$$\int \frac{\cos A\,dx}{1+\cos A} = -\frac{1}{b}\tan\frac{A}{2}+x$$

$$\int \frac{dx}{\cos A\,(1+\cos A)} = -\frac{1}{b}\tan\frac{A}{2} + \frac{1}{b}\ln\left[\tan\left(\frac{\pi}{4}+\frac{A}{2}\right)\right]$$

$$\int \frac{dx}{1-\cos A} = -\frac{1}{b}\cot\frac{A}{2}$$

$$\int \frac{x\,dx}{1-\cos A} = -\frac{x}{b}\cot\frac{A}{2} + \frac{2}{b^2}\ln\left(\cos\frac{A}{2}\right)$$

$$\int \frac{\cos A\,dx}{1-\cos A} = -\frac{1}{b}\cot\frac{A}{2} - x$$

$$\int \frac{dx}{\cos A\,(1-\cos A)} = -\frac{1}{b}\cot\frac{A}{2} + \frac{1}{b}\ln\left[\tan\left(\frac{\pi}{4}+\frac{A}{2}\right)\right]$$

$$\int \frac{dx}{1+\cos^2 A} = \frac{1}{2b\sqrt{2}}\sin^{-1}\frac{1-3\cos^2 A}{1+\cos^2 A} \qquad \int \frac{dx}{1-\cos^2 A} = -\frac{\cot A}{b}$$

$$\int \frac{dx}{(1+\cos A)^2} = \frac{1}{2b}\tan\frac{A}{2} + \frac{1}{6b}\tan^3\frac{A}{2}$$

$$\int \frac{\cos A\,dx}{(1+\cos A)^2} = \frac{1}{2b}\tan\frac{A}{2} - \frac{1}{6b}\tan^3\frac{A}{2}$$

$$\int \frac{dx}{(1-\cos A)^2} = -\frac{1}{2b}\cot\frac{A}{2} - \frac{1}{6b}\cot^3\frac{A}{2}$$

$$\int \frac{\cos A\,dx}{(1-\cos A)^2} = \frac{1}{2b}\cot\frac{A}{2} - \frac{1}{6b}\cot^3\frac{A}{2}$$

Ii. Indefinite Integrals Involving $f(x) = f(1 \pm a\sin^2 A) \quad A = bx$

$$\int \frac{\sin A\,dx}{1+a\sin A} = \frac{x}{a} - \frac{1}{a}\int \frac{dx}{1+a\sin A}$$

$$\int \frac{dx}{(1+a\sin A)\sin A} = \frac{1}{b}\ln\left(\tan\frac{A}{2}\right) - a\int \frac{dx}{1+a\sin A}$$

$$\int \frac{dx}{1+a\sin^2 A} = \frac{1}{b\sqrt{1+a}}\tan^{-1}[(\sqrt{1+a})\tan A] \qquad a > 0$$

$$\int \frac{dx}{1-a\sin^2 A} = \begin{cases} \dfrac{1}{b\sqrt{1-a}}\tan^{-1}[(\sqrt{1-a})\tan A] & a < 1 \\[2ex] \dfrac{1}{b\sqrt{a-1}}\ln\dfrac{(\sqrt{a-1})\tan A + 1}{(\sqrt{a-1})\tan A - 1} & a > 1 \end{cases}$$

$$\int \frac{\sin A\, dx}{(1+a\sin A)^2} = \frac{\cos A}{b\,(a^2-1)\,(1+a\sin A)} + \frac{a}{a^2-1}\int \frac{dx}{1+a\sin A} \qquad a^2 > 1$$

$$\int \sin A\sqrt{1+a\sin^2 A}\, dx = -\frac{\cos A}{2b}\sqrt{1+a\sin^2 A} - \frac{1+a}{2b\sqrt{a}}\sin^{-1}\frac{\sqrt{a}\cos A}{\sqrt{1+a}}$$

$$\int \sin A\sqrt{1-a\sin^2 A}\, dx = -\frac{\cos A}{2b}\sqrt{1-a\sin^2 A} - \frac{1-a}{2b\sqrt{a}}\ln(\sqrt{a}\cos A + \sqrt{1-a\sin^2 A})$$

$$\int \frac{\sin A\, dx}{\sqrt{1+a\sin^2 A}} = \frac{-1}{b\sqrt{a}}\sin^{-1}\frac{\sqrt{a}\cos A}{\sqrt{1+a}}$$

$$\int \frac{\sin A\, dx}{\sqrt{1-a\sin^2 A}} = \frac{-1}{b\sqrt{a}}\ln(\sqrt{a}\cos A + \sqrt{1-a\sin^2 A}) \qquad a > 0$$

iii. Indefinite Integrals Involving $f(x) = f(1 \pm a\cos^2 A) \quad A = bx$

$$\int \frac{\cos A\, dx}{1+a\cos A} = \frac{x}{a} - \frac{1}{a}\int \frac{dx}{1+a\cos A}$$

$$\int \frac{dx}{(1+a\cos A)\cos A} = \frac{1}{b}\ln\left[\tan\left(\frac{A}{2}+\frac{\pi}{4}\right)\right] - a\int \frac{dx}{1+a\cos A}$$

$$\frac{dx}{1+a\cos^2 A} = \frac{1}{b\sqrt{1+a}}\tan^{-1}\frac{\tan A}{\sqrt{1+a}} \qquad a > 0$$

$$\int \frac{dx}{1-a\cos^2 A} = \begin{cases} \dfrac{1}{b\sqrt{1-a}}\tan^{-1}\dfrac{\tan A}{\sqrt{1-a}} & a < 1 \\ \dfrac{1}{b\sqrt{a-1}}\ln\dfrac{\tan A - \sqrt{a-1}}{\tan A + \sqrt{a-1}} & a > 1 \end{cases}$$

$$\int \frac{\cos A\, dx}{(1+a\cos A)^2} = \frac{-\sin A}{b\,(a^2-1)\,(1+a\cos A)} + \frac{a}{a^2-1}\int \frac{dx}{1+a\cos A} \qquad a^2 > 1$$

$$\int \cos A\sqrt{1+a\sin^2 A}\, dx = \frac{\sin A}{2b}\sqrt{1+a\sin^2 A} + \frac{1}{2b\sqrt{a}}\ln(\sqrt{a}\sin A + \sqrt{1+a\sin^2 A})$$

$$\int \cos A\sqrt{1-a\sin^2 A}\, dx = \frac{\sin A}{2b}\sqrt{1-a\sin^2 A} + \frac{1}{2b\sqrt{a}}\sin^{-1}(\sqrt{a}\sin A)$$

$$\int \frac{\cos A\, dx}{\sqrt{1+a\sin^2 A}} = \frac{1}{b\sqrt{a}}\ln(\sqrt{a}\sin A + \sqrt{1+a\sin^2 A})$$

$$\int \frac{\cos A\, dx}{\sqrt{1-a\sin^2 A}} = \frac{1}{b\sqrt{a}}\sin^{-1}(\sqrt{a}\sin A) \qquad a > 0$$

liii. Indefinite Integrals Involving $f(x) = \sin^m A \cos^n A \quad A = bx$

$$\int \sin A \cos A \, dx = -\frac{\cos^2 A}{2b}$$

$$\int \sin A \cos^n A \, dx = -\frac{\cos^{n+1} A}{(n+1)\,b} \quad n \neq -1 \qquad \int \sin^m A \cos A \, dx = \frac{\sin^{m+1}}{(m+1)b} \quad m \neq -1$$

$$\int \sin^2 A \cos A \, dx = \frac{\sin^3 A}{3b}$$

$$\int \sin^2 A \cos^2 A \, dx = \frac{x}{8} - \frac{\sin 4A}{32b}$$

$$\int \sin^2 A \cos^3 A \, dx = -\frac{\sin^3 A \cos^2 A}{5b} + 2\frac{\sin^3 A}{15b}$$

$$\int \sin^2 A \cos^n A \, dx = -\frac{\sin A \cos^{n+1} A}{(n+2)\,b} + \int \frac{\cos^n A \, dx}{n+2} \qquad n \neq -2$$

$$\int \sin^3 A \cos A \, dx = \frac{\sin^4 A}{4b}$$

$$\int \sin^3 A \cos^n A \, dx = -\frac{\cos^{n+1} A}{(n+1)\,b} + \frac{\cos^{n+3} A}{(n+3)\,b} \qquad n \neq -1, -3$$

$$\int \sin^4 A \cos A \, dx = -\frac{\sin^5 A}{5b}$$

$$\int \sin^4 A \cos^2 A \, dx = \frac{1}{192b}(\sin 6A - 3\sin 4A - 3\sin 2A + 12A)$$

$$\int \sin^4 A \cos^3 A \, dx = -\frac{\sin^5 A}{5b} - \frac{\sin^7 A}{7b}$$

$$\int \sin^m A \cos^n A \, dx = \frac{\sin^{m+1} A \cos^{n-1} A}{b\,(m+n)} + \frac{n-1}{m+n}\int \sin^m A \cos^{n-2} A \, dx$$

$$= -\frac{\sin^{m-1} A \cos^{n+1}}{b\,(m+n)} + \frac{m-1}{m+n}\int \sin^{m-2} A \cos^n A \, dx \qquad m > 0, \; n > 0$$

liv. Indefinite Integrals Involving $f(x) = \dfrac{1}{\sin^m A \cos^n A} \quad A = bx$

$$\int \frac{dx}{\sin A \cos A} = \frac{1}{b} \ln(\tan A)$$

$$\int \frac{dx}{\sin A \cos^2 A} = \frac{1}{b} \ln\left(\tan \frac{A}{2}\right) + \frac{1}{b \cos A}$$

$$\int \frac{dx}{\sin A \cos^n A} = \frac{1}{b(n-1)\cos^{n-1} A} + \int \frac{dx}{\sin A \cos^{n-2} A} \qquad n \neq 1$$

$$\int \frac{dx}{\sin^2 A \cos A} = -\frac{1}{b \sin A} + \frac{1}{b} \ln\left[\tan\left(\frac{\pi}{4} + \frac{A}{2}\right)\right]$$

$$\int \frac{dx}{\sin^2 A \cos^2 A} = -\frac{2}{b} \cot 2A$$

$$\int \frac{dx}{\sin^2 A \cos^n A} = \frac{1 - n\cos^2 A}{b(n-1)\sin A \cos^{n-1} A} + \frac{n(n-2)}{n-1} \int \frac{dx}{\cos^{n-2} A} \qquad n \neq 1$$

$$\int \frac{dx}{\sin^3 A \cos A} = \frac{-1}{2b \sin^2 A} + \frac{1}{b} \ln(\tan A)$$

$$\int \frac{dx}{\sin^3 A \cos^2 A} = \frac{1}{b \cos A} - \frac{\cos A}{2b \sin^2 A} + \frac{3}{2b} \ln\left(\tan \frac{A}{2}\right)$$

$$\int \frac{dx}{\sin^3 A \cos^n A} = \frac{2 - (n+1)\cos^2 A}{2b(n-1)\sin^2 A \cos^{n-1} A} + \int \frac{(n+1)\,dx}{2 \sin A \cos^{n-2} A} \qquad n \neq 1$$

$$\int \frac{dx}{\sin^m A \cos A} = \frac{-1}{b(m-1)\sin^{m-1} A} + \int \frac{dx}{\sin^{m-2} A \cos A} \qquad m \neq 1$$

$$\int \frac{dx}{\sin^m A \cos^n A} = \begin{cases} \dfrac{-1}{b(m-1)\sin^{m-1} A \cos^{n-1} A} + \dfrac{m+n-2}{m-1} \displaystyle\int \frac{dx}{\sin^{m-2} A \cos^n A} & m \neq 1 \\[2ex] \dfrac{1}{b(n-1)\sin^{m-1} A \cos^{n-1} A} + \dfrac{m+n+2}{n-1} \displaystyle\int \frac{dx}{\sin^m A \cos^{n-2} A} & n \neq 1 \end{cases}$$

IV. Indefinite Integrals Involving $f(x)\ \dfrac{\sin^m A}{\cos^n A} \qquad A = bx$

$$\int \frac{\sin A\, dx}{\cos A} = -\frac{\ln(\cos A)}{b} \qquad \int \frac{\sin A\, dx}{\cos^n A} = \frac{1}{b(n-1)\cos^{n-1} A} \qquad n \neq 1$$

$$\int \frac{\sin^2 A\, dx}{\cos A} = -\frac{\sin A}{b} + \frac{1}{b} \ln\left[\tan\left(\frac{\pi}{4} + \frac{A}{2}\right)\right]$$

$$\int \frac{\sin^2 A\, dx}{\cos^2 A} = -x + \frac{1}{b} \tan A$$

$$\int \frac{\sin^2 A\,dx}{\cos^3 A} = \frac{\sin A}{2b\cos^2 A} - \frac{1}{2b}\ln\left[\tan\left(\frac{\pi}{4}+\frac{A}{2}\right)\right]$$

$$\int \frac{\sin^2 A\,dx}{\cos^n A} = -\frac{\sin A}{b(n-1)\cos^{n-1} A} - \frac{1}{n-1}\int \frac{dx}{\cos^{n-2} A} \qquad n \neq 1$$

$$\int \frac{\sin^3 A\,dx}{\cos A} = -\frac{\sin^2 A}{2b} - \frac{1}{b}\ln(\cos A)$$

$$\int \frac{\sin^3 A\,dx}{\cos^2 A} = \frac{\cos A}{b} + \frac{1}{b\cos A}$$

$$\int \frac{\sin^3 A\,dx}{\cos^3 A} = \frac{\tan^2 A}{2b} + \frac{1}{b}\ln(\cos A)$$

$$\int \frac{\sin^3 A\,dx}{\cos^n A} = -\frac{1}{b(n-1)\cos^{n-1} A} - \frac{1}{b(n-3)\cos^{n-3} A} \qquad n \neq 1, 3$$

$$\int \frac{\sin^m A\,dx}{\cos A} = -\frac{\sin^{m-1} A}{b(m-1)} + \int \frac{\sin^{m-2} A\,dx}{\cos A} \qquad m \neq 1$$

$$\int \frac{\sin^m A\,dx}{\cos^n A} = \begin{cases} -\dfrac{\sin^{m-1} A}{b(m-n)\cos^{n-1} A} + \dfrac{m-1}{m-n}\displaystyle\int \frac{\sin^{m-2} A\,dx}{\cos^n A} & m \neq n \\[2ex] \dfrac{\sin^{m-1} A}{b(n-1)\cos^{n-1} A} - \dfrac{m-1}{n-1}\displaystyle\int \frac{\sin^{m-2} A\,dx}{\cos^{n-2} A} & n \neq 1 \end{cases}$$

---

lvi. Indefinite Integrals Involving $f(x) = \dfrac{\cos^n A}{\sin^m A}$ $\quad A = bx$

$$\int \frac{\cos A\,dx}{\sin A} = \frac{\ln(\sin A)}{b} \qquad \int \frac{\cos A\,dx}{\sin^m A} = \frac{-1}{b(m-1)\sin^{m-1} A} \qquad m \neq 1$$

$$\int \frac{\cos^2 A\,dx}{\sin A} = \frac{\cos A}{b} + \frac{1}{b}\ln\left(\tan\frac{A}{2}\right)$$

$$\int \frac{\cos^2 A\,dx}{\sin^2 A} = -x - \frac{1}{b}\cot A$$

$$\int \frac{\cos^2 A\,dx}{\sin^3 A} = -\frac{\cos A}{2b\sin^2 A} - \frac{1}{2b}\ln\left(\tan\frac{A}{2}\right)$$

$$\int \frac{\cos^2 A\,dx}{\sin^m A} = \frac{-\cos A}{b\,(m-1)\sin^{m-1} A} - \frac{1}{m-1}\int \frac{dx}{\sin^{m-2} A} \qquad m \neq 1$$

$$\int \frac{\cos^3 A\,dx}{\sin A} = \frac{\cos^2 A}{2b} + \frac{1}{b}\ln(\sin A)$$

$$\int \frac{\cos^3 A\,dx}{\sin^2 A} = -\frac{\sin A}{b} - \frac{1}{b\sin A}$$

$$\int \frac{\cos^3 A\,dx}{\sin^3 A} = \frac{-\cot^2 A}{2b} - \frac{1}{b}\ln(\sin A)$$

$$\int \frac{\cos^3 A\,dx}{\sin^m A} = \frac{-1}{b\,(m-1)\sin^{m-1} A} + \frac{1}{b\,(m-3)\sin^{m-3} A} \qquad m \neq 1, 3$$

$$\int \frac{\cos^n A\,dx}{\sin A} = \frac{\cos^{n-1} A}{b\,(n-1)} + \int \frac{\cos^{n-2} A\,dx}{\sin A} \qquad n \neq 1$$

$$\int \frac{\cos^n A\,dx}{\sin^m A} = \begin{cases} -\dfrac{\cos^{n-1} A}{b\,(m-n)\sin^{m-1} A} - \dfrac{n-1}{m-n}\displaystyle\int \frac{\cos^{n-2} A\,dx}{\sin^m A} & m \neq n \\ -\dfrac{\cos^{n-1} A}{b\,(m-1)\sin^{m-1} A} - \dfrac{n-1}{m-1}\displaystyle\int \frac{\cos^{n-2} A\,dx}{\sin^{m-2} A} & n \neq 1 \end{cases}$$

---

lvii. Indefinite Integrals Involving $f(x) = f(\sin A, 1 \pm a\cos A)$ $\quad A = bx$

$$\int \frac{\sin A\,dx}{1 \pm a\cos A} = \mp \frac{\ln(1 \pm \cos A)}{ab}$$

$$\int \frac{\sin A\,dx}{(1 \pm a\cos A)^2} = \pm \frac{1}{ab\,(1 \pm a\cos A)}$$

$$\int \frac{\sin A\,dx}{(1 \pm a\cos A)^n} = \pm \frac{1}{ab\,(n-1)\,(1 \pm a\cos A)^{n-1}}$$

$$\int \frac{dx}{(1 \pm \cos A)\sin A} = \frac{\mp 1}{2b\,(1 \pm \cos A)} + \frac{1}{2b}\ln\left(\tan\frac{A}{2}\right)$$

$$\int \frac{\cos A\,dx}{(1 \pm \cos A)\sin A} = \frac{-1}{2b\,(1 \pm \cos A)} \pm \frac{1}{2b}\ln\left(\tan\frac{A}{2}\right)$$

$$\int \frac{\sin A\,dx}{(1 \pm \cos A)\cos A} = \frac{1}{b}\ln\frac{1 \pm \cos A}{\cos A}$$

lviii. Indefinite Integrals Involving $f(x) = f(\cos A, 1 \pm a \sin A) \qquad A = bx$

$$\int \frac{\cos A\, dx}{1 \pm a \sin A} = \pm \frac{\ln (1 \pm a \sin A)}{ab}$$

$$\int \frac{\cos A\, dx}{(1 \pm a \sin A)^2} = \pm \frac{1}{ab\,(1 \pm a \sin A)}$$

$$\int \frac{\cos A\, dx}{(1 \pm a \sin A)^n} = \mp \frac{1}{ab\,(n-1)\,(1 \pm a \sin A)^{n-1}}$$

$$\int \frac{dx}{(1 \pm \sin A) \cos A} = \frac{\mp 1}{2b\,(1 \pm \sin A)} + \frac{1}{2b} \ln \left[\tan \left(\frac{\pi}{4} + \frac{A}{2} \right) \right]$$

$$\int \frac{\sin A\, dx}{(1 \pm \sin A) \cos A} = \frac{1}{2b\,(1 \pm \sin A)} \pm \frac{1}{2b} \ln \left[\tan \left(\frac{\pi}{4} + \frac{A}{2} \right) \right]$$

$$\int \frac{\cos A\, dx}{(1 \pm \sin A) \sin A} = -\frac{1}{b} \ln \frac{1 \pm \sin A}{\sin A}$$

lix. Indefinite Integrals Involving $f(x) = f(p \sin A \pm q \cos A) \qquad A = bx$

$$\int \frac{dx}{p \sin A \pm q \cos A} = \frac{1}{b\sqrt{p^2 + q^2}} \ln \left[\tan \frac{A \pm \tan^{-1}(q/p)}{2} \right]$$

$$\int \frac{dx}{(p \sin A \pm q \cos A)^2} = \frac{1}{b\,(p^2 + q^2)} \tan \frac{A \pm \tan^{-1}(q/p)}{2}$$

$$\int \frac{\sin A\, dx}{p \sin A \pm q \cos A} = \frac{p}{b\,(p^2 + q^2)} [A \mp \ln \sqrt[p]{(p \sin A \pm q \cos A)^q}\,]$$

$$\int \frac{\cos A\, dx}{p \sin A \pm q \cos A} = \frac{\pm p}{b\,(p^2 + q^2)} [A \pm \ln \sqrt[p]{(p \sin A \pm q \cos A)^q}\,]$$

$$\int \frac{dx}{1 + p \sin A + q \cos A} = \frac{2}{b\sqrt{1 - p^2 - q^2}} \tan^{-1} \frac{p + (1 \mp q) \tan (A/2)}{\sqrt{1 - p^2 - q^2}}$$

$$\int \frac{dx}{p \sin A + q\,(1 + \cos A)} = \frac{1}{bp} \ln \left(q + p \tan \frac{A}{2} \right)$$

lx. Indefinite Integrals Involving $f(x) = f(p^2 \sin^2 A \pm q^2 \cos^2 A) \qquad A = bx$

$$\int \frac{dx}{p^2 \sin^2 A + q^2 \cos^2 A} = \frac{1}{bpq} \tan^{-1} \frac{p \tan A}{q}$$

$$\int \frac{dx}{p^2 \sin^2 A - q^2 \cos^2 A} = \frac{1}{2bpq} \ln \frac{p \tan A - q}{p \tan A + q}$$

$$\int \frac{dx}{(p^2 \sin^2 A + q^2 \sin A \cos A + r^2 \cos^2 A)} = \frac{1}{b} \int \frac{dz}{p^2 z^2 + q^2 z + r^2} \qquad \tan A = z$$

$$\int \frac{dx}{(p^2 \sin^2 x + q^2 \cos^2 x)^2} = \frac{(p^2 + q^2)}{2p^3 q^3} - \frac{(q^2 - p^2) \sin z \cos z}{2p^3 q^3} \qquad \tan z = \frac{p}{q} \tan x$$

lxi. Indefinite Integrals Involving $f(x) = f(x, \tan A)$ $A = bx$

$$\int \tan A \, dx = -\frac{\ln(\cos A)}{b}$$

$$\int \tan^2 A \, dx = \frac{\tan A}{b} - x$$

$$\int \tan^3 A \, dx = \frac{\tan^2 A}{2b} + \frac{\ln(\cos A)}{b}$$

$$\int \tan^n A \, dx = \frac{\tan^{n-1} A}{(n-1) b} - \int \tan^{n-2} A \, dx \qquad n > 1$$

$$\int \frac{dx}{\tan A} = \frac{\ln(\sin A)}{b} \qquad \int \frac{dx}{\cos^2 A \tan A} = \frac{\ln(\tan A)}{b}$$

$$\int \frac{dx}{\tan^n A} = \frac{-1}{(n-1) b \tan^{n-1} A} - \int \frac{dx}{\tan^{n-2} A} \qquad n > 1$$

$$\int x \tan A \, dx = \frac{1}{b^2} \left[\frac{A^3}{3} + \frac{A^2}{15} + \frac{2A^3}{105} + \dots + \frac{2^{2k} (2^{2k} - 1) B_k A^{2k+1}}{(2k+1)!} + \dots \right]$$

$$\int \frac{\tan A \, dx}{x} = A \left[1 + \frac{A^2}{9} + \frac{2A^4}{75} + \dots + \frac{2^{2k} (2^{2k} - 1) B_k A^{2k-2}}{(2k-1)(2k)!} + \dots \right]$$

(B_k = Bernoulli's number)

$$\int \frac{dx}{p + q \tan A} = \frac{1}{b(p^2 + q^2)} [pA + q \ln(q \sin A + p \cos A)]$$

$$\int \frac{dx}{\sqrt{p + q \tan^2 A}} = \frac{1}{b\sqrt{p - q}} \sin^{-1} \left(\sqrt{\frac{p - q}{p}} \sin A \right) \qquad p > q$$

lxii. Indefinite Integrals Involving $f(x) = f(x, \cot A)$ $A = bx$

$$\int \cot A\, dx = \frac{\ln(\sin A)}{b}$$

$$\int \cot^2 A\, dx = \frac{\cot A}{b} - x$$

$$\int \cot^3 A\, dx = \frac{-\cot^2 A}{2b} - \frac{\ln(\sin A)}{b}$$

$$\int \cot^n A\, dx = -\frac{\cot^{n-1} A}{(n-1)\, b} - \int \cot^{n-2} A\, dx \qquad n > 1$$

$$\int \frac{dx}{\cot A} = -\frac{\ln(\cos A)}{b} \qquad \int \frac{dx}{\sin^2 A \cot A} = -\frac{\ln(\cot A)}{b}$$

$$\int \frac{dx}{\cot^n A} = \frac{1}{(n-1)\, b \cot^{n-1} A} - \int \frac{dx}{\cot^{n-2} A} \qquad n > 1$$

$$\int x \cot A\, dx = \frac{1}{b^2}\left[A - \frac{A^3}{9} - \frac{A^5}{225} - \dots - \frac{2^{2k} B_k A^{2k+1}}{(2k+1)!} - \dots\right]$$

$$\int \frac{\cot A\, dx}{x} = -\frac{1}{A}\left[1 + \frac{A^2}{3} - \frac{A^4}{135} + \dots + \frac{2^{2k} B_k A^{2k}}{(2k-1)(2k)!} + \dots\right]$$

(B_k is Bernoulli's number)

$$\int \frac{dx}{p + q \cot A} = \frac{1}{b(p^2 + q^2)}[pA - q \ln(p \sin A + q \cos A)]$$

$$\int \frac{dx}{\sqrt{p + q \cot^2 A}} = \frac{1}{b\sqrt{p-q}} \cos^{-1}\left(\sqrt{\frac{p-q}{p}} \cos A\right) \qquad p > q$$

---

lxiii. Indefinite Integrals Involving $f(x) = f(x, \sin^{-1} B)$ $B = \frac{x}{b}$

$$\int \sin^{-1} B\, dx = x \sin^{-1} B + \sqrt{b^2 - x^2}$$

$$\int x \sin^{-1} B\, dx = \left(\frac{x^2}{2} - \frac{b^2}{4}\right) \sin^{-1} B + \frac{x\sqrt{b^2 - x^2}}{4}$$

$$\int x^2 \sin^{-1} B\, dx = \frac{x^3}{3} \sin^{-1} B + \frac{(2b^2 + x^2)\sqrt{b^2 - x^2}}{9}$$

$$\int x^m \sin^{-1} B\,dx = \frac{x^{m+1}}{m+1}\sin^{-1} B - \frac{1}{m+1}\int \frac{x^{m+1}}{\sqrt{b^2 - x^2}}\,dx \qquad m \neq -1$$

$$\int \frac{\sin^{-1} B\,dx}{x} = B + \frac{B^3}{(2)(3)(3)} + \frac{(1)(3)B^5}{(2)(4)(5)(5)} + \frac{(1)(3)(5)B^7}{(2)(4)(6)(7)(7)} + \ldots$$

$$\int \frac{\sin^{-1} B\,dx}{x^2} = -\frac{\sin^{-1} B}{x} - \frac{1}{b}\ln\frac{b + \sqrt{b^2 - x^2}}{x}$$

$$\int (\sin^{-1} B)^2\,dx = x(\sin^{-1} B)^2 - 2x + 2\sqrt{b^2 - x^2}\sin^{-1} B$$

lxiv. Indefinite Integrals Involving $f(x) = f(x, \tan^{-1} B)$ $\quad B = \dfrac{x}{b}$

$$\int \tan^{-1} B\,dx = x\tan^{-1} B - b\ln\sqrt{b^2 + x^2}$$

$$\int x\tan^{-1} B\,dx = \frac{b^2 + x^2}{2}\tan^{-1} B - \frac{bx}{2}$$

$$\int x^m \tan^{-1} B\,dx = \frac{x^{m+1}}{m+1}\tan^{-1} B - \frac{b}{m+1}\int \frac{x^{m+1}\,dx}{b^2 + x^2} \qquad m \neq -1$$

$$\int \frac{\tan^{-1} B\,dx}{x} = B - \frac{B^3}{3^2} + \frac{B^5}{5^2} - \frac{B^7}{7^2} + \ldots.$$

$$\int \frac{\tan^{-1} B\,dx}{x^2} = -\frac{1}{b}\left(\frac{\tan^{-1} B}{B} + \ln\frac{\sqrt{1 + B^2}}{B}\right)$$

lxv. Indefinite Integrals Involving $f(x) = f(x, \cos^{-1} B)$ $\quad B = \dfrac{x}{b}$

$$\int \cos^{-1} B\,dx = x\cos^{-1} B - \sqrt{b^2 - x^2}$$

$$\int x\cos^{-1} B\,dx = \left(\frac{x^2}{2} - \frac{b^2}{4}\right)\cos^{-1} B - \frac{x\sqrt{b^2 - x^2}}{4}$$

$$\int x^2\cos^{-1} B\,dx = \frac{x^3}{3}\cos^{-1} B - \frac{(2b^2 + x^2)\sqrt{b^2 - x^2}}{9}$$

$$\int x^m \cos^{-1} B\,dx = \frac{x^{m+1}}{m+1}\cos^{-1} B + \frac{1}{m+1}\frac{x^{m+1}}{\sqrt{b^2 - x^2}}\,dx \qquad m \neq -1$$

$$\int \frac{\cos^{-1} B\, dx}{x} = \frac{\pi}{2} \ln x - B - \frac{B^3}{(2)(3)(3)} - \frac{(1)(3)B^5}{(2)(4)(5)(5)} - \frac{(1)(3)(5)B^7}{(2)(4)(6)(7)(7)} + \ldots$$

$$\int \frac{\cos^{-1} B\, dx}{x^2} = -\frac{\cos^{-1} B}{x} + \frac{1}{b} \ln \frac{b + \sqrt{b^2 - x^2}}{x}$$

$$\int (\cos^{-1} B)^2\, dx = x(\cos^{-1} B)^2 - 2x - 2\sqrt{b^2 - x^2} \cos^{-1} B$$

lxvi. Indefinite Integrals Involving $f(x) = f(x, \cot^{-1} B)$ $B = \frac{x}{b}$

$$\int \cot^{-1} B\, dx = x \cot^{-1} B + b \ln \sqrt{b^2 + x^2}$$

$$\int x \cot^{-1} B\, dx = \frac{b^2 + x^2}{2} \cot^{-1} B + \frac{bx}{2}$$

$$\int x^m \cot^{-1} B\, dx = \frac{x^{m+1}}{m+1} \cot^{-1} B + \frac{b}{m+1} \int \frac{x^{m+1}}{b^2 + x^2} dx \qquad m \neq -1$$

$$\int \frac{\cot^{-1} B\, dx}{x} = \frac{\pi}{2} \ln x - B + \frac{B^3}{3^2} - \frac{B^5}{5^2} - \frac{B^7}{7^2} + \ldots.$$

$$\int \frac{\cot^{-1} B\, dx}{x^2} = -\frac{1}{b}\left(\frac{\cot^{-1} B}{B} - \ln \frac{\sqrt{1 + B^2}}{B}\right)$$

lxvii. Indefinite Integrals Involving $f(x) = f(x^m, e^{\pm A})$ $A = bx$

$$\int e^A\, dx = \frac{e^A}{b} \qquad \int xe^A\, dx = \frac{(A-1)e^A}{b^2}$$

$$\int x^2 e^A\, dx = \frac{(A^2 - 2A + 2)e^A}{b^3}$$

$$\int x^m e^A\, dx = \frac{x^m e^A}{b} - \frac{m}{b} \int x^{m-1} e^A\, dx$$

$$\int \frac{e^A}{x} dx = \ln x + \frac{A}{(1)(1!)} + \frac{A^2}{(2)(2!)} + \frac{A^3}{(3)(3!)} + \ldots$$

$$\int \frac{e^A}{x^m} dx = \frac{-e^A}{(m-1)x^{m-1}} + \frac{b}{m-1} \int \frac{e^A}{x^{m-1}} dx$$

$$\int \frac{dx}{1 + ae^A} = \frac{A}{b} - \frac{1}{b} \ln(1 + ae^A)$$

$$\int \frac{dx}{pe^{A} + qe^{-A}} = \begin{cases} \dfrac{1}{b\sqrt{pq}} \tan^{-1} e^{A} \sqrt{\dfrac{p}{q}} & pq > 0 \\[2ex] \dfrac{1}{2b\sqrt{-pq}} \ln \dfrac{q + e^{A}\sqrt{-pq}}{q - e^{A}\sqrt{-pq}} & pq < 0 \end{cases}$$

lxviii. Indefinite Integrals Involving $f(x) = f(x^m, a^A)$ $A = bx$

$$\int a^{A}\, dx = \frac{a^{A}}{b \ln a}$$

$$\int x^{m} a^{A}\, dx = \frac{x^{m} a^{A}}{b \ln a} - \frac{m}{b \ln a} \int x^{m-1} a^{A}\, dx$$

$$\int \frac{a^{A}}{x}\, dx = \ln x + \frac{A \ln a}{(1)(1!)} + \frac{a^{2} (\ln a)^{2}}{(2)(2!)} + \frac{A^{3} (\ln a)^{3}}{(3)(3!)} + \ldots.$$

$$\int \frac{a^{A}}{x^{m}}\, dx = \frac{-a^{A}}{(m-1)\, x^{m-1}} + \frac{b \ln a}{m-1} \int \frac{a^{A}}{x^{m-1}}\, dx \qquad m > 1$$

lxix. Indefinite Integrals Involving $f(x) = f(x, e^{\lambda x}, \sin \alpha x, \cos \alpha x, \sin \beta x, \cos \beta x)$

$$\int e^{\lambda x} \sin \alpha x\, dx = \frac{e^{\lambda x} (\lambda \sin \alpha x - \alpha \cos \alpha x)}{\lambda^{2} + \alpha^{2}}$$

$$\int e^{\lambda x} \cos \alpha x\, dx = \frac{e^{\lambda x} (\lambda \cos \alpha x + \alpha \sin \alpha x)}{\lambda^{2} + \alpha^{2}}$$

$$\int x e^{\lambda x} \sin \alpha x\, dx = \frac{x e^{\lambda x} (\lambda \sin \alpha x - \alpha \cos \alpha x)}{\lambda^{2} + \alpha^{2}} - \frac{e^{\lambda x} [(\lambda^{2} - \alpha^{2}) \sin \alpha x - 2\alpha\lambda \cos \alpha x]}{(\lambda^{2} + \alpha^{2})^{2}}$$

$$\int x e^{\lambda x} \cos \alpha x\, dx = \frac{x e^{\lambda x} (\lambda \cos \alpha x + \alpha \sin \alpha x)}{\lambda^{2} + \alpha^{2}} - \frac{e^{\lambda x} [(\lambda^{2} - \alpha^{2}) \cos \alpha x + 2\alpha\lambda \sin \alpha x]}{(\lambda^{2} + \alpha^{2})^{2}}$$

$$\int e^{\lambda x} \sin \alpha x \sin \beta x\, dx = \frac{e^{\lambda x} [(\alpha - \beta) \sin (\alpha - \beta) x + \lambda \cos (\alpha - \beta) x]}{2[\lambda^{2} + (\alpha - \beta)^{2}]} - \frac{e^{\lambda x} [(\alpha + \beta) \sin (\alpha + \beta) x + \lambda \cos (\alpha + \beta) x]}{2[\lambda^{2} + (\alpha + \beta)^{2}]}$$

$$\int e^{\lambda x} \sin \alpha x \cos \beta x\, dx = \frac{-e^{\lambda x} [(\alpha - \beta) \cos (\alpha - \beta) x - \lambda \sin (\alpha - \beta) x]}{2[\lambda^{2} + (\alpha - \beta)^{2}]}$$

$$-\frac{e^{\lambda x}[(\alpha+\beta)\cos(\alpha+\beta)x-\lambda\sin(\alpha+\beta)x]}{2[\lambda^2+(\alpha+\beta)^2]}$$

$$\int e^{\lambda x}\cos\alpha x\cos\beta x\,dx=\frac{e^{\lambda x}[(\alpha-\beta)\sin(\alpha-\beta)x+\lambda\cos(\alpha-\beta)x]}{2[\lambda^2+(\alpha-\beta)^2]}$$

$$+\frac{e^{\lambda x}[(\alpha+\beta)\sin(\alpha+\beta)x+\lambda\cos(\alpha+\beta)x]}{2[\lambda^2+(\alpha+\beta)^2]}$$

$$\int e^{\lambda x}\sin^n\alpha x\,dx=\frac{e^{\lambda x}\sin^{n-1}\alpha x(\lambda\sin\alpha x-n\alpha\cos\alpha x)}{\lambda^2+n^2\alpha^2}+\frac{n(n-1)\alpha^2}{\lambda^2+n^2\alpha^2}\int e^{\lambda x}\sin^{n-2}\alpha x\,dx$$

$$\int e^{\lambda x}\cos^n\alpha x\,dx=\frac{e^{\lambda x}\cos^{n-1}\alpha x(\lambda\cos\alpha x+n\alpha\sin\alpha x)}{\lambda^2+n^2\alpha^2}+\frac{n(n-1)\alpha^2}{\lambda^2+n^2\alpha^2}\int e^{\lambda x}\cos^{n-2}\alpha x\,dx$$

---

lxx. Indefinite Integrals Involving $f(x)=f(x^m,\sinh A)$ $A=bx$

$$\int\sinh A\,dx=\frac{\cosh A}{b}$$

$$\int\sinh^2 A\,dx=\frac{\sinh A\cosh A}{2b}-\frac{x}{2}$$

$$\int\sinh^n A\,dx=\frac{\sinh^{n-1}A\cosh A}{bn}-\frac{n-1}{n}\int\sinh^{n-2}A\,dx$$

$$\int x\sinh A\,dx=\frac{x\cosh A}{b}-\frac{\sinh A}{b^2}$$

$$\int x^2\sinh A\,dx=\left(\frac{x^2}{b}+\frac{2}{b^3}\right)\cosh A-\frac{2x}{b^2}\sinh A$$

$$\int x^m\sinh A\,dx=\frac{x^m\cosh A}{b}-\frac{m}{b}\int x^{m-1}\cosh A\,dx$$

$$\int\frac{\sinh A\,dx}{x}=A+\frac{A^3}{(3)(3!)}+\frac{A^5}{(5)(5!)}+\dots.$$

$$\int\frac{\sinh A\,dx}{x^2}=-\frac{\sinh A}{x}+b\left(\ln x+\frac{A^2}{(2)(2!)}+\frac{A^4}{(4)(4!)}+\dots\right)$$

$$\int\frac{\sinh A\,dx}{x^m}=-\frac{-\sinh A}{(m-1)x^{m-1}}+\frac{b}{m-1}\int\frac{\cosh A}{x^{m-1}}\qquad m\neq 1$$

$$\int \frac{dx}{\sinh A} = \frac{\ln[\tanh(A/2)]}{b}$$

$$\int \frac{dx}{\sinh^2 A} = -\frac{\coth A}{b}$$

$$\int \frac{dx}{\sinh^n A} = \frac{-\cosh A}{b(n-1)\sinh^{n-2} A} - \frac{n-2}{n-1}\int \frac{dx}{\sinh^{n-2} A} \qquad n \neq 1$$

$$\int \frac{x\,dx}{\sinh^n A} = \frac{-x\cosh A}{b(n-1)\sinh^{n-1} A} - \frac{1}{b^2(n-1)(n-2)\sinh^{n-2} A} - \frac{n-2}{n-1}\int \frac{x\,dx}{\sinh^{n-2} A} \qquad n \neq 1, 2$$

---

lxxi. Indefinite Integrals Involving $f(x) = f(x^m, \cosh A)$ $A = bx$

$$\int \cosh A\,dx = \frac{\sinh A}{b}$$

$$\int \cosh^2 A\,dx = \frac{\sinh A \cosh A}{2b} + \frac{x}{2}$$

$$\int \cosh^n A\,dx = \frac{\cosh^{n-1} A \sinh A}{bn} + \frac{n-1}{n}\int \cosh^{n-2} A\,dx$$

$$\int x\cosh A\,dx = \frac{x\sinh A}{b} - \frac{\cosh A}{b^2}$$

$$\int x^2 \cosh A\,dx = \left(\frac{x^2}{b} + \frac{2}{b^3}\right)\sinh A - \frac{2x}{b^2}\cosh A$$

$$\int x^m \cosh A\,dx = \frac{x^m \sinh A}{b} - \frac{m}{b}\int x^{m-1}\sinh A\,dx$$

$$\int \frac{\cosh A\,dx}{x} = \ln x + \frac{A^2}{(2)(2!)} + \frac{A^4}{(4)(4!)} + ...$$

$$\int \frac{\cosh A\,dx}{x^2} = -\frac{\cosh A}{x} + b\left[A + \frac{A^3}{(3)(3!)} + \frac{A^5}{(5)(5!)} + ...\right]$$

$$\int \frac{\cosh A\,dx}{x^m} = -\frac{\cosh A}{(m-1)x^{m-1}} + \frac{b}{m-1}\int \frac{\sinh A}{x^{m-1}}dx \qquad m \neq 1$$

$$\int \frac{dx}{\cosh A} = \frac{\tan^{-1}(\sinh A)}{b}$$

$$\int \frac{dx}{\cosh^2 A} = \frac{\tanh A}{b}$$

$$\int \frac{dx}{\cosh^n A} = \frac{\sinh A}{b(n-1)\cosh^{n-1} A} + \frac{n-2}{n-1}\int \frac{dx}{\cosh^{n-2} A} \qquad n \neq 1$$

$$\int \frac{x\,dx}{\cosh^n A} = \frac{x \sinh A}{b(n-1)\cosh^{n-1} A} + \frac{1}{b^2(n-1)(n-2)\cosh^{n-2} A} + \frac{n-2}{n-1}\int \frac{x\,dx}{\cosh^{n-2} A} \qquad n \neq 1, 2$$

lxxii. Indefinite Integrals Involving $f(x) = f(\sinh A, \cosh A)$ $A = bx$

$$\int \sinh A \cosh A\, dx = \frac{\sinh^2 A}{2b}$$

$$\int \sinh^2 A \cosh^2 A\, dx = \frac{\sinh 4A}{32b} - \frac{x}{8}$$

$$\int \sinh^n A \cosh A\, dx = \frac{\sinh^{n+1} A}{(n+1)b} \qquad n \neq -1$$

$$\int \sinh A \cosh^n A\, dx = \frac{\cosh^{n+1} A}{(n+1)b} \qquad n \neq -1$$

$$\int \frac{dx}{\sinh A \cosh A} = \frac{\ln(\tanh A)}{b}$$

$$\int \frac{dx}{\sinh^2 A \cosh A} = -\frac{\tan^{-1}(\sinh A) + \operatorname{csch} A}{b}$$

$$\int \frac{dx}{\sinh A \cosh^2 A} = \frac{\ln[\tanh(A/2) + \operatorname{sech} A}{b}$$

$$\int \frac{dx}{\sinh^2 A \cosh^2 A} = -\frac{2\coth 2A}{b}$$

$$\int \frac{\sinh^2 A}{\cosh A}\, dx = \frac{\sinh A - \tan^{-1}(\sinh A)}{b}$$

$$\int \frac{\cosh^2 A}{\sinh A}\, dx = \frac{\cosh A + \ln[\tanh(A/2)]}{b}$$

$$\int \frac{\sinh A}{\cosh^n}\, dx = -\frac{1}{(n-1)b\cosh^{n-1} A} \qquad n \neq 1$$

$$\int \frac{\cosh A}{\sinh^n}\, dx = -\frac{1}{(n-1)b\sinh^{n-1} A} \qquad n \neq 1$$

$$\int \frac{\sinh A}{\cosh A \pm 1}\, dx = \frac{1}{b} \ln (\cosh A \pm 1)$$

lxxiii. Indefinite Integrals Involving $f(x) = f(\sinh \alpha x, \cosh \alpha x, \sinh \beta x, \cosh \beta x)$

$$\left.\begin{aligned}
\int \sinh \alpha x \sinh \beta x &= \frac{\sinh (\alpha + \beta)x}{2(\alpha + \beta)} - \frac{\sinh (\alpha - \beta)x}{2(\alpha - \beta)} \\
\int \sinh \alpha x \cosh \beta x &= \frac{\cosh (\alpha + \beta)x}{2(\alpha + \beta)} + \frac{\cosh (\alpha - \beta)x}{2(\alpha - \beta)} \\
\int \cosh \alpha x \cosh \beta x &= \frac{\sinh (\alpha + \beta)x}{2(\alpha + \beta)} + \frac{\sinh (\alpha - \beta)x}{2(\alpha - \beta)}
\end{aligned}\right] \quad \alpha^2 \neq \beta^2$$

lxxiv. Indefinite Integrals Involving $f(x) = f(\sinh \alpha x, \cosh \alpha x, \sin \beta x, \cos \beta x)$

$$\int \sinh \alpha x \sinh \beta x = \frac{\alpha \cosh \alpha x \sin \beta x - \beta \sinh \alpha x \cos \beta x}{\alpha^2 + \beta^2}$$

$$\int \sinh \alpha x \cosh \beta x = \frac{\alpha \cosh \alpha x \cos \beta x + \beta \sinh \alpha x \sin \beta x}{\alpha^2 + \beta^2}$$

$$\int \cosh \alpha x \sin \beta x = \frac{\alpha \sinh \alpha x \sin \beta x - \beta \cosh \alpha x \cos \beta x}{\alpha^2 + \beta^2}$$

$$\int \cosh \alpha x \cos \beta x = \frac{\alpha \sinh \alpha x \cos \beta x + \beta \cosh \alpha x \sin \beta x}{\alpha^2 + \beta^2}$$

lxxv. Indefinite Integrals Involving $f(x) = f(\tanh A, \coth A)$ $A = bx$

$$\int \tanh A\, dx = \frac{\ln (\cosh A)}{b} \qquad \int \tanh^2 A\, dx = x - \frac{\tanh A}{b}$$

$$\int \tanh^n A\, dx = -\frac{\tanh^{n-1} A}{b(n-1)} + \int \tanh^{n-2} A\, dx \qquad n \neq 1$$

$$\int \coth A\, dx = \frac{\ln (\sinh A)}{b} \qquad \int \coth^2 A\, dx = x - \frac{\coth A}{b}$$

$$\int \coth^n A\, dx = -\frac{\coth^{n-1} A}{b(n-1)} + \int \coth^{n-2} A\, dx \qquad n \neq 1$$

lxxvi. Indefinite Integrals Involving $f(x) = f(x, \sinh^{-1} B)$ $B = \frac{x}{b}$

$$\int \sinh^{-1} B\, dx = x \sinh^{-1} B - \sqrt{x^2 + b^2}$$

$$\int x \sinh^{-1} B\, dx = \left(\frac{x^2}{2} + \frac{b^2}{4}\right) \sinh^{-1} B - \frac{x\sqrt{x^2 + b^2}}{4}$$

$$\int x^2 \sinh^{-1} B\, dx = \frac{x^3}{3} \sinh^{-1} B + \frac{(2b^2 - x^2)\sqrt{x^2 + b^2}}{9}$$

$$\int x^m \sinh^{-1} B\, dx = \frac{x^{m+1}}{m+1} \sinh^{-1} B - \frac{1}{m+1} \int \frac{x^{m+1}}{\sqrt{x^2 + b^2}}\, dx \qquad m \neq -1$$

$$\int \frac{\sinh^{-1} B\, dx}{x} = B - \frac{B^3}{(2)(3)(3)} + \frac{(1)(3)B^5}{(2)(4)(5)(5)} - \frac{-(1)(3)(5)B^7}{(2)(4)(6)(7)(7)} + \ldots \qquad x^2 < b^2$$

$$\int \frac{\sinh^{-1} B\, dx}{x^2} = -\frac{\sinh^{-1} B}{x} - \frac{1}{b} \ln \frac{b + \sqrt{x^2 + b^2}}{x}$$

---

lxxvii. Indefinite Integrals Involving $f(x) = f(x, \tanh^{-1} B)$ $\qquad B = \frac{x}{b}$

$$\int \tanh^{-1} B\, dx = x \tanh^{-1} B + b \ln \sqrt{b^2 - x^2}$$

$$\int x \tanh^{-1} B\, dx = \frac{x^2 - b^2}{2} \tanh^{-1} B + \frac{bx}{2}$$

$$\int x^m \tanh^{-1} B\, dx = \frac{x^{m+1}}{m+1} \tanh^{-1} B - \frac{b}{m+1} \int \frac{x^{m+1}\, dx}{b^2 - x^2} \qquad m \neq 1$$

$$\int \frac{\tanh^{-1} B\, dx}{x} = B + \frac{B^3}{3^2} + \frac{B^5}{5^2} + \frac{B^7}{7^2} + \ldots.$$

$$\int \frac{\tanh^{-1} B\, dx}{x^2} = -\frac{1}{b}\left(\frac{\tanh^{-1} B}{B} + \ln \frac{\sqrt{1 - B^2}}{B}\right)$$

---

lxxviii. Indefinite Integrals Involving $f(x) = f(x, \cosh^{-1} B)$ $\qquad B = \frac{x}{b}$

$$\int \cosh^{-1} B\, dx = x \cosh^{-1} B \mp \sqrt{x^2 - b^2}$$

$$\int x \cosh^{-1} B\, dx = \left(\frac{x^2}{2} + \frac{b^2}{2}\right) \cosh^{-1} B \mp \frac{x\sqrt{x^2 - b^2}}{4}$$

$$\int x^2 \cosh^{-1} B\, dx = \frac{x^3}{3} \cosh^{-1} B \mp \frac{(2b^2 + x^2)\sqrt{x^2 - b^2}}{9}$$

$$\int x^m \cosh^{-1} B\, dx = \frac{x^{m+1}}{m+1} \cosh^{-1} B \mp \frac{1}{m+1} \int \frac{x^{m+1}\, dx}{\sqrt{x^2 - b^2}} \qquad m \neq -1$$

$-$ if $\cosh^{-1} B > 0$
$+$ if $\cosh^{-1} B < 0$

$$\int \frac{\cosh^{-1} B\,dx}{x} = \mp\left[\frac{1}{2}(\ln 2B)^2 + \frac{B^2}{(2)(2)(2)} + \frac{(1)(3)B^4}{(2)(4)(4)(4)} + \frac{(1)(3)(5)B^6}{(2)(4)(6)(6)(6)} + \ldots\right] \quad \begin{matrix} -\text{ if } \cosh^{-1} B < 0 \\ +\text{ if } \cosh^{-1} B > 0 \end{matrix}$$

$$\int \frac{\cosh^{-1} B\,dx}{x^2} = -\frac{\cosh^{-1} B}{x} \mp \frac{1}{b}\ln\frac{b+\sqrt{x^2+b^2}}{x}$$

lxxix. Indefinite Integrals Involving $f(x) = f(x, \coth^{-1} B)$ $B = \dfrac{x}{b}$

$$\int \coth^{-1} B\,dx = x\coth^{-1} B + b\ln\sqrt{x^2-b^2}$$

$$\int x\coth^{-1} B\,dx = \frac{x^2-b^2}{2}\coth^{-1} B + \frac{bx}{2}$$

$$\int x^m \coth^{-1} B\,dx = \frac{x^{m+1}}{m+1}\coth^{-1} B - \frac{b}{m+1}\int \frac{x^{m+1}}{x^2-b^2}\,dx \qquad m \neq -1$$

$$\int \frac{\coth^{-1} B\,dx}{x} = -B - \frac{B^3}{3^2} - \frac{B^5}{5^2} - \frac{B^7}{7^2}\ldots.$$

$$\int \frac{\coth^{-1} B\,dx}{x^2} = -\frac{1}{b}\left(\frac{\coth^{-1} B}{B} + \ln\frac{\sqrt{B^2-1}}{B}\right)$$

lxxx. Indefinite Integrals Involving $f(x) = f(x, \ln A)$ $A = bx$

$$\int \ln A\,dx = x(\ln A - 1)$$

$$\int x\ln A\,dx = \left(\frac{x}{2}\right)^2 [\ln(A^2) - 1]$$

$$\int x^m \ln A\,dx = \frac{x^{m+1}}{(m+1)^2}(\ln A^{m+1} - 1) \qquad m \neq -1$$

CHAPTER

22

Digital Logic

22.1 AND GATE

A *logic gate* is a device that controls the flow of information, usually in the form of pulses. The symbol for an AND gate is shown in Fig. 22.1. A· B is read "A AND B." As indicated in the truth table, an output appears only when there are inputs at A AND B.

22.2 OR GATE

The symbol for an OR gate is shown in Fig. 22.2, where A + B is read "A OR B." As indicated in the truth table, the output is 1 if input A OR input B is 1. For no input, the output is zero (0).

22.3 NOT GATE

The logic NOT is represented by the symbol in Fig. 22.3, where $\bar{A}$ is read "NOT A." As indicated in the truth table the NOT element is an *inverter*; the output is the *complement* of the single input.

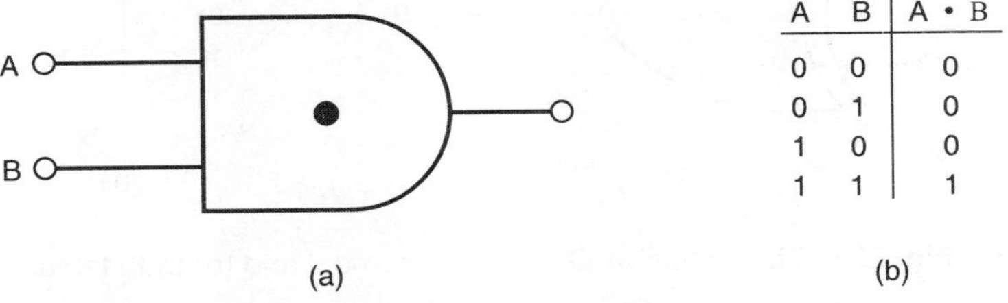

A	B	A • B
0	0	0
0	1	0
1	0	0
1	1	1

Fig. 22.1 (a) Symbol and (b) truth table for the AND gate

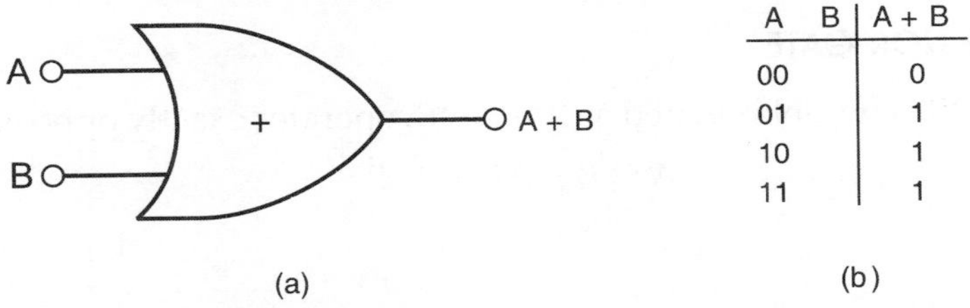

A B	A + B
00	0
01	1
10	1
11	1

Fig. 22.2 A two-input OR gate (a) symbol and (b) truth table

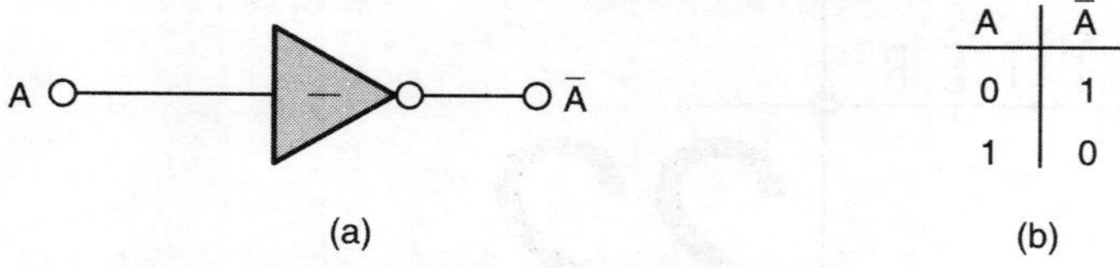

A	$\bar{A}$
0	1
1	0

(a) (b)

Fig. 22.3 A NOT gate (a) symbol and (b) truth table

22.4 NAND GATE

The NAND gate is defined by the truth table of Fig. 22.4. The circle on the NAND element symbol and the bar on the $\overline{A \cdot B}$ output indicate the inversion process.

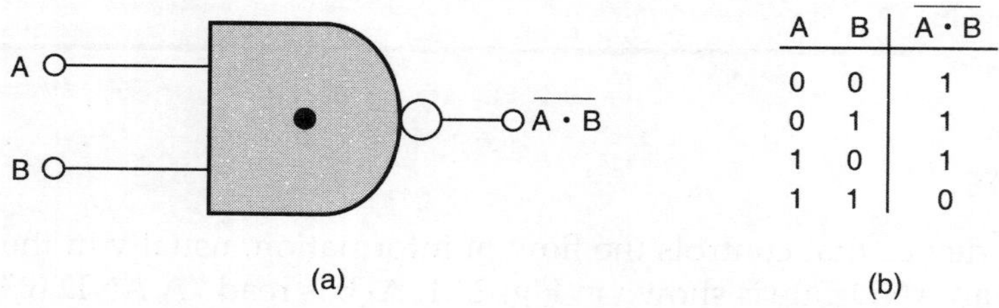

A	B	$\overline{A \cdot B}$
0	0	1
0	1	1
1	0	1
1	1	0

(a) (b)

Fig. 22.4 The NAND gate (a) symbol and (b) truth table

22.5 EXCLUSIVE-OR GATE

The Exclusive-OR operation is $(A + B)\ \overline{AB}$ As shown in Fig. 22.5. The Exclusive-OR gate is used so free

$$A \text{ XOR } B = A \oplus B = (A + B)\ \overline{AB}$$

A B + A ⊕ B

A B	A ⊕ B
00	0
01	1
10	1
11	0

(a) (b)

Fig. 22.5 The exclusive-OR gate (a) symbol and (b) truth table

quently that it is represented by the special symbol ⊕ defined by

22.6 EXCLUSIVE NOR GATE

The exclusive NOR Gate, abbreviated as EX- NOR, operates exactly opposite to EX- OR gate.

$$Y = A \oplus B = AB + \bar{A} \cdot \bar{B}$$

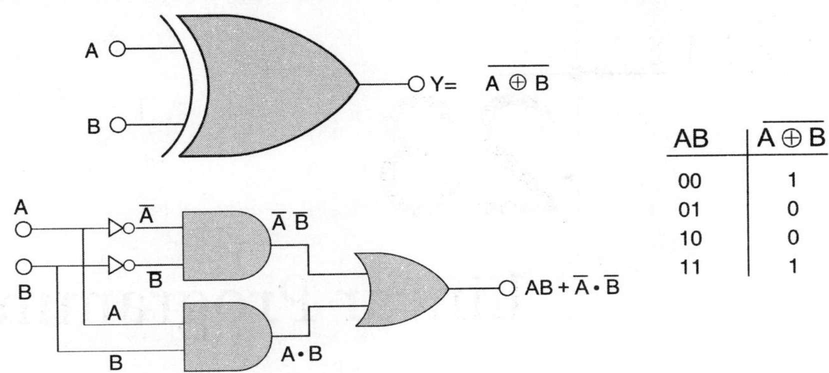

AB	$\overline{A \oplus B}$
00	1
01	0
10	0
11	1

Fig. 22.6 The Exclusive-NOR gate (a) symbol and (b) truth table.

This circuit is also called an equality detector as its output is 1 only when the inputs are equal.

22.7 DEMORGAN'S THEOREM

De Morgans Theorem states that "to obtain the inverse of any Boolean function, invert all variables and replace all ORs by ANDs and all ANDs by ORs."

The first De Morgan theorem says that a NOR gate ($\overline{A+B}$) is equivalent to an AND gate with NOT circuits in the inputs ($\bar{A} \cdot \bar{B}$). The second says that a NAND gate ($\bar{A} \cdot \bar{B}$) is equivalent to an OR gate with NOT circuits in the inputs ($\bar{A} + \bar{B}$).

CHAPTER

23

Linear Programming

LINEAR PROGRAMMING

Mathematical programming consists of methods for solving optimization problems with constraints i.e. for finding a maximum or (a minimum) of the objective function $z = f(x_1 ..., x_n)$ satisfying the constraints.

Linear Programming or linear optimization means mathematical programming in which the objective function is a linear function

$$z = f(x_1 ... x_n) = a_1 x_1 + a_2 x_2 + ... a_n x_n.$$

and the constraints are linear inequalities.

DEFINITIONS

Solution

A set of values of the decision variables which satisfy the constraints of a linear programming problems (*LPP*) is called a solution of the *LPP*.

Feasible Solution

A solution of a *LPP* which also satisfy the non-negativity restrictions of the problems is called its Feasible Solution. The set of all Feasible Solutions of a *LPP* is called the Feasible region.

Optical Solution

A feasible solution which optimises (maximise or minimise) the objective function of a *LPP* is called Optical Solution of *LPP*. A linear programming problem may have many optical solutions.

Linear Inequations

If $a, b, C \in R$ then the equation $ax \oplus by = c$ is called linear equation in two variables.

The inequalilies of the form $ax + by \leq c$, $ax + by \geq c$, $ax + by < c$ and $ax + by > c$ are called linear inequations in two variables.

A general linear optimization problem can be represented by normal form.

Maximize,

$$f = C_1 x_1 + C_2 x_2 \dots C_n x_n$$

Subject to the constraints

$$a_{11} x_1 + \dots + a_{1n} x_n = b_1$$
$$a_{21} x_1 + \dots + a_{2n} x_n = b_2$$
$$\dots\dots\dots\dots\dots\dots\dots\dots\dots\dots$$
$$a_{m1} x_1 + \dots + a_{mn} x_n = b_m$$
$$x_1 \geq 0 \qquad (i = 1 \dots n),$$

with all b_j non-negative.

The same problem also includes the minimization of an objective function f since this correspond to maximizing $-f$ and thus needs no separate consideration. An n-tuple $(x_1 \dots x_n)$ that satisfy all the constants is called feasible point or feasible solution if for the objective function f becomes maximum, compared with the values of f at all feasible solutions.

Simplex Method

A linear optimization problem can be written in the normal for, i.e.

maximise

$$z = f(x) = C_1 x_1 + \dots C_n x_n$$

Subject to the constraints

$$a_{11} x_1 + \dots + a_{1n} x_n = b_1$$
$$a_{21} x_1 + \dots + a_{2n} x_n = b_2$$
$$\dots\dots\dots\dots\dots\dots\dots\dots\dots\dots$$
$$a_{m1} x_1 + \dots + a_{mn} x_n = b_m$$
$$x_1 \geq 0 \qquad (i = 1 \dots n)$$

For finding an optical solution of this problem, consider basic feasible solutions known as simplex method.

Appendix A

Numerical Tables

A.1 DECIMAL OF AN INCH WITH MILLIMETER EQUIVALENTS

Function	*Decimal*	*Millimeters*
1/64	0.015625	0.397
1/32 or 2/64	0.031250	0.794
3/64	0.46875	1.191
1/16 or 2/32 or 4/64	0.062500	1.588
5/64	0.078125	1.984
3/32 or 6/64	0.093750	2.381
7/64	0.109375	2.778
1/8 or 4/32 or 8/64	0.125000	3.175
9/64	0.140625	3.572
5/32 or 10/64	0.156250	3.969
11/64	0.171875	4.366
3/16 or 6/32 or 12/64	0.187500	4.763
13/64	0.203125	5.159
7/32 or 14/64	0.218750	5.556
15/64	0.234375	5.953
1/4 or 8/32 or 16/64	0.250000	6.350
17/64	0.265625	6.747
9/32 or 18/64	0.281250	7.144
19/64	0.296875	7.541
5/16 or 10/32 or 20/64	0.312500	7.938
21/64	0.328125	8.334
11/32 or 22/64	0.343750	8.731
23/64	0.359375	9.128
3/8 or 12/32 or 24/64	0.375000	9.525

Function					*Decimal*	*Millimeters*
				25/64	0.406250	100.319
				27/64	0.421875	10.716
7/16	or	14/32	or	28/64	0.437500	11.113
				29/64	0.453125	11.509
		15/32	or	30/64	0.468750	11.906
				31/64	0.484375	12.303
1/2	or	16/32	or	32/64	0.500000	12.700
				33/64	0.515625	13.097
		17/32	or	34/64	0.531250	13.491
				35/64	0.546875	13.891
9/16	or	18/32	or	36/64	0.562500	14.288
				37/64	0578125	14.684
		19/32	or	38/64	0.593750	15.081
				39/64	0.609375	15.478
5/8	or	20/32	or	40/64	0.625000	15.875
				41/64	0.640625	16.272
		21/32	or	42/64	0.656250	16.669
				43/64	0.671875	17.066
11/16	or	22/32	or	44/64	0.687500	17.463
				45/64	0.703125	17.859
		23/32	or	46/64	0.718750	18.256
				47/64	0.734375	18.653
3/4	or	24/32	or	48/64	0.750000	19.050
				49/64	0.765625	19.447
		25/32	or	50/64	0.781250	19.844
				51/64	0.796875	20.241
13/16	or	26/32	or	52/64	0.812500	20.638
				53/64	0.828125	21.034
		27/32	or	54/64	0.843750	21.431
				55/64	0.859375	21.828
7/8	or	28/32	or	56/64	0.875000	22.225
				57/64	0.890625	22.622
		29/32	or	58/64	0.906250	23.019
				59/64	0.921875	23.416
15/16	or	30/32	or	60/64	0.937500	23.813
				61/64	0.953125	24.209
		31/32	or	62/64	0.968750	24.606
				63/64	0.984375	25.003
1	or	31/32	or	64/64	1.000000	25.400

A.2 POWERS OF N^n FOR N RANGING FROM $N = 2$ to 9 and $n = 2$ to 9

n	2^n	3^n	4^n	5^n	n
2	4	9	16	25	2
3	8	27	64	125	3
4	16	81	256	625	4
5	32	243	1,024	3,125	5
6	64	729	4,096	15,625	6
7	128	2,187	16,384	78,125	7
8	256	6,561	65,536	390,625	8
9	512	19,683	262,144	1,953,125	9

n	6^n	7^n	8^n	9^n	n
2	36	49	64	81	2
3	216	343	512	729	3
4	1,296	2,401	4,096	6,561	4
5	7,776	16,807	32,768	59,049	5
6	49,656	117,649	262,144	531,441	6
7	279,936	823,543	2,097,152	4,782,969	7
8	1,679,616	5,764,801	16,777,216	43,046,721	8
9	10,077,696	40,353,607	134,217,728	387,420,489	9

A.3 SQUARE ROOTS OF N/n RANGING FROM $N = 1$ to 9 and $n = 2$ to 9

N	$\sqrt[n]{N/2}$	$\sqrt{N/3}$	$\sqrt{N/4}$	$\sqrt{N/5}$	$\sqrt{N/6}$	$\sqrt{N/7}$	$\sqrt{N/8}$	$\sqrt{N/9}$	N
1	0.7071	0.5773	0.5000	0.4471	0.4083	0.3779	0.3536	0.3333	1
2		0.8165	0.7071	0.6325	0.5773	0.5345	0.5000	0.4714	2
3	1.2247		0.8660	0.7746	0.7071	0.6547	0.6124	0.5773	3
4	1.4142	1.1546		0.8944	0.8165	0.7559	0.7071	0.6667	4
5	1.5811	1.2911	1.1180		0.9129	0.8452	0.7906	0.7454	5
6	1.7321	1.4142	1.2247	1.0955	1.0000	0.9258	0.8660	0.8165	6
7	1.8708	1.5278	1.3229	1.1832	1.0801	1.0000	0.9354	0.8819	7
8	2.0000	1.6330	1.4142	1.2649	1.1546	1.0690	1.0000	0.9428	8
9	2.1213	1.7321	1.5000	1.3416	1.2247	1.1337	1.0606	1.0000	9

A.4 POWERS; ROOTS AND LOGARITHM FUNCTIONS OF NUMBER *N*, RANGING FROM 1 to 100

N	N^2	N^3	$\sqrt{N}$	$\sqrt[3]{N}$	$\log N$	$\ln N$	$1{,}000/N$	$N\pi$	N
1	1	1	1.0000	1.0000	0.00000	0.00000	1000.000	3.142	1
2	4	8	1.4142	1.2599	0.30103	0.69315	500.000	6.283	2
3	9	27	1.7321	1.4423	0.47712	1.09861	333.333	9.425	3
4	16	64	2.0000	1.5874	0.60206	1.38629	250.000	12.566	4
5	25	125	2.2361	1.7100	0.69897	1.60944	200.000	15.708	5
6	36	216	2.4495	1.8171	0.77815	1.79176	166.667	18.850	6
7	49	343	2.6458	1.9129	0.84510	1.94591	142.857	21.991	7
8	64	512	2.8284	2.0000	0.90309	2.07944	125.000	25.133	8
9	81	729	3.0000	2.0801	0.95424	2.19722	111.111	28.274	9
10	100	1000	3.1623	2.1544	1.00000	2.30259	100.000	31.416	10
11	121	1331	3.3166	2.2240	1.04139	2.39790	90.9091	34.558	11
12	144	1728	3.4641	2.2894	1.07918	2.48491	83.3333	37.699	12
13	169	2197	3.6056	2.3513	1.11394	2.59495	76.9231	40.841	13
14	196	2744	3.7417	2.4101	1.14613	2.63906	71.4286	43.982	14
15	225	3375	3.8730	2.4662	1.17609	2.70805	66.6667	47.124	15
16	256	4096	4.0000	2.5198	1.20412	2.77259	62.5000	50.265	16
17	289	4913	4.1231	2.5713	1.23045	2.83321	58.8235	53.407	17
18	324	5832	4.2426	2.6207	1.25527	2.89037	55.5556	56.549	18
19	361	6859	4.3589	2.6684	1.27875	2.94444	52.6316	59.690	19
20	400	8000	4.4721	2.7144	1.30103	2.99573	50.0000	62.832	20
21	441	9261	4.5826	2.7589	1.32222	3.04452	47.6190	65.973	21
22	484	10648	4.6904	2.8020	1.34242	3.09104	45.4545	69.115	22
23	529	12167	4.7958	2.8439	1.36173	3.13549	43.4783	72.257	23
24	576	13824	4.8990	2.8845	1.38021	3.17805	41.6667	75.398	24
25	625	15625	5.0000	2.9240	1.39794	3.21888	40.0000	78.540	25
26	676	17576	5.0990	2.9625	1.41497	3.25810	38.4615	81.381	26
27	729	19683	5.1962	3.0000	1.43136	3.29584	37.0370	84.823	27
28	784	21952	5.2915	3.0366	1.44716	3.33220	35.7143	87.965	28
29	841	24389	5.2915	3.0366	1.44716	3.33220	35.7143	87.965	29
30	900	27000	5.4772	3.1072	1.47712	3.40120	33.3333	94.248	30
31	961	29791	5.5678	3.1414	1.49136	3.43399	32.2581	97.389	31
32	1024	32768	5.6569	3.1748	1.50515	3.46574	31.2500	100.531	32
33	1089	35937	5.7446	3.2075	1.51851	3.49651	30.3030	103.673	33
34	1156	39304	5.8310	3.2396	1.531148	3.52636	29.4118	106.814	34
35	1125	42875	5.9161	3.2711	1.54407	3.55535	28.5714	109.956	35
36	1296	46656	6.0000	3.3019	1.55630	3.58352	27.7778	113.097	36

(contd...)

N	N^2	N^3	$\sqrt{N}$	$\sqrt[3]{N}$	$\log N$	$\ln N$	$1{,}000/N$	$N\pi$	N
37	1369	50653	6.0828	3.3322	1.56820	3.61092	27.0270	116.239	37
38	1444	54872	6.1644	3.3620	1.57978	3.63759	26.3158	119.381	38
39	1521	59319	6.2450	3.3912	1.59106	3.66356	25.6410	122.522	39
40	1600	64000	6.3246	3.4200	1.60206	3.68888	25.0000	125.664	40
41	1681	68921	6.4031	3.4482	1.61278	3.71357	24.3902	128.805	41
42	1764	74088	6.4807	3.4760	1.62325	3.73767	23.8095	131.946	42
43	1849	79507	6.5574	3.5034	1.63347	3.76120	23.2558	135.089	43
44	1936	85184	6.6332	3.5303	1.64345	3.78419	22.7273	138.230	44
45	2025	91125	6.7082	3.5569	1.65321	3.80667	22.2222	141.372	45
46	2116	97336	6.7823	3.5830	1.66276	3.82864	21.7391	144.513	46
47	2209	103823	6.8557	3.6088	1.67210	3.85015	21.2766	147.655	47
48	2304	110592	6.9282	3.6342	1.68124	3.87120	20.8333	150.896	48
49	2401	117649	7.0000	3.6593	1.69020	3.89182	20.4082	153.938	49
50	2500	125000	7.0711	3.6840	1.69897	3.91202	20.0000	157.079	50
51	2601	132651	7.1414	3.7084	1.70757	3.93183	19.6078	160.221	51
52	2704	140608	7.2111	3.7325	1.71600	3.95124	19.2308	163.363	52
53	2809	14887	7.2801	3.7563	1.72428	3.97029	18.8679	166.504	53
54	2916	157464	7.3485	3.7798	1.73239	3.98828	18.5185	169.646	54
55	3025	166375	7.4162	3.8030	1.74036	4.00733	18.1818	172.788	55
56	3136	175616	7.4833	3.8259	1.74819	4.02535	17.8571	175.929	56
57	3249	185193	7.5498	3.84851	1.75587	4.04305	17.5439	179.071	57
58	3364	195112	7.6158	3.8709	1.76343	4.06044	17.2414	182.212	58
59	3481	205379	7.6811	3.8930	1.77085	4.07754	16.9492	185.354	59
60	3600	216000	7.7460	3,9149	1.77815	4.09434	16.6667	188.496	60
61	3721	226981	7.8102	3.9365	1.78533	4.11087	16.3934	191.637	61
62	3844	238328	7.8740	3.9579	1.79239	4.12713	16.1290	194.778	62
63	3969	250047	7.9373	3.9791	1.79934	4.14313	15.8730	197.920	63
64	4096	262144	8.0000	4.0000	1.80618	4.15888	15.6250	201.062	64
65	4225	274625	8.0623	4.0207	1.81291	4.17439	15.3846	204.204	65
66	4356	287496	8.1240	4.0412	1.81954	4.18965	15.1515	207.345	66
67	4489	300763	8.1854	4.0615	1.82607	4.20469	14.9254	210.487	67
68	4624	314432	8.2462	4.0817	1.83251	4.21951	14.7059	213.628	68
69	4761	328509	8.3066	4.1016	1.83885	4.23411	14.4928	216.700	69
70	4900	343000	8.3666	4.1213	1.84510	4.24850	14.2857	219.912	70
71	5041	357911	8.4261	4.1408	1.85126	4.26268	14.0845	223.053	71
72	5184	373248	8.4853	4.1602	1.85733	4.27667	13.8889	226.195	72
73	5329	389017	8.5440	4.1793	1.86332	4.29046	13.6986	229.336	73
74	5476	405224	8.6023	4.1983	1.86923	4.30407	13.5135	232.478	74

(contd...)

N	N^2	N^3	$\sqrt{N}$	$\sqrt[3]{N}$	log N	ln N	1,000/N	$N\pi$	N
75	5625	421875	8.6603	4.2172	1.87506	4.31749	13.3333	235.619	75
76	5776	438976	8.7178	4.2358	1.88081	4.33073	13.1579	238.761	76
77	5929	456533	8.7750	4.2543	1.88649	4.34381	12.9870	241.903	77
78	6048	474552	8.8318	4.2727	1.89209	4.35671	12.8205	245.044	78
79	6241	493039	8.8882	4.2908	1.89763	4.36945	12.6582	245.186	79
80	6400	512000	8.9443	4.3089	1.90309	4.38203	12.5000	251.337	80
81	6561	531441	9.0000	4.3267	1.90849	4.39445	12.3457	254.469	81
82	6724	551368	9.0554	4.3445	1.91381	4.40672	12.1951	257.611	82
83	6889	571787	9.1104	4.3621	1.91908	4.41884	12.0482	260.752	83
84	7056	592704	9.1652	4.3795	1.92428	4.43082	11.9048	263.894	84
85	7225	614125	9.2195	4.3968	1.92942	4.44265	11.7647	267.035	85
86	7396	636056	9.2736	4.4140	1.93450	4.45435	11.627	270.177	86
87	7569	658503	9.3274	4.4310	1.93952	4.46591	11.4943	273.319	87
88	7744	681472	9.3808	4.4480	1.94448	4.47734	11.3636	276.460	88
89	7921	704969	9.4340	4.4647	1.94939	4.48864	11.2360	279.602	89
90	8100	729000	9.4868	4.4814	1.95424	4.49981	11.1111	282.743	90
91	8281	753571	9.5394	4.4979	1.95904	4.51086	10.9880	285.885	91
92	8464	778688	9.5917	4.5144	1.96379	4.52179	10.8696	289.027	92
93	8649	804357	9.6437	4.5307	1.96848	4.53260	10.7527	292.168	93
94	8836	830584	9.6954	4.5468	1.97313	4.54329	10.6383	295.310	94
95	9025	857375	9.7468	4.5629	1.97772	4.55388	10.5263	298.451	95
96	9216	884736	9.7980	4.5789	1.98227	4.56435	10.4167	301.593	96
97	9409	912673	9.8489	4.5947	1.98677	4.57471	10.3093	304.735	97
98	9604	941192	9.8995	4.6104	1.99123	1.58497	10.2041	307.876	98
99	9801	970299	9.9499	4.6261	1.99564	4.59512	10.1010	311.017	99
100	10000	1000000	10.0000	4.6416	2.00000	4.60517	10.0000	314.159	100

A.5 NATURAL TRIGNOMETRIC FUNCTIONS OF sin ω (ω = 0° to 90°)

<·············· sin ω ··············>			ω = 0° – 45°					
Degree	0′	10′	20′	30′	40′	50′	60′	Degree
0	0.00000	00.00291	0.00582	0.00873	0.01164	0.01454	0.04745	89
1	0.01745	0.02036	0.02327	0.02618	0.02908	0.03199	0.03490	88
2	0.03490	0.03781	0.04071	0.04362	0.04653	0.04948	0.05234	87
3	0.05234	0.05524	0.05814	0.06105	0.06385	0.06685	0.06976	86
4	0.06976	0.07266	0.07556	0.07847	0.08136	0.08426	0.08716	85
5	0.08716	0.09005	0.09295	0.09585	0.09874	0.10164	0.10453	84
6	0.10453	0.10742	0.11031	0.11320	0.11609	0.11898	0.12187	83

(contd...)

<…………sin ω…………>				ω = 0° – 45°			
Degree	0′	10′	20′	30′	40′	50′	60′	Degree
7	0.12187	0.12476	0.12764	0.13053	0.13341	0.13629	0.13917	82
8	0.13917	0.14205	0.14493	0.14781	0.15069	0.15356	0.15643	81
9	0.15643	0.15931	0.16218	0.16505	0.16792	0.17078	0.17365	80
10	0.17365	0.17651	0.17937	0.18224	0.18509	0.18795	0.19081	79
11	0.19081	0.19366	0.19652	0.19937	0.20222	0.20507	0.20791	78
12	0.20791	0.21076	0.21360	0.21644	0.21928	0.22212	0.22495	77
13	0.22495	0.22778	0.23062	0.23345	0.23627	0.23910	0.24192	76
14	0.24192	0.24474	0.24756	0.25038	0.25320	0.25601	0.25882	75
15	0.25882	0.26163	0.26443	0.26724	0.27004	0.27284	0.27564	74
16	0.27564	0.27843	0.28123	0.28402	0.28680	0.28959	0.29237	73
17	0.29237	0.29515	0.29793	0.30071	0.30348	0.30625	0.30902	72
18	0.30902	0.31178	0.31454	0.31730	0.32006	0.22882	0.32557	71
19	0.32557	0.32832	0.33106	0.33381	0.33655	0.33929	0.34202	70
20	0.34202	0.34475	0.54748	0.35021	0.35293	0.35565	0.35837	69
21	0.35837	0.36108	0.36379	0.36650	0.36921	0.37191	0.37461	68
22	0.37461	0.37730	0.37999	0.38268	0.38537	0.38805	0.39073	67
23	0.39073	0.39341	0.39608	0.39575	0.40141	0.40408	0.40674	66
24	0.40674	0.40939	0.41204	0.41469	0.41734	0.41998	0.42262	65
25	0.42262	0.72525	0.42788	0.43051	0.43513	0.43575	0.43837	64
26	0.43837	0.44098	0.44359	0.44620	0.44880	0.45140	0.45399	63
27	0.45399	0.45658	0.45917	0.46175	0.46433	0.46690	0.46947	62
28	0.46947	0.47204	0.47460	0.47716	0.47971	0.48226	0.48481	64
29	0.48481	0.48735	0.48989	0.49242	0.49495	0.49748	0.50000	60
30	.0.50000	0.50252	0.50503	0.50754	0.51004	0.51254	0.51504	59
31	0.51504	0.51753	0.52002	0.52250	0.52498	0.52745	0.52992	58
32	0.52992	0.53238	0.55484	0.53730	0.53975	0.54220	0.54464	57
33	0.54464	0.54708	0.54951	0.55194	0.55436	0.55678	0.55919	56
34	0.55919	0.56160	0.56401	0.56641	0.56880	0.57119	0.57358	55
35	0.57358	0.57596	0.57833	0.58070	0.58307	0.58543	0.58779	54
36	0.58779	0.59014	0.59248	0.59482	0.59716	0.59949	0.60182	53
37	0.60182	0.60414	0.60645	0.60876	0.61107	0.61337	0.61566	52
38	0.61566	0.61795	0.62024	0.62251	0.62479	0.62706	0.62932	51
39	0.62932	0.63158	0.63383	0.63608	0.63832	0.64056	0.64279	50
40	0.64279	0.64501	0.64723	0.64945	0.65166	0.65386	0.65606	49
41	0.65606	0.65825	0.66044	0.66262	0.66480	0.66697	0.66913	48
42	0.66913	0.67129	0.67344	0.67559	0.67773	0.67987	0.68200	47
43	0.68200	0.68412	0.68624	0.38835	0.69046	0.69256	0.69466	46

(contd...)

<·············· sin ω ··············>			ω = 0° – 45°				
Degree	0′	10′	20′	30′	40′	50′	60′	Degree
44	0.69466	0.69675	0.69883	0.70091	0.70298	0.70505	0.70711	45
	60′	50′	40′	30′	20′	10′	0′	ω
ω = 45° to 90°						<·············· cos ω ··············>		

A.6 NATURAL TRIGONOMETRIC FUNCTIONS OF cos ω (ω = 45° to 90°)

<·············· sin ω ··············>			ω = 45° – 90°					
Degree ω°	0′	10′	20′	30′	40′	50′	60′	Degree
45	0.70711	0.70916	0.71121	0.71325	0.71529	0.71732	0.71934	44
46	0.71934	0.72136	0.72337	0.72537	0.72737	0.72937	0.73135	43
47	0.73135	0.73333	0.73531	0.73728	0.73924	0.74120	0.74314	42
48	0.74314	0.74509	0.74703	0.74896	0.75088	0.75280	0.75471	41
49	.75471	0.75661	0.75851	0.76041	0.76229	0.76417	0.76604	40
50	0.76604	0.76791	0.76977	0.77162	0.77347	0.77531	0.77715	39
51	0.77715	0.77897	0.78079	0.78261	0.78442	0.78622	0.78801	38
52	0.78801	0.78980	0.79158	0.79335	0.79512	0.79688	0.79864	37
53	0.79864	0.80038	0.80212	0.80386	0.80558	0.80730	0.80902	36
54	0.80902	0.81072	0.81242	0.81412	0.81580	0.81748	0.81915	35
55	0.81915	0.82082	0.82248	0.82413	0.82577	0.82741	0.82904	34
56	0.82904	0.83066	0.83228	0.83389	0.83549	0.83708	0.83867	33
57	0.83867	0.84025	0.84182	0.84339	0.84495	0.84650	0.84805	32
58	0.84805	0.84959	0.85112	0.85264	0.85416	0.85567	0.86603	31
59	0.85717	0.85866	0.86015	0.86163	0.86310	0.86457	0.86603	30
60	0.86603	0.86748	0.86892	0.87036	0.87178	0.87321	0.87462	29
61	0.87462	0.87603	0.87743	0.87882	0.88020	0.88158	0.88295	28
62	0.88295	0.88431	0.88566	0.88701	0.88835	0.88968	0.89101	27
63	0.89101	0.89232	0.89363	0.89493	0.89623	0.89752	0.89879	26
64	0.89879	0.90007	0.90133	0.90259	0.90383	0.90507	0.90631	25
65	0.90631	0.90753	0.90875	0.90996	0.91116	0.91236	0.91355	24
66	0.91355	0.91472	0.91590	0.91706	0.91822	0.91936	0.92050	23
67	0.92050	0.92164	0.92276	0.92388	0.92499	0.92609	0.92718	22
68	0.92718	0.92827	0.92935	0.93042	0.93148	0.93253	0.93358	21
69	0.93358	0.93462	0.93565	0.93667	0.93769	0.93869	0.93969	20
70	0.93969	0.94068	0.94167	0.94264	0.94361	0.94457	0.94552	19
71	0.94552	0.94646	0.94740	0.94832	0.94924	0.95015	0.95106	18
72	0.95106	0.95195	0.95284	0.95372	0.95459	0.95545	0.95630	17
73	0.95630	0.95715	0.95799	0.95882	0.95964	0.96046	0.96126	16
74	0.96126	0.96206	0.96285	0.96363	0.96440	0.96517	0.96593	15

(contd...)

<……sin ω……>			ω = 45° – 90°					
Degree ω°	0′	10′	20′	30′	40′	50′	60′	Degree
75	0.96593	0.96667	0.96742	0.96815	0.96887	0.96959	0.97030	14
76	0.97030	0.97100	0.97169	0.97237	0.97304	0.97371	0.97437	13
77	0.97437	0.97502	0.97566	0.97630	0.97692	0.97754	0.97815	12
78	0.97815	0.97875	0.97934	0.97992	0.98050	0.98107	0.98163	11
79	0.98163	0.98218	0.98272	0.98325	0.98378	0.98430	0.98481	10
80	0.98481	0.98531	0.98580	0.98629	0.98676	0.98723	0.98769	9
81	0.98769	0.98814	0.98858	0.98902	0.98944	0.98986	0.99027	8
82	0.99027	0.99067	0.99106	0.99144	0.99182	0.99219	0.99255	7
83	0.99255	0.99290	0.99324	0.99357	0.99390	0.99421	0.99452	6
84	0.99452	0.99482	0.99511	0.99540	0.99567	0.99594	0.99619	5
85	0.99619	0.99644	0.99668	0.99692	0.99714	0.99736	0.99756	4
86	0.99756	0.99776	0.99795	0.99813	0.99831	0.99847	0.99863	3
87	0.99863	0.99878	0.99892	0.99905	0.99917	0.99929	0.99939	2
88	0.99939	0.99949	0.99958	0.99966	0.99973	0.99979	0.99985	1
89	0.99985	0.99989	0.99993	0.99996	0.99998	1.00000	1.00000	0
	60′	50′	40′	30′	20′	10′	0′	ω
			ω = 0° – 90°			<……cos ω……>		

A.7 NATURAL TRIGONOMETRIC FUNCTIONS OF tan ω (ω = 0° to 45°)

<……tan ω……>			ω = 0° to 45°					
Degree ω	0′	10′	20′	30′	40′	50′	60′	Degree
0	0.00000	00.00291	0.00582	0.00873	0.01164	0.01455	0.01746	89
1	0.01746	0.02036	0.02328	0.02619	0.02910	0.03201	0.03492	88
2	0.03492	0.03783	0.04075	0.04366	0.04658	0.04949	0.05241	87
3	0.05241	0.05533	0.05824	0.06116	0.06408	0.06700	0.06993	86
4	0.06993	0.07285	0.07578	0.07870	0.08163	0.08456	0.08749	85
5	0.08749	0.09042	0.09335	0.09629	0.09923	0.010216	0.10510	84
6	0.10510	0.10805	0.11099	0.11394	0.11688	0.11983	0.12278	83
7	0.12278	0.12574	0.12869	0.13165	0.13461	0.13758	0.14054	82
8	0.14054	0.14351	0.14648	0.14945	0.15243	0.15540	0.15838	81
9	0.15838	0.16137	0.16435	0.16734	0.17033	0.17333	0.17633	80
10	0.17633	0.17933	0.18233	0.18534	0.18835	0.19136	0.19438	79
11	0.19438	0.19740	0.20042	0.29345	0.20648	0.20952	0.21256	78
12	0.21256	0.21560	0.21864	0.22169	0.22475	0.22781	0.23087	77
13	0.23087	0.23393	0.23700	0.24008	0.24316	0.24624	0.24933	76
14	0.24933	0.25242	0.25552	0.25862	0.26172	0.26483	0.26795	75
15	0.26795	0.27107	0.27419	0.27732	0.28046	0.28360	0.28675	74

(contd...)

<……tan ω……>			ω = 0° to 45°					
Degree ω	0′	10′	20′	30′	40′	50′	60′	Degree
16	0.28675	0.28990	0.29305	0.29621	0.29938	0.30255	0.30573	73
17	0.30573	0.30891	0.31210	0.31530	0.31850	0.32171	0.32492	72
18	0.32492	0.32814	0.33136	0.33460	0.33783	0.34108	0.34433	71
19	0.34433	0.34758	0.35085	0.35412	0.35740	0.36068	0.36397	70
20	0.36397	0.36727	0.37057	0.37388	0.37720	0.38053	0.38386	69
21	0.38386	0.38721	0.39055	0.39391	0.39727	0.40065	0.40403	68
22	0.40403	0.40741	0.41081	0.41421	0.41763	0.42105	0.42447	67
23	0.42447	0.42791	0.43136	0.43481	0.43828	0.44175	0.44523	66
24	0.44523	0.44872	0.45222	0.45573	0.45924	0.46277	0.46631	65
25	0.46631	0.46985	0.47341	0.47698	0.48055	0.48414	0.48773	64
26	0.48773	0.48134	0.49495	0.49858	0.50222	0.50587	0.50953	63
27	0.50953	0.51320	0.51688	0.52057	0.52427	0.52798	0.53171	62
28	0.53171	0.53545	0.53920	0.54296	0.54673	0.55051	0.55431	61
29	0.55431	0.55812	0.56194	0.56577	0.56962	0.57348	0.57735	60
30	0.57735	0.58124	0.58513	0.58905	0.59297	0.59691	0.60086	59
31	0.60086	0.60483	0.60881	0.61280	0.61681	0.62083	0.62487	58
32	0.62487	0.62892	0.63299	0.63707	0.64117	0.64528	0.64941	57
33	0.64941	0.65355	0.65771	0.66189	0.66608	0.67028	0.67451	56
34	0.67451	0.67875	0.68301	0.68728	0.69157	0.69588	0.70021	55
35	0.70021	0.70455	0.70891	0.71329	0.71769	0.72211	0.72654	54
36	0.72654	0.73100	0.73547	0.73996	0.74447	0.74900	0.75355	53
37	0.75355	0.75812	0.76272	0.76733	0.77196	0.77661	0.78129	52
38	0.78129	0.78598	0.79070	0.79544	0.80020	0.80498	0.80978	51
39	0.80978	0.81461	0.80946	0.82434	0.82923	0.83415	0.83910	50
40	0.83910	0.84407	0.84906	0.85408	0.85912	0.86419	0.86929	49
41	0.86929	0.87441	0.87955	0.88473	0.88992	0.89515	0.90040	48
42	0.90040	0.90569	0.91099	0.91633	0.92170	0.92709	0.93252	47
43	0.93252	0.93797	0.94345	0.94896	0.95451	0.96008	0.96569	46
44	0.86569	0.97133	0.97700	0.98270	0.98843	0.99420	1.00000	45
	60′	50′	40′	30′	20′	10′	0′	ω
ω = 45° – 90°						<……cot ω……>		

A.8 NATURAL TRIGONOMETRIC FUNCTIONS OF tan (ω = 45° to 90°)

<……tan ω……>			ω = 45° to 90°					
ω	0′	10′	20′	30′	40′	50′	60′	
45	1.00000	1.00583	1.01170	1.01761	1.02355	1.02952	1.03553	44
46	1.03553	1.04158	1.04766	1.05378	1.05994	1.06613	1.07237	43

(contd...)

<··············· **tan ω** ···············>			**ω = 45° to 90°**					
ω	0′	10′	20′	30′	40′	50′	60′	
47	1.07237	1.07864	1.08496	1.09131	1.09770	1.10414	1.11061	42
48	1.11061	1.11713	1.12369	1.13029	1.13694	1.14363	1.15037	41
49	1.15037	1.15715	1.16398	1.17085	1.17777	1.18474	1.91175	40
50	1.19175	1.19882	1.20593	1.21310	1.22031	1.22758	1.23490	39
51	1.23490	1.24227	1.24969	1.25717	1.26471	1.27230	1.27994	38
52	1.27994	1.28764	1.29541	1.30323	1.31110	1.31904	1.32704	37
53	1.32704	1.33511	1.34323	1.35142	1.35968	1.36800	1.37638	36
54	1.37638	1.38484	1.39336	1.40195	1.41061	1.41934	1.42815	35
55	1.42815	1.43703	1.44598	1.45501	1.46411	1.47330	1.48256	34
56	1.48256	1.49190	1.50133	1.51084	1.52043	1.53010	1.53987	33
57	1.53987	1.54972	1.55966	1.56969	1.57981	1.59002	1.60033	32
58	1.60033	1.61074	1.62125	1.63185	1.64256	1.65337	1.66428	31
59	1.66428	1.67530	1.68643	1.69766	1.70901	1.72047	1.73205	30
60	1.73205	1.74375	1.75556	1.76749	1.77955	1.79174	1.80405	29
61	1.80405	1.81649	1.82906	1.84177	1.85462	1.86760	1.88073	28
62	1.88073	1.89400	1.90741	1.92098	1.93470	1.94858	1.96261	27
63	1.96261	1.97681	1.99116	2.00569	2.02039	2.03526	2.05030	26
64	2.05030	2.06553	2.08094	2.09654	2.11233	2.12832	2.14451	25
65	2.14451	2.16090	2.17749	2.19430	2.21132	2.22857	2.24604	24
66	2.24604	2.26374	2.28167	2.29984	2.31826	2.33693	2.35585	23
67	2.35585	2.37504	2.39449	2.41421	2.43422	2.45451	0.47509	22
68	2.47509	2.49397	2.51715	2.53865	2.56046	2.58261	2.60509	21
69	2.60509	2.62791	2.65109	2.67462	2.69853	2.72281	2.74748	20
70	2.74748	2.77254	2.79802	2.82391	2.85023	2.87700	2.90421	19
71	2.90421	2.93189	2.96004	2.98869	3.01783	3.04749	3.07768	18
72	3.07768	3.10842	3.13972	3.17159	3.20406	3.23714	3.27085	17
73	3.27085	3.30521	3.34023	3.37594	3.41236	3.44951	3.48741	16
74	3.48741	3.52609	3.56557	3.60588	3.64705	3.68909	3.73205	15
75	3.73205	3.77595	3.82083	3.86671	3.91364	3.96165	4.01078	14
76	4.01078	4.06107	4.11256	4.16530	4.21933	4.27471	4.33148	13
77	4.33148	4.38969	4.44942	4.51071	4.57363	4.63825	4.70463	12
78	4.70463	4.77286	4.84300	4.91516	4.98940	5.06584	5.14455	11
79	5.14455	5.22566	5.30928	5.39552	5.48451	5.57638	5.67128	10
80	5.57128	5.76937	5.87080	5.97576	6.08444	6.19703	6.31375	9
81	6.31375	6.43484	6.56055	6.69116	6.82694	6.96823	7.11537	8
82	7.11537	7.26873	7.42871	7.59575	7.77035	7.95302	8.14435	7
83	8.14435	8.34496	8.55555	8.77689	9.00983	9.285530	9.51436	6

(contd...)

<…………… tan ω ……………>			ω = 45° to 90°					
ω	0′	10′	20′	30′	40′	50′	60′	
84	9.51436	9.78817	10.07803	10.38540	10.71191	11.05943	11.43005	5
85	11.43005	11.82617	12.25051	12.70621	13.19688	13.72674	14.30067	4
86	14.30067	14.92442	15.60478	16.34986	17.16934	18.07498	19.08114	3
87	19.08114	20.20555	21.47040	22.90377	24.54176	26.43160	28.63625	2
88	28.63625	31.24158	34.36777	38.18846	42.96408	49.10388	57.28996	1
89	57.28996	68.75009	85.93979	114.58865	171.88540	343.77371	Infinite	0
	60′	50′	40′	30′	20′	10′	0′	ω
		ω = 0° – 40°				<…………… cot ω ……………>		

A.9 NATURAL TRIGONOMETRIC FUNCTIONS OF sec ω (ω = 0° to 45°)

<…………… sec ω ……………>			ω = 0° to 45°					
ω	0′	10′	20′	30′	40′	50′	60′	
0	1.00000	1.00001	1.00002	1.00004	1.00007	1.00011	1.00015	89
1	1.00015	1.00021	1.00027	1.00034	1.00042	1.00051	1.00061	88
2	1.00061	1.00072	1.00083	1.00095	1.00108	1.00122	1.00137	87
3	1.00137	1.00153	1.00169	1.00187	0.00205	1.00224	1.00244	86
4	1.00244	1.00265	1.00287	1.00309	1.00333	1.00357	1.00382	85
5	1.00382	1.00408	1.00435	1.00463	1.00491	1.00521	1.00551	84
6	1.00551	1.00582	1.00614	1.00647	1.00681	1.00715	1.00751	83
7	1.00751	1.00787	1.00825	1.00863	1.00902	1.00942	1.00983	82
8	1.00983	1.01024	1.01067	1.01111	1.01155	1.01200	1.01247	81
9	1.01247	1.01294	1.01342	1.01391	1.01440	1.01491	1.01543	80
10	1.01543	1.01595	1.01649	1.01703	1.01758	1.01815	1.01872	79
11	1.01872	1.01930	1.01989	1.02049	1.02110	1.02171	1.02234	78
12	1.02234	1.02298	1.02362	1.02428	1.02494	1.02562	1.02630	77
13	1.02630	1.02700	1.02770	1.02842	1.02914	1.02987	1.03061	76
14	1.03061	1.03137	1.03213	1.03290	1.03368	1.03447	1.03528	75
15	1.03528	1.03609	1.03691	1.03774	1.03858	1.03944	1.04030	74
16	1.04030	1.04117	1.04206	1.04295	1.04385	1.04477	1.04569	73
17	1.04569	1.04663	1.04757	1.04853	1.04950	1.05047	1.05146	72
18	1.05146	1.05246	1.05347	1.05449	1.05552	1.05657	1.05762	71
19	1.05762	1.05869	1.05976	1.06085	1.06195	1.06306	1.06418	70
20	1.06418	1.06531	1.06645	1.06761	1.06878	1.06995	1.07115	69
21	1.07115	1.07235	1.07356	1.07479	1.07602	1.07727	1.07853	68
22	1.07853	1.07981	1.08109	1.08239	1.08370	1.08503	1.08636	67
23	1.08636	1.08771	1.08907	1.09044	1.09183	1.09323	1.09464	66
24	1.09464	1.09606	1.09750	1.09895	1.10041	1.10189	1.10338	65

(contd...)

<··············· sec ω ···············>			ω = 0° to 45°				
ω	0′	10′	20′	30′	40′	50′	60′	
25	1.10338	1.10488	1.10640	1.10793	1.10947	1.11103	1.11260	64
26	1.11260	1.11419	1.11579	1.11740	1.11903	1.12067	1.12233	63
27	1.12233	1.12400	1.12568	1.12738	1.12910	1.13083	1.13257	62
28	1.13257	1.13433	1.13610	1.13789	1.13970	1.14152	1.14335	61
29	1.14335	1.14521	1.14707	1.14896	1.15085	1.15277	1.15470	60
30	1.15470	1.15665	1.15861	1.16059	1.16259	1.16460	1.16663	59
31	1.16663	1.16868	1.17075	1.17283	1.17493	1.17704	1.17918	58
32	1.17918	1.18133	1.18350	1.18569	1.18790	1.19012	1.19236	57
33	1.19236	1.19463	1.19691	1.19920	1.20152	1.20386	1.20622	56
34	1.20622	1.20859	1.21099	1.21341	1.21854	1.21830	1.22077	55
35	1.22077	1.22327	1.22579	1.22833	1.23089	1.23347	1.23607	54
36	1.23607	1.23869	1.24134	1.24400	1.24669	1.24940	1.25214	53
37	1.25214	1.25489	1.25767	1.26047	1.26330	1.26615	1.26902	52
38	1.26902	1.27191	1.27483	1.27778	1.28075	1.28374	1.28676	51
39	1.28676	1.28980	1.29287	1.29597	1.29909	1.30223	1.30541	50
40	1.30541	1.30861	1.31183	1.31509	1.31837	1.32168	1.32501	49
41	1.32501	1.32838	1.33177	1.33519	1.33864	1.34212	1.34563	48
42	1.34563	1.34917	1.35274	1.35634	1.35997	1.36363	1.36733	47
43	1.36733	1.37105	1.37481	1.37860	1.38242	1.38628	1.39016	46
44	1.39016	1.39409	1.39804	1.40203	1.40606	1.41012	1.41421	45
	60′	50′	40′	30′	20′	10′	0′	ω
	ω = 45° – 90°					<··············· csc ω ···············>		

A.10 NATURAL TRIGONOMETRIC FUNCTIONS OF sec ω (ω = 45° to 90°)

<··············· sec ω ···············>			ω = 45° to 90°					
Degree (ω)	0′	10′	20′	30′	40′	50′	60′	Degree
45°	1.41421	1.41835	1.42251	1.42672	1.43096	1.43524	1.43956	44°
46	1.43956	1.44391	1.44831	1.45274	1.45721	1.46173	1.46628	43
47	1.46628	1.47087	1.47551	1.48019	1.48491	1.48967	1.49448	42
48	1.49448	1.49933	1.50422	1.50916	1.51415	1.51918	1.52425	41
49	1.52425	1.52938	1.53455	1.53977	1.54504	1.55036	1.55572	40
50	1.55572	1.56114	1.56661	1.575213	1.57771	1.58333	1.58902	39
51	1.58902	1.59475	1.60054	1.60639	1.61229	1.61825	1.62427	38
52	1.62427	1.63035	1.63648	1.64268	1.64894	1.65526	1.66164	37
53	1.66164	1.66809	1.67460	1.68117	1.68782	1.69452	1.70130	36
54	1.70130	1.70815	1.71506	1.72205	1.72911	1373624	1.74345	35
55	1.74345	1.75073	1.75808	1.76552	1.77303	1.78062	1.78829	34

(contd...)

<…………… sec ω ……………>			ω = 45° to 90°					
Degree (ω)	0′	10′	20′	30′	40′	50′	60′	Degree
56	1.78829	1.79604	1.80388	1.81180	1.81981	1.82790	1.83608	33
57	1.83608	1.84435	1.85271	1.86116	1.86970	1.87834	1.88708	32
58	1.88708	1389591	1.90485	1.91388	1.92302	1.93226	1.94160	31
59	1.94160	1.95106	1.96062	1.97029	1.98008	1.98998	2.00000	30
60	2.00000	2.01014	2.02039	2.03077	2.04128	2.05191	2.06267	29
61	2.06267	2.07356	2.08458	2.09574	2.10704	2.11847	2.13005	28
62	2.13005	2.14178	2.15366	2.16568	2.17786	2.19019	2.20269	27
63	2.20269	2.21535	2.22817	2.24116	2.25432	2.26766	2.28117	26
64	2.28117	2.29487	2.30875	2.32282	2.33708	2.35154	2.36620	25
65	2.36620	2.38107	2.39614	2.41142	2.42692	2.44264	2.45859	24
66	2.45859	2.47477	2.49119	2.50784	2.52474	2.54190	2.55930	23
67	2.55930	2.57693	2.59491	2.61313	2.63162	2.65040	0.66947	22
68	2.66947	2.68884	2.70851	2.72850	2.74881	2.76945	2.79043	21
69	2.79043	2.81175	2.83342	2.85545	2.87785	2.90063	2.92380	20
70	2.92380	2.94737	2.97135	2.99574	3.02057	304584	3.07155	19
71	3.07155	3.09774	3.12440	3.15155	3.17920	3.20737	3.23607	18
72	3.23607	3.26531	3.29512	3.32551	3.35649	3.38808	3.42030	17
73	3.42030	3.45317	3.48671	3.52094	3.55587	3.59154	3.62796	2
74	3.62796	3.66515	3.70315	3.74198	3.78166	3.82223	3.86370	15
75	3.86370	3.90613	3.94952	3.99393	4.03938	4.08591	4.13357	14
76	4.13357	4.18238	4.23239	4.28366	4.33622	4.39012	4.44541	13
77	4.44541	4.50216	4.56041	4.62023	4.68167	4.74482	4.80973	12
78	4.80973	4.87649	4.94517	5.01585	5.08863	5.16359	5.24084	11
79	5.24084	5.32049	5.40263	5.48740	5.57493	5.66533	5.75877	10
89	5.75877	5.85539	5.95536	6.05886	6.16607	6.27719	6.39245	9
81	6.39245	6.51208	6.63633	6.76547	6.89979	7.03962	7.18530	8
82	7.18530	7.33719	7.49571	7.66130	7.83445	8.01565	8.20551	7
83	8.20551	8.40466	8.61379	8.83367	9.06515	9.30917	9.56677	6
84	9.56677	9.83912	10.12752	10.43343	10.75849	11.10455	11.47371	5
85	11.47371	11.86837	12.29125	12.74550	13.23472	13.76312	14.33559	4
86	14.33559	14.95788	15.63679	16.38041	17.19843	18.10262	19.10732	3
87	19.10732	20.23028	21.49368	22.92559	24.56212	26.45051	28.65371	2
88	28.65371	31.25758	34.38232	38.20155	42.97571	49.11406	57.29869	1
89	57.29869	68.75736	85.94561	114.59301	171.8831	343.77516	Infinite	0
	60′	50′	40′	30′	20′	10′	0′	ω
		ω = 45° – 40°				<…………… csc ω ……………>		

A.11 TRIGONOMETRIC FUNCTIONS RANGING FROM (x = 0.01 to 5.0 radians) or (x = 0.57 to 2.86.48 degree)

x, rad	sin x	cos x	tan x	e^x	e^{-x}	sinh x	cosh x	tanh x	x, deg
0.01	0.01000	0.99995	0.01000	1.01005	0.99005	0.01000	1.00005	0.01000	0.57
0.02	0.02000	0.99980	0.02000	1.02020	0.98020	0.02000	1.00020	0.02000	1.15
0.03	0.03000	0.99955	0.03001	1.03045	0.97045	0.03000	1.00045	0.02999	1.72
0.04	0.03999	0.99920	0.04002	1.04081	0.96079	0.04001	1.00080	0.03998	2.29
0.05	0.04998	0.99875	0.05004	1.05127	0.95123	0.05002	1.00125	0.04996	2.86
0.06	0.05996	0.99820	0.06007	1.06184	0.94176	0.06004	1.00180	0.05993	3.44
0.07	0.06994	0.99755	0.07011	1.07251	0.93239	0.07006	1.00245	0.06989	4.01
0.08	0.07991	0.99680	0.08017	1.08329	0.92312	0.08009	1.00320	0.07983	4.58
0.09	0.08988	0.99595	0.09024	1.09417	0.91393	0.09012	1.00405	0.08976	5.16
0.10	0.09983	0.99500	0.10033	1.10517	0.90484	0.10017	1.00500	0.09967	5.73
0.11	0.10978	0.99396	0.11045	1.11628	0.89583	0.11022	1.00606	0.10956	6.30
0.12	0.11971	0.99281	0.12058	1.12750	0.88692	0.12029	1.00721	0.11943	6.88
0.13	0.12963	0.99156	0.13074	1.13883	0.87810	0.13037	1.00846	0.12927	7.45
0.14	0.13954	0.99022	0.14092	1.15027	0.86936	0.14046	1.00982	0.13909	8.02
0.15	0.14944	0.98877	0.15114	0.16183	0.86071	0.15056	0.011127	0.14889	8.59
0.16	0.15932	0.98723	0.16138	1.17351	0.85214	0.16068	1.01283	0.15865	9.17
0.17	0.16918	0.98558	0.17165	1.18530	0.84366	0.17082	1.01448	0.16838	9.74
0.18	0.17903	0.98384	0.18197	1.19722	1.83527	0.18097	1.01624	0.17808	10.31
0.19	0.18886	0.98200	0.19232	1.20925	0.82696	0.19115	0.01810	0.18775	10.89
0.20	0.19867	0.98007	0.20271	1.22140	0.81873	0.20134	1.02007	0.19738	11.46
0.21	0.20846	0.97803	0.21314	1.23368	0.81058	0.21155	1.02213	0.20697	12.03
0.22	0.21823	0.97590	0.22362	1.24608	0.80252	0.22178	1.02430	0.21652	12.61
0.23	0.22798	0.97367	0.23414	1.25860	0.79453	0.23203	1.02657	0.22603	13.18
0.24	0.23770	0.97134	0.24472	1.27125	0.78663	0.24231	1.02894	0.23550	13.75
0.25	0.24740	0.96891	0.25534	1.28403	0.77880	0.25261	1.03141	0.24492	14.32
0.26	0.25708	0.96639	0.26602	1.29693	0.77105	0.26294	1.03399	0.25430	14.90
0.27	0.26673	0.96377	0.27676	1.30996	0.76338	0.27329	1.03667	0.26362	15.47
0.28	0.27636	0.96106	0.28755	1.32313	0.75578	0.28367	1.03946	0.27291	16.04
0.29	0.28595	0.95824	0.29841	1.33643	0.74826	0.29408	1.04235	0.28213	16.62
0.30	0.29552	0.95534	0.30934	1.34986	0.74082	0.30452	1.04534	0.29131	17.19
0.31	0.30506	0.85233	0.32033	1.36343	0.73345	0.31499	1.04844	0.30044	17.76
0.32	0.31457	0.94924	0.33139	1.37713	0.72615	0.32549	1.05164	0.30951	18.33
0.33	0.32404	0.94604	0.34252	1.39097	0.71892	0.33602	1.05495	0.31852	18.91
0.34	0.33349	0.94275	0.35374	1.40495	0.71177	0.34659	1.05836	0.32748	19.48
0.35	0.34290	0.93937	0.36503	1.41907	0.70469	0.35719	1.06188	0.33638	20.05
0.36	0.35227	0.93590	0.37640	1.43333	0.69768	0.36783	1.06550	0.34521	20.63

(contd...)

x, rad	$\sin x$	$\cos x$	$\tan x$	e^x	e^{-x}	$\sinh x$	$\cosh x$	$\tanh x$	x, deg
0.37	0.36162	0.93233	0.38786	1.44773	0.69073	0.37850	1.06923	0.35399	21.20
0.38	0.37092	0.92866	039941	1.46228	0.68386	0.38921	1.07307	0.36271	21.27
0.39	0.38019	0.92491	0.41105	1.17698	0.67706	0.3996	1.07702	0.37136	22.35
0.40	0.38942	0.92106	0.42279	1.49182	0.67032	0.41075	1.08107	0.37995	22.92
0.41	0.39861	0.91712	0.43463	1.50682	0.66365	0.42158	1.08523	0.38847	23.49
0.42	0.40776	0.91309	0.44657	1.52196	0.65705	0.43246	1.08950	0.39693	24.06
0.43	0.41687	0.90897	0.45862	1.53726	0.65051	0.44337	1.09388	0.40532	24.64
0.44	0.42594	0.90475	0.47078	1.55271	0.64404	0.45434	1.09837	0.41364	25.21
0.45	0.43497	0.90045	0.48306	1.56831	0.63763	0.46534	1.10297	0.42190	25.78
0.46	0.44395	0.89605	0.49545	1.58407	0.63128	0.47640	1.10768	0.43008	26.36
0.47	0.45289	0.89157	0.50797	1.59999	0.62500	0.48750	1.11250	0.43820	26.93
0.48	0.46178	0.88699	0.52061	1.61607	0.61878	0.49865	1.11743	0.44624	27.50
0.49	0.47063	0.88233	0.53339	1.63232	1.61263	0.50984	1.12247	0.45422	28.07
0.50	0.47943	0.87758	0.54630	1.64872	0.60653	0.52110	1.12763	0.46212	28.65
0.51	0.48818	0.87274	0.55936	1.66529	0.60050	0.53240	1.13289	0.46995	29.22
0.52	0.49688	0.86782	0.57256	1.68203	0.59452	0.54375	1.13827	0.47770	29.79
0.53	0.50553	0.86281	0.58592	1.69893	0.58860	0.55516	1.14377	0.48538	30.37
0.54	0.51414	0.85771	0.59943	1.71601	0.58275	0.56663	1.14938	0.49299	30.94
0.55	0.52269	0.85252	0.61311	1.73325	0.57695	0.57515	1.15510	0.50052	31.51
0.56	0.53119	0.84726	0.62695	1.75067	0.57121	0.58973	1.16094	0.50798	32.09
0.57	0.53963	0.84190	0.64097	1.76827	0.86553	0.60137	1.16690	0.51536	32.66
0.58	0.54802	0.83646	0.65517	1.78604	0.55990	0.61307	1.17297	0.52267	33.23
0.59	0.55636	0.83094	0.66956	1.80399	0.55433	0.62483	1.17916	0.52990	33.80
0.60	0.56464	0.82534	0.68414	1.82212	0.54881	0.63665	1.18547	0.53705	34.38
0.61	0.57287	0.81965	0.69892	1.84043	0.54335	0.64854	1.19189	0.54413	34.95
0.62	0.58104	0.81388	0.71391	1.85893	0.53794	0.66049	1.19844	0.55113	35.52
0.63	0.58914	0.80803	0.7211	1.87761	0.53259	0.67251	1.20510	0.55805	36.10
0.64	0.59120	0.80210	0.74454	1.89648	0.52729	0.68459	1.21189	0.56490	37.24
0.65	0.60519	0.79608	0.76020	1.91554	0.52205	0.69675	1.21879	0.57167	37.24
0.66	0.61312	0.78999	0.77610	1.93479	0.51685	0.70887	1.22582	0.57836	37.82
0.67	0.62099	0.78382	0.79225	1.95424	0.51171	0.73126	1.23297	0.58498	38.39
0.68	0.62879	0.77757	0.80866	1.97388	0.50662	0.73363	1.24025	0.59152	38.96
0.69	0.63654	0.77125	0.82534	1.99372	0.50158	0.74607	1.24765	0.59798	39.53
0.70	0.64422	0.76484	0.84229	2.01375	0.49659	0.75858	1.25517	0.60437	40.11
0.71	0.65183	0.75836	0.85953	2.03399	0.49164	0.77117	1.26282	0.61068	40.68
0.72	0.65938	0.75181	0.87707	2.05443	0.48675	0.78384	1.27059	0.61691	41.25
0.73	0.66687	0.74517	0.89492	2.07508	0.48191	0.79659	1.27849	0.62307	41.83
0.74	0.67429	0.73847	0.91309	2.09594	0.47711	0.80941	1.28652	0.62913	42.40

(contd...)

x, rad	$\sin x$	$\cos x$	$\tan x$	e^x	e^{-x}	$\sinh x$	$\cosh x$	$\tanh x$	x, deg
0.75	0.68164	0.73169	0.93160	2.11700	0.47237	0.82237	1.29468	0.63515	42.47
0.76	0.68892	0.72484	0.95045	2.13828	0.46767	0.83530	1.30297	0.64108	43.54
0.77	0.69614	0.71791	0.96967	2.15977	0.46301	0.84838	1.31139	0.64693	44.12
0.78	0.70328	0.71091	0.98926	2.18147	0.45841	0.86153	1.31994	0.65271	44.69
0.79	0.71035	0.70385	1.00925	2.20340	0.45384	0.87478	1.32862	0.65841	45.26
0.80	0.71736	0.69671	1.02964	2.22554	0.44933	0.88811	1.33743	0.66404	45.84
0.81	0.72429	0.68950	1.05046	2.24791	0.44486	0.90152	1.34638	0.66959	46.41
0.82	0.73115	0.68222	1.07171	2.27050	0.44043	0.91503	1.35547	0.67507	46.98
0.83	0.73793	0.67488	1.09343	2.29332	0.43605	0.92863	1.36468	0.68048	47.56
0.84	0.74464	0.66746	1.11563	2.31637	0.43171	0.94233	1.37404	0.68581	48.13
0.85	0.75128	0.68998	1.13833	2.33965	0.42741	0.95612	1.38353	0.60107	48.70
0.86	0.75784	0.65244	1.16156	2.36316	0.43216	0.97000	1.39306	0.69626	49.27
0.87	0.76433	0.64483	1.18532	2.38691	0.41895	0.98398	1.40293	0.70137	49.85
0.88	0.77074	0.63715	1.20966	2.41090	0.41478	0.99806	1.41284	0.70642	50.42
0.89	0.77707	0.62941	1.23460	2.43513	0.41066	1.01224	1.42289	0.71139	50.99
0.90	0.78333	0.62161	1.26016	2.45960	0.40657	1.02652	1.43309	0.71630	51.57
0.91	0.78950	0.61375	1.28637	2.48432	0.40252	1.04090	1.44342	0.72113	52.14
0.92	0.79560	0.60582	1.31326	2.50929	0.39852	1.05539	1.45390	0.72590	52.71
0.93	0.80162	0.59783	1.34087	2.53451	0.39455	1.06998	1.46453	0.73059	53.29
0.94	0.80756	0.58979	1.36923	2.55998	0.39063	1.08468	1.47530	0.73522	53.86
0.95	0.81342	0.58168	1.39838	2.58571	0.38674	1.09948	1.48623	0.73978	54.43
0.96	0.81919	0.57352	1.42836	2.61170	0.38289	1.11440	1.49729	0.74428	55.00
0.97	0.82489	0.56530	1.45920	2.63794	0.37908	1.12943	1.50851	0.74870	55.58
0.98	0.83050	0.55702	1.49096	2.66446	0.37531	1.14457	1.51988	0.75307	56.15
0.99	0.83603	0.54869	1.52368	2.69123	0.37158	1.15983	1.83141	0.75736	56.72
1.00	0.84147	0.54030	1.55741	2.71828	0.36788	1.17520	1.54308	0.76159	57.30
1.01	0.84683	0.53186	1.59221	2.74560	0.36422	1.19069	1.55491	0.76576	57.87
1.02	0.85211	0.52337	1.62813	2.77319	0.36059	1.20630	1.56689	0.76987	58.44
1.03	0.85730	0.51482	1.66524	2.80107	0.35701	1.22203	1.57904	0.77391	59.01
1.04	0.86240	0.50622	1.70361	2.80122	0.35345	1.23788	1.59134	0.77789	59.59
1.05	0.86742	0.49757	1.74332	2..85765	0.34994	1.25386	1.60379	0.78181	60.10
1.06	0.87236	0.48887	1.78442	2.88637	0.34646	1.26996	1.61641	0.78566	60.73
1.07	0.87720	0.48012	1.82703	2.91538	0.34301	1.28619	1.62919	0.78946	61.31
1.08	0.88196	0.47133	1.87122	2.94468	0.33960	1.30254	1.64214	0.79320	61.88
1.09	0.88663	0.46249	1.91709	2.97427	0.33622	1.31903	1.65525	0.79688	62.45
1.10	0.89121	0.45360	1.96476	3.00417	0.33287	1.33565	1.66852	0.80050	63.03
1.11	0.89570	0.44466	2.01434	3.03436	0.32956	1.35240	1.68196	0.80406	63.60
1.12	0.90010	0.43568	2.06596	3.06485	0.32628	1.36929	1.69557	0.80757	64.17

(contd...)

x, rad	$\sin x$	$\cos x$	$\tan x$	e^x	e^{-x}	$\sinh x$	$\cosh x$	$\tanh x$	x, deg
1.13	0.90441	0.42666	2.11975	3.09566	0.32303	1.38631	1.70934	0.81102	64.74
1.14	0.90863	0.41759	2.17588	3.12677	0.31982	1.40347	1.72329	0.81441	65.32
1.15	0.91276	0.40849	2.23450	3.15819	0.31664	1.42078	1.73741	0.81775	65.89
1.16	0.91680	0.39934	2.29580	3.18993	0.31349	1.43822	1.75171	0.82104	66.46
1.17	0.92075	0.39015	2.35998	3.22199	0.31037	1.45581	1.76618	0.82427	67.04
1.18	0.92461	0.38092	2.42727	3.25437	0.30728	1.47355	1.78083	0.82745	67.61
1.19	0.92837	0.37166	2.49760	3.28708	0.30422	1.49143	1.97565	0.83058	68.18
1.20	0.93204	0.36236	2.57215	3.32012	0.30119	1.50946	1.81066	0.83365	68.75
1.21	0.93562	0.35302	2.65032	3.35348	0.29820	1.52764	1.82584	0.83668	69.33
1.22	0.93910	0.34365	2.73275	3.38719	0.29523	1.54598	1.84121	0.83965	69.90
1.23	0.94249	0.33424	2.81982	3.42123	0.29229	1.56447	1.85676	0.84258	70.47
1.24	0.94578	0.32480	2.91193	3.45561	0.28938	1.58311	1.87250	0.84546	71.05
1.25	0.94898	0.31532	3.00957	3.49034	0.28650	1.60192	1.88842	0.84828	71.62
1.26	0.95209	0.30582	3.11327	3.52542	0.28365	1.62088	1.90454	0.85106	72.19
1.27	0.95510	0.29628	3.22363	3.56085	0.28083	1.64001	1.92084	0.85380	72.77
1.28	0.95802	0.28672	3.34135	3.59664	0.27804	1.65930	1.93734	0.85648	73.34
1.29	0.96084	0.27712	3.46721	3.63279	0.27527	1.67876	1.95403	0.85913	73.91
1.30	0.96356	0.26750	3.60210	3.66930	0.27253	1.69838	1.97091	0.86172	74.48
1.31	0.96618	0.25785	3.74708	3.70617	0.26982	1.71818	1.98800	0.86428	75.06
1.32	0.96872	0.24818	3.90335	3.74342	0.26714	1.73814	2.00528	0.86678	75.63
1.33	0.97115	0.23848	4.07231	3.78104	0.26448	1.75828	2.02276	0.86925	76.20
1.34	0.97348	0.22875	4.25562	3.81904	0.26185	1.77860	2.04044	0.87167	76.78
1.35	0.97572	0.21901	4.45522	3.85743	0.25924	1.79909	2.05833	0.87405	77.35
1.36	0.97786	0.20924	4.67344	3.89619	0.25666	1.81977	2.07463	0.87639	77.92
1.37	0.97991	0.19945	4.91306	3.93535	0.25411	1.84062	2.09473	0.87869	78.50
1.38	0.98185	0.18964	5.17744	5.97490	0.25158	1.86289	2.11324	0.88095	79.07
1.39	0.98370	0.17981	5.47069	4.01485	0.24908	1.88289	2.13196	0.88317	79.64
1.40	0.98545	0.16997	5.79788	4.05520	0.24660	1.90430	2.15090	0.88535	80.21
1.41	0.98710	0.16010	6.16536	4.09596	0.24414	1.92591	2.17005	0.88749	80.79
1.42	0.98865	0.15023	6.58112	4.13712	0.24171	1.94770	2.18942	0.88960	81.36
1.43	0.99010	0.14033	7.05546	4.17870	0.23931	1.96970	2.20900	0.89167	81.93
1.44	0.99146	0.13042	7.60183	4.22070	0.23693	1.99188	2.22881	0.89370	82.51
1.45	0.99271	0.12050	8.23809	4.26311	0.23457	2.01427	2.24884	0.89569	83.08
1.46	0.99387	0.11057	8.98861	4.30596	0.23224	2.03686	2.26910	0.89765	83.65
1.47	0.99492	0.10063	9.88737	4.34924	0.22993	2.05965	2.28958	0.89958	84.22
1.48	0.99588	0.09067	10.98338	4.39295	0.22764	2.08265	2.31029	0.90147	84.80
1.49	0.99647	0.08071	12.34986	4.43710	0.22537	2.10586	2.33123	0.90332	85.37
1.50	0.99749	0.07074	14.10142	4.48169	0.22313	2.12928	2.35241	0.90515	85.94

(contd...)

x, rad	sin x	cos x	tan x	e^x	e^{-x}	sinh x	cosh x	tanh x	x, deg
1.51	0.99815	0.06076	16.42809	4.52673	0.22091	2.15291	2.37382	0.90694	86.52
1.52	0.99871	0.05077	19.66953	4.57223	0.21871	2.17676	2.39547	0.90870	87.09
1.53	0.99917	0.04079	24.49841	4.61818	0.21654	2.20082	2.41736	0.91042	87.66
1.54	0.99953	0.03079	32.46114	4.66459	0.21438	2.22510	2.43949	0.91212	88.24
1.55	0.99978	0.02079	48.07848	4.71147	2.21225	2.24961	2.46186	0.91379	88.81
1.56	0.99994	0.01080	92.62050	4.75882	0.21014	2.27434	2.48448	0.91542	89.38
1.57	1.00000	0.00080	1255.76559	4.80665	0.20805	2.29930	2.50735	0.91703	89.95
1.58	0.99996	– 0.00920	– 108.64920	4.85496	0.20598	2.32449	2.53047	0.91860	90.53
1.59	0.99982	– 0.01920	– 52.06697	4.90375	0.20393	2.34991	2.55384	0.92015	91.10
1.60	0.99957	– 0.02920	– 34.23253	4.95303	0.20190	2.37557	2.57746	0.92167	91.67
1.61	0.99923	– 0.03919	– 25.49474	5.00281	0.19989	2.40146	2.60135	0.92316	92.25
1.62	0.99879	– 0.04918	– 20.30728	5.05309	0.19790	2.42760	2.62549	0.92462	92.82
1.63	0.99825	– 0.05917	– 16.87110	5.10387	0.19593	2.45397	2.64990	0.92606	93.39
1.64	0.99761	– 0.06915	– 14.42702	5.15517	0.19398	2.48059	2.67457	0.92747	93.97
1.65	0.99687	– 0.07912	– 12.59926	5.20698	0.19208	2.50746	2.69951	0.92886	94.54
1.66	0.99602	– 0.08909	– 11.18055	5.25931	0.19014	2.53459	2.72472	0.93022	95.11
1.67	0.99508	– 0.09904	– 10.04718	5.31217	0.18825	2.56196	2.75021	0.93155	95.68
1.68	0.99404	– 0.10899	– 9.12077	5.36556	0.18637	2.58959	2.77596	0.93286	96.26
1.69	0.99290	– 0.11892	– 8.34923	5.41948	0.18452	2.61748	2.80200	0.93415	96.83
1.70	0.99166	– 0.12884	– 7.69660	5.47395	0.18268	2.64563	2.82832	0.93541	97.40
1.71	0.99033	– 0.13875	– 7.13726	5.52896	0.18087	2.67405	2.84591	0.93665	97.98
1.72	0.98889	– 0.14865	– 6.65244	5.58453	0.17907	2.70273	2.88180	0.93786	98.55
1.73	0.98735	– 0.15853	– 6.22810	5.64065	0.17728	2.73168	2.90897	0.93906	99.12
1.74	0.98572	– 0.16840	– 5.85353	5.69734	0.17552	3.76091	2.93643	0.94023	99.69
1.75	0.98399	– 0.17825	– 5.52038	5.75460	0.17377	2.79041	2.96419	0.94138	100.27
1.76	0.98215	– 0.18808	– 5.22209	5.81244	0.17204	2.82020	2.99224	0.94250	100.84
1.77	0.98022	– 0.19789	– 4.95341	5.87085	0.17033	2.85026	3.02059	0.94361	101.41
1.78	0.97820	– 0.20768	– 4.71009	5.92986	0.16864	2.88061	3.04925	0.94470	101.99
1.79	0.97607	– 0.21745	– 4.48866	5.98945	0.16696	2.91125	3.07821	0.94576	102.56
1.80	0.97385	– 0.22720	– 4.28626	6.04965	0.16530	2.94217	3.10747	0.94681	103.13
1.81	0.97153	– 0.23693	– 4.10050	6.11045	0.16365	2.97340	3.13705	0.94783	103.71
1.82	0.96911	– 0.24663	– 3.92937	6.17186	0.16203	3.00492	3.16694	0.94884	104.28
1.83	0.96659	– 0.25631	– 3.77118	6.23389	0.16041	3.03674	3.19715	0.94983	104.85
1.84	0.96398	– 0.26596	– 3.62449	6.29654	0.15882	1.06886	3.22768	0.95080	105.42
1.85	0.96128	– 0.27559	– 3.48806	6.35982	0.15724	3.10129	3.25853	0.95175	106.00
1.86	0.95847	– 0.28519	– 3.36083	6.42374	0.15567	3.13403	3.28970	0.95268	106357
1.87	0.95557	– 0.29476	– 3.24187	6.48830	0.15412	3.16700	3.32121	0.95359	107.14
1.88	0.95258	– 0.30430	– 3.13038	6.55350	0.15259	3.20046	3.35305	0.95449	107.72

(contd...)

x, rad	$\sin x$	$\cos x$	$\tan x$	e^x	e^{-x}	$\sinh x$	$\cosh x$	$\tanh x$	x, deg
1.89	0.94949	– 0.31381	– 3.02566	6.61937	0.15107	3.23415	3.38522	0.95537	108.29
1.90	0.94630	– 0.32329	– 2.92710	6.68589	0.14957	3.26816	3.41773	0.95624	108.86
1.91	0.94302	– 0.33274	– 2.83414	6.75309	0.14808	3.30250	3.45058	0.95709	109.43
1.92	0.93965	– 0.34215	– 2.74630	6.82096	0.14661	3.33718	3.48378	0.95792	110.01
1.93	0.93618	– 0.35153	– 2.66316	6.88951	0.14515	3.37218	3.51733	0.95873	110.58
1.94	0.93262	– 0.36087	– 2.58433	6.95875	0.14370	3.40752	3.55123	0.95953	111.15
1.95	0.92896	– 0.37018	– 2.50948	7.02869	0.14227	3.44321	3.58548	0.96032	111.73
1.96	0.92521	– 0.37945	– 2.43828	7.09933	0.14086	3.47923	3.62009	0.96109	112.30
1.97	0.92137	– 0.38868	– 2.37048	7.17068	0.13946	3.51561	3.05507	0.96185	112.87
1.98	0.91744	– 0.39788	– 2.30582	7.24274	0.13807	3.55234	3.69041	0.96185	113.45
1.99	0.91341	– 0.40703	– 2.24408	7.31553	0.13670	3.58942	3.72611	0.96331	114.02
2.00	0.90930	– 2.41615	– 2.18504	7.38906	0.13534	3.62686	3.76220	0.96403	114.59
2.01	0.90509	– 0.42522	– 2.12853	7.46332	0.13399	3.66466	3.79865	0.96473	115.16
2.02	0.90079	– 0.43425	– 2.07437	7.53832	0.13266	3.70283	3.83549	0.96541	115.74
2.03	0.89641	– 0.44323	– 2.02242	7.61409	0.13134	3.74138	3.87271	0.96609	116.31
2.04	0.89193	– 0.45218	– 1.97252	7.69061	0.13003	3.78029	3.91032	0.96675	116.88
2.05	0.88736	– 0.46107	– 1.92456	7.76790	0.12873	3.81958	3.94832	0.96740	117.46
2.06	0.88271	– 0.46992	– 1.87841	7.84597	0.12745	3.85926	3.98671	0.96803	118.03
2.07	0.87796	– 0.47873	– 1.83396	7.92482	0.12619	3.89932	4.02550	0.96865	118.60
2.08	0.87313	– 0.48748	– 1.79111	8.00447	0.12493	3.93977	4.06470	0.96926	119.18
2.09	0.86821	– 0.49619	– 1.74927	8.08492	0.12369	3.98061	4.10430	0.96986	119.75
2.10	0.86321	– 0.50485	– 1.70985	8.16617	0.12246	4.02186	4.14431	0.97045	120.32
2.11	0.85812	– 0.51345	– 1.67127	8.24824	0.12124	4.06350	4.18474	0.97103	120.89
2.12	0.85294	– 0.52201	– 1.63396	8.33114	0.12003	4.10555	4.22558	0.97159	121.47
2.13	0.84768	– 0.53051	– 1.59785	8.41487	0.11884	4.14801	4.26685	0.97215	122.04
2.14	0.84233	– 0.53896	– 1.56288	8.49944	0.11765	4.19089	4.30855	0.97269	122.61
2.15	0.83690	– 0.54736	– 1.52898	8.58486	0.11648	4.23419	4.35067	0.97323	123.19
2.16	0.83138	– 0.55570	– 1.49610	8.67114	0.11533	4.27791	4.39323	0.97375	123.76
2.17	0.82578	– 0.56399	– 1.46420	8.75828	0.11418	4.32205	4.43623	0.97426	124.33
2.18	0.82010	– 0.57221	– 1.43321	8.84631	0.11304	4.36663	4.47967	0.97477	124.90
2.19	0.81434	– 0.58039	– 1.40310	8.93521	0.11192	4.41165	4.52356	0.97526	125.48
2.20	0.80850	– 0.58850	– 1.37382	9.02501	0.11080	4.45711	4.56791	0.97574	126.05
2.21	0.80257	– 0.59656	– 1.34534	9.11572	0.10970	4.50301	4.61271	0.97622	126.62
2.22	0.79657	– 0.60455	– 1.31761	9.20733	0.10861	4.54936	4.65797	0.97668	127.20
2.23	0.79048	– 0.61249	– 1.29061	9.29987	0.10753	4.59617	4.70370	0.97714	127.77
2.24	0.78432	– 0.62036	– 1.26429	9.39333	0.10646	4.64344	4.74989	0.97759	128.34
2.25	0.77807	– 0.62817	– 1.23863	9.48774	0.10540	4.69117	4.79657	0.97803	128.92
2.26	0.77175	– 0.63592	– 1.21359	9.58309	0.10435	4.73937	4.84872	0.97846	129.49

(contd...)

x, rad	$\sin x$	$\cos x$	$\tan x$	e^x	e^{-x}	$\sinh x$	$\cosh x$	$\tanh x$	x, deg
2.27	0.76535	– 0.64361	– 1.18916	9.67940	0.10331	4.78804	4.89136	0.97888	130.06
2.28	0.75888	– 0.65123	– 1.16530	9.77668	0.10228	4.83720	4.93948	0.97929	130.63
2.29	0.75233	– 0.65879	– 1.14200	9.87494	0.10127	4.88684	4.98810	0.97970	131.21
2.30	0.74541	– 0.66628	– 1.11921	9.97418	0.10026	4.93696	5.03722	0.98010	131.78
2.31	0.73901	– 0.67370	– 1.09694	10.07442	0.09926	4.98758	5.08684	0.98049	132.35
2.32	0.73223	– 0.68106	– 1.07514	10.17567	0.09827	5.03870	5.13697	0.90087	132.93
2.33	0.72538	– 0.68834	– 1.05381	10.27794	0.09730	5.09032	5.18762	0.98124	133.50
2.34	0.78146	– 0.69556	– 1.03293	10.38124	0.09633	5.14245	5.23878	0.98161	134.07
2.35	0.71147	– 0.70271	– 1.01247	10.48557	0.09537	5.19510	5.29047	0.98197	134.65
2.36	0.70441	– 0.70979	– 0.99242	10.59095	0.09442	5.24827	5.34269	0.98233	135.22
2.37	0.69728	– 0.71680	– 9.97276	10.69739	0.09348	5.30196	5.39544	0.98267	135.79
2.38	0.69007	– 0.72374	– 0.95349	10.80490	0.09255	5.35618	5.44873	0.98301	136.36
2.39	0.68280	– 0.73060	– 0.93438	10.91349	0.09163	5.41093	5.50256	0.98335	136.94
2.40	0.67546	– 0.73739	– 0.91601	11.02318	0.09072	5.46623	5.55695	0.98367	137.51
2.41	0.66806	– 0.74411	– 0.89779	11.13396	0.08982	5.52207	5.61189	0.98400	138.08
2.42	0.66058	– 0.75075	– 0.87989	11.24586	0.08892	5.57847	5.66739	0.98431	138.66
2.43	0.65304	– 0.75732	– 0.86230	11.35888	0.08804	5.63542	5.72346	0.98462	139.23
2.44	0.64543	– 0.76382	– 0.84501	11.47304	0.08716	5.69294	5.78010	0.98492	139.80
2.45	0.63776	– 0.77023	– 0.82802	11.58835	0.08629	5.75103	5.83732	0.98522	140.37
2.46	0.63003	– 0.77657	– 0.81130	11.70481	0.08543	5.80969	5.89512	0.98551	140.95
2.47	0.62223	– 0.78283	– 0.79485	11.82245	0.08458	5.86893	5.95352	0.98379	141.52
2.48	0.61437	– 0.78901	– 0.77866	11.94126	0.08374	5.92876	6.01250	0.98607	142.09
2.49	0.60645	– 0.79512	– 0.76272	12.06128	0.08291	5.98918	6.07209	0.98635	142.67
2.50	0.59847	– 0.80114	– 0.74702	12.18249	0.08208	6.05020	6.13229	0.98661	143.24
2.51	0.59043	– 0.80709	– 0.73156	12.30493	0.08127	6.11183	6.19310	0.98688	143.81
2.52	0.58233	– 0.81295	– 0.71632	12.42860	0.08046	6.17407	6.25453	0.98714	144.39
2.53	0.57417	– 0.81873	– 0.70129	12.55351	0.07966	6.23692	6.31658	0.98739	144.96
2.54	0.56596	– 0.82444	– 0.68648	12.67967	0.07887	6.30040	6.37927	0.98764	145.53
2.55	0.55768	– 0.83005	– 0.67186	12.80710	0.07808	6.36451	6.44259	0.98788	146.10
2.56	0.54936	– 0.83559	– 0.63745	12.93582	0.07730	0.42926	6.30656	0.98812	146.68
2.57	0.54097	– 0.84104	– 0.64322	13.06582	0.07654	6.49464	6.57118	0.98835	147.25
2.58	0.53253	– 0.84641	– 0.62917	13.19714	0.07577	6.56068	6.63646	0.98858	147.82
2.59	0.52404	– 0.85169	– 0.61530	13.32977	0.07502	6.62738	6.70240	0.98881	148.40
2.60	0.51550	– 0.85689	– 0.60160	13.46374	0.07427	6.69473	6.76901	0.98903	148.97
2.61	0.50691	– 0.86200	– 0.58806	13.59905	0.07353	0.76276	6.83629	0.98924	149.54
2.62	0.49826	– 0.86703	– 0.57468	13.73572	0.07280	6.83146	6.90426	0.98946	150.11
2.63	0.48957	– 0.87197	– 0.56145	13.87377	0.07208	6.90085	6.97292	0.98966	150.69
2.64	0.48082	– 0.87682	– 0.54837	14.01320	0.07136	6.97092	7.04228	0.98987	151.26

(contd...)

x, rad	sin x	cos x	tan x	e^x	e^{-x}	sinh x	cosh x	tanh x	x, deg
2.65	0.47203	– 0.88158	– 0.53544	14.15404	0.07065	7.04169	7.11234	0.99007	151.83
2.66	0.46319	– 0.88626	– 0.52264	14.29629	0.06995	7.11317	7.18312	0.99026	152.41
2.67	0.45431	– 0.89085	– 0.50997	14.43997	0.06925	7.18536	7.25461	0.99045	152.98
2.68	0.44537	– 0.89534	– 0.49743	14.58509	0.06856	7.25827	7.32683	0.99064	153.55
2.69	0.43640	– 0.89975	– 0.48502	14.73168	0.06788	7.33190	7.39978	0.99083	154.13
2.70	0.42738	– 0.90407	– 0.47273	14.87973	0.6721	7.40626	7.47347	0.99101	154.70
2.71	0.41832	– 0.90830	– 0.46055	15.02928	0.06654	7.48137	7.54791	0.99118	155.27
2.72	0.40921	– 0.91244	– 0.44848	15.18032	0.06587	7.55722	7.62310	0.99136	155.84
2.73	0.40007	– 0.91648	– 0.43653	15.33289	0.06522	7.63383	7.69905	0.99153	156.42
2.74	0.39088	– 0.92044	– 0.42467	15.48699	0.06457	7.71121	7.77578	0.99170	156.99
2.75	0.38166	– 0.92430	– 0.41292	15.64263	0.06393	7.78935	7.85328	0.99186	157.56
2.76	0.37240	– 0.92807	– 0.40126	15.79984	0.06329	7.86928	7.93157	0.99202	158.14
2.77	0.36310	– 0.93175	– 0.38970	15.95863	0.06266	7.94799	8.01065	0.98218	158.71
2.78	0.36310	– 0.93175	– 0.38970	15.95863	0.06266	7.94799	8.01065	0.98218	158.71
2.78	0.35376	– 0.93533	– 0.37822	16.11902	0.06204	8.02840	8.09053	0.99233	159.28
2.79	0.34439	– 0.93883	– 0.36683	16.28102	0.06142	8.10980	8.17122	0.99248	159.86
2.80	0.33499	– 0.94222	– 0.35553	16.44465	0.06081	8.19192	8325273	0.99263	160.43
2.81	0.32555	– 0.94553	– 0.34431	16.60992	0.06020	8.27486	8.33506	9.99278	161.00
2.82	0.31608	– 0.94873	– 0.33316	16.77685	0.05961	8.35862	8.41823	0.99292	161.57
2.83	0.30657	– 0.95185	– 0.32208	16.94546	0.05901	8.44322	8.50224	0.99306	162.15
2.84	0.29704	– 0.95486	– 0.31108	17.11577	0.05843	0.52867	8.58710	0.99320	162.72
2.85	0.28748	– 0.95779	– 0.30015	17.28778	0.05784	8.61497	8.67281	0.99333	163.29
2.86	0.27789	– 0.96061	– 0.28928	17.46153	0.05727	8.70213	8.75940	0.99346	164.44
2.87	0.26827	– 0.96334	– 0.27847	17.63702	0.05670	8.79016	8.84886	0.99359	164.44
2.88	0.25862	– 0.96598	– 0.26773	17.81427	0.05613	8.87907	8.93520	0.99372	165.01
2.89	0.24895	– 0.96852	– 0.25704	17.99331	0.05558	8.96887	9.02444	0.99384	165.58
2.90	0.23925	– 0.97096	– 0.24641	18.17415	0.05502	9.05956	9.11458	0.99396	166.16
2.91	0.22953	– 0.97330	– 0.23582	18.35680	0.05448	9.15116	9.20564	0.99408	166.73
2.92	0.21978	– 0.97555	– 0.22529	18.54129	0.05393	9.24368	9.29761	0.99420	167.30
2.93	0.21002	– 0.97770	– 0.21481	18.72763	0.05340	9.33712	9.39051	0.99431	167.88
2.94	0.20023	– 0.97975	– 0.20437	18.91585	9.43149	9.43149	9.48436	0.99443	168.45
2.95	0.19042	– 0.98170	– 0.19397	19.10595	0.05234	9.52681	9.57915	0.99494	169.02
2.96	0.18060	– 0.98356	– 0.18362	19.29797	0.05182	9.62308	9.67490	0.99464	169.60
2.97	0.17075	– 0.98531	– 0.17330	19.49192	0.05130	9.72031	9.77161	0.99475	170.17
2.98	0.16089	– 0.98697	– 0.16301	19.68782	0.05079	9.81851	9.86930	0.99485	170.74
2.99	0.15101	– 0.98853	– 0.15276	19.88568	0.05029	9.91770	9.96789	0.99496	171.31
3.00	0.14112	– 0.98999	– 0.14255	20.08554	0.04979	10.01787	10.06766	0.99505	171.89

(contd...)

x, rad	$\sin x$	$\cos x$	$\tan x$	e^x	e^{-x}	$\sinh x$	$\cosh x$	$\tanh x$	x, deg
3.05	0.09146	– 0.99581	– 0.09185	21.115534	0.04736	10.53399	10.58135	0.99552	174.75
3.10	0.04158	– 0.99914	– 0.04162	22.19795	0.04505	11.07645	11.12150	0.99595	177.62
3.15	– 0.00841	– 0.99996	0.00841	23.33606	0.04285	11.64661	11.68946	0.99633	180.48
3.20	– 0.05837	– 0.99829	0.05847	24.53253	0.04076	12.24588	12.28665	0.99668	183.35
3.25	– 0.10820	– 0.99413	0.10883	25.79034	0.03877	12.87578	12.91456	0.99700	186.21
3.30	– 0.15775	– 0.98748	0.15975	27.11264	0.03688	13.53788	13.57476	0.99728	189.08
3.35	– 0.20690	– 0.97836	0.21148	28.50273	0.03508	14.23382	14.26891	0.99754	191.94
3.40	– 0.25554	– 0.96680	0.26432	29.96410	0.03337	14.96536	14.99874	0.99777	194.81
3.45	– 0.30354	– 0.95282	0.31857	31.50039	0.03175	15.73432	15.76607	0.99799	197.67
3.50	– 0.35078	– 0.93646	0.37459	33.11545	0.03020	16.54263	16.57282	0.99818	200.54
3.55	– 0.39715	– 0.91775	0.43274	34.81332	0.02872	17.39230	17.42102	0.99835	203.40
3.60	– 0.44252	– 0.89676	0.49347	36.59523	0.02732	18.28546	18.31278	0.90851	206.26
3.65	– 0.48679	– 0.87352	0.55727	38.47467	0.02599	19.22434	19.25033	0.99865	209.13
3.70	– 0.52984	– 0.84810	0.62473	40.44730	0.02472	20.21129	20.23601	0.98878	211.99
3.75	– 0.57156	– 0.82056	0.69655	42.52108	0.02352	21.24878	21.27230	0.99889	214.86
3.80	– 0.61186	– 0.79097	0.77356	44.70118	0.02237	22.33941	22.36178	0.99900	217.72
3.85	– 0.65–63	– 0.75940	0.85676	46.99306	0.02128	23.48589	23.50717	0.99909	220.59
3.90	– 0.68777	– 0.72593	0.94742	49.40245	0.02024	24.69110	24.71135	0.99918	223.45
3.95	– 0.72319	– 0.69065	1.04711	51.93537	0.01925	25.95806	25.97731	0.99926	226.32
4.00	– 0.75680	– 0.65364	1.15782	54.59815	0.01832	27.28992	27.30823	0.99933	229.18
4.05	– 0.78853	– 0.61500	1.28215	57.39746	0.01742	28.69002	28.70744	0.99939	232.05
4.10	– 0.81828	– 0.57482	1.42353	60.34029	0.01637	30.16186	30.17843	0.99945	234.91
4.15	– 0.84598	– 0.53321	1.58659	63.43400	0.01576	31.70912	31.72488	0.99950	237.78
4.20	– 0.37158	– 0.49026	1.77778	66.68633	0.01500	33.33567	33.3566	0.99955	240.64
4.25	– 0.89499	– 0.44609	2.00631	70.10541	0.01426	35.04557	35.05984	0.99959	243.51
4.30	– 0.91617	– 0.40080	2.28585	73.69979	0.01357	36.84311	36.85668	0.99963	246.37
4.35	– 0.93505	– 0.35451	2.63760	77.47846	0.01291	38.73278	38.74568	0.99967	249.24
4.40	– 0.95160	– 0.30733	3.09632	81.45087	0.01228	40.71930	40.73157	0.99970	252.10
4.45	– 0.96577	– 0.25939	3.72327	85.62694	0.01168	42.80763	42.81913	0.99973	254.97
4.50	– 0.97753	– 0.21080	4.63733	90.01713	0.01111	45.00301	45.01412	0.99975	257.83
4.55	– 0.98684	– 0.16168	6.10383	94.63241	0.01057	47.31092	47.32149	0.99978	260.70
4.60	– 0.99369	– 0.11215	8.86017	99.48432	0.01005	49.73713	49.74718	0.99980	263.56
4.65	– 0.99805	– 0.06235	16.00767	104.58499	0.00956	52.28771	52.29727	0.99982	266.43
4.70	– 0.99992	– 0.01239	80.71276	109.94717	0.00910	54.96904	54.97813	0.99983	269.29
4.75	– 0.99929	0.03760	– 26.57541	115.58428	0.90865	57.78782	57.79647	0.99985	272.15
4.80	– 0.99616	0.08750	– 11.38487	121.51042	0.00823	60.75109	60.75932	0.99986	275.02
4.85	– 0.99055	0.13718	– 7.22093	127.74039	0.00783	63.86628	63.87411	0.99988	277.88
4.90	– 0.98245	0.18651	– 5.26749	134.28978	0.00745	07.14117	67.14861	0.99987	280.75
4.95	– 0.97190	0.23538	– 4.12906	141.17496	0.00708	70.58394	70.59102	0.99990	283.61
5.00	– 0.95892	0.28366	– 3.38052	148.41316	0.00674	74.20321	74.20995	0.99991	286348

Appendix B

Conversion Tables

B.1 LENGTHS

i. Metric System

1 meter (m) = 10 decimeters (dm) = 10^2 centimeters (cm) = 10^3 millimeters (mm)

1 kilometer (km) = 10 hectrometers (hm) = 10^2 decameters (dkm) = 10^3 meters (m)

ii. FPS System

	Inches (in)	Feet (ft)	Yards	Miles
Inches	1.0	0.083	0.027	0.00001578
Feet	12.0	1.0	0.333	0.00018939
Yards	36.0	3.0	1.0	0.00056818
Miles	63,360.0	5,280.0	1,760.0	1.0

iii. Conversion Factors

1 centimeter (cm) = 0.3937 in = 0.0328083 ft

1 meter (m) = 39.37 in = 3.280833 ft

1 kilometer (km) = 0.6214 miles = 3,280.833 ft

1 inch = 2.540 cm   1 foot = 30.48 cm

1 yard = 0.91440 m   1 mile = 1.609 km

B.2 AREAS

i. Metric System

1 square kilometer = 100 hectares = 100^2 acres = 100^3 square meters

ii. FPS System

	Square inches	Square feet	Acres	Square miles
Squares inches	1.0	0.006944		
Square feet	144.0	1.0	0.000023	
Acres		43,560.0	1.0	0.0015625
Square miles			640.0	1.0

iii. Conversion Factors

1 m^2 = 1,550 in^2 = 10,76387 ft^2

1 km^2 = 0.3861006 sq. mile = 2,47,104 acres

1 in^2 = 6.4516258 cm^2

1 ft^2 = 929.0341 cm^2

1 $yard^2$ = 0.8361 m^2

1 $mile^2$ = 2.590 km^2

B.3 VOLUMES

i. Metric System

For pure water at 4°C (39.2°F), 1 cubic decimeter = 1 litre = 1 kilogram

1 liter (l) = 10 decilitres (dl) = 10^2 centiliters (cl) = 10^3 milliliters (ml)

1 kiloliter (kl) = 10 hectoliters (hl) = 10^2 dekaliters (dkl) = 10^3 liters (l)

ii. FPS System

	Gills	Pints	Quarts	Gallons
Gil's	1.0	0.25	0.125	0.03125
Pints	4.0	1.0	0.5	0.125
Quarts	8.0	2.0	1.0	0.250
Gallons	32.0	8.0	4.0	1.0

iii. Conversion Factors

1 liter = 1.057 qt = 0.26417762 gal

1 litre = 61.02 in^3 = 0.03532 ft^3

1 ft^3 = 7.481 gal = 28.32 liters

1 gal = 0.13368 ft^3 = 3.785 liters

B.4 WEIGHTS

i. Metric System

1 gram (g) = 10 decigrams (dg) = 10^2 centigrams (cg) = 10^3 milligrams (mg)

1 kilogram (kg) = 10 hectograms (hg) = 10^2 decagrams (dkg) = 10^3 grams (g)

1 ton = 1,000 kg

ii. FPS System

	Grains	Ounces	Pounds	Tons
Grains	1.0	0.002286	0.000143	714×10^{-10}
Ounces	437.5	1.0	0.0625	1.95×10^{-8}
Pounds	7,000.0	16.0	1.0	0.0005
Tons	14×10^{6}	32×10^{3}	2,000.0	1.0

iii. Conversion Factors

1 kg = 35.2740 oz = 2.20462 lb

1 ton (MS) = 1.10231 tons (US) = 2,240 lb

1 grain = 64.798918 mg

1 lb = 0.45359242 kg

1 oz = 28.349527 g

1 ton (IS) = 0.984207 tons (MS)

B.5 PHYSICAL CONSTANTS

i. Temperature

1° Centigrade (C) = 1.8° Fahrenheit (F) = 0.8° Reaumur (Re)
= 1.8° Rankine (Ra) = 1° Kelvin (K)

0°C = 32°F = 0° Re = 492°Ra = 273°K

100°C = 212°F = 80° Re = 672° Ra = 373°K

ii. Density

1 lb/cu ft = 5.787×10^{-4} lb/cu in = 16.018 kg/cu meter
= 1.6018×10^{-2} grams/cu cm

1 gram/cu cm = 0.03613 lb/cu in = 62.43 lb/cu ft

iii. Pressure

1 atmoshpere (atm) = 1.0133 bars = 14.696 lb/sq in = 1.013246×10^{6} dynes/sq cm
= 1,033.2 grams/sq cm (0°C) = 760 mm Hg (0°C)
= 29.921 in. Hg (0°C) = 33.903 ft water (0°C)

1 dynes/sq cm = 1.01971×10^{-3} gram/sq cm = 1.4504×10^{-5} lb/sq in

1 bar = 1.0 × 106 dynes/sq cm = 0.98692 atm

1 lb wt/sq in = 70.307 grams/sq cm = 68,947 dynes/sq cm

iv. Force

1 newton = 1×10^{5} dynes = 0.22481 lb wt

1 dyne = 2.2481×10^{-6} lb wt = 7.2330×10^{-5} poundal = 0.0010197 gram wt

1 gram wt = 0.07932 poundal = 980.665 dynes = 2.2046×10^{-3} lb wt

1 lb wt = 32.17 poundals = 453.59 gram wt = 4.4482 newtons

1 poundal = 0.031081 lb wt = 14.098 gram wt = 1.3825×10^{4} dynes

v. Energy

1 absolute (abs) joule = 1 newton.meter = 1×10^7 ergs = 1×10^7 dyne.cm
= 1 watt.sec = 1 volt.coulomb = 0.73756 ft.lb
= 2.3889×10^{-4} kg.cal (mean) = 9.4805×10^{-4} BTU (mean)
= 23.730 ft.poundal = 2.778×10^{-7} kw.hr = 3.725×10^{-7} hp.h

1 gram.cal (mean) = 4.186 joules (abs)

1 gram.cal (15°C) = 4.1855 joules (abs) = 0.003968 BTU

1 joule (International) = 1.000165 joules (abs)

1 kw/hr = 3413.0 BTU (mean) = 2.6552×106 ft.lb = 1.3410 hp.h

1 liter atm (normal) = 3.7745×10^{-5} hp.h = 24.206 g.cal (mean)
= 101.328 joules (abs)

1 atomic mass unit (amu) = 931.16 MeV = 1.65983×10^{-24} gram mass (energy equiv)

1 electron volt = 1.60207×10^{-19} joule (abs)

vi. Power

1 BTU (mean)/min = 0.023575 hp = 17.580 watts (abs) = 778.0 ft.lb/min

1 erg/sec = 10^{-10} kw = 1.3412×10^{-10} hp = 10^{-7} watt
= 5.688×10^{-9} BTU (mean)/min
= 1.4333×10^{-90} kg.cal (mean)/min = 7.3756×10^{-8} ft.lb/sec

1 hp (mech) = 0.70696 BTU (mean)/sec = 550 ft.lb/sec = 745.70 watts
= 10.688 kg.cal (mean)/min

1 hp (elec) = 746.00 watts

Appendix C

Glossary of Symbols

C.1 RELATIONSHIPS

Symbol	Meaning	Symbol	Meaning
$=$ or $::$	Equals	$\neq$ or $\neq$	Does not equal
$>$	Greater than	$<$	Less than
$\geq$	Greater than or equal	$\leq$	Less than or equal
$\cong$	Identical	$\approx$	Approximately equal

C.2 ALGEBRA

Symbol	Meaning	Symbol	Meaning
$+$	Plus or positive	$-$	Minus or negative
$\pm$	Plus or minus Positive or negative	$\mp$	Minus or plus Negative or positive
$\times$	Multiplied by	+ or :	divided by
a^n	nth power of a	$\sqrt[n]{a}$	nth root of a
log $\log_{10}$	Common logarithms or Brigg's logarithm	ln $\log_e$	Natural logarithm or Napier's logarithm

( ) parentheses [ ] Brackets { } Braces

$$\begin{vmatrix} a_1 & a_2 & \dots \\ b_1 & b_2 & \dots \\ & & \dots \end{vmatrix}$$ Determinant

$$\begin{bmatrix} a_1 & a_2 & \dots \\ b_1 & b_2 & \dots \\ & & \dots \end{bmatrix}$$ Matrix

Symbol	Meaning	Symbol	Meaning
I	Unit matrix	Adj	Adjoint matrix
A^{-1}	Inverse of the A matrix	A^T	Transpose of the A matrix
A^+	Complex matrix	$A^{\neq}$	Conjugate complex matrix
$n!$	n factorial	$\begin{pmatrix} n \\ k \end{pmatrix}$	Binomial coefficient

$k^p n$ The number of all possible permutations of n elements, among which there are k elements of equal value.

$k^V n$ The number of all possible permutations of n elements taken k at a time.

k_n^c The number of all possible permutations (without repetition) of n elements taken k at a time.

C.3 COMPLEX NUMBERS

$i = \sqrt{-1}$	Unit imaginary number	$z = x + iy$	Complex variable
$\lvert z \rvert$	Absolute value of z	$\bar{z}$	Conjugate of z
Re (z), R(z)	Real part of z	Im (z)	Imaginary part of z
$\vert\vert$	Parallel to	$\alpha°$	α in degrees
$\perp$	Perpendicular to	α'	α in minutes
$\angle$	Angle	α''	α in seconds
$\cong$	Congruent to	$\sim$	Similar to
Δ	Triangle	○	Circle
▱	Parallelogram	□	Square
$\overline{AB}$	The line segment between A and B	$\overset{\frown}{AB}$	The arc segment between A and B

P (x, y, z) Point P given by the cartesian coordinates x, y, z

P (r, θ, z) Point P given by the cylindrical coordinates r, θ, z

P (ρ, θ, ϕ) Point P given by the spherical coordinates ρ, θ, ϕ

C.4 CIRCULAR AND HYPERBOLIC FUNCTIONS

sin	Sine	sinh	Hyperbolic sine
cos	Cosine	cosh	Hyperbolic cosine
tan	Tangent	tanh	Hyperbolic tangent
cot	Cotangent	coth	Hyperbolic cotangent
sec	Secant	sech	Hyperbolic secant
csc	Cosecant	csch	Hyperbolic cosecant
vers	Versine	covers	Coversine
$\sin^{-1}$	Inverse sine	$\sinh^{-1}$	Inverse hyperbolic sine
$\cos^{-1}$	Inverse cosine	$\cosh^{-1}$	Inverse hyperbolic cosine
$\tan^{-1}$	Inverse tangent	$\tanh^{-1}$	Inverse hyperbolic tangent
$\cot^{-1}$	Inverse cotangent	$\coth^{-1}$	Inverse hyperbolic cotangent
$\sec^{-1}$	Inverse secant	sech^{-1}	Inverse hyperbolic secant
$\csc^{-1}$	Inverse cosecant	csch^{-1}	Inverse hyperbolic cosecant

C.5 VECTOR ANALYSIS

i, j, k	Unit vectors, cartensian system of coordinates	e_s	Unit vector in s direction

$x = ix + jy + kz$	Position vector, cartesian coordinates
$r = r_a e_a + r_\theta e_\theta + r_z e_z$	Position vector, spherical coordinates
$r = r_b e_b + r_\theta e_\theta + r_\phi e_\phi$	Positive vector, spherical coordinates
$r_1 \bullet r_2$	Scalar product
$r_1 \times r_2$	Vector product
∇	Vector differential operator
∇^2	Laplacian operator

C.6 ANALYSIS

(a, b)	The bounded open interval
$[a, b]$	The bounded closed interval
$f(x), F(x)$	The function of x
$f^{-1}(x), F^{-1}(x)$	The inverse function of x
$\sum_{i=1}^{n} u_i$	The sum of n terms
$\prod_{i=1}^{n} u_i$	The product of n terms
Δu	The increment of u
du	The differential of u

$$\left.\begin{array}{l}\dfrac{dy}{dx}, y', D_x y \\ \dfrac{df(x)}{dx}, f'(x), D_x f(x)\end{array}\right\}$$ The first-order derivative of $y = f(x)$ with repect to x

$$\left.\begin{array}{l}\dfrac{d^n y}{dx^n}, y^{(n)}, D_x^{(n)} y \\ \dfrac{d^n f(x)}{dx}, f^{(n)}(x), D_x^{(n)} f(x)\end{array}\right\}$$ The nth-order derivative of $y = f(x)$ with respect to x

$$\dfrac{dw}{dx}, w_x, D_x w$$

$$\dfrac{\partial f}{\partial x}, f_x, F_x$$ The first-order partial derivative of $w = f(x, y \ldots)$ with respect to x

$$\left.\begin{array}{l}\dfrac{\partial^2 w}{\partial x \partial y} w_{xy}, D_{xy} w \\ \dfrac{\partial^2 f}{\partial x \partial y} f_{xy}, F_{xy}\end{array}\right\}$$ The second-order partial derivative of $w = f(x, y)$ with respect to x and then with respect to y

$f'(a+)$	The derivative on the right of $x = a$
$f'(a-)$	The derivative on the left of $x = a$
$\dfrac{\partial(f_1, f_2, \cdots f_n)}{\partial(x_1, x_2, \ldots, x_n)}, \; J\dfrac{(f_1, f_2, \cdots f_n)}{(x_1, x_2, \ldots, xn)}$	Jocobian determinant
$\int f(x)\, dx$	The indefinite integral of $y = f(x)$
$\int_a^b f(x)\, dx$	The definite integral of $y = f(x)$ between limits a and b
$\iint$	The double integral
$\iiint$	The triple integral

$\int_C$ The line integral		$\int_s$ The surface integral	$\int_V$ The volume integral
A	Area	M	Static moment
V	Volume	I	Moment of inertia
k	Radius of gyration	J	Polar moment of inertia
ρ	Radius of curvature	k	Curvature
L	Length of surve	x_c, y_c, z_c	Coordinates of centroid

C.7 SPECIAL CONSTANTS

π	Ludolf's number (pi)	g	Acceleration due to gravity
e	Euler's number	C, γ	Euler's constant
E_k	Euler's number	B_k	Bernoulli's number

C.8 SPECIAL FUNCTIONS

$\Gamma(x)$	Gamma function	$\Pi(x)$	Pi function
$Z(X)$	Riemann's zeta function	erf (x)	Error integral

Si (x), Ci (x) Ei (x)	Integral functions
S(x), C(x)	Fresnel integrals
$\left.\begin{matrix} E(k,x), F(k,x), \Pi(n,k,x) \\ E(k,\phi,), F(k,\phi), \Pi(n,k,\phi) \end{matrix}\right\}$	Eilliptic integrals
snu, cnu, dnu	Jacobi's elliptic functions
$£ f(t) = f(s)$	Laplace transform
$D(x-t)$	Dirac delta functions

C.9 BESSEL FUNCTIONS

J_n	Bessel function of the first kind of order n
Y_n	Bessel function of the second kind of order n
I_n	Modified Bessel function of the first kind of order n
K_n	Modified Bessel function of the second kind of order n
$H_n^{(1)}$	Hankel function of the first kind of order n
$H_n^{(2)}$	Hankel function of the second kind of order n
Ber_n	Ber function of order n
Bei_n	Bei function of order n
Ker_n	Ker function of order n
Kei_n	Kei function of order n

C.10 ORTHOGONAL POLYNOMIALS

$p_n(x)$	Orthogonal polynomial in $nw(x)$ weight function
$F(\alpha, \beta, \gamma, x)$	Hypergeometric series
$F(\beta, \gamma, x)$	Cofluent hypergeometric series
$P_n(x)$	Legendre polynomial in n
$T_n(x)$	Chebyshev polynomial in n
$L_n(x)$	Laguerre polynomial in n
$H_n(x)$	Hernite polynomail in n

C.11 NUMERICAL METHODS

ε	Absolute error	$\bar{\varepsilon}$	Relative error
ε_T	Truncation error	$\bar{y}$	Substitute function
r_{ij}	Carryover value	m_j	Starting value
Δy_n	Forward difference	∇y_n	Backward difference
δy_n	Central difference	Δx_n	Divided difference

C.12 PROBABILITY AND STATISTICS

$\cup$	Union	$\cap$	Intersection
$P(E)$	Probability of occurrence	$P(\bar{E})$	Probability of nonoccurrence
$\phi(x)$	Probability density	$\phi(x, y)$	Joint probability density
$\bar{X}$	Arithmetic means	$\bar{G}$	Geometric means
$\bar{H}$	Harmonic means	$\bar{Q}$	Quadratic means
D	Deviation	$\bar{D}$	Mean deviation
σ	Standard deviation	σ^2	Variance
μ_k	Moment of degree k	β^2	Kurtosis
γ_1	Coefficient of skewness	γ_2	Coefficient of excess
$\phi_N(t)$	Ordinate of the standard normal curve	$F_N(t)$	Area under the standard normal curve

C.13 GREEK ALPHABET

A	α	Alpha	I	ι	Iota	P	ρ	Rho
B	β	Beta	K	κ	Kappa	Σ	σ	Sigma
Γ	γ	Gamma	A	$\wedge$	lambda	T	τ	Tau
Δ	δ	Delta	M	μ	Mu	Y	υ	Upsilon
E	ε	Epsilon	N	ν	Nu	Φ	ϕ	Phi
Z	ζ	Zeta	Ξ	ξ	Xi	X	χ	Chi
H	η	Eta	O	o	Omicron	Ψ	ψ	Psi
Θ	θ	Theta	Π	π	Pi	Ω	ω	Omega

Appendix D

Tables of Physical Constants

Table of Physical Constants

Constant	*symbol*	*Value*	*Systeme International (MKSA)*	*Centimeter-gram-second (CGS)*
Speed of light in vacuum	c	2.997925	$\times 10^{8}$ m s^{-1}	$\times 10^{10}$ cm s^{-1}
Elementary charge	e	1.60210	10^{-19} C	10^{-20} cm$^{1/2}$ g$^{1/2}$
Avogadro constant	NA	6.02252	10^{23} mol^{-1}	10^{23} mol^{-1}
Electron rest mass	m_e	9.1091	10^{-31} Kg	10^{-28} g
		5.48597	10^{-4} u	10^{-4} u
Proton rest mass	m_p	1.67252	10^{-27} kg	10^{-24} g
		1.00727663	10^{0} u	10^{0} u
Neutron rest mass	m_n	1.67482	10^{-27} kg	10^{-24} g
		1.0086654	10^{0} u	10^{0} u
Faraday constant	F	9.64870	10^{4} C mol^{-1}	10^{3} cm$^{1/2}$g$^{1/2}$ mol^{1}
		2.89261	....	10^{14} cm$^{1/2}$g$^{1/2}$ mol^{-1}
Planck constant	h	6.6256	10^{-34} J s	10^{-27} erg s
	h	1.05450	10^{-34} J s	10^{-27} erg s
Fine structure constant	α	7.29720	10^{-3} ...	10^{-3}
	$1/\alpha$	1.370388	10^{2} ...	10^{2}
	$\alpha/2\pi$	1.161385	10^{-3} ...	10^{-3}
	α^2	5.32492	10^{-5} ...	10^{-5}
Charge-to-mass ratio for electron	e/m_e	1.758796	10^{11} C kg^{-1}	10^{7} cm$^{1/2}$ g$^{-1/2}$
		5.27274	...	10^{17} cm$^{3/2}$ g$^{-1/2}$ s^{-1}

Constant	*symbol*	*Value*	*Systeme Interna-tional (MKSA)*	*Centimeter-gram-second (CGS)*
Quantum-charge ratio	h/e	4.13556	10^{-15} J.s C^{-1}	10^{-7} $cm^{3/2}$ $g^{1/2}$ s^{-1}
		1.37947	...	10^{-17} $cm^{1/2}$ $g^{1/2}$
Compton wavelength of electron	λ_c	2.42621	10^{-12} m	10^{-10} cm
	$\lambda_c/2\pi$	3.86144	10^{-13} m	10^{-11} cm
		1.32140	10^{-15} m	10^{-13} cm
Compton wavelength of proton	$\lambda_{c,p}\,2\pi$	2.10307	10^{-16} m	10^{-14} cm
Rydberg constant	R_∞	1.0973731	10^7 m^{-1}	$10\ n^5$ cm^{-1}
Bohr radius	α_0	5.29167	10^{-11} m	10^{-9} cm
Electron radius	r_e	2.81777	10^{-15} m	10^{-13} cm
	r_e^2	7.9398	10^{-30} m	10^{-26} cm^2
Thomson cross section	$8\pi^{2/3}$	6.6516	10^{-29} m^2	10^{-25} cm^2
Gyromagnetic ratio of proton	γ	2.67519	10^8 rad s^{-1} T^{-1}	10^4 rad s^{-1} G^{-1}
	$\gamma/2\pi$	4.25770	10^7 Hz T^{-1}	10^3 s^{-1} G^{-1}
(Uncorrected for diamag-	γ'	2.67512	10^8 rad s^{-1} T^{-1}	10^4 rad s^{-1} G^{-1}
netism, H_2O)	$y'/2\pi$	4.25759	10^7 Hz T^{-1}	10^3 s^{-1} G^{-1}
Bohr magneton	μ_N	9.2732	10^{-24} J T^{-1}	10^{-21} erg G^{-1}
Nuclear magneton	μ_N	5.0505	10^{-27} J T^{-1}	10^{-24} erg G^{-1}
Proton moment	μ_p	1.41049	10^{-26} J T^{-1}	10^{-23} erg G^{-1}
				10^7 Hz T^{-1}
	μ_p/μ_S	2.79276	10^0 ...	10^0
(uncorrected for diamag-netism, H_2O)	μ'_p/μ_N	2.79268	10^0 ...	10^0
Anomalous electron moment correction	$(\mu_e/\mu_0)-1$	1.159615	10^{-3} ...	10^{-3}
Zeeman splitting constant	u_B/h_C	4.66858	10^1 m^{-1} T^{-1}	10^{-5} cm^{-1} G^{-1}
Gas constant	R	8.3143	100 J° K^{-1} mol^{-1}	10^7 erg $°K^{-1}$ mol^{-1}
Normal volume perfect gas	V_0	2.24136	10^{-2} m^3 mol^{-1}	10^4 cm^3 mol^{-1}
Boltzmann constant	k	1.38054	10^{-23} J $°K^{-1}$	10^{-16} erg $°K^{-1}$
First radiation constant ($2\pi hc^2$)	c_1	3.7415	10^{-16} W m^2	10^{-5} erg cm^2 s^{-1}
Second radiation constant	c^2	1.43879	10^{-2} m °K	10^6 cm °K
Wien displacement constant	b	2.8978	10^{-3} m °K	10^{-1} cm °K
Stefan–Boltzmann constant	σ	5.6697	10^{-8} W m^{-2} $°K^{-4}$	10^{-5} erg $cm^{-2}s^{-1}°K^{-4}$
Gravitational constant	G	6.670	10^{-11} N m^2 kg^{-2}	10^{-8} dyn cm^2 g^{-2}

C–coulomb, J–joule, Hz–hertz, W–watt, N–newton, T–tesla, G–gauss

Appendix E

Table E.1 Fundamental, Mechanical, Electrical, and Magnetic Units

Name of dimension or quantity	*Symbol*	*Description*	*S.I. unit and abbreviation*	*Equivalent unit*	*Dimensions*
			Fundamental units		
Current	I	$\frac{\text{charge}}{\text{time}}$	ampere (A)	6.25×10^{18} electron charges per second $= \frac{C}{s}$	I
Length	L, l		metre (m)	1,000 mm = 100 cm	L
Mass	M, m		kilogram (kg)	1,000 g	M
Time	T, t		second (s)	$\frac{1}{60}$ min $= \frac{1}{360}$ h $= \frac{1}{86400}$ day	T

(*Contd...*)

Table E.1 Fundamental, Mechanical, Electrical, and Magnetic Units (Contd.)

Name of dimension or quantity	*Symbol*	*Description*	*S.I. unit and abbreviation*	*Equivalent unit*	*Dimensions*
			Mechanical units		
Acceleration	a	$\frac{\text{velocity}}{\text{time}} = \frac{\text{length}}{\text{time}^2}$	$\frac{\text{metre}}{\text{second}^2}$ $(\text{m} \cdot \text{s}^{-2})$		$\frac{L}{T^2}$
Area	A, a, s	length^2	metre^2 (m^2)		L^2
Energy or work	W	force x length = power x time	joule (J)	$\text{N} \cdot \text{m} = \text{W} \cdot \text{s} = \text{V} \cdot \text{C}$ $= 10^7$ ergs $= 10^8$ dynes · mm	$\frac{ML^2}{T^2}$
Energy density	w	$\frac{\text{energy}}{\text{volume}}$	$\frac{\text{joule}}{\text{metre}^3}$ $(\text{J} \cdot \text{m}^{-3})$	$\frac{\frac{1}{100}\text{erg}}{\text{mm}^3}$	$\frac{ML}{T^2}$
Force	F	mass × acceleration	newton (N)	$\frac{\text{kg} \cdot \text{m}}{\text{s}^2} = \frac{\text{J}}{\text{m}}$ = 105 dynes	$\frac{ML}{T^2}$

(*Contd...*)

Table E.1 Fundamental, Mechanical, Electrical, and Magnetic Units (Contd.)

Name of dimension or quantity	*Symbol*	*Description*	*S.I. unit and abbreviation*	*Equivalent unit*	*Dimensions*
Frequency	f	$\frac{\text{force}}{\text{time}}$	hertz (Hz)	$\frac{\text{cycle}}{\text{s}}$	$\frac{1}{T}$
Impedance	Z	$\frac{\text{force}}{\text{mass} \times \text{velocity}}$	$\frac{\text{newton} \cdot \text{second}}{\text{kilogram} \cdot \text{metre}}$	$\frac{\text{N} \cdot \text{s}}{\text{kg} \cdot \text{m}}$	$\frac{1}{T}$
Length	L, l		metre (m)	1,000 mm = 100 cm	L
Mass	M, m		kilogram (kg)	1,000 g	M
Moment (torque)		force × length	newton · metre (N · m)	$\frac{\text{kg} \cdot \text{m}^2}{\text{s}^2} = \text{J}$	$\frac{ML^2}{T^2}$
Momentum	mv	mass × velocity = force × time $= \frac{\text{energy}}{\text{velocity}}$	newton · second (N · s)	$\frac{\text{kg} \cdot \text{m}}{\text{s}} = \frac{\text{J} \cdot \text{s}}{\text{m}}$	$\frac{ML}{T}$

(Contd...)

Table E.1 Fundamental, Mechanical, Electrical, and Magnetic Units (Contd.)

Name of dimension or quantity	*Symbol*	*Description*	*S.I. unit and abbreviation*	*Equivalent unit*	*Dimensions*
Period	T	$\frac{1}{\text{frequency}}$	second (s)	T	
Power	P	$\frac{\text{force} \times \text{length}}{\text{time}}$ $= \frac{\text{energy}}{\text{time}}$	watt (W)	$\frac{\text{J}}{\text{s}} = \frac{\text{N} \cdot \text{m}}{\text{s}}$ $= \frac{\text{kg} \cdot \text{m}^2}{\text{s}^3}$	$\frac{ML^2}{T^3}$
Time	T, t		second (s)	$\frac{1}{60}$ min $= \frac{1}{3600}$ h $= \frac{1}{86400}$ day	T
Velocity (velocity of light in vacuum = 300 ms^{-1})	v	$\frac{\text{length}}{\text{time}}$	$\frac{\text{metre}}{\text{second}}$ (ms^{-1})		$\frac{L}{T}$
Volume	V	length^3	metre^3 (m^3)		L^3

(Contd...)

Table E.1 Fundamental, Mechanical, Electrical, and Magnetic Units (Contd.)

Name of dimension or quantity	Symbol	Description	S.I. unit and abbreviation	Equivalent unit	Dimensions
			Electrical units		
Admittance	Y	$\frac{1}{\text{impedance}}$	mho (℧)	$\frac{A}{V} = \frac{C^2}{J \cdot s} = S\dagger$	$\frac{I^2T^3}{ML^2}$
Capacitance	C	$\frac{\text{charge}}{\text{potential}}$	farad (F)	$\frac{Q}{V} = \frac{C^2}{J} = \frac{A \cdot s}{V}$ = 9 × 1011 cm esu (egs)	$\frac{I^2T^4}{ML^2}$
Charge	Q, q	current × time	coulomb (C)	6.25×10^{18} electron charges $= A \cdot s$ $= 3 \times 10^9$ esu (cgs) = 0.1 emu (cgs)	IT
Charge density	ρ	$\frac{\text{charge}}{\text{volume}} = \nabla \cdot \boldsymbol{D}$	$\frac{\text{coulomb}}{\text{metre}^3}$ (Cm^{-3})	$\frac{A \cdot s}{m^3}$	$\frac{IT}{L^3}$
Conductance	G	$\frac{1}{\text{resistance}}$	mho (℧)	$\frac{A}{V} = \frac{C^2}{J \cdot s} = S\dagger$	$\frac{I^2T^3}{ML^2}$

(Contd...)

Table E.1 Fundamental, Mechanical, Electrical, and Magnetic Units (Contd.)

Name of dimension or quantity	*Symbol*	*Description*	*S.I. unit and abbreviation*	*Equivalent unit*	*Dimensions*
Conductivity	σ	$\frac{1}{\text{resistivity}}$	$\frac{\text{mho}}{\text{metre}}$ (℧ m^{-1})	$\frac{1}{\Omega m}$	$\frac{I^2T^3}{ML^3}$
Current	I, i	$\frac{\text{charge}}{\text{time}}$	ampere (A)	$\frac{\text{C}}{\text{s}} = 3 \times 10^9$ esu (cgs) $= 0.1$ emu (cgs)	I
Current	J	$\frac{\text{current}}{\text{area}}$	$\frac{\text{ampere}}{\text{metre}^2}$ ($\text{A} \cdot \text{m}^{-2}$)	$\frac{\text{C}}{\text{s} \cdot \text{m}^2}$	$\frac{I}{L^2}$
Dipole moment	$p\ (= ql)$	charge × length	coulomb · metre (C · m)	A · s · m	LIT
Emf	V	$\int \mathbf{E} \cdot dl$	volt (V)	$\frac{\text{Wb}}{\text{s}} = \frac{\text{J}}{\text{C}}$	$\frac{ML^2}{IT^3}$
Energy density (electric)	w_e	$\frac{\text{energy}}{\text{volume}}$	$\frac{\text{joule}}{\text{metre}}$ (Jm^{-3})	$\frac{\frac{1}{100}\text{erg}}{\text{mm}^3}$	$\frac{M}{LT^2}$

(Contd...)

Table E.1 Fundamental, Mechanical, Electrical, and Magnetic Units (Contd.)

Name of dimension or quantity	*Symbol*	*Description*	*S.I. unit and abbreviation*	*Equivalent unit*	*Dimensions*
Field intensity	$\boldsymbol{E}$	$\frac{\text{potential}}{\text{length}} = \frac{\text{force}}{\text{charge}}$	$\frac{\text{volt}}{\text{metre}}$ (V m^{-1})	$\frac{\text{N}}{\text{C}} = \frac{\text{J}}{\text{C} \cdot \text{m}}$ $= \frac{1}{3} \times 10^{-4}$ esu (cgs) $= 10^6$ emu (cgs)	$\frac{ML}{IT^3}$
Flux	ψ	charge $= \iint \boldsymbol{D} \cdot ds$	coulomb (C)	A . s	IT
Flux density (displacement)	$\boldsymbol{D}$	$\frac{\text{charge}}{\text{area}}$	$\frac{\text{coulomb}}{\text{metre}^2}$ (C · m^{-2})	$\frac{\text{A} \cdot \text{s}}{\text{m}^2} = \frac{\text{A}}{\text{m}^2\,\text{s}^{-1}}$	$\frac{IT}{L^2}$
Impedance	Z	$\frac{\text{potential}}{\text{current}}$	ohm (Ω)	$\frac{\text{V}}{\text{L}}$	$\frac{ML^2}{I^2T^3}$
Linear charge desity	ρ_L	$\frac{\text{charge}}{\text{length}}$	$\frac{\text{coulomb}}{\text{metre}}$ (C · m^{-1})	$\frac{\text{A} \cdot \text{s}}{\text{m}}$	$\frac{IT}{L}$

(Contd...)

Table E.1 Fundamental, Mechanical, Electrical, and Magnetic Units (Contd.)

Name of dimension or quantity	*Symbol*	*Description*	*S.I. unit and abbreviation*	*Equivalent unit*	*Dimensions*
Permittivity (dielectric constant) (for vacuum, $\varepsilon_0 = 8.85\ \text{pF} \cdot \text{m}^{-1}$ $= 10^{-9}/36\pi\ \text{F} \cdot \text{m}^{-1}$	ε	$\frac{\text{capacitance}}{\text{length}}$	$\frac{\text{farad}}{\text{metre}}$ $(\text{F} \cdot \text{m}^{-1})$	$\frac{\text{C}}{\text{Vm}}$	$\frac{I^2T^4}{ML^3}$
Polarization	P	$\frac{\text{dipole moment}}{\text{volume}}$	$\frac{\text{coloumb}}{\text{metre}^2}$ $(\text{C} \cdot \text{m}^{-2})$	$\frac{\text{A} \cdot \text{s}}{\text{m}^2}$	$\frac{IT}{L^2}$
Potential	V	$\frac{\text{work}}{\text{charge}}$	volt (V)	$\frac{\text{J}}{\text{C}} = \frac{\text{N} \cdot \text{m}}{\text{C}} = \frac{\text{W} \cdot \text{s}}{\text{C}}$ $= \frac{\text{W}}{\text{A}} = \frac{\text{Wb}}{\text{s}}$ $= \frac{1}{300}$ esu (cgs) $= 10^8$ emu (cgs)	$\frac{ML^2}{IT^3}$
Poynting vector	S	$\frac{\text{power}}{\text{area}}$	$\frac{\text{watt}}{\text{metre}^2}$ $(\text{W} \cdot \text{m}^{-2})$	$\frac{\text{J}}{\text{s} \cdot \text{m}^2}$	$\frac{M}{T^3}$

(Contd...)

Table E.1 Fundamental, Mechanical, Electrical, and Magnetic Units (Contd.)

Name of dimension or quantity	*Symbol*	*Description*	*S.I. unit and abbreviation*	*Equivalent unit*	*Dimensions*
Radiation intensity	P	$\frac{\text{power}}{\text{unit solid angle}}$	$\frac{\text{watt}}{\text{steradian}}$ $(\text{W} \cdot \text{sr}^{-1})$		$\frac{ML^2}{T^3}$
Reactance	X	$\frac{\text{potential}}{\text{current}}$	ohm (Ω)	$\frac{\text{V}}{\text{A}}$	$\frac{ML^2}{I^2T^3}$
Relative permittivity	ε_r	ratio $\frac{\varepsilon_0}{\varepsilon}$			dimensionless
Resistance	R	$\frac{\text{potential}}{\text{current}}$	ohm (Ω)	$\frac{\text{V}}{\text{A}} = \frac{\text{J} \cdot \text{s}}{\text{C}^2}$ $= \frac{1}{9} \times 10^{-11}$ esu (cgs) $= 10^{-9}$ emu (cgs)	$\frac{ML^2}{I^2T^3}$
Resistivity	S	resistance × length $= \frac{1}{\text{conductivity}}$	ohm · metre (Ω · m)	$\frac{\text{V}_\text{m}}{\text{A}}$	$\frac{ML^3}{I^2T^3}$

(Contd...)

Table E.1 Fundamental, Mechanical, Electrical, and Magnetic Units (Contd.)

Name of dimension or quantity	*Symbol*	*Description*	*S.I. unit and abbreviation*	*Equivalent unit*	*Dimensions*
Sheet-current density	K	$\frac{\text{current}}{\text{length}}$	$\frac{\text{ampere}}{\text{metre}}$ $(\text{A} \cdot \text{m}^{-1})$	$\frac{\text{A}}{\text{m}^2} \times \text{m}$	$\frac{I}{L}$
Susceptance	B	$\frac{1}{\text{reactance}}$	ohm (℧)	$\frac{\text{A}}{\text{V}} = \text{S}$	$\frac{I^2T^3}{ML^2}$
wavelength	λ	length	metre (m)		L
			Magnetic units		
Dipole moment (magnetic)	m $(= Qm\, I)$	pole strength × length = current × area $= \frac{\text{torque}}{\text{magnetic flux density}}$	ampere · metre² $(\text{A} \cdot \text{m}^2)$	$\frac{\text{C} \cdot \text{m}^2}{\text{s}}$	IL^2
Energy density (magnetic)	w_m	$\frac{\text{energy}}{\text{volume}}$	$\frac{\text{joule}}{\text{metre}}$ $(\text{J} \cdot \text{m}^{-3})$	$\frac{\frac{1}{100}\text{erg}}{\text{mm}^3}$	$\frac{M}{LT^2}$

(Contd...)

Table E.1 Fundamental, Mechanical, Electrical, and Magnetic Units (Contd.)

Name of dimension or quantity	*Symbol*	*Description*	*S.I. unit and abbreviation*	*Equivalent unit*	*Dimensions*
Flux (magnetic)	Ψ_m	$\iint \boldsymbol{B} \cdot ds$	weber (Wb)	$\text{V} \cdot \text{s} = \frac{\text{N} \cdot \text{m}}{\text{A}}$ $= 10^8$ Mx cgs (emu)	$\frac{ML^2}{IT^2}$
Flux density	$\boldsymbol{B}$	$\frac{\text{force}}{\text{pole}} = \frac{\text{force}}{\text{current moment}}$ $= \frac{\text{magnetic flux}}{\text{area}}$	tesla (T) $\frac{\text{weber}}{\text{metre}^2}$ $(\text{Wb} \cdot \text{m}^{-2})$	$\frac{\text{V} \cdot \text{s}}{\text{m}^2} = \frac{\text{N}}{\text{A} \cdot \text{m}}$ $= 10^4$ G‡ emu (cgs)	$\frac{M}{IT^2}$
Flux linkage	A	flux × turns	weber-turn (Wb-turn)		$\frac{ML^2}{IT^2}$
H field	$\boldsymbol{H}$	$\frac{\text{mmf}}{\text{length}}$	$\frac{\text{ampere}}{\text{metre}}$ $(\text{A} \cdot \text{m}^{-1})$	$\frac{\text{N}}{\text{Wb}} = \frac{\text{W}}{\text{V} \cdot \text{m}}$ $= 4\pi \times 10^{-3}$ Oe [cgs] (emu) $= 400\pi$ gammas	$\frac{I}{L}$

(Contd...)

Table E.1 Fundamental, Mechanical, Electrical, and Magnetic Units (Contd.)

Name of dimension or quantity	*Symbol*	*Description*	*S.I. unit and abbreviation*	*Equivalent unit*	*Dimensions*
Inductance	L	$\frac{\text{magnetic flux linkage}}{\text{current}}$	henry (H)	$\frac{\text{Wb}}{\text{A}} = \frac{\text{J}}{\text{A}^2} = \Omega \cdot \text{s}$ $= \frac{1}{9} \times 10^{-11}$ esu (cgs) $= 10^9$ cm emu (cgs)	$\frac{ML^2}{I^2T^2}$
Magnetization (magnetic polarization)	M	$\frac{\text{magnetic moment}}{\text{volume}}$	$\frac{\text{ampere}}{\text{metre}}$ $(\text{A} \cdot \text{m}^{-1})$	$\frac{\text{A} \cdot \text{m}^2}{\text{m}^3} = \frac{\text{A} \cdot \text{m}}{\text{m}^2}$	$\frac{I}{L}$
Mmf	F	$\int \boldsymbol{H} \cdot dl$	ampere-turn (A-turn)	$\frac{\text{C}}{\text{s}}$	I
Permeability (for vacuum, $\mu_0 = 400\pi$ nH m^{-1})	μ	$\frac{\text{inductance}}{\text{length}}$	$\frac{\text{henry}}{\text{metre}}$ (Hm^{-1})	$\frac{\text{Wb}}{\text{A} \cdot \text{m}} = \frac{\text{V} \cdot \text{s}}{\text{A} \cdot \text{m}}$	$\frac{ML}{I^2T^2}$

(*Contd...*)

Table E.1 Fundamental, Mechanical, Electrical, and Magnetic Units (Contd.)

Name of dimension or quantity	*Symbol*	*Description*	*S.I. unit and abbreviation*	*Equivalent unit*	*Dimensions*
Permeance	P	$\frac{\text{magnetic flux}}{\text{mmf}} = \frac{1}{\text{reluctance}}$	henry (H)	$\frac{\text{Wb}}{\text{A}}$	$\frac{ML^2}{I^2T^2}$
Pole density	ρ_m	$\frac{\text{pole strength}}{\text{volume}} = \frac{\text{current}}{\text{area}} = \nabla \cdot H = -\nabla \cdot M$	$\frac{\text{ampere}}{\text{metre}^2}$ $(\text{A} \cdot \text{m}^{-2})$		$\frac{I}{L^2}$
Pole strength	Q_m, ρ_m	current × length $= \iiint \rho_m \, dv$	ampere-turn $(\text{A} \cdot \text{m})$	$\frac{\text{C.m}}{\text{s}}$	IL
Potential (magnetic ***H***)	U	$\int \boldsymbol{H} \cdot dl$	ampere (A)	$\frac{\text{J}}{\text{Wb}} = \frac{\text{W}}{\text{V}} = \frac{\text{C}}{\text{s}} = \frac{4\pi}{10}$ Gb§ emu (cgs)	I

(*Contd...*)

Table E.1 Fundamental, Mechanical, Electrical, and Magnetic Units (Contd.)

Name of dimension or quantity	*Symbol*	*Description*	*S.I. unit and abbreviation*	*Equivalent unit*	*Dimensions*
Relative permeability	μ_r	ratio $\frac{\mu}{\mu_0}$			Dimensionless
Reluctance	R	$\frac{\text{mmf}}{\text{magnetic flux}} = \frac{1}{\text{permeance}}$	$\frac{1}{\text{henry}}$ (H^{-1})	$\frac{\text{A}}{\text{Wb}}$	$\frac{I^2T^2}{ML^2}$
Vector potential	A	current × permeability	$\frac{\text{Weber}}{\text{henry}}$ $(\text{Wb} \cdot \text{m}^{-1})$	$\frac{\text{HA}}{\text{m}} = \frac{\text{N}}{\text{A}}$	$\frac{ML}{IT^2}$

†S is the SI abbreviation for Siemens, used often for mho.

‡Mx, G and Oe are SI abbreviations for maxwell, gauss, and oersted.

§Gb is the SI abbreviation for gilbert.

Appendix F

Table F.1 Trigonometric, Hyperbolic, Logarithmic, and Other Relations

Trigonometric Relations

$$\sin (x \pm y) = \sin x \cos y \pm \cos x \sin y$$

$$\cos (x \pm y) = \cos x \cos y \pm \sin x \sin y$$

$$\sin (x + y) + \sin (x - y) = 2 \sin x \cos y$$

$$\cos (x + y) + \cos (x - y) = 2 \cos x \cos y$$

$$\sin (x + y) - \sin (x - y) = 2 \cos x \sin y$$

$$\cos (x + y) - \cos (x - y) = -2 \sin x \sin y$$

$$\sin 2x = 2 \sin x \cos x$$

$$\cos 2x = \cos^2 x - \sin^2 x = 2 \cos^2 x - 1 = 1 - 2 \sin^2 x$$

$$\cos x = 2\cos^2 \frac{1}{2}x - 1 = 1 - 2\sin^2 \frac{1}{2}x$$

$$\sin x = 2\sin \frac{1}{2}x \cos \frac{1}{2}x$$

$$\sin^2 x + \cos^2 x = 1$$

$$\tan (x + y) = \frac{\tan x + \tan y}{1 - \tan x \tan y}$$

$$\tan 2x = \frac{2 \tan x}{1 - \tan^2 x}$$

$$\sin x = x - \frac{x^3}{3!} + \frac{x^5}{5!} - \frac{x^7}{7!} + \ldots$$

$$\cos x = 1 - \frac{x^2}{2!} + \frac{x^4}{4!} - \frac{x^6}{6!} + \ldots$$

$$\tan x = x + \frac{x^3}{3} + \frac{2x^5}{15} + \frac{17x^7}{315} + \frac{62x^9}{2835} + \ldots$$

Hyperbolic Relations

$$\sinh x = \frac{e^x - e^{-x}}{2} = x + \frac{x^3}{3!} + \frac{x^5}{5!} + \frac{x^7}{7!} + \ldots$$

$$\cosh x = \frac{e^x + e^{-x}}{2} = 1 + \frac{x^2}{2!} + \frac{x^4}{4!} + \frac{x^6}{6!} + \ldots$$

$$\tanh x = \frac{\sinh x}{\cosh x}$$

$$\coth x = \frac{\cosh x}{\sinh x} = \frac{1}{\tanh x}$$

$$\sinh (x \pm jy) = \sinh x \cos y \pm j \cosh x \sin y$$

$$\cosh (x \pm jy) = \cosh x \cos y \pm j \sinh x \sin y$$

$$\left.\begin{aligned} \cosh(jx) &= \frac{1}{2}\left(e^{+jx} + e^{-jx}\right) = \cos x \\ \sinh(jx) &= \frac{1}{2}\left(e^{+jx} - e^{-jx}\right) = j \sin x \end{aligned}\right\} \text{de Moivre's theorem}$$

$$e^{\pm jx} = \cos x \pm j \sin x$$

$$e^{\pm jx} = 1 \pm jx - \frac{x^2}{2!} \mp j\frac{x^3}{3!} + \frac{x^4}{4!} \pm j\frac{x^5}{5} - \ldots$$

$$e^x = \cosh x + \sinh x$$

$$e^{-x} = \cosh x - \sinh x$$

$$e^x = 1 + x + \frac{x^2}{2!} + \frac{x^3}{3!} + \frac{x^4}{4!} + \ldots$$

$$\cosh x = \cos jx$$

$$j \sinh x = \sin jx$$

$$\tanh (x \pm jy) = \frac{\sinh 2x}{\cosh 2x + \cos 2y} \pm j\frac{\sin 2y}{\cosh 2x + \cos 2y}$$

$$\coth (x \pm jy) = \frac{\sinh 2x}{\cosh 2x - \cos 2y} \pm j\frac{\sin 2y}{\cosh 2x - \cos 2y}$$

Logarithmic Relations

$\log_{10} x = \log x$ common logarithm

$\log_e x = \ln x$ natural logarithm

$$\log_{10} x = 0.4343 \log_e x = 0.4343 \ln x$$

$$\ln x = \log_e x = 2.3026 \log_{10} x$$

$$e = 2.71828$$

dB = 10 log (power ratio) = 20 log (voltage ratio)

$$1\text{ Np (voltage attenuation)} = \frac{1}{e} = 0.368\text{ (voltage)} = -8.68\text{ dB}$$

Approximation Formulas for Small Quantities

(δ is a small quantity compared with unit)

$(1 \pm \delta)^2 = 1 \pm 2\delta$

$(1 \pm \delta)^n = 1 \pm n\delta$

$$\sqrt{1+\delta} = 1 + \frac{1}{2}\delta$$

$$\frac{1}{\sqrt{1+\delta}} = 1 - \frac{1}{2}\delta$$

$e^{\delta} = 1 + \delta$

$\ln(1 + \delta) = \delta$

$$J_n(\delta) = \frac{\delta^n}{n!\,2^n} \text{ for } |\delta| << 1$$

where J_n is Bessel function of order n. Thus,

$$J_1(\delta) = \frac{\delta}{2}$$

Series

Binomial:

$$(x + y)^n = x^n + nx^{n-1}y + \frac{n(n-1)}{2!}x^{n-2}y^2 + \frac{n(n-1)(n-2)}{3!}x^{(n-3)}y^3 + \ldots$$

Taylor's:

$$f(x + y) = f(x) + \frac{df(x)}{dx}\frac{y}{1} + \frac{d^2f(x)}{dx^2}\frac{y^2}{2!} + \frac{d^3f(x)}{dx^3}\frac{y^3}{3!} + \ldots$$

Solution of Quadratic Equation

If $ax^2 + bx + c = 0$, then

$$x = \frac{-b \pm \sqrt{b^2 - 4ac}}{2a}$$

Gradient, Divergence, and Curl in Rectangular, Cylindrical, and Spherical Coordinates

Rectangular Coordinates

$$\nabla f = \hat{x}\frac{\partial f}{\partial x} + \hat{y}\frac{\partial f}{\partial x} + \hat{z}\frac{\partial f}{\partial x}$$

$$\nabla \cdot A = \frac{\partial A_x}{\partial x} + \frac{\partial A_y}{\partial y} + \frac{\partial A_z}{\partial z}$$

$$\nabla \times \boldsymbol{A} = \hat{\boldsymbol{x}}\left(\frac{\partial A_z}{\partial y} - \frac{\partial A_y}{\partial z}\right) + \hat{\boldsymbol{y}}\left(\frac{\partial A_x}{\partial z} - \frac{\partial A_z}{\partial x}\right) + \hat{\boldsymbol{z}}\left(\frac{\partial A_y}{\partial x} - \frac{\partial A_x}{\partial y}\right)\begin{vmatrix} \hat{\boldsymbol{x}} & \hat{\boldsymbol{y}} & \hat{\boldsymbol{z}} \\ \frac{\partial}{\partial x} & \frac{\partial}{\partial y} & \frac{\partial}{\partial z} \\ A_x & A_y & A_z \end{vmatrix}$$

Cylindrical Coordinates

$$\nabla f = \hat{\boldsymbol{r}}\frac{\partial f}{\partial r} + \hat{\Phi}\frac{1}{r}\frac{\partial f}{\partial \phi} + \hat{\boldsymbol{z}}\frac{\partial f}{\partial z}$$

$$\nabla \cdot \boldsymbol{A} = \frac{1}{r}\frac{\partial}{\partial r} r A_r + \frac{1}{r}\frac{\partial A_\phi}{\partial \phi} + \frac{\partial A_z}{\partial z}$$

$$\nabla \times \boldsymbol{A} = \hat{\boldsymbol{r}}\left(\frac{1}{r}\frac{\partial A_z}{\partial \phi} - \frac{\partial A_\phi}{\partial z}\right) + \hat{\Phi}\left(\frac{\partial A_r}{\partial z} - \frac{\partial A_z}{\partial r}\right) + \hat{\boldsymbol{z}}\frac{1}{r}\left(\frac{\partial}{\partial r} r A_\phi - \frac{\partial A_r}{\partial \phi}\right)$$

$$= \begin{vmatrix} \hat{\boldsymbol{r}}\frac{1}{r} & \hat{\Phi} & \hat{\boldsymbol{z}}\frac{1}{r} \\ \frac{\partial}{\partial r} & \frac{\partial}{\partial \phi} & \frac{\partial}{\partial z} \\ A_r & r A_\phi & A_z \end{vmatrix}$$

Spherical Coordinates

$$\nabla f = \hat{\boldsymbol{r}}\frac{\partial f}{\partial r} + \hat{\theta}\frac{1}{r}\frac{\partial f}{\partial \theta} + \hat{\Phi}\frac{1}{r\sin\theta}\frac{\partial f}{\partial \phi}$$

$$\nabla \cdot \boldsymbol{A} = \frac{1}{r^2}\frac{\partial}{\partial r} r^2 A_r + \frac{1}{r\sin\theta}\frac{\partial}{\partial \theta}(A_\theta \sin\theta) + \frac{1}{r\sin\theta}\frac{\partial A_\phi}{\partial \phi}$$

$$\nabla \times \boldsymbol{A} = \hat{\boldsymbol{r}}\frac{1}{r\sin\theta}\left(\frac{\partial}{\partial \theta}(A_\phi \sin\theta) - \frac{\partial A_\theta}{\partial \phi}\right) + \hat{\theta}\frac{1}{r}\left(\frac{1}{\sin\theta}\frac{\partial A_r}{\partial \phi} - \frac{\partial}{\partial r} r A_\phi\right) + \hat{\Phi}\frac{1}{r}\left(\frac{\partial}{\partial r} r A_\theta - \frac{\partial A_r}{\partial \theta}\right)$$

Appendix G

Table G.1 Glyphs (Nonalphabetic Pictograph Symbols)

Symbol	*Definition*
$=$	Equal to
$\sim$ or $\approx$	Approximately equal to
$\cong$	Nearly equal to
$\equiv$	Identical with or by definition
$\neq$	Not equal to
$\propto$	Proportional to
%	Percent
$\rightarrow$	Approaches
$<$	Less than
$>$	Greater than
$\leq$	Less than or equal to
$\geq$	Greater than or equal to
$<<$	Much less than
$>>$	Much greater than
∞	Infinity
$\therefore$	Therefore
!	Factorial
$\sqrt{\ }$	Square root
$\mid\ \mid$	Absolute value
$\sum$	Summation sign

$\int$	Integral sign
$\oint$	Line integral around a closed path
$\iint$ or $\int_s$	Surface integral
$\oiint$ or $\oint_s$	Surface integral completely enclosing a volume
$\iiint$ or $\int_v$	Volume integral